ITERATIVE SOLUTION OF LARGE LINEAR SYSTEMS

This is a volume in
COMPUTER SCIENCE AND APPLIED MATHEMATICS
A Series of Monographs and Textbooks

Editor: WERNER RHEINBOLDT

A complete list of titles in this series appears at the end of this volume.

ITERATIVE SOLUTION OF LARGE LINEAR SYSTEMS

David M. Young

THE UNIVERSITY OF TEXAS AT AUSTIN

ACADEMIC PRESS, INC.

Harcourt Brace Jovanovich, Publishers

Orlando San Diego New York
Austin Boston London Sydney
Tokyo Toronto

ACADEMIC PRESS, INC.
Orlando, Florida 32887

United Kingdom Edition published by
ACADEMIC PRESS, INC. (LONDON) LTD.
24/28 Oval Road, London NW1 7DX

LIBRARY OF CONGRESS CATALOG CARD NUMBER: 73-170124

AMS (MOS) Subject Classification (1970): 65F10

PRINTED IN THE UNITED STATES OF AMERICA

87 88 89 9 8 7 6 5 4 3

512.94
Y84

91-549

TO

GARRETT BIRKHOFF

CLASS NO.
CUST. ACCT. #
NO. COPIES 1
LIST PRICE 89.50
COST PRICE 91.20
BINDING
DATE ORDERED 3/20/91
DEALER
DEALER FILE NO.
AUTHOR'S FULL NAME Young, David M.
TITLE Iterative Solution of Large Linear Systems
PUBLISHER Academic Press
PLACE
PUB. DATE 1971
EDITION
ACCESSION NO. 91-549
FUND CHARGED Math
SPECIAL INSTRUCTIONS
LIBRARY NAME
SETS L.C. H WI X
ISBN 0-12-773050-8
PURCHASE ORD. # 5570
L. C. CARD NO.
F NF
CARD SETS ORDERED 3C 3C/LC LC LJ OCLC OTHER
3/20/91

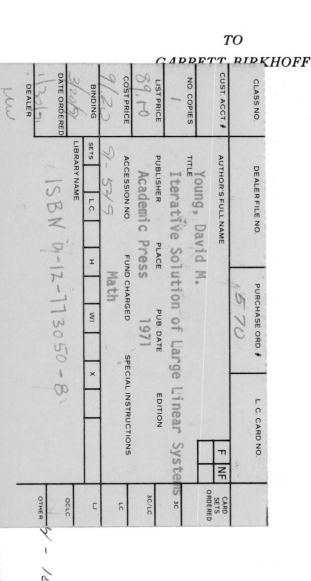

CONTENTS

Preface xiii
Acknowledgments xvii
Notation xix
List of Fundamental Matrix Properties xxi
List of Iterative Methods xxiii

1. Introduction 1

1.1. The Model Problem 2
Supplementary Discussion 6
Exercises 6

2. Matrix Preliminaries 7

2.1. Review of Matrix Theory 7
2.2. Hermitian Matrices and Positive Definite Matrices 18
2.3. Vector Norms and Matrix Norms 25
2.4. Convergence of Sequences of Vectors and Matrices 34
2.5. Irreducibility and Weak Diagonal Dominance 36
2.6. Property A 41
2.7. *L*-Matrices and Related Matrices 42
2.8. Illustrations 48
Supplementary Discussion 53
Exercises 55

3. Linear Stationary Iterative Methods . . . 63

3.1. Introduction . . . 63
3.2. Consistency, Reciprocal Consistency, and Complete Consistency . . . 65
3.3. Basic Linear Stationary Iterative Methods . . . 70
3.4. Generation of Completely Consistent Methods . . . 75
3.5. General Convergence Theorems . . . 77
3.6. Alternative Convergence Conditions . . . 80
3.7. Rates of Convergence . . . 84
3.8. The Jordan Condition Number of a 2×2 Matrix . . . 89
Supplementary Discussion . . . 94
Exercises . . . 95

4. Convergence of the Basic Iterative Methods . . . 106

4.1. General Convergence Theorems . . . 106
4.2. Irreducible Matrices with Weak Diagonal Dominance . . . 107
4.3. Positive Definite Matrices . . . 108
4.4. The SOR Method with Varying Relaxation Factors . . . 118
4.5. L-Matrices and Related Matrices . . . 120
4.6. Rates of Convergence of the J and GS Methods for the Model Problem 127
Supplementary Discussion . . . 132
Exercises . . . 133

5. Eigenvalues of the SOR Method for Consistently Ordered Matrices . . . 140

5.1. Introduction . . . 140
5.2. Block Tri-Diagonal Matrices . . . 141
5.3. Consistently Ordered Matrices and Ordering Vectors . . . 144
5.4. Property A . . . 148
5.5. Nonmigratory Permutations . . . 153
5.6. Consistently Ordered Matrices Arising from Difference Equations . . . 157
5.7. A Computer Program for Testing for Property A and Consistent Ordering 159
5.8. Other Developments of the SOR Theory . . . 162
Supplementary Discussion . . . 163
Exercises . . . 163

6. Determination of the Optimum Relaxation Factor . . . 169

6.1. Virtual Spectral Radius . . . 170
6.2. Analysis of the Case Where All Eigenvalues of B Are Real . . . 171

6.3. Rates of Convergence: Comparison with the Gauss–Seidel Method . . . 188
6.4. Analysis of the Case Where Some Eigenvalues of B Are Complex 191
6.5. Practical Determination of ω_b: General Considerations 200
6.6. Iterative Methods of Choosing ω_b 209
6.7. An Upper Bound for $\bar{\mu}$ 211
6.8. A Priori Determination of $\bar{\mu}$: Exact Methods 216
6.9. A Priori Determination of $\bar{\mu}$: Approximate Values 222
6.10. Numerical Results 224
Supplementary Discussion 227
Exercises 228

7. Norms of the SOR Method 233

7.1. The Jordan Canonical Form of $\mathscr{L}_\omega$ 234
7.2. Basic Eigenvalue Relation 239
7.3. Determination of $\| \mathscr{L}_\omega \|_{D^{1/2}}$ 245
7.4. Determination of $\| \mathscr{L}_{\omega_b}^m \|_{D^{1/2}}$ 248
7.5. Determination of $\| \mathscr{L}_\omega \|_{A^{1/2}}$ 255
7.6. Determination of $\| \mathscr{L}_{\omega_b}^m \|_{A^{1/2}}$ 258
7.7. Comparison of $\| \mathscr{L}_{\omega_b}^m \|_{D^{1/2}}$ and $\| \mathscr{L}_{\omega_b}^m \|_{A^{1/2}}$ 264
Supplementary Discussion 265
Exercises 266

8. The Modified SOR Method: Fixed Parameters 271

8.1. Introduction 271
8.2. Eigenvalues of $\mathscr{L}_{\omega,\omega'}$ 273
8.3. Convergence and Spectral Radius 277
8.4. Determination of $\| \overline{\mathscr{L}_{\omega,\omega'}} \|_{D^{1/2}}$ 283
8.5. Determination of $\| \overline{\mathscr{L}_{\omega,\omega'}} \|_{A^{1/2}}$ 288
Supplementary Discussion 291
Exercises 291

9. Nonstationary Linear Iterative Methods 295

9.1. Consistency, Convergence, and Rates of Convergence 295
9.2. Periodic Nonstationary Methods 300
9.3. Chebyshev Polynomials 301
Supplementary Discussion 304
Exercises 304

10. The Modified SOR Method: Variable Parameters 306

10.1. Convergence of the MSOR Method 307
10.2. Optimum Choice of Relaxation Factors 307
10.3. Alternative Optimum Parameter Sets 311
10.4. Norms of the MSOR Method: Sheldon's Method 315
10.5. The Modified Sheldon Method 319
10.6. Cyclic Chebyshev Semi-Iterative Method 321
10.7. Comparison of Norms 327
Supplementary Discussion 340
Exercises 341

11. Semi-Iterative Methods 344

11.1. General Considerations 345
11.2. The Case Where G Has Real Eigenvalues 347
11.3. J, JOR, and RF Semi-Iterative Methods 355
11.4. Richardson's Method 361
11.5. Cyclic Chebyshev Semi-Iterative Method 365
11.6. GS Semi-Iterative Methods 367
11.7. SOR Semi-Iterative Methods 374
11.8. MSOR Semi-Iterative Methods 376
11.9. Comparison of Norms 383
Supplementary Discussion 385
Exercises 386

12. Extensions of the SOR Theory: Stieltjes Matrices . . . 391

12.1. The Need for Some Restrictions on A 391
12.2. Stieltjes Matrices 395
Supplementary Discussion 401
Exercises 401

13. Generalized Consistently Ordered Matrices 404

13.1. Introduction 404
13.2. CO(q, r)-Matrices, Property $A_{q,r}$, and Ordering Vectors 405
13.3. Determination of the Optimum Relaxation Factor 413
13.4. Generalized Consistently Ordered Matrices 418
13.5. Relation between GCO(q, r)-Matrices and CO(q, r)-Matrices 419

13.6. Computational Procedures: Canonical Forms . . . 422
13.7. Relation to Other Work . . . 428
Supplementary Discussion . . . 429
Exercises . . . 430

14. Group Iterative Methods . . . 434

14.1. Construction of Group Iterative Methods . . . 435
14.2. Solution of a Linear System with a Tri-Diagonal Matrix . . . 441
14.3. Convergence Analysis . . . 445
14.4. Applications . . . 452
14.5. Comparison of Point and Group Iterative Methods . . . 454
Supplementary Discussion . . . 456
Exercises . . . 457

15. Symmetric SOR Method and Related Methods . . . 461

15.1. Introduction . . . 461
15.2. Convergence Analysis . . . 463
15.3. Choice of Relaxation Factor . . . 464
15.4. SSOR Semi-Iterative Methods: The Discrete Dirichlet Problem . . . 471
15.5. Group SSOR Methods . . . 474
15.6. Unsymmetric SOR Method . . . 476
15.7. Symmetric and Unsymmetric MSOR Methods . . . 478
Supplementary Discussion . . . 480
Exercises . . . 481

16. Second-Degree Methods . . . 486

Supplementary Discussion . . . 493
Exercises . . . 493

17. Alternating Direction Implicit Methods . . . 495

17.1. Introduction: The Peaceman-Rachford Method . . . 495
17.2. The Stationary Case: Consistency and Convergence . . . 498
17.3. The Stationary Case: Choice of Parameters . . . 503
17.4. The Commutative Case . . . 514
17.5. Optimum Parameters . . . 518

17.6. Good Parameters 525
17.7. The Helmholtz Equation in a Rectangle 531
17.8. Monotonicity 534
17.9. Necessary and Sufficient Conditions for the Commutative Case 535
17.10. The Noncommutative Case 545
Supplementary Discussion 547
Exercises 548

18. Selection of Iterative Method 553

Bibliography 556

Index 565

PREFACE

The availability of very high-speed computers with large, fast memories has made it possible to obtain accurate numerical solutions of mathematical problems which, although algorithms for handling them were well known previously, could not be used in practice because the number of calculations required would have been prohibitive. A problem for which this is particularly true is that of solving a large system of linear algebraic equations where the matrix of the system is very "sparse," that is, most of its elements vanish. Such systems frequently arise in the numerical solution of elliptic partial differential equations by finite difference methods. These problems, in turn, arise in studies in such areas as neutron diffusion, fluid flow, elasticity, steady-state heat flow, and weather prediction. The solution of a large system with a sparse matrix is usually obtained by iterative methods instead of by direct methods such as the Gauss elimination method. An extensive theory has been developed which is interesting mathematically while at the same time is of practical use.

The purpose of the book is to provide a systematic development of a substantial portion of the theory of iterative methods for solving large linear systems. The emphasis is on practical techniques, and the treatment has been made as elementary as possible without unnecessary generalizations. Every effort has been made to give a mathematically rigorous and self-contained treatment with the exception of some basic theorems of matrix theory and analysis which are stated without proof in Chapter 2. Only in later chapters has it been necessary to depart from this objective to any appreciable extent. The material includes published and unpublished work of the author as well as results of others as indicated. Particular reference has been made to the works of Richard Varga and Eugene Wachspress.

In order to make the book as widely useful as possible a minimum amount of mathematical background is assumed. In addition to a knowledge of the fundamentals of matrix theory, at a level indicated by Section 2.1, the reader should be familiar with the elementary aspects of real analysis as covered in a good course in "advanced calculus" or "elementary analysis." He should also be acquainted with some elementary complex variable theory. In addition, a general background in numerical analysis, especially matrix methods and finite difference methods for solving partial differential equations, as well as in computer programming would be highly desirable.

Chapters 1 and 2 are introductory in nature. Chapter 1 shows how the solution of a certain partial differential equation by finite difference methods leads to a large linear system with a sparse matrix. Chapter 2 provides a general review of matrix theory and the definition and study of various matrix properties which are used later. At the beginning of Chapter 2 several theorems of matrix theory are stated without proof. From then on, with some noted exceptions, all subsequent results are either proved or given as exercises which can be worked out from the presentation given.

The focal point of the book is an analysis of the convergence properties of the successive overrelaxation method (SOR method) as applied to a linear system where the matrix is "consistently ordered." In Chapters 3 and 4 we consider a number of iterative methods, including the SOR method, without assuming that the matrix is consistently ordered. In Chapter 3 the concepts of consistency, reciprocal consistency, and complete consistency are given for linear stationary iterative methods together with convergence conditions and various measures of convergence rates. In Chapter 4 convergence theorems are given for various iterative methods under certain assumptions on the matrix A of the system.

Chapters 5–8 are concerned with the SOR method and the modified SOR method when A is consistently ordered. In Chapter 5 it is shown how, if A is "consistently ordered," the eigenvalues of the matrix corresponding to the SOR method can be determined in terms of the eigenvalues of the matrix B corresponding to the Jacobi method of iteration. From this relation, in Chapter 6, we are able, under various assumptions on the location of the eigenvalues of B, to give necessary and sufficient conditions for convergence of the SOR method and also to determine an optimum value of the relaxation factor. In Chapter 7 we study the effectiveness of the successive overrelaxation method in terms of various norms of the associated matrix. As an outgrowth of the study of Chapter 7 we are led to consider, in Chapter 8, the modified SOR method which is a variant of the SOR method and which can be defined when the matrix has a 2×2 block form where the diagonal blocks are diagonal matrices.

Chapters 9–11 are concerned with nonstationary iterative methods including the modified SOR method with variable iteration parameters and semi-iterative methods based on the stationary methods considered previously.

Chapters 12 and 13 are concerned with studies of the convergence properties of the SOR method when the matrix is not consistently ordered.

Chapter 14 and 15 are concerned with modifications of the SOR method which are intended to make the SOR theory applicable where it might not otherwise be, as in the use of group iterative methods which are treated in Chapter 14, or to improve the convergence as in the case of the symmetric SOR method which is treated in Chapter 15.

Second-degree methods, sometimes referred to as "second-order methods," are discussed in Chapter 16. Alternating direction implicit methods, especially the Peaceman–Rachford method, are discussed in Chapter 17. An overall discussion of the various methods considered is given in Chapter 18.

Exercises are given for most of the sections. These are designed to enhance the reader's understanding of the material presented, to fill in details not covered in the text, and to supplement and extend the theory.

With judicious choice of topics the book could be used in connection with a one-semester first-year graduate-level course. Such a course could include: Chapter 1, Chapter 2 (the amount of coverage needed depending on the backgrounds of the students), Chapters 3, 4, 5, and part of Chapter 6. Selections could be made from the remaining chapters. It would certainly be desirable to at least define semi-iterative methods (Chapter 11), block methods (Chapter 14), the symmetric successive overrelaxation method (Chapter 15), and the Peaceman–Rachford alternating direction method (Chapter 17). One possible approach in a first course might be to examine all iterative methods in terms of the spectral radii of the corresponding matrices and to omit all material relating to matrix norms appearing after Chapter 4. Much of the material in the book not covered in a first course could be treated in subsequent seminars.

Decimal notation is used for the numbering of sections and chapters. Thus the third section of Chapter 5 is numbered 5.3. The 15th numbered equation in Section 3 of Chapter 5 is numbered (3.15); the second numbered equation not in any section is designated (2). The equations are referenced in another chapter by (5-3.15) and (5-2), respectively. A similar numbering system is used for theorems, figures, etc.

ACKNOWLEDGMENTS

In the preparation of the manuscript, I received encouragement and useful suggestions from many sources. I would particularly like to acknowledge the encouragement of the publisher, Academic Press, Inc., and its Consulting Editor, Dr. Werner C. Rheinboldt, who offered many helpful suggestions. I would also like to acknowledge useful suggestions of Dr. William F. Ames, Dr. Louis W. Ehrlich, Dr. Yasuhiko Ikebe, Mr. John Dauwalder, and Mr. David Kincaid. Invaluable assistance on proofreading was provided by John Dauwalder, David Kincaid, and Vitalius Benokraitis. They also, together with Jerry Webb and Harold Eidson, assisted in numerical calculations and in the exercises. Additional numerical calculations were done by Belinda Wilkinson and Alkis Mouradoglou. All of this assistance is gratefully acknowledged. Finally, I wish to thank Mrs. Dorothy Baker and Mrs. Marjorie Dragoo for their extremely careful and competent work in the preparation of the manuscript.

Much of the material included in the book is an outgrowth of work done over the past several years by the author and supported in part by research grants including Grants GP 217, 5253, 8442 from the National Science Foundation, and Grants DA-ARO(D)-31-124-G388, 721, and 1050 from the U.S. Army Research Office (Durham) with The University of Texas at Austin. This support is gratefully acknowledged.

I wish to dedicate this book to Garrett Birkhoff, who introduced me to this area of study and who has provided continued interest and encouragement over the years.

NOTATION

I. Matrices

Symbol	*Meaning*	*First used in section*
A^{T}	transpose	2.1
A^{*}	(complex) conjugate	2.1
A^{H}	conjugate transpose	2.1
A^{-1}	inverse	2.1
trace A	trace of A	2.1
det A	determinant of A	2.1
I	identity matrix	2.1
S_A	eigenvalue spectrum	2.1
$S(A)$	spectral radius	3.7
$R(A)$	(asymptotic) rate of convergence	3.7
$\| A \| = \| A \|_2$	spectral norm	2.3
$\| A \|_L = \| LAL^{-1} \|$	L-norm	2.3
$\bar{S}(A)$	virtual spectral radius	6.1
$\bar{R}(A)$	virtual (axymptotic) rate of convergence	9.1
$\overline{\| A \|}_L$	virtual L-norm	7
diag A	the diagonal matrix with the same diagonal elements as A	2.1
$\nu(A)$	spectral condition number	3.7
$\mathscr{J}(A)$	Jordan condition number	3.7
$K(A)$	condition	11.3

Frequently, given a matrix A, we define the following matrices

Symbol	*Meaning*	*First used in section*
$D = \operatorname{diag} A$		4.3
$C = D - A$		4.3
C_{L}:	a strictly lower triangular matrix with the same elements below the main diagonal as C	4.3
C_{U}:	a strictly upper triangular matrix with the same elements above the main diagonal as C	4.3
$B = D^{-1}C$		4.3
L:	a strictly lower triangular matrix with the same elements below the main diagonal as B	4.3
U:	a strictly upper triangular matrix with the same elements above the main diagonal as B	4.3

For any H, if D is positive definite

$\tilde{H} = D^{1/2}HD^{-1/2}$		4.3
$\hat{H} = D^{-1/2}HD^{-1/2}$		4.3

If A is positive definite

$H' = A^{1/2}HA^{-1/2}$		4.4

II. Vectors

E^N	the set of all $N \times 1$ matrices	2.1
(x, y)	inner product	2.1
$\| x \| = (x, x)^{1/2}$	Euclidean norm	2.1
$\| x \|_L = \| Lx \|$	L-norm	2.2

III. Miscellaneous

$z^* = x - iy$ where $z = x + iy$ and x and y are real

$T_m(x) = \frac{1}{2}\{[x + (x^2 - 1)^{1/2}]^m + [x^2 + (x^2 - 1)^{1/2}]^{-m}\}$

Chebyshev polynomial of degree m (Chap. 9)

$\mathscr{S}(A, b)$ The set of solutions of the linear system $Au = b$ (Chap. 3)

LIST OF FUNDAMENTAL MATRIX PROPERTIES

Property	*First used in section*
positive ($A > 0$)	2.1
nonnegative ($A \geq 0$)	2.1
nonsingular	2.1
orthogonal	2.1
Hermitian	2.1
normal	2.1
positive definite	2.2
nonnegative definite	2.2
irreducible	2.5
weak diagonal dominance	2.5
Property A	2.6
L-matrix	2.7
M-matrix	2.7
Stieltjes	2.7
monotone	2.7
stable	3.6
N-stable	3.6
T-matrix	5.2
consistently ordered (CO)	5.3
generalized consistently ordered (GCO)	13.2
(q, r)-consistently ordered (CO(q, r))	13.2
generalized (q, r)-consistently ordered (GCO(q, r))	13.4
π-consistently ordered (π-CO)	13.4
generalized π-consistently ordered (π-GCO)	14.3
Property $A^{(\pi)}$	14.3

LIST OF ITERATIVE METHODS

Linear Stationary Methods of First Degree

$$u^{(n+1)} = Gu^{(n)} + k$$

Abbreviation	*Name*	*First appears in section*	*G*
J	Jacobi	3.3	$B = D^{-1}C$
JOR	simultaneous overrelaxation	3.3	$B_\omega = \omega B + (1 - \omega)I$
GS	Gauss–Seidel	3.3	$\mathscr{L} = (I - L)^{-1}U$
SOR	successive overrelaxation	3.3	$\mathscr{L}_\omega$ (forward) $= (I - \omega L)^{-1}(\omega U + (1 - \omega)I)$
	successive overrelaxation	7.5	$\mathscr{U}_\omega$ (backward) $= (I - \omega U)^{-1}(\omega L + (1 - \omega)I)$
RF	—	3.3	$R_p = I + pA$
GRF	—	3.3	$R_P = I + PA$
MSOR	modified SOR	8.1	$\mathscr{L}_{\omega,\omega'}$ (forward)
	modified SOR	8.4	$\mathscr{U}_{\omega,\omega'}$ (backward)
SSOR	symmetric SOR	15.1	$\mathscr{S}_\omega = \mathscr{U}_\omega \mathscr{L}_\omega$
USSOR	unsymmetric SSOR	15.6	$\mathscr{C}_{\omega,\bar{\omega}} = \mathscr{U}_{\bar{\omega}} \mathscr{L}_\omega$
SMSOR	symmetric MSOR	15.7	$\mathscr{S}_{\omega,\omega'} = \mathscr{U}_{\omega,\omega'} \mathscr{L}_{\omega,\omega'}$
USMSOR	unsymmetric MSOR	15.7	$\mathscr{W}_{\omega,\omega',\bar{\omega},\bar{\omega}'} = \mathscr{U}_{\bar{\omega},\bar{\omega}'} \mathscr{L}_{\omega,\omega'}$
PR	Peaceman–Rachford	17.1	$T_{\varrho,\varrho'} = (V + \varrho' I)^{-1}(H - \varrho' I)(H + \varrho I)^{-1}(V - \varrho I)$

Nonstationary Methods

$$u^{(n+1)} = \mathscr{G}_n u^{(0)} + k_n$$

Semi-iterative methods (Chapter 11) are a special case. Semi-iterative methods based on the J method, JOR method, etc. are referred to as the J–SI method, the JOR–SI method, etc.

Abbreviation	*Name*	*First appears in section*	$\mathscr{G}_n$
CCSI	cyclic Chebyshev semi-iterative	10.6	$\mathscr{C}_n$
—	Sheldon's method	10.4	$\mathscr{H}_n = \mathscr{L}_{\omega_b}^{n-1}\mathscr{L}$
—	modified Sheldon	10.5	$\mathscr{K}_n = \mathscr{L}_{\omega_b}^{n-1}\mathscr{L}_{1,\omega_b}$
J–SI	—	11.3	$\mathscr{B}_n$
RF–SI	—	11.4	$\mathscr{Q}_n$
GS–SI	—	11.6	$\mathscr{L}_n$
GS–SSI	Sheldon's modification of the GS–SI method	11.6	$\mathscr{D}_n$
—	Richardson's method	11.4	—
SOR–SI	—	11.7	—
MSOR–SI	—	11.8	—

Properties of Iterative Methods

Property	*Section*
consistent	3.3
reciprocally consistent	3.3
completely consistent	3.3
convergent	3.5
weakly convergent	3.5
strongly convergent	6.1

Special Notation

$\bar{\mu} = S(B)$	
$\omega_b = 2[1 + (1 - \bar{\mu}^2)^{1/2}]^{-1}$	optimum relaxation factor
$r = \omega_b - 1$	$S(\mathscr{L}_{\omega_b})$

ITERATIVE SOLUTION OF LARGE LINEAR SYSTEMS

Chapter 1 / INTRODUCTION

In this book we study numerical methods for solving linear systems of the form

$$Au = b \tag{1}$$

where A is a given real $N \times N$ matrix and b is a given real column vector of order N. It is desired to determine the unknown column vector u. For the case $N = 3$, the system (1) may be written in the alternative forms

$$\begin{pmatrix} a_{1,1} & a_{1,2} & a_{1,3} \\ a_{2,1} & a_{2,2} & a_{2,3} \\ a_{3,1} & a_{3,2} & a_{3,3} \end{pmatrix} \begin{pmatrix} u_1 \\ u_2 \\ u_3 \end{pmatrix} = \begin{pmatrix} b_1 \\ b_2 \\ b_3 \end{pmatrix} \tag{2}$$

or

$$\begin{aligned} a_{1,1}u_1 + a_{1,2}u_2 + a_{1,3}u_3 &= b_1 \\ a_{2,1}u_1 + a_{2,2}u_2 + a_{2,3}u_3 &= b_2 \\ a_{3,1}u_1 + a_{3,2}u_2 + a_{3,3}u_3 &= b_3 \end{aligned} \tag{3}$$

We shall be primarily concerned with cases where N is large, say in the range 10^3 to 10^6, and where the matrix A is "sparse," i.e., has only a few nonzero elements as compared to the total number of elements of A. We shall assume that the matrix has one or more of a number of properties, described in Chapter 2, such as weak diagonal dominance, irreducibility, symmetry, etc. In nearly every case the assumed properties will be sufficient to guarantee that the matrix A is nonsingular.

We shall study various iterative methods for solving (1). Such methods appear to be ideally suited for problems involving large sparse matrices, much more so in most cases than direct methods such as the Gauss elimination method. A typical iterative method involves the selection of an initial approximation $u^{(0)}$ to the solution of (1) and the determination of a sequence $u^{(1)}, u^{(2)}, \ldots$ according to some algorithm, which, if the method is properly chosen, will converge to the exact solution $\bar{u}$ of (1). The use of such an algorithm has the advantage that the matrix A is not altered during the computation. Hence, though the computation may be long, the problem of the accumulation of rounding errors is less serious than for those methods, such as most direct methods, where the matrix is changed during the computation process.

1.1. THE MODEL PROBLEM

In order to illustrate our discussion we shall frequently consider the following *model problem*. Let $G(x, y)$ and $g(x, y)$ be continuous functions defined in R and S, respectively, where R is the interior and S is the boundary of the unit square $0 \leq x \leq 1$, $0 \leq y \leq 1$. We seek a function $u(x, y)$ continuous in $R + S$, which is twice continuously differentiable in R and which satisfies *Poisson's equation*

$$\frac{\partial^2 u}{\partial x^2} + \frac{\partial^2 u}{\partial y^2} = G(x, y) \tag{1.1}$$

On the boundary, $u(x, y)$ satisfies the condition

$$u(x, y) = g(x, y) \tag{1.2}$$

If $G(x, y) \equiv 0$, then (1.1) reduces to Laplace's equation

$$\frac{\partial^2 u}{\partial x^2} + \frac{\partial^2 u}{\partial y^2} = 0 \tag{1.3}$$

and the model problem is a special case of the *Dirichlet problem*.

We shall primarily be concerned with the linear system arising from the numerical solution of the model problem using a five-point difference equation. We superimpose a mesh of horizontal and vertical lines over the region with a uniform spacing $h = M^{-1}$, for some integer M. We seek to determine approximate values of $u(x, y)$ at the mesh points, i.e., at the intersections of these lines. For a given mesh point (x, y) we use the usual central difference quotients to approximate the partial derivatives. Thus we use the approximations

$$\begin{aligned} \partial^2u/\partial x^2 &\sim [u(x+h, y) + u(x-h, y) - 2u(x, y)]/h^2 \\ \partial^2u/\partial y^2 &\sim [u(x, y+h) + u(x, y-h) - 2u(x, y)]/h^2 \end{aligned} \tag{1.4}$$

Replacing the partial derivatives by difference quotients in (1.1) and multiplying by $-h^2$ we obtain the difference equation

$$\begin{aligned} 4u(x, y) - u(x+h, y) - u(x-h, y) - u(x, y+h) - u(x, y-h) \\ = -h^2G(x, y) \end{aligned} \tag{1.5}$$

This equation together with the boundary condition (1.2) defines a discrete analog of the model problem. Evidently from (1.5) we obtain one linear algebraic equation for each interior mesh point, i.e., one equation for each unknown value of u.

To be specific, if $G(x, y) \equiv 0$ and if $h = \frac{1}{3}$, we have the situation shown in Figure 1.1. We seek approximate values of $u_1 = u(x_1, y_1)$, u_2, u_3, and u_4. The values of u at the points labeled 5, 6, ... are determined by (1.2). Thus we have $u_5 = g_5$, $u_6 = g_6$, etc. From (1.5) we obtain

$$\begin{aligned} 4u_1 - u_2 - u_3 - u_6 - u_{16} &= 0 \\ 4u_2 - u_{13} - u_4 - u_1 - u_{15} &= 0 \\ 4u_3 - u_4 - u_9 - u_7 - u_1 &= 0 \\ 4u_4 - u_{12} - u_{10} - u_3 - u_2 &= 0 \end{aligned} \tag{1.6}$$

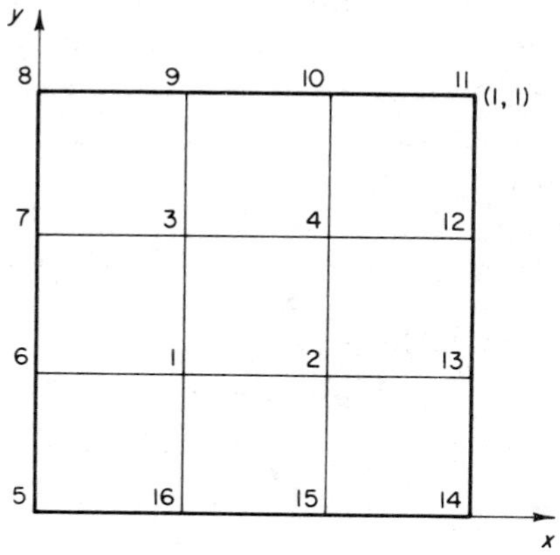

Figure 1.1. The model problem with $h = \frac{1}{3}$.

Moving the known values to the right-hand side, we have the linear system

$$\begin{aligned}
4u_1 - u_2 - u_3 \qquad\qquad &= u_6 + u_{16} \\
-u_1 + 4u_2 \qquad\qquad - u_4 &= u_{13} + u_{15} \\
-u_1 \qquad + 4u_3 - u_4 &= u_7 + u_9 \\
- u_2 - u_3 - 4u_4 &= u_{10} + u_{12}
\end{aligned} \tag{1.7}$$

which can be written in the form (1),

$$\begin{pmatrix} 4 & -1 & -1 & 0 \\ -1 & 4 & 0 & -1 \\ -1 & 0 & 4 & -1 \\ 0 & -1 & -1 & 4 \end{pmatrix} \begin{pmatrix} u_1 \\ u_2 \\ u_3 \\ u_4 \end{pmatrix} = \begin{pmatrix} u_6 + u_{16} \\ u_{13} + u_{15} \\ u_7 + u_9 \\ u_{10} + u_{12} \end{pmatrix} \tag{1.8}$$

Suppose now that $g(x, y) = 0$ on all sides of the square except where $y = 1$, where $g(x, y) = 4500x(1 - x)$. All boundary values vanish except $u_9 = u_{10} = 1000$. The system (1.8) becomes

$$\begin{pmatrix} 4 & -1 & -1 & 0 \\ -1 & 4 & 0 & -1 \\ -1 & 0 & 4 & -1 \\ 0 & -1 & -1 & 4 \end{pmatrix} \begin{pmatrix} u_1 \\ u_2 \\ u_3 \\ u_4 \end{pmatrix} = \begin{pmatrix} 0 \\ 0 \\ 1000 \\ 1000 \end{pmatrix} \tag{1.9}$$

We can solve (1.9) directly, obtaining

$$u_1 = 125, \qquad u_2 = 125, \qquad u_3 = 375, \qquad u_4 = 375$$

While the matrix of (1.9) is not particularly sparse, we note that no matter how small h is, there will not be more than five nonzero elements in any row or column of the matrix. Thus, if $h = \frac{1}{20}$ the matrix has order $19^2 = 361$. However, out of $361^2 = 130{,}321$ elements in the matrix, fewer than 1800 are nonzero. Thus, by any reasonable definition the matrix is indeed sparse.

We now apply a very simple iterative method to the system (1.9). We simply solve the first equation for u_1, the second for u_2, etc., obtaining

$$\begin{aligned}
u_1 &= \qquad \tfrac{1}{4}u_2 + \tfrac{1}{4}u_3 \\
u_2 &= \tfrac{1}{4}u_1 \qquad\qquad + \tfrac{1}{4}u_4 \\
u_3 &= \tfrac{1}{4}u_1 \qquad\qquad + \tfrac{1}{4}u_4 + 250 \\
u_4 &= \qquad \tfrac{1}{4}u_2 + \tfrac{1}{4}u_3 \qquad + 250
\end{aligned} \tag{1.10}$$

At any stage, given $u_1^{(n)}, u_2^{(n)}, u_3^{(n)}, u_4^{(n)}$, we determine $u_1^{(n+1)}, u_2^{(n+1)}, \ldots$ by

the above equations using $u_1^{(n)}, u_2^{(n)}, \ldots$ in the right-hand sides. Thus, we have

$$\begin{aligned} u_1^{(n+1)} &= \tfrac{1}{4}u_2^{(n)} + \tfrac{1}{4}u_3^{(n)} \\ u_2^{(n+1)} &= \tfrac{1}{4}u_1^{(n)} + \tfrac{1}{4}u_4^{(n)} \\ u_3^{(n+1)} &= \tfrac{1}{4}u_1^{(n)} + \tfrac{1}{4}u_4^{(n)} + 250 \\ u_4^{(n+1)} &= \tfrac{1}{4}u_2^{(n)} + \tfrac{1}{4}u_3^{(n)} + 250 \end{aligned} \tag{1.11}$$

This method is known as the *Jacobi* method. Thus, letting $u_1^{(0)} = u_2^{(0)} = u_3^{(0)} = u_4^{(0)} = 0$, we have the values shown in the accompanying tabulation.

$u^{(n)}$ \ n	0	1	2	3	4	5	6
$u_1^{(n)}$	0	0	62.5	93.75	109.375	117.175	121.0875
$u_2^{(n)}$	0	0	62.5	93.75	109.375	117.175	121.0875
$u_3^{(n)}$	0	250	312.5	343.75	359.375	367.175	371.0875
$u_4^{(n)}$	0	250	312.5	343.75	359.325	367.175	371.0875

$u^{(n)}$ \ n	7	8	9	10	11	12
$u_1^{(n)}$	123.04	124.02	124.51	124.75	124.88	124.94
$u_2^{(n)}$	123.04	124.02	124.51	124.75	124.88	124.94
$u_3^{(n)}$	373.04	374.02	374.51	374.75	374.88	374.94
$u_4^{(n)}$	373.04	374.02	374.51	374.75	374.88	374.94

While this method converges fairly rapidly in the simple case shown, it is very slow indeed if the mesh size h is small. For example, if the mesh size $h = \frac{1}{80}$ is used, it would require approximately 18,000 iterations to obtain convergence to six significant figures. This number can be reduced by approximately 50% by using the Gauss–Seidel method which involves using values of $u_i^{(n+1)}$ whenever available in place of $u_i^{(n)}$. Thus, for instance, in the second equation of (1.11) one would use the formula

$$u_2^{(n+1)} = \tfrac{1}{4}u_1^{(n+1)} + \tfrac{1}{4}u_4^{(n)}$$

Neither the Jacobi nor the Gauss–Seidel method is satisfactory even for the model problem unless h is fairly large. By a slight modification of the Gauss–Seidel method one can dramatically improve the convergence in certain cases. Still further improvements can also be obtained by other methods in certain cases. Our purpose in this book is to study a number of iterative methods from the standpoints of speed of convergence and suitability for use on high-speed computers. In order to do so we shall make extensive use of matrix theory. In Chapter 2 we shall review matrix theory and define certain matrix properties which we shall use in our studies.

SUPPLEMENTARY DISCUSSION

The use of direct methods, even for solving very large problems, has received increasing attention recently (see, for instance, Angel, 1970). In some cases their use is quite appropriate. However, there is the danger that if one does not properly apply iterative methods in certain cases one will incorrectly conclude that they are not effective and that direct methods must be used. It is hoped that this book will provide some guidance for the proper application of iterative methods so that a decision on whether or not to use them can be made on a sound basis.

Section 1.1. The model problem has been used frequently in the study of the behavior of iterative methods for solving large linear systems. The term "model problem" for the problem (1.2)–(1.3) was introduced by Varga [1962].

EXERCISES

Section 1.1

1. Solve (1.10) using the Gauss–Seidel method. Use the same starting values as were used for the Jacobi method.

2. For the numerical results given for the Jacobi method, determine $\varepsilon_i^{(n)} = u_i^{(n)} - u_i$ where the $\{u_i\}$ are exact values. Show that the ratios $\varepsilon^{(n+1)}/\varepsilon^{(n)}$ approach a limit. Do the same for the Gauss–Seidel method.

3. Consider the discrete analog of the model problem for (1.1) where $G(x, y) = x + y + 1$ and $g(x, y) = 1 + x^2$. Determine the numerical solution for the case $h = \frac{1}{3}$ by: direct elimination; by the Jacobi method; by the Gauss–Seidel method. Use starting values of zero and iterate until the iterants do not change in the third significant figure.

4. For the model problem with $h = \frac{1}{40}$ what is the ratio of nonvanishing elements of A to the total number?

Chapter 2 / **MATRIX PRELIMINARIES**

In this chapter we shall present basic facts about matrix theory which we shall later use in the study of iterative methods. We assume that the reader is already familiar with the general theory of matrices as presented, for instance, in Faddeev and Faddeeva [1963] and Birkhoff and MacLane [1953]. We shall review some of this theory in Section 2.1 and shall also state, often without proof, some of the less elementary facts which will be used later. Special attention is given to Hermitian and positive definite matrices in Section 2.2. Sections 2.3 and 2.4 are devoted to vector and matrix norms and to the convergence of sequences of vectors and matrices. Sections 2.5–2.7 are devoted to matrices with certain properties. Finally, in Section 2.8 we show how one obtains one or more of these properties when one uses finite difference methods to obtain numerical solutions of certain problems involving partial differential equations.

2.1. REVIEW OF MATRIX THEORY

A *rectangular* $N \times M$ *matrix* A is a rectangular array of complex numbers

$$A = (a_{i,j}) = \begin{pmatrix} a_{1,1} & a_{1,2} & \cdots & a_{1,M} \\ a_{2,1} & a_{2,2} & \cdots & a_{2,M} \\ \vdots & \vdots & \cdots & \vdots \\ a_{N,1} & a_{N,2} & \cdots & a_{N,M} \end{pmatrix} \tag{1.1}$$

If $N = M$, then the matrix is said to be a *square matrix of order N*. We shall use capital letters for matrices, and, unless otherwise specified, a matrix indicated by a capital letter is assumed to be square. The elements of a matrix will be represented by lowercase letters with two subscripts.

If $N = 1$, the matrix is a *row matrix* or *row vector*. Similarly, if $M = 1$ the matrix is a *column matrix* or *column vector*. We shall denote a column vector by a small letter and the elements of the vector will be denoted by lowercase letters with a single subscript. Thus we have

$$v = \begin{pmatrix} v_1 \\ v_2 \\ \cdot \\ \cdot \\ \cdot \\ v_N \end{pmatrix} \tag{1.2}$$

The *diagonal elements* of a matrix A are the elements $a_{1,1}, a_{2,2}, \ldots, a_{N,N}$. For a *diagonal matrix*, all off-diagonal elements vanish. The *identity matrix* of order N, usually denoted by I, is a diagonal matrix all of whose diagonal elements are unity. If $a_{i,j} = 0$ for $i < j$ ($i \leq j$), then A is a *lower triangular* (*strictly lower triangular*) matrix. If $a_{i,j} = 0$ for $i > j$ ($i \geq j$), then A is an *upper triangular* (*strictly upper triangular*) matrix.

The *sum* of two $N \times M$ matrices $A = (a_{i,j})$ and $B = (b_{i,j})$ is an $N \times M$ matrix $C = (c_{i,j})$ where $c_{i,j} = a_{i,j} + b_{i,j}$. The product of an $N \times J$ matrix A and a $J \times M$ matrix B is an $N \times M$ matrix C where

$$c_{i,j} = \sum_{k=1}^{J} a_{i,k} b_{k,j}, \qquad i = 1, 2, \ldots, N, \qquad j = 1, 2, \ldots, M \tag{1.3}$$

The *trace* of a matrix $A = (a_{i,j})$ is given by

$$\text{trace } A = \sum_{i=1}^{N} a_{i,i} \tag{1.4}$$

We define the matrix obtained from A by replacing all off-diagonal elements of A by zero as "diag A." Thus we have

$$\text{diag } A = \begin{pmatrix} a_{1,1} & 0 & 0 & \cdots & 0 \\ 0 & a_{2,2} & 0 & \cdots & 0 \\ 0 & 0 & a_{3,3} & \cdots & 0 \\ \vdots & \vdots & \vdots & \cdots & \vdots \\ 0 & 0 & 0 & \cdots & a_{N,N} \end{pmatrix} \tag{1.5}$$

Given an $N \times M$ matrix $A = (a_{i,j})$ we obtain the *transpose of A*, denoted by A^{T} by interchanging the rows and columns of A. Thus if A is given by (1.1), then

$$A^{\mathrm{T}} = \begin{pmatrix} a_{1,1} & a_{2,1} & \cdots & a_{N,1} \\ a_{1,2} & a_{2,2} & \cdots & a_{N,2} \\ \vdots & \vdots & \cdots & \vdots \\ a_{1,M} & a_{2,M} & \cdots & a_{N,M} \end{pmatrix} \tag{1.6}$$

Given a complex number z we denote the complex conjugate of z by z^* (rather than $\bar{z}$ as is more usual). The *conjugate* of A, denoted by A^* is obtained from A by replacing each element by its complex conjugate. Thus

$$A^* = \begin{pmatrix} a_{1,1}^* & a_{1,2}^* & \cdots & a_{1,M}^* \\ a_{2,1}^* & a_{2,2}^* & \cdots & a_{2,M}^* \\ \vdots & \vdots & \cdots & \vdots \\ a_{N,1}^* & a_{N,2}^* & \cdots & a_{N,M}^* \end{pmatrix} \tag{1.7}$$

The *conjugate transpose* of A, denoted by A^{H}, is the transpose of A^* (also the conjugate of A^{T}). Thus we have

$$A^{\mathrm{H}} = (A^*)^{\mathrm{T}} = (A^{\mathrm{T}})^* = \begin{pmatrix} a_{1,1}^* & a_{2,1}^* & \cdots & a_{N,1}^* \\ a_{1,2}^* & a_{2,2}^* & \cdots & a_{N,2}^* \\ \vdots & \vdots & \cdots & \vdots \\ a_{1,M}^* & a_{2,M}^* & \cdots & a_{N,M}^* \end{pmatrix} \tag{1.8}$$

Evidently we have

$$(A^{\mathrm{H}})^{\mathrm{H}} = (A^{\mathrm{T}})^{\mathrm{T}} = (A^*)^* = A \tag{1.9}$$

and for any two rectangular matrices A and B such that AB is defined we have

$$(AB)^{\mathrm{H}} = B^{\mathrm{H}}A^{\mathrm{H}}, \qquad (AB)^{\mathrm{T}} = B^{\mathrm{T}}A^{\mathrm{T}}, \qquad (AB)^* = A^*B^* \tag{1.10}$$

If $A = A^{\mathrm{T}}$, then A is *symmetric* (usually the property of symmetry is useful only for real matrices). If $A = A^{\mathrm{H}}$, then A is *Hermitian*. If A is real and $AA^{\mathrm{T}} = I$, then A is *orthogonal*. If $AA^{\mathrm{H}} = I$, then A is *unitary*. If $AA^{\mathrm{H}} = A^{\mathrm{H}}A$, then A is *normal*.

If A is square and if there exists a square matrix B such that $AB = I$ or $BA = I$, then A is *nonsingular* and we let $B = A^{-1}$. If A is nonsingular, then the *inverse matrix*, denoted by A^{-1}, is unique and $AA^{-1} = A^{-1}A = I$. Moreover, if A and B are nonsingular, then AB is nonsingular and

$$(AB)^{-1} = B^{-1}A^{-1} \tag{1.11}$$

A *permutation matrix* $P = (p_{i,j})$ is a matrix with exactly one nonzero element, namely unity, in each row and each column. Thus, for example,

$$P = \begin{pmatrix} 0 & 1 & 0 & 0 \\ 0 & 0 & 0 & 1 \\ 0 & 0 & 1 & 0 \\ 1 & 0 & 0 & 0 \end{pmatrix}$$

is a permutation matrix. A permutation matrix corresponds to a permutation function, $\sigma(i)$, defined for $i = 1, 2, \ldots, N$ by

$$\sigma(i) = j$$

where $p_{i,j} = 1$. In general, a permutation function σ is a single-valued function defined for $i = 1, 2, \ldots, N$ and with a single-valued inverse σ^{-1} also defined for $i = 1, 2, \ldots, N$. Thus in the above example we have

$$\sigma(1) = 2, \qquad \sigma(2) = 4, \qquad \sigma(3) = 3, \qquad \sigma(4) = 1$$

and

$$\sigma^{-1}(1) = 4, \qquad \sigma^{-1}(2) = 1, \qquad \sigma^{-1}(3) = 3, \qquad \sigma^{-1}(4) = 2$$

It is easy to verify that if P corresponds to σ, then $P^{-1} = P^{\mathrm{T}}$ corresponds to σ^{-1}. Thus in the example we have

$$P^{-1} = P^{\mathrm{T}} = \begin{pmatrix} 0 & 0 & 0 & 1 \\ 1 & 0 & 0 & 0 \\ 0 & 0 & 1 & 0 \\ 0 & 1 & 0 & 0 \end{pmatrix}$$

which corresponds to σ^{-1}.

Vector Spaces

Let E^N denote the set of all $N \times 1$ column matrices, which we shall henceforth refer to as *vectors*. The reader is assumed to be familiar with the notions of vector spaces and subspaces, linear independence, etc. A *basis* for E^N is a set of N linearly independent vectors of E^N. Given such a basis, say $v^{(1)}, v^{(2)}, \ldots, v^{(N)}$, then any vector $v \in E^N$ can be expressed uniquely as a linear combination of basis vectors. Thus there exists a unique set of numbers $c_1, c_2, \ldots, c_N$ such that

$$v = \sum_{i=1}^{N} c_i v^{(i)} \tag{1.12}$$

For any subspace W of E^N, every basis for W has the same number of vectors. This number, which is independent of the choice of bases, is known as the *dimension* of W.

Given any two vectors $v = (v_1, v_2, \ldots, v_N)^T$ and $w = (w_1, w_2, \ldots, w_N)^T$ we define the *inner product* of v and w by

$$(v, w) = v^H w = \sum_{i=1}^{N} v_i^* w_i \tag{1.13}$$

Evidently we have for any v and w

$$(v, w) = (w, v)^* \tag{1.14}$$

and for any matrix A

$$(v, Aw) = (A^H v, w) \tag{1.15}$$

We define the *Euclidean length* of a vector $v = (v_1, v_2, \ldots, v_N)^T$ as

$$\| v \|_2 = (v, v)^{1/2} = \left(\sum_{i=1}^{N} | v_i |^2 \right)^{1/2} \tag{1.16}$$

It is easy to verify the Schwarz inequality which states that

$$| (v, w) | \leq \| v \|_2 \| w \|_2 \tag{1.17}$$

Two vectors v and w are *orthogonal* if $(v, w) = 0$. A set of vectors $v^{(1)}, v^{(2)}, \ldots, v^{(p)}$ is *pairwise orthogonal* if $(v^{(i)}, v^{(j)}) = 0$ for $i \neq j$. They are *orthonormal* if

$$(v^{(i)}, v^{(j)}) = \begin{cases} 1 & \text{if } i = j \\ 0 & \text{if } i \neq j \end{cases} \tag{1.18}$$

Any set of nonvanishing pairwise orthogonal vectors is linearly independent. The columns of an orthogonal matrix form an orthonormal set of vectors as well as a basis for E^N. The same is true for the columns of a unitary matrix.

Determinants

The *determinant* of an $N \times N$ matrix A, denoted by $\det A$, is the sum of $N!$ terms of the form

$$t(\sigma) = s(\sigma)a_{1,\sigma(1)}a_{2,\sigma(2)} \cdots a_{N,\sigma(N)} \tag{1.19}$$

where σ is a permutation of the integers $1, 2, \ldots, N$. Here $s(\sigma) = 1$ or $s(\sigma) = -1$ if the permutation σ is *even* or *odd*, respectively. A permutation σ is even if the sequence $\sigma(1), \sigma(2), \ldots, \sigma(N)$ can be put in the form $1, 2, \ldots, N$ by an even number of *transpositions*, i.e., interchanges of any pair of distinct elements of the sequence. Otherwise the permutation is odd. Thus the permutation $\sigma(1) = 2$, $\sigma(2) = 4$, $\sigma(3) = 5$, $\sigma(4) = 1$, $\sigma(5) = 3$, $\sigma(6) = 6$ is odd since the sequence 2, 4, 5, 1, 3, 6 can be put in the form 1, 2, 3, 4, 5, 6 by the three transpositions: $1 \leftrightarrow 2$, $2 \leftrightarrow 4$, $3 \leftrightarrow 5$. In practice it is usually more convenient to count the number of *inversions* in the array $\sigma(1), \sigma(2), \ldots, \sigma(N)$. An inversion occurs when $\sigma(j) < \sigma(i)$ even though $i < j$. Thus in the example we have the five inversions (2,1), (4,1), (4,3), (5,1), (5,3). Using either procedure we find that the sign of the term $a_{1,2}a_{2,4}a_{3,5}a_{4,1}a_{5,3}a_{6,6}$ in the expansion of the determinant of the 6×6 matrix $A = (a_{i,j})$ is negative.

We shall assume that the basic properties of determinants are known. We now give an alternative method for determining the sign of a term in a determinant which will be used in Chapter 10. A *cycle* is a permutation σ such that $\sigma(i) = i$ except that for s distinct integers $i_1, i_2, \ldots, i_s$ we have

$$\sigma(i_1) = i_2, \qquad \sigma(i_2) = i_3, \ldots, \sigma(i_{s-1}) = i_s, \qquad \sigma(i_s) = i_1$$

We designate such a cycle by $(i_1, i_2, \ldots, i_s)$. A cycle is *even* or *odd* depending on whether s is even or odd. Any permutation can be written as the product of disjoint cycles, and the permutation is even or odd depending on whether or not there are an even number of even cycles. Thus in the example given above, σ is the product of the cycles (1, 2, 4), (3, 5), (6). Since there is only one even cycle, the permutation is odd.

We now state without proof

Theorem 1.1. If A and B are $N \times N$ matrices, then

$$\det AB = \det A \det B = \det BA \tag{1.20}$$

Systems of Linear Equations

Of fundamental importance is the following theorem.

Theorem 1.2. If $\det A \neq 0$, then for given b the linear system

$$Au = b \tag{1.21}$$

has a unique solution which is given by

$$u_k = \Delta_k / \det A, \qquad k = 1, 2, \ldots, N \tag{1.22}$$

where $\Delta_k = \det A_k$ and where A_k is the matrix obtained from A by replacing the kth column of A by b.

The above representation of the solution of a linear system is known as *Cramer's rule.* While important theoretically, it is not practical to actually carry out unless the order of A is very small. If one wishes to use a direct method, the Gauss elimination method is much superior. As already stated, we shall be primarily concerned with iterative methods as applied to very large systems with sparse matrices.

Theorem 1.3. A matrix A is nonsingular if and only if $\det A \neq 0$.

One can in principle determine A^{-1} by solving the N linear systems $Av^{(k)} = e^{(k)}$, $k = 1, 2, \ldots, N$, where $e_i^{(k)} = 1$ if $i = k$ and $e_i^{(k)} = 0$ otherwise. The matrix whose columns are the $v^{(k)}$ is A^{-1}. Once one has determined A^{-1} one can easily solve (1.21) for any b by multiplying both sides by A^{-1} obtaining

$$u = A^{-1}b \tag{1.23}$$

Theorem 1.4. There exists a vector $u \neq 0$ such that $Au = 0$ if and only if A is singular.

The fact that $Au = 0$ has only the trivial solution $u = 0$ if A is nonsingular follows from Theorem 1.2. We also have

Theorem 1.5. The linear system (1.21) has a unique solution if and only if A is nonsingular. If A is singular, then (1.21) either has no solution or else it has an infinite number.

Proof. Suppose that A is singular and (1.21) has a solution, say $\bar{u}$. By Theorem 1.4 there exists a solution $v \neq 0$ of $Av = 0$. Hence for any constant c the vector $\bar{u} + cv$ satisfies (1.21). Hence the solution of (1.21) is not unique and, in fact, there are an infinite number of solutions. The rest of the proof follows from Theorem 1.2.

The *null space* $\mathcal{N}(A)$ of an $N \times M$ matrix A is the set of all $v \in E^M$ such that $Av = 0$. The *range* $\mathcal{R}(A)$ of A is the set of all $w \in E^N$ such that for some $x \in E^M$ we have $Ax = w$. Evidently, if A is an $N \times M$ matrix, then (1.21) has a solution if and only if $b \in \mathscr{R}(A)$.

The *column rank* of an $N \times M$ matrix A is the dimension of the subspace of E^N spanned by the columns of A. Similarly, the *row rank* is the dimension of the subspace of E^M spanned by the rows of A. The row rank and the column rank are equal, and their common value equals the *rank*. If P is a nonsingular $N \times N$ matrix, then PA has the same rank as A. For a square matrix of order N the sum of the row rank and the *nullity* (i.e., the dimension of the null space) is equal to the order of the matrix. The system (1.21), where A is an $N \times M$ matrix, has a solution if and only if the rank of the augmented matrix $(A\ b)$ is the same as the rank of A.

Eigenvalues and Eigenvectors

An *eigenvalue* of a matrix A is a real or complex number λ such that for some $v \neq 0$ we have

$$Av = \lambda v \tag{1.24}$$

An *eigenvector* of A is a vector v such that $v \neq 0$ and for some λ (1.24) holds. From Theorem 1.4 we have

Theorem 1.6. The number λ is an eigenvalue of A if and only if

$$\det(A - \lambda I) = 0 \tag{1.25}$$

Evidently (1.25) is a polynomial equation, referred to as the *characteristic equation of* A, of the form

$$(-1)^N \det(A - \lambda I) = \lambda^N - (\text{trace } A)\lambda^{N-1} + \cdots + (-1)^N \det A = 0 \tag{1.26}$$

Since the sum and product of the roots of (1.26) are trace A and det A, respectively, we have

Theorem 1.7. If A is a square matrix of order N with eigenvalues $\lambda_1, \lambda_2, \ldots, \lambda_N$, then

$$\det A = \prod_{i=1}^{N} \lambda_i, \qquad \text{trace } A = \sum_{i=1}^{N} \lambda_i \tag{1.27}$$

Theorem 1.8. If $\lambda_1, \lambda_2, \ldots, \lambda_k$ are distinct eigenvalues of A and if $v^{(1)}, v^{(2)}, \ldots, v^{(k)}$ are associated eigenvectors, then the $\{v^{(k)}\}$ are linearly independent.

Two matrices A and B of order N are *similar* if $B = P^{-1}AP$ for some nonsingular matrix P. Similar matrices have the same rank.

Theorem 1.9. If A and B are similar, then they have the same eigenvalues. Moreover, λ is an eigenvalue of A of multiplicity k if and only if λ is an eigenvalue of B of multiplicity k.

We now define the *spectral radius* of a matrix A as

$$S(A) = \max_{\lambda \in S_A} |\lambda| \tag{1.28}$$

where S_A is the set of all eigenvalues of A (i.e., the *spectrum* of A).

The eigenvalues of a matrix are continuous functions of the elements of the matrix. Thus we have

Theorem 1.10. Given a matrix $A = (a_{i,j})$ and any $\varepsilon > 0$ there exists $\delta > 0$ such that if $A' = (a'_{i,j})$ is any matrix with $|a'_{i,j} - a_{i,j}| < \delta$ for all i and j, the following holds: If $\lambda_1, \lambda_2, \ldots, \lambda_N$ are the eigenvalues of A and $\lambda_1', \lambda_2', \ldots, \lambda_N'$ are the eigenvalues of A', then for some permutation $\sigma(i)$ of the integers $1, 2, \ldots, N$ we have

$$|\lambda_i - \lambda'_{\sigma(i)}| < \varepsilon, \qquad i = 1, 2, \ldots, N \tag{1.29}$$

(A proof of this result can be found in Appendix A of Ostrowski [1960].)

As an application of Theorem 1.10 we have

Theorem 1.11. If A and B are square matrices, then AB and BA have the same eigenvalues. Moreover, λ is an eigenvalue of AB of multiplicity k if and only if λ is an eigenvalue of BA of multiplicity k.

Proof. If B is nonsingular, then B^{-1} exists and BA is similar to $B^{-1}(BA)B = AB$, and the result follows from Theorem 1.9. A similar argument holds if A is nonsingular.

If both A and B are singular, then 0 is an eigenvalue of B. Let δ be the modulus of the nonzero eigenvalue of B of smallest modulus. If $0 < \varepsilon < \delta$, then $B + \varepsilon I$ is nonsingular and the eigenvalues of $A(B + \varepsilon I)$ are the same as those of $(B + \varepsilon I)A$. As $\varepsilon \to 0$ both sets of eigenvalues converge to the eigenvalues of AB and BA, respectively, by Theorem 1.10.

A somewhat weaker version of Theorem 1.11 can be proved without using the continuity argument.

Theorem 1.12. If A is an $N \times M$ matrix and if B is an $M \times N$ matrix, then $\lambda \neq 0$ is an eigenvalue of AB if and only if λ is an eigenvalue of BA. If $M = N$, then the conclusion is true even for $\lambda = 0$.

Proof. If $\lambda \neq 0$ is an eigenvalue of AB and if v is an eigenvector, then $ABv = \lambda v \neq 0$ since $\lambda \neq 0$ and $v \neq 0$. Hence $Bv \neq 0$. Therefore, $BABv = \lambda Bv$ and λ is an eigenvalue of BA.

If A and B are square matrices and $\lambda = 0$ is an eigenvalue of AB with eigenvector v, then, by Theorems 1.1 and 1.4 we have $\det AB = \det BA = 0$. Hence $\lambda = 0$ is an eigenvalue of BA.

As an example, if $A^{\mathrm{T}} = (\frac{1}{2}, \frac{1}{2}) = B$, then

$$AB = \begin{pmatrix} \frac{1}{4} & \frac{1}{4} \\ \frac{1}{4} & \frac{1}{4} \end{pmatrix}, \qquad BA = (\tfrac{1}{2})$$

The eigenvalues of AB are $\frac{1}{2}$ and 0 while $\frac{1}{2}$ is the only eigenvalue of BA.

We remark that the conclusion of Theorem 1.12 is weaker than that of Theorem 1.11 in the case of square matrices since we do not assert that the eigenvalues of AB, together with their multiplicities, are the same as those of BA.

Canonical Forms

By means of similarity transformations involving matrices with certain properties one can reduce a given matrix to one or more special forms. In some cases one can obtain a diagonal matrix by a similarity transformation involving a unitary transformation. Thus we have

Theorem 1.13. If A is a normal matrix then there exists a unitary matrix P such that $P^{-1}AP$ is a diagonal matrix. If A is Hermitian, then $P^{-1}AP$ is real. If A is real and symmetric, then P can be taken to be an orthogonal matrix.

We remark that in each case the columns of P form an orthogonal set of eigenvectors of A.

In the general case one cannot always reduce A to diagonal form by means of unitary similarity transformations or even with more general transformations. However, one can always obtain an upper triangular form by unitary similarity transformations and a special type of bidiagonal form by more general transformations. Thus we have

Theorem 1.14. For any matrix A there exists a unitary matrix P such that $P^{-1}AP$ is an upper triangular matrix whose diagonal elements are the eigenvalues of A.

This result is due to Schur [1909]. An elementary proof is given by Rheinboldt [1969] using a result of Householder [1958]. The proof of the following theorem, on the other hand, is somewhat long and complicated.

Theorem 1.15. For any matrix A there exists a nonsingular matrix P such that $P^{-1}AP$ has the form

$$P^{-1}AP = J = \begin{pmatrix} J_1 & 0 & \cdots & 0 \\ 0 & J_2 & \cdots & 0 \\ 0 & 0 & \cdots & 0 \\ \vdots & \vdots & \cdots & \vdots \\ 0 & 0 & \cdots & J_p \end{pmatrix} \tag{1.30}$$

where each block J_i has the form

$$J_i = \begin{pmatrix} \lambda_i & 1 & 0 & \cdots & 0 & 0 \\ 0 & \lambda_i & 1 & \cdots & 0 & 0 \\ \vdots & \vdots & \vdots & \cdots & \vdots & \vdots \\ 0 & 0 & 0 & \cdots & \lambda_i & 1 \\ 0 & 0 & 0 & \cdots & 0 & \lambda_i \end{pmatrix} \tag{1.31}$$

and where the λ_i are the eigenvalues of A. Moreover, if J' is any matrix of the form of J such that for some nonsingular matrix Q we have $Q^{-1}AQ = J'$, then J and J' are identical except for possible permutation of the diagonal blocks.

The matrix J is said to be a *Jordan canonical form* of A. It is unique except as noted above. As an immediate corollary we note that if the eigenvalues of A are distinct, then the Jordan canonical form of A is diagonal. We also note that two similar matrices have the same Jordan canonical form, up to permutations of the Jordan blocks.

It is easy to see that for each block J_i there is one column of P which is an eigenvector of A. The remaining columns of P associated with J_i are said to be *principal vectors* of A. We say that the *grade* of a principal vector v is the smallest integer k such that

$$(A - \lambda I)^k v = 0 \tag{1.32}$$

where λ is the eigenvalue associated with v. An eigenvector is thus a principal vector of grade one. For a block J_i of size q there corresponds one principal vector for each grade $1, 2, \ldots, q$.

Nonnegative Matrices

If every element of A is real and nonnegative (positive), we say that A is *nonnegative* and $A \geq 0$ (*positive* and $A > 0$). If $A \geq 0$, $B \geq 0$, and $A - B \geq 0$ we say that $A \geq B$. Similarly, if every element of a vector v is nonnegative (positive) we say that $v \geq 0$ ($v > 0$). We let $|A|$ denote the matrix whose elements are the moduli of the corresponding elements of A.

The Perron–Frobenius theory of nonnegative matrices provides many theorems concerning the eigenvalues and eigenvectors of nonnegative matrices. (See, for instance, Varga [1962], Chapter 2.) We shall use the following results.

Theorem 1.16. If $A \geq |B|$, then $S(A) \geq S(B)$.

Theorem 1.17. If $A \geq 0$, then $S(A)$ is an eigenvalue of A, and there exists a nonnegative eigenvector of A associated with $S(A)$.

Proofs of Theorems 1.16 and 1.17 are given by Oldenberger [1940] and by Frobenius [1908], respectively. Many other proofs are given in the literature. In Section 2.7 we shall prove a stronger form of Theorem 1.17 for the case where A is real and symmetric.

2.2. HERMITIAN MATRICES AND POSITIVE DEFINITE MATRICES

We first prove the following characteristic property of Hermitian matrices.

Theorem 2.1. A is an Hermitian matrix if and only if (v, Av) is real for all v.

Proof. Suppose that (v, Av) is real for all v. Let $A = B + iC$, where B and C are real. If v is real, then

$$(v, Av) = (v, Bv) + i(v, Cv) \tag{2.1}$$

and hence

$$(v, Cv) = 0 \tag{2.2}$$

for all real v. Moreover, since C is real we have

$$(v, C^T v) = (C^T v, v)^* = (v, Cv)^* = 0 \tag{2.3}$$

by (2.2). Therefore we have

$$(v, (C + C^T)v) = 0 \tag{2.4}$$

for all real v. But this implies

$$C + C^T = 0 \tag{2.5}$$

To see this, consider for each i and j the vector $v^{(i,j)}$ such that $v_k^{(i,j)} = 0$ unless $k = i$ or $k = j$ and such that $v_i^{(i,j)} = v_j^{(i,j)} = 1$. In order that (2.4) be satisfied for $v^{(i,j)}$, the (i, j) element of $C + C^T$ must vanish.

Next, let $v = u + iw$ where u and w are real. Then the imaginary part of (v, Av) is

$$\begin{aligned} J &= -(w, Bu - Cw) + (u, Cu + Bw) \\ &= (u, Bw) - (w, Bu) = (u, (B - B^T)w) \end{aligned} \tag{2.6}$$

by (2.2). We now show that

$$B - B^T = 0 \tag{2.7}$$

Given i and j let

$$u_k^{(i,j)} = \begin{cases} 1, & k = i \\ 0, & k \neq i \end{cases} \qquad w_k^{(i,j)} = \begin{cases} 1, & k = j \\ 0, & k \neq j \end{cases} \tag{2.8}$$

Evidently $J = (u^{(i,j)}, (B - B^T)w^{(i,j)})$ is the (i, j) element of $B - B^T$. Since J vanishes for all real u and w, it follows that this element must vanish and (2.7) holds. By (2.5) and (2.7) it follows that A is Hermitian since

$$A^H = (B + iC)^H = B^T - iC^T = B + iC = A.$$

If A is Hermitian, then (v, Av) is real for all v since by (1.15) we have

$$(v, Av)^* = (Av, v) = (v, A^H v) = (v, Av)$$

This completes the proof of Theorem 2.1.

From Theorem 1.13 the eigenvalues of an Hermitian matrix are real. We now give a method for determining bounds on these eigenvalues.

Theorem 2.2. If A is an Hermitian matrix and if $\bar{\lambda}$ and $\underline{\lambda}$ are the largest and the smallest eigenvalues of A, respectively, then

$$\begin{aligned} \bar{\lambda} &= \max_{v \neq 0} \frac{(v, Av)}{(v, v)} = \frac{(\bar{v}, A\bar{v})}{(\bar{v}, \bar{v})} \\ \underline{\lambda} &= \min_{v \neq 0} \frac{(v, Av)}{(v, v)} = \frac{(\underline{v}, A\underline{v})}{(\underline{v}, \underline{v})} \end{aligned} \tag{2.9}$$

where $\bar{v}$ and $\underline{v}$ are eigenvectors of A corresponding to $\bar{\lambda}$ and $\underline{\lambda}$, respectively.

Proof. Let the eigenvalues of A be $\lambda_1, \lambda_2, \ldots, \lambda_N$ where $\underline{\lambda} = \lambda_1 \leq \lambda_2 \leq \cdots \leq \lambda_N = \bar{\lambda}$. By Theorem 1.13 there exists an orthonormal set of eigenvectors $v^{(1)}, v^{(2)}, \ldots, v^{(N)}$ corresponding to $\lambda_1, \lambda_2, \ldots, \lambda_N$. Since the $\{v^{(i)}\}$ form a basis of E^N, then for any $v \in E^N$ we have

$$v = \sum_{i=1}^{N} c_i v^{(i)} \tag{2.10}$$

for suitable constants $c_1, c_2, \ldots, c_N$. If $v \neq 0$, then

$$(v, v) = \sum_{i=1}^{N} | c_i |^2 \neq 0 \tag{2.11}$$

and

$$(v, Av) = \sum_{i=1}^{N} | c_i |^2 \lambda_i \tag{2.12}$$

Evidently

$$(v, Av)/(v, v) = \sum_{i=1}^{N} w_i \lambda_i \tag{2.13}$$

where

$$w_i = | c_i |^2 \Big/ \sum_{j=1}^{N} | c_j |^2 \geq 0 \tag{2.14}$$

and

$$\sum_{i=1}^{N} w_i = 1 \tag{2.15}$$

It follows that $(v, Av)/(v, v)$ is a weighted average of the λ_i with non-negative weights. Thus we have for all v

$$\frac{(v^{(1)}, Av^{(1)})}{(v^{(1)}, v^{(1)})} = \lambda_1 \leq \frac{(v, Av)}{(v, v)} \leq \lambda_N = \frac{(v^{(N)}, Av^{(N)})}{(v^{(N)}, v^{(N)})}$$

and (2.9) follows.

There are many definitions given for positive definite matrices. We shall use the following

Definition 2.1. A matrix A is *positive definite* if A is Hermitian and

$$(v, Av) > 0 \tag{2.16}$$

for all $v \neq 0$. If $(v, Av) \geq 0$ for all v, then A is *nonnegative definite.*

Similar definitions can be given for *negative definite* and *nonpositive definite* matrices.

From Theorem 2.2 we have

Theorem 2.3. A matrix A is positive definite (nonnegative definite) if and only if it is Hermitian and all of its eigenvalues are positive (nonnegative).

Theorem 2.4. The real part of a positive definite (nonnegative definite) matrix is positive definite (nonnegative definite).

Proof. Let $A = B + iC$ where B and C are real be a positive definite matrix and let $v = u + iw$ where u and w are real. Since $B = B^{\mathrm{T}}$ the imaginary part of $(v, Bv) = 0$ by (2.6). Therefore

$$(v, Bv) = (u, Bu) + (w, Bw)$$

But $(x, Bx) > 0$ for all real $x > 0$ since

$$(x, Ax) = (x, (B + iC)x) = (x, Bx) + i(x, Cx)$$

is real and positive. Therefore (v, Bv) is real and positive for all $v \neq 0$ and, by Theorem 2.1, B is positive definite.

Theorem 2.5. An Hermitian matrix of order N is positive definite (nonnegative definite) if and only if every subdeterminant formed by deleting any m rows and the corresponding columns is positive (nonnegative), where $0 \leq m < N$. In particular, $\det A$ and all diagonal elements of A are positive (nonnegative).

Proof. Let $\tilde{A}$ be any matrix formed by deleting certain rows and the corresponding columns of A. Let v be any vector such that $v_i = 0$ if the ith row and column have been deleted, and let $\tilde{v}$ be the vector formed from v by deleting the (zero) elements corresponding to deleted rows of A. Evidently, we have

$$(v, Av) = (\tilde{v}, \tilde{A}\tilde{v})$$

If A is positive definite and if $\tilde{v} \neq 0$, then $v \neq 0$ and $(v, Av) > 0$. Hence $(\tilde{v}, \tilde{A}\tilde{v}) > 0$, for all $\tilde{v}$, and $\tilde{A}$ must be positive definite. Since all eigenvalues of $\tilde{A}$ are thus positive and since the product of the eigenvalues is $\det \tilde{A}$, we have $\det \tilde{A} > 0$.

Suppose, on the other hand, that all subdeterminants are positive. Then the coefficients of the characteristic polynomial of A have alternating signs. Thus, for example, in the case $N = 4$, the characteristic polynomial is

$$\begin{aligned} &\det \begin{pmatrix} a_{1,1} - \lambda & a_{1,2} & a_{1,3} & a_{1,4} \\ a_{2,1} & a_{2,2} - \lambda & a_{2,3} & a_{2,4} \\ a_{3,1} & a_{3,2} & a_{3,3} - \lambda & a_{3,4} \\ a_{4,1} & a_{4,2} & a_{4,3} & a_{4,4} - \lambda \end{pmatrix} \\ &\quad = \lambda^4 - \lambda^3(\Delta_{1,2,3} + \Delta_{1,2,4} + \Delta_{1,3,4} + \Delta_{2,3,4}) \\ &\qquad + \lambda^2(\Delta_{1,2} + \Delta_{1,3} + \Delta_{1,4} + \Delta_{2,3} + \Delta_{2,4} + \Delta_{3,4}) \\ &\qquad + \lambda(\Delta_1 + \Delta_2 + \Delta_3 + \Delta_4) + \Delta \end{aligned}$$

where $\Delta = \det A$, Δ_i is the determinant formed by deleting the ith row and column, etc. Since all Δ's are positive, it follows that for $\lambda < 0$ the characteristic polynomial is positive; hence there can be no negative roots. Therefore, by Theorem 2.3, A is positive definite.

For any matrix L we have

$$(v, LL^{\mathrm{H}}v) = (L^{\mathrm{H}}v, L^{\mathrm{H}}v) \geq 0$$

for all v. Moreover, if $(v, LL^{\mathrm{H}}v) = 0$, and if L is nonsingular, then $L^{\mathrm{H}}v = 0$ and $v = 0$. Thus we have

Theorem 2.6. For any matrix L the matrix LL^{H} is Hermitian and nonnegative definite. If L is nonsingular, then LL^{H} is positive definite.

We now show that given a positive definite matrix A we can find a positive definite "square root." Thus we have

Theorem 2.7. If A is a positive definite (nonnegative definite) matrix, then there exists a unique positive definite (nonnegative definite) matrix B such that

$$B^2 = A \tag{2.17}$$

Moreover, if A is real, then B is real. The matrix B satisfying (2.17) is usually denoted by $A^{1/2}$.

Proof. Since A is Hermitian, by Theorem 1.13 there exists a unitary matrix P such that $P^{-1}AP = \Lambda$ where Λ is a diagonal matrix with positive diagonal elements, say λ_i. If $\Lambda^{1/2}$ is the diagonal matrix with diagonal elements $(\lambda_i)^{1/2}$, then $(\Lambda^{1/2})^2 = \Lambda$. If $B = P\Lambda^{1/2}P^{-1}$, (2.17) holds. If A is real, then P may be taken to be orthogonal, by Theorem 1.13, and hence B is real.

To prove the uniqueness let us assume that the diagonal elements of Λ are arranged in ascending order. Let C be any positive definite matrix such that $C^2 = A$. There exists a unitary matrix R such that

$$R^{-1}CR = \Gamma$$

where Γ is a diagonal matrix. Moreover,

$$\Gamma^2 = R^{-1}C^2R = R^{-1}AR$$

so that Γ^2 is a Jordan canonical form of A. Hence by Theorem 1.1 for some permutation matrix Q we have

$$\Gamma^2 = Q^{-1}\Lambda Q$$

and

$$\Lambda = Q\Gamma^2Q^{-1} = (Q\Gamma Q^{-1})^2$$

Since $Q\Gamma Q^{-1}$ is a diagonal matrix it follows that the diagonal elements of $Q\Gamma Q^{-1}$ are the (positive) square roots of the corresponding elements of Λ, i.e.,

$$Q\Gamma Q^{-1} = \Lambda^{1/2}$$

Evidently we have

$$Q^{-1}\Lambda Q = \Gamma^2 = R^{-1}C^2R = R^{-1}AR = R^{-1}P\Lambda P^{-1}R$$

or

$$E\Lambda = \Lambda E$$

where

$$E = P^{-1}RQ^{-1}$$

Since Λ is a diagonal matrix all elements of E vanish except for those elements in diagonal blocks associated with equal eigenvalues of A.

Thus we have

$$E = \begin{pmatrix} K_1 & 0 & 0 & \cdots & 0 \\ 0 & K_2 & 0 & \cdots & 0 \\ \cdot & \cdot & \cdot & \cdots & \cdot \\ 0 & 0 & 0 & \cdots & K_p \end{pmatrix}, \quad \Lambda^{1/2} = \begin{pmatrix} \lambda_1^{1/2} I_1 & 0 & 0 & \cdots & 0 \\ 0 & \lambda_2^{1/2} I_2 & 0 & \cdots & 0 \\ \cdot & \cdot & \cdot & \cdots & \cdot \\ 0 & 0 & 0 & \cdots & \lambda_p^{1/2} I_p \end{pmatrix}$$

where the K_i and I_i are diagonal blocks and $\lambda_1, \lambda_2, \ldots, \lambda_p$ are the distinct eigenvalues of A. Evidently

$$C = R\Gamma R^{-1} = RQ^{-1}\Lambda^{1/2}QR^{-1} = PE\Lambda^{1/2}E^{-1}P^{-1} = P\Lambda^{1/2}P^{-1} = B$$

and the uniqueness follows.

Corollary 2.8. If A is a positive definite matrix, then for any nonsingular matrix L the matrix M given by

$$M = LAL^{\mathrm{H}} \tag{2.18}$$

is positive definite.

Proof. Let $A^{1/2}$ be the positive definite matrix whose square is A. Since $M = LA^{1/2}(LA^{1/2})^{\mathrm{H}}$ and since $LA^{1/2}$ is nonsingular, it follows from Theorem 2.6 that M is positive definite. The corollary also can be proved directly by noting that $(u, LAL^{\mathrm{H}}u) = (L^{\mathrm{H}}u, AL^{\mathrm{H}}u) > 0$ if $u \neq 0$.

Wachspress [1966] defined a class of matrices such that (v, Av) is real and positive, not necessarily for all v, but when v is real.

Definition 2.2. A matrix A is *positive real* (*nonnegative real*) if $(v, Av) > 0$ (≥ 0), for all real $v \neq 0$.

While any positive definite matrix is, of course, positive real, the converse is not true. For example, the matrix

$$A = \begin{pmatrix} 2 & 0 \\ 1 & 2 \end{pmatrix} \tag{2.19}$$

is positive real but not positive definite.

Positive real matrices are characterized by the following theorem.

Theorem 2.9. A matrix A is positive real (nonnegative real) if and only if $A + A^{\mathrm{T}}$ is real and positive definite (nonnegative definite).

Proof. Let $A = B + iC$ where B and C are real. In the proof of Theorem 2.1 we showed that in order for (v, Av) to be real for all real v, we must have (2.5). Hence $A + A^T$ is real if A is positive real. If v is real, then

$$(v, Av) = (v, Bv) = (Bv, v) = (v, B^T v)$$

so that

$$(v, Av) = \tfrac{1}{2}(v, (B + B^T)v) \tag{2.20}$$

Suppose $B + B^T$ is not positive definite. Then since $B + B^T$ is Hermitian, by Theorem 2.3, it must have a nonpositive eigenvalue. Hence there must exist a vector $z = x + iy$ such that

$$(B + B^T)(x + iy) = \lambda(x + iy)$$

where $\lambda \leq 0$ and where x and y are real. Since x and y are not both zero, this implies that $(B + B^T)v = \lambda v$ for some real $v \neq 0$. But by (2.20) this implies that $(v, Av) \leq 0$ for some real $v \neq 0$ and we have a contradiction.

If, on the other hand, $A + A^T$ is real and positive definite, then $C + C^T = 0$ and $B + B^T$ is positive definite; hence the matrix is positive real.

2.3. VECTOR NORMS AND MATRIX NORMS

In dealing with vectors and matrices it is convenient to have some measure of their sizes. Such a measure is provided by any one of several real-valued, nonnegative functions, known as *norms*, of the elements of the vector or matrix.

Let us first consider vector norms. A *vector norm* is a real-valued function whose domain is the set of all vectors of E^N. For any $v \in E^N$ we denote the *α-norm* of v by $\| v \|_\alpha$. We require that any vector norm, $\| \cdot \|_\alpha$, have the following properties:

$$\| v \|_\alpha > 0 \quad \text{if} \quad v \neq 0 \tag{3.1a}$$

$$\| v \|_\alpha = 0 \quad \text{if} \quad v = 0 \tag{3.1b}$$

$$\| cv \|_\alpha = | c | \, \| v \|_\alpha \quad \text{for any complex number } c \tag{3.1c}$$

$$\| v + w \|_\alpha \leq \| v \|_\alpha + \| w \|_\alpha \tag{3.1d}$$

We have already defined the Euclidean length $\| v \|_2$ of a vector $v \in E^N$ by (1.16). The first three properties of (3.1) are trivially satisfied for $\| v \|_2$. We also note that, by (1.16) and (1.17),

$$\begin{aligned}\| v + w \|_2^2 &= (v + w, v + w) = (v, v) + (v, w) + (w, v) + (w, w) \\ &\le \| v \|_2^2 + | (v, w) | + | (w, v) | + \| w \|_2^2 \\ &\le \| v \|_2^2 + 2 \| v \|_2 \| w \|_2 + \| w \|_2^2 = (\| v \|_2 + \| w \|_2)^2\end{aligned}$$

Hence (3.1d) is satisfied and $\| v \|_2$ defines a vector norm. Other examples of vector norms are

$$\| v \|_\infty = \max | v_i |, \qquad i = 1, 2, \ldots, N \tag{3.2}$$

$$\| v \|_1 = \sum_{i=1}^{N} | v_i | \tag{3.3}$$

Let us now consider matrix norms. A *matrix norm* is a real-valued function whose domain is the set of all $N \times N$ matrices. Thus for any $N \times N$ matrix A we let $\| A \|_\beta$ denote the *β-norm* of the matrix A. We require that the matrix norm $\| \cdot \|_\beta$ have the following properties.

$$\| A \|_\beta > 0 \qquad \text{if} \quad A \ne 0 \tag{3.4a}$$

$$\| A \|_\beta = 0 \qquad \text{if} \quad A = 0 \tag{3.4b}$$

$$\| cA \|_\beta = | c | \, \| A_\beta \| \qquad \text{for any complex number} \quad c \tag{3.4c}$$

$$\| A + B \|_\beta \le \| A \|_\beta + \| B \|_\beta \tag{3.4d}$$

$$\| AB \|_\beta \le \| A \|_\beta \| B \|_\beta \tag{3.4e}$$

Some examples of matrix norms are

$$\| A \|_\infty = \max \sum_{j=1}^{N} | a_{i,j} |, \qquad i = 1, 2, \ldots, N \tag{3.5}$$

$$\| A \|_1 = \max \sum_{i=1}^{N} | a_{i,j} |, \qquad j = 1, 2, \ldots, N \tag{3.6}$$

$$\| A \|_2 = [S(A^{\mathrm{H}} A)]^{1/2} \tag{3.7}$$

$$\| A \|_M = N \max | a_{i,j} |, \qquad i, j = 1, 2, \ldots, N \tag{3.8}$$

$$\| A \|_T = \Big[\sum_{i,j=1}^{N} | a_{i,j} |^2 \Big]^{1/2} \tag{3.9}$$

We shall postpone showing that (3.5)–(3.7) are matrix norms. For (3.8) the verification of (3.4a)–(3.4d) is trivial. To show (3.4e) we have

$$\begin{aligned}\| AB \|_M &= N \max_{i,j} \left| \sum_{k=1}^{N} a_{i,k} b_{k,j} \right| \\ &\leq N^2 \max_{i,j,k} (| a_{i,k} | \, | b_{k,j} |) \\ &\leq N^2 (\max_{i,j} | a_{i,j} |)(\max_{i,j} | b_{i,j} |) \leq \| A \|_M \, \| B \|_M\end{aligned}$$

For (3.9) the verification of (3.4a)–(3.4c) is trivial. We verify (3.4d) as for the vector norm $\| \cdot \|_2$. To prove (3.4e) we have

$$\| AB \|_T^2 = \sum_{i,j=1}^{N} \left| \sum_{k=1}^{N} a_{i,k} b_{k,j} \right|^2 \leq \sum_{i,j=1}^{N} \left(\sum_{k=1}^{N} | a_{i,k} |^2 \right) \left(\sum_{s=1}^{N} | b_{s,j} |^2 \right)$$

by the Schwarz inequality (1.17). Hence

$$\| AB \|_T^2 \leq \left(\sum_{i=1}^{N} \sum_{k=1}^{N} | a_{i,k} |^2 \right) \left(\sum_{j=1}^{N} \sum_{s=1}^{N} | b_{s,j} |^2 \right) = \| A \|_T^2 \, \| B \|_T^2$$

In order to compare various vector norms we shall use two theorems in analysis. First, we observe that the set E^N is a *metric space* with the distance function

$$d(v, w) = \| v - w \|_\infty$$

Given any subset E of E^N the *closure* $\bar{E}$ of E is the set of all $v \in E^N$ such that for any $\varepsilon > 0$ there exists $w \in E$ with $d(v, w) < \varepsilon$. A subset E is *closed* if $E = \bar{E}$. We assume without proof the following well-known result.

Any continuous real-valued function defined on a closed and bounded subset E of E^N assumes minimum and maximum values on E.

It is easy to show that the "unit sphere" E^* of E^N, i.e., the set of all v such that $d(0, v) = 1$ is closed and bounded. Hence any continuous function defined on E^* assumes maximum and minimum values on E^*.

Let us now show that all vector norms are in a sense equivalent.

Theorem 3.1. If $\| \cdot \|_\alpha$ and $\| \cdot \|_{\alpha'}$ are any two vector norms, then there exist positive constants $c(\alpha, \alpha')$ and $C(\alpha, \alpha')$ such that for all $v \in E^N$ we have

$$c(\alpha, \alpha') \| v \|_\alpha \leq \| v \|_{\alpha'} \leq C(\alpha, \alpha') \| v \|_\alpha \tag{3.10}$$

Proof. It suffices to prove that for any vector norm $\| \cdot \|_\alpha$ there exist positive constants c and C depending on α such that

$$c \| v \|_\infty \leq \| v \|_\alpha \leq C \| v \|_\infty \tag{3.11}$$

for all $v \in E^N$. For each $i = 1, 2, \ldots, N$ let $e^{(i)}$ denote the unit vector such that $e_j^{(i)} = 0$ if $j \neq i$ and $e_i^{(i)} = 1$. If $v = (v_1, v_2, \ldots, v_N)^T$, then $v = \sum_{i=1}^N v_i e^{(i)}$ and

$$\| v \|_\alpha \leq \sum_{i=1}^N | v_i | \, \| e^{(i)} \|_\alpha \leq C \max_i | v_i | = C \| v \|_\infty \tag{3.12}$$

where

$$C = \sum_{i=1}^N \| e^{(i)} \|_\alpha \tag{3.13}$$

The function

$$f(v) = \left\| \sum_{i=1}^N v_i e^{(i)} \right\|_\alpha \tag{3.14}$$

where $v = (v_1, v_2, \ldots, v_N)^T$ is a continuous function of v, with respect to the norm $\| \cdot \|_\infty$, since

$$| f(v) - f(v') | \leq \left\| \sum_{i=1}^N (v_i - v_i')e^{(i)} \right\|_\alpha \leq C \| v \|_\infty \tag{3.15}$$

where $v' \in E^N$ and C is given by (3.13). The first inequality follows from (3.1d) since we have in general for any $w, w' \in E^N$

$$| \, \| w \|_\alpha - \| w' \|_\alpha \, | \leq \| w - w' \|_\alpha \tag{3.16}$$

Since $f(v)$ is a continuous function defined on the unit sphere E^* it follows that f attains a minimum value $c \geq 0$ on E^*. Thus for some $v^{(0)} \in E^*$ we have $f(v^{(0)}) = c$. If $c = 0$, then

$$\left\| \sum_{i=1}^N v_i^{(0)} e^{(i)} \right\|_\alpha = 0$$

and by (3.1a) and (3.1b)

$$\sum_{i=1}^N v_i^{(0)} e^{(i)} = 0$$

Unless the $v_i^{(0)}$ are all zero, this relation implies that the $e^{(i)}$ are linearly independent and we have a contradiction. Since $\| v^{(0)} \|_\infty = 1$, not all the $v_i^{(0)}$ vanish. Therefore we have $c > 0$. Hence for any $v \in E^N$ we have

$$\| v \|_\alpha = \left\| \sum_{i=1}^{N} v_i e^{(i)} \right\|_\alpha = \| v \|_\infty \left\| \sum_{i=1}^{N} (v_i/\| v \|_\infty) e^{(i)} \right\|_\alpha \geq c \| v \|_\infty \tag{3.17}$$

since

$$\max_i | v_i/\| v \|_\infty | = 1$$

This completes the proof of Theorem 3.1.

We can also show that all matrix norms are in a sense equivalent.

Theorem 3.2. If $\| \cdot \|_\beta$ and $\| \cdot \|_{\beta'}$ are any two matrix norms, then there exist positive constants $d(\beta, \beta')$ and $D(\beta, \beta')$ such that for all A we have

$$d(\beta, \beta') \| A \|_\beta \leq \| A \|_{\beta'} \leq D(\beta, \beta') \| A \|_\beta \tag{3.18}$$

Proof. As in the case of Theorem 3.1, it suffices to prove that for any matrix norm $\| \cdot \|_\beta$ there exist positive constants d and D depending on β such that

$$d \| A \|_M \leq \| A \|_\beta \leq D \| A \|_M \tag{3.19}$$

The proof is similar to that of Theorem 3.1 and will be omitted.

For a given vector norm $\| \cdot \|_\alpha$ let us consider the function

$$g(v) = \| Av \|_\alpha \tag{3.20}$$

Since $v = \sum_{i=1}^{N} v_i e^{(i)}$ and $Av = \sum_{i=1}^{N} v_i A e^{(i)}$, we have by (3.16)

$$\begin{aligned} | g(v) - g(v') | &\leq \sum_{i=1}^{N} | v_i - v_i' | \, \| Ae^{(i)} \|_\alpha \\ &\leq \left(N \max_i \| Ae^{(i)} \|_\alpha \right) \| v - v' \|_\infty \end{aligned} \tag{3.21}$$

Hence $g(v)$ is a continuous function of v.

Let $E^{(\alpha)}$ be the set of all $v \in E^N$ such that $\| v \|_\alpha = 1$. By Theorem 3.1 the set $E^{(\alpha)}$ is bounded since $E^{(\infty)} = E^*$ is bounded. Since, as can be easily shown, $E^{(\alpha)}$ is closed, it follows from the continuity of $g(v)$ that $g(v)$ assumes a maximum value on $E^{(\alpha)}$, i.e., there exists $v^{(0)}$ such that $\| v^{(0)} \| = 1$ and

$$\| Av^{(0)} \|_\alpha = \max_{\|v\|_\alpha = 1} \| Av \|_\alpha \tag{3.22}$$

For any $v \neq 0$ we have $\| \, \| v \|_\alpha^{-1} v \, \|_\alpha = 1$ and hence

$$\| Av^{(0)} \|_\alpha \geq \| A(\| v \|_\alpha^{-1} v) \|_\alpha = \| Av \|_\alpha / \| v \|_\alpha \tag{3.23}$$

Therefore, we have

$$\| Av^{(0)} \|_\alpha / \| v^{(0)} \|_\alpha = \max_{v \neq 0} \| Av \|_\alpha / \| v \|_\alpha \tag{3.24}$$

Given a vector norm $\| \cdot \|_\alpha$ we define the *induced matrix norm* $\| \cdot \|_{\beta(\alpha)}$ by

$$\| A \|_{\beta(\alpha)} = \max_{v \neq 0} \| Av \|_\alpha / \| v \|_\alpha \tag{3.25}$$

It is obvious that (3.4b), (3.4c), and (3.4d) are satisfied. To show that (3.4a) is satisfied let $A \neq 0$. If $a_{i,j} \neq 0$, choose v such that $v_k = 0$ if $k \neq j$ and such that $v_j = 1$. Evidently $(Av)_i = a_{i,j}$ so that $Av \neq 0$. Since $\| Av \|_\alpha \neq 0$ we have (3.4a). To prove (3.4e) we have

$$\| AB \|_{\beta(\alpha)} = \max_{v \neq 0} \| ABv \|_\alpha / \| v \|_\alpha = \max_{v \in S_1} \| ABv \|_\alpha / \| v \|_\alpha$$

where S_1 is the set of all v such that $v \neq 0$ and $Bv = 0$. If S_1 is empty, then $Bv = 0$ for all v and $\| AB \|_{\beta(\alpha)} = 0$ so that (3.4e) holds. If S_1 is not empty we have

$$\begin{aligned} \| AB \|_{\beta(\alpha)} &= \max_{v \in S_1} \big((\| ABv \|_\alpha / \| Bv \|_\alpha)(\| Bv \|_\alpha / \| v \|_\alpha) \big) \\ &\leq \max_{v \in S_1} (\| ABv \|_\alpha / \| Bv \|_\alpha) \max_{v \in S_1} (\| Bv \|_\alpha / \| v \|_\alpha) \\ &\leq \| A \|_{\beta(\alpha)} \| B \|_{\beta(\alpha)} \end{aligned}$$

We can now show that (3.5), (3.6), and (3.7) are matrix norms by showing that they are the induced matrix norms corresponding to the vector norms $\| \cdot \|_\infty$, $\| \cdot \|_1$, and $\| \cdot \|_2$, respectively. For (3.5) we have

$$\begin{aligned} \| Av \|_\infty / \| v \|_\infty &= \max_i \left| \sum_{j=1}^N a_{i,j} (v_j / \| v \|_\infty) \right| \leq \max_i \sum_{j=1}^N | a_{i,j} | \\ &= \| A \|_\infty \end{aligned} \tag{3.26}$$

On the other hand, let us define k by

$$\sum_{j=1}^N | a_{k,j} | = \max \sum_{j=1}^N | a_{i,j} |, \qquad i = 1, 2, \ldots, N \tag{3.27}$$

and let $w = (w_1, w_2, \ldots, w_N)^{\mathrm{T}}$ where

$$w_j = \begin{cases} |a_{k,j}|/a_{k,j} & \text{if} \quad a_{k,j} \neq 0 \\ 0 & \text{if} \quad a_{k,j} = 0 \end{cases} \tag{3.28}$$

Evidently

$$(Aw)_k = \sum_{j=1}^{N} a_{k,j} w_j = \sum_{j=1}^{N} |a_{k,j}| \tag{3.29}$$

(We note that $w \neq 0$; otherwise, $a_{k,j} = 0$ for all j and $A = 0$). Moreover, $\| w \|_\infty = 1$. Thus we have

$$\| Aw \|_\infty \geq \| A \|_\infty \| w \|_\infty \tag{3.30}$$

and (3.5) follows.

It is left to the reader to verify that (3.6) is the induced matrix norm corresponding to the vector norm $\| \cdot \|_1$. For the induced matrix norm corresponding to $\| \cdot \|_2$ we have

$$\max_{v \neq 0} \| Av \|_2 / \| v \|_2 = \max_{v \neq 0} [(Av, Av)/(v, v)]^{1/2} = \max_{v \neq 0} [(v, A^{\mathrm{H}}Av)/(v, v)]^{1/2}$$

But since $A^{\mathrm{H}}A$ is an Hermitian nonnegative definite matrix, by Theorem 2.6, it follows from Theorem 2.2 that the last expression is equal to $(S(A^{\mathrm{H}}A))^{\frac{1}{2}}$.

In our future discussions we shall frequently use the vector norm $\| \cdot \|_2$ and the corresponding induced matrix norm given by (3.7). This latter norm is sometimes referred to as the *spectral norm*. When no confusion will arise we shall omit the norm labels. Thus we define

$$\| v \| = \| v \|_2 = (v, v)^{1/2} \tag{3.31}$$

$$\| A \| = \| A \|_2 = [S(A^{\mathrm{H}}A)]^{1/2} \tag{3.32}$$

A vector norm $\| \cdot \|_\alpha$ and a matrix norm $\| \cdot \|_\beta$ are said to be *consistent*, or *compatible*, if for all $v \in E^N$ we have

$$\| Av \|_\alpha \leq \| A \|_\beta \| v \|_\alpha \tag{3.33}$$

Evidently, any vector norm and the induced matrix norm are consistent. If the matrix norm $\| \cdot \|_\beta$ and the vector norm are consistent and if for some $v \in E^N$ and $v \neq 0$, we have

$$\| Av \|_\alpha = \| A \|_\beta \| v \|_\alpha \tag{3.34}$$

then the matrix norm is *subordinate* to the vector norm. Evidently, the induced matrix norm corresponding to a vector norm is subordinate to that vector norm.

Given a matrix norm $\| \cdot \|_\beta$ one can define a consistent vector norm $\| \cdot \|_\alpha$ as follows

$$\| v \|_\alpha = \| B \|_\beta \tag{3.35}$$

where B is a matrix whose first column is v and all other elements vanish. One can easily verify that $\| \cdot \|_\alpha$ is a norm and that $\| \cdot \|_\alpha$ and $\| \cdot \|_\beta$ are consistent. Thus we have

Theorem 3.3. Given a matrix norm $\| \cdot \|_\beta$ there exists a vector norm $\| \cdot \|_\alpha$ such that $\| \cdot \|_\alpha$ and $\| \cdot \|_\beta$ are consistent.

We now prove

Theorem 3.4. For any matrix norm $\| \cdot \|_\beta$ we have

$$S(A) \leq \| A \|_\beta \tag{3.36}$$

Proof. By Theorem 3.3 there exists a vector norm $\| \cdot \|_\alpha$ such that for all A and v we have

$$\| Av \|_\alpha \leq \| A \|_\beta \| v \|_\alpha \tag{3.37}$$

Suppose λ is an eigenvalue of A and v is an associated eigenvector. Then $Av = \lambda v$ and hence, by (3.1c), $| \lambda | \leq \| A \|_\beta$.

Given a matrix norm $\| \cdot \|_\beta$ and any nonsingular matrix L we define the "β, L-norm" of a matrix A by

$$\| A \|_{\beta,L} = \| LAL^{-1} \|_\beta \tag{3.38}$$

In the case of the matrix norm $\| \cdot \|_2 = \| \cdot \|$, we denote the L-norm by $\| A \|_L$ and we have, by (3.32),

$$\| A \|_L = \| LAL^{-1} \| \tag{3.39}$$

Also, for a vector norm $\| \cdot \|_\alpha$ we can define the "α,L-norm" of a vector v by

$$\| v \|_{\alpha,L} = \| Lv \|_\alpha \tag{3.40}$$

In the case of the vector norm $\| \cdot \|_2 = \| \cdot \|$, we let

$$\| v \|_L = \| Lv \| \tag{3.41}$$

Evidently, if $\|\cdot\|_\beta$ is the induced matrix norm of $\|\cdot\|_\alpha$, then $\|\cdot\|_{\beta,L}$ is the induced matrix norm of $\|\cdot\|_{\alpha,L}$. Moreover, if $\|\cdot\|_\alpha$ and $\|\cdot\|_\beta$ are consistent, then so are $\|\cdot\|_{\alpha,L}$ and $\|\cdot\|_{\beta,L}$. If $\|\cdot\|_\beta$ is subordinate to $\|\cdot\|_\alpha$, then $\|\cdot\|_{\beta,L}$ is subordinate to $\|\cdot\|_{\alpha,L}$.

Theorem 3.5.† For any matrix A and any $\varepsilon > 0$ there exists a nonsingular matrix L such that

$$\| A \|_L \leq S(A) + \varepsilon \tag{3.42}$$

Proof. Let $A' = \varepsilon^{-1}A$ and let V be any matrix which reduces A' to Jordan canonical form. Evidently,

$$V^{-1}AV = J^{(\epsilon)} \tag{3.43}$$

where $J^{(\epsilon)}$ is the same as the Jordan canonical form J of A except that the off-diagonal elements of J are multiplied by ε. Thus a typical block of $J^{(\epsilon)}$ is

$$J_p^{(\epsilon)} = \begin{pmatrix} \lambda & \varepsilon & 0 & \cdots & 0 & 0 \\ 0 & \lambda & \varepsilon & \cdots & 0 & 0 \\ \vdots & \vdots & \vdots & \cdots & \vdots & \vdots \\ 0 & 0 & 0 & \cdots & \lambda & \varepsilon \\ 0 & 0 & 0 & \cdots & 0 & \lambda \end{pmatrix} = \lambda I + \varepsilon E \tag{3.44}$$

Since EE^{T} has the form

$$EE^{\mathrm{T}} = \begin{pmatrix} 1 & 0 & 0 & \cdots & 0 & 0 \\ 0 & 1 & 0 & \cdots & 0 & 0 \\ \vdots & \vdots & \vdots & \cdots & \vdots & \vdots \\ 0 & 0 & 0 & \cdots & 1 & 0 \\ 0 & 0 & 0 & \cdots & 0 & 0 \end{pmatrix} \tag{3.45}$$

it follows that $S(EE^{\mathrm{T}}) = 1$ and $\| E \| = 1$. Therefore by (3.44) we have

$$\| J_p^{(\epsilon)} \| \leq |\lambda| + \varepsilon \tag{3.46}$$

† This theorem is given by Householder [1964]. The proof is based on that of Householder but with the use of a technique of Ortega and Rheinboldt [1970].

and hence

$$\| V^{-1}AV \| \leq S(A) + \varepsilon \tag{3.47}$$

If we let $L = V^{-1}$, we obtain (3.42).

We remark that given W such that $W^{-1}AW = J$ one can construct V as follows. Let $w_1, w_2, \ldots, w_s$ be vectors corresponding to the Jordan block J_p, given by (3.44) with ε replaced by one. The columns of V corresponding to $J_p^{(\varepsilon)}$ are

$$v_i = \varepsilon^{i-1} w_i, \qquad i = 1, 2, \ldots, ps \tag{3.48}$$

From Theorems 3.5 and 3.2 it follows that given any A and any $\varepsilon > 0$ there exists a matrix norm $\| \cdot \|_\beta$ such that $\| A \|_\beta \leq S(A) + \varepsilon$. This can also be proved using Schur's theorem (Theorem 1.14). (See Rheinboldt [1969].) One can show that a nonsingular matrix Q can be found such that $\| A \|_{\infty,Q} \leq S(A) + \varepsilon$. An advantage of this procedure is the fact that, already noted, Schur's theorem is much easier to prove than the corresponding theorem on the Jordan canonical form.

If we apply Theorem 3.5 with $\varepsilon = (1 - S(A))/2$ we have

Corollary 3.6. If $S(A) < 1$, then there exists a nonsingular matrix L such that

$$\| A \|_L < 1 \tag{3.49}$$

In Section 3.6 we show that L can be taken to be positive definite.

2.4. CONVERGENCE OF SEQUENCES OF VECTORS AND MATRICES

A sequence of vectors $v^{(1)}, v^{(2)}, v^{(3)}, \ldots$ is said to converge to a limit v if

$$\lim_{n\to\infty} v_i^{(n)} = v_i, \qquad i = 1, 2, \ldots, N \tag{4.1}$$

Similarly, a sequence of matrices $A^{(1)}, A^{(2)}, A^{(3)}, \ldots$ is said to converge to a limit A if

$$\lim_{n\to\infty} a_{i,j}^{(n)} = a_{i,j}, \qquad i, j = 1, 2, \ldots, N \tag{4.2}$$

Theorem 4.1. The sequence $v^{(1)}, v^{(2)}, \ldots$ converges to v if and only if for any vector norm $\| \cdot \|_\alpha$

$$\lim_{n\to\infty} \| v^{(n)} - v \|_\alpha = 0 \tag{4.3}$$

Proof. Evidently $v^{(n)} \to v$ if and only if $\| v^{(n)} - v \|_\infty \to 0$. From Theorem 3.1 it follows that $\| v^{(n)} - v \|_\infty \to 0$ if and only if $\| v^{(n)} - v \|_\alpha \to 0$.

In a similar manner we can prove

Theorem 4.2. The sequence $A^{(1)}, A^{(2)}, \ldots$ converges to A if and only if for every matrix norm $\| \cdot \|_\beta$

$$\lim_{n\to\infty} \| A^{(n)} - A \|_\beta = 0 \tag{4.4}$$

Theorem 4.3. $A^n \to 0$ as $n \to \infty$ if and only if

$$S(A) < 1 \tag{4.5}$$

Proof. If $S(A) \geq 1$, then $S(A^n) \geq 1$ for all n, and, by Theorem 3.4, $\| A^n \|_\beta \geq S(A^n) \geq 1$ for any norm $\| \cdot \|_\beta$. On the other hand, if $A^n \to 0$, then $\| A^n \|_\beta \to 0$ which is impossible if $S(A) \geq 1$. Hence (4.5) is a necessary condition for convergence.

If $S(A) < 1$, then by Corollary 3.6 there exists a nonsingular matrix L such that $\| A \|_L < 1$. Since $\| A^n \|_L \leq \| A \|_L^n$ it follows that $\| A^n \|_L \to 0$. Hence, by Theorem 4.1 it follows that $A^n \to 0$.

An infinite series of matrices $A^{(1)} + A^{(2)} + A^{(3)} + \cdots$ is said to converge to a limit A if the sequence of partial sums $S^{(1)} = A^{(1)}$, $S^{(2)} = A^{(1)} + A^{(2)}$, etc., converges to A. If convergence holds, then we write

$$A = A^{(1)} + A^{(2)} + A^{(3)} + \cdots = \sum_{i=1}^{\infty} A^{(i)}$$

We now prove

Theorem 4.4. The matrix $I - B$ is nonsingular and the series $I + B + B^2 + \cdots$ converges if and only if $S(B) < 1$. Moreover if $S(B) < 1$, then

$$(I - B)^{-1} = I + B + B^2 + \cdots = \sum_{i=0}^{\infty} B^i \tag{4.6}$$

Proof. Let $S^{(n)} = I + B + B^2 + \cdots + B^n$. Evidently

$$(I - B)S^{(n)} = I - B^{n+1} \tag{4.7}$$

If $S(B) < 1$, then $I - B$ is nonsingular and by (4.7)

$$S^{(n)} - (I - B)^{-1} = -(I - B)^{-1}B^{n+1} \tag{4.8}$$

Since $S(B) < 1$, then by Theorems 4.3 and 4.2 it follows that $\| B^{n+1} \| \to 0$. Therefore since

$$\| S^{(n)} - (I - B)^{-1} \| \leq \| (I - B)^{-1} \| \, \| B^{n+1} \| \tag{4.9}$$

it follows from Theorem 4.2 that $S^{(n)} \to (I - B)^{-1}$ and (4.6) holds.

If on the other hand the series $I + B + B^2 + \cdots$ converges, then one can easily show that $B^n \to 0$. By Theorem 4.3 this implies that $S(B) < 1$.

Theorem 4.5. The condition

$$\lim_{n\to\infty} A^n v = 0 \tag{4.10}$$

holds for all $v \in E^N$ if and only if $S(A) < 1$.

Proof. By Theorem 4.3 the proof is an immediate consequence of the following lemma.

Lemma 4.6. The condition (4.10) holds if and only if $A^n \to 0$ as $n \to \infty$.

Proof. If $A^n \to 0$, then by Theorem 4.2 we have $\| A^n \| \to 0$. Since $\| A^n v \| \leq \| A^n \| \, \| v \|$ it follows that $\| A^n v \| \to 0$ and, by Theorem 4.1, that $A^n v \to 0$. On the other hand, suppose A^n does not converge to zero. Then $S(A) \geq 1$ by Theorem 4.3 and for some vector $v \neq 0$ and some λ with $|\lambda| \geq 1$ we have $Av = \lambda v$. Hence $\| A^n v \| = |\lambda|^n \| v \|$ which does not converge to zero.

2.5. IRREDUCIBILITY AND WEAK DIAGONAL DOMINANCE

In this section we shall consider two important classes of matrices and show that any matrix belonging to both classes is nonsingular.

Definition 5.1. A matrix $A = (a_{i,j})$ of order N is *irreducible* if $N = 1$ or if $N > 1$ and given any two nonempty disjoint subsets S and T of W, the set of the first N positive integers, such that $S + T = W$, there exist $i \in S$ and $j \in T$ such that $a_{i,j} \neq 0$.

Definition 5.2. A matrix $A = (a_{i,j})$ of order N has *weak diagonal dominance* if

$$| a_{i,i} | \geq \sum_{\substack{j=1 \\ j \neq i}}^{N} | a_{i,j} |, \qquad i = 1, 2, \ldots, N \tag{5.1}$$

and for at least one i

$$| a_{i,i} | > \sum_{\substack{j=1 \\ j \neq i}}^{N} | a_{i,j} | \tag{5.2}$$

If a matrix is irreducible, then one cannot extract a subsystem of the system (1.21) which preserves the correspondence between the equations and the unknowns, and which can be solved independently of the larger system. This follows from the next theorem.

Theorem 5.1. *A is irreducible if and only if there does not exist a permutation matrix P such that $P^{-1}AP$ has the form,*

$$P^{-1}AP = \begin{pmatrix} F & 0 \\ G & H \end{pmatrix} \tag{5.3}$$

where F and H are square matrices and where all elements of 0 vanish.

Proof. If A is irreducible and such a permutation exists, let S be the set of all i corresponding to the first p rows where p is the order of F and let $T = W - S$. But there exists $i \in S$ and $j \in T$, such that $a_{i,j} \neq 0$. But this contradicts the fact that all elements of 0 vanish.

If A is not irreducible, then a permutation satisfying P clearly exists. For there exist sets S and T such that $a_{i,j} = 0$ for all $i \in S$ and $j \in T$. If we choose P so that all rows and columns corresponding to S come first in $P^{-1}AP$, then we clearly obtain the form (5.3). Thus if no such permutation exists, A is irreducible.

If A is nonsingular but not irreducible, we can solve (1.21) by reordering the equations and relabelling the unknowns to obtain a matrix of the form (5.3). We then can solve the subsystem with the matrix F and and then another subsystem with the matrix H.

We remark that if one does not insist on preserving the correspondence between equations and unknowns one can consider all matrices of the form $P^{-1}AQ$ where P and Q are permutation matrices. Thus Forsythe and Moler [1967] use the term *irreducible* to mean that $P^{-1}AQ$ cannot be put into the form (5.3).

Evidently, Theorem 5.1 could be used as an alternative definition of irreducibility. Another alternative definition could be based on the following

Theorem 5.2. A matrix A of order N is irreducible if and only if $N = 1$ or, given any two distinct integers i and j with $1 \leq i \leq N$, $1 \leq j \leq N$, then $a_{i,j} \neq 0$ or there exist $i_1, i_2, \ldots, i_s$ such that

$$a_{i,i_1} a_{i_1,i_2} \cdots a_{i_s,j} \neq 0 \tag{5.4}$$

Proof. If the above condition holds, let S and T be any two nonempty disjoint subsets of $W = \{1, 2, \ldots, N\}$ such that $S + T = W$. Let i be any element of S and let j be any element of T. By hypothesis $a_{i,j} \neq 0$ or else there exist $i_1, i_2, \ldots, i_s$ such that (5.4) holds. Thus, either $a_{i,j} \neq 0$ or else for at least one of the factors $a_{i,i_1}, a_{i_1,i_2}, \ldots, a_{i_s,j}$ the first subscript is in S and the second is in T. Since the factor does not vanish, it follows that A is irreducible.

If A is irreducible and $N > 1$, for each i let X_i be the set of all j such that $i \neq j$ and either $a_{i,j} \neq 0$ or else there exist $i_1, i_2, \ldots, i_s$ such that (5.4) holds. Let $S_1 = \{i\}$ and $T_1 = W - S_1$. By the irreducibility there exists $i_1 \in T_1$ such that $a_{i,i_1} \neq 0$; hence $i_1 \in X_i$. If $N = 2$, then $X_i = W - \{i\}$. Otherwise, let $S_2 = \{i, i_1\}$ and let $T_2 = W - S_2$. Again, by the irreducibility there exists $i_2 \in T_2$ such that $a_{i,i_2} \neq 0$ or else $a_{i_1,i_2} \neq 0$. Hence since $a_{i,i_2} \neq 0$ or $a_{i,i_1} a_{i_1,i_2} \neq 0$, we have $i_2 \in X_i$. Continuing in this way we can show that $X_i = W - \{i\}$. Hence the alternative condition is satisfied, and the theorem is proved.

The concept of irreducibility can be illustrated graphically. Given a matrix A we construct the *directed graph* of A as follows. Label any distinct N points in the plane as $1, 2, \ldots, N$. For each i and j such that $a_{i,j} \neq 0$ draw an arrow from i to j. If $a_{i,j} \neq 0$ and $a_{j,i} \neq 0$, there will be an arrow from i to j and one from j to i. (If $a_{i,i} \neq 0$, we can also draw a small loop containing the point i, but this has no effect on the irreducibility.) The matrix is irreducible if and only if either $N = 1$ or else the graph is *connected* in the following sense. Given any two distinct points i and j

either there is an arrow from i to j or else there is a path of arrows from i to i_1, i_1 to i_2, ..., i_s to j.

For example, the directed graph of the matrix

$$\begin{pmatrix} 0 & 1 & 0 & 1 \\ 1 & 0 & 1 & 0 \\ 0 & 1 & 0 & 1 \\ 1 & 0 & 0 & 1 \end{pmatrix}$$

is given in Figure 5.1. One can easily verify that the graph is connected.

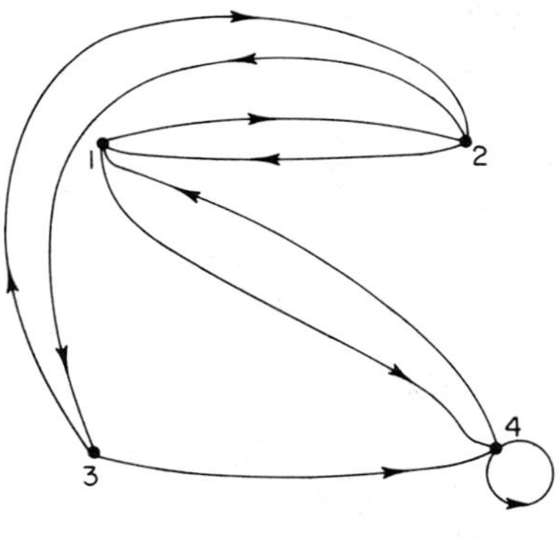

Figure 5.1.

The matrix

$$A = \begin{pmatrix} 1 & 1 & 1 \\ 0 & 0 & 1 \\ 0 & 0 & 1 \end{pmatrix}$$

is not irreducible. Its directed graph shown in Figure 5.2 is not connected since we cannot reach the point 1 starting from 3.

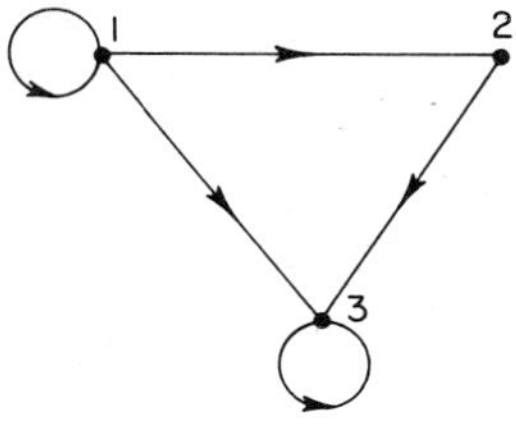

Figure 5.2.

We now prove the following fundamental theorem.

Theorem 5.3. If A is an irreducible matrix with weak diagonal dominance, then $\det A \neq 0$ and none of the diagonal elements of A vanishes.

Proof. If $N = 1$, then $a_{1,1} > 0$ by weak diagonal dominance and $\det A \neq 0$. Suppose that $N > 1$ and $a_{i,i} = 0$ for some i. By the weak diagonal dominance we have $a_{i,j} = 0$ for all j. But this contradicts the irreducibility. For if $S = \{i\}$ and $T = W - \{i\}$, we have $a_{i,j} = 0$ for all $i \in S$, $j \in T$.

By Theorem 1.4, if $\det A = 0$, then there exists a solution $u \neq 0$ of the homogeneous system $Au = 0$. Since $a_{i,i} \neq 0$ for all i we can solve the ith equation for u_i obtaining

$$u_i = \sum_{j=1}^{N} b_{i,j} u_j, \qquad i = 1, 2, \ldots, N \tag{5.5}$$

where

$$\begin{aligned} b_{i,i} &= 0 \\ b_{i,j} &= -a_{i,j}/a_{i,i}, \qquad i \neq j, \qquad i, j = 1, 2, \ldots, N \end{aligned} \tag{5.6}$$

Evidently by (5.1) and (5.6) we have

$$\sum_{j=1}^{N} | b_{i,j} | \leq 1 \tag{5.7}$$

for all i and for some i we have

$$\sum_{j=1}^{N} | b_{i,j} | < 1 \tag{5.8}$$

Let $M = \max | u_i |$, $i = 1, 2, \ldots, N$. Since $u \neq 0$, $M > 0$. Let k be any value of i such that $| u_k | = M$. By (5.5) we have

$$M = | u_k | = \left| \sum_{j=1}^{N} b_{k,j} u_j \right| \leq \sum_{j=1}^{N} | b_{k,j} | \, | u_j |$$

On the other hand, by (5.7) we have

$$\sum_{j=1}^{N} | b_{k,j} | \, (| u_j | - M) \geq 0 \tag{5.9}$$

Since $| u_j | \leq M$ for all j it follows that if $b_{k,j} \neq 0$, then $| u_j | = M$.

By Theorem 5.2, for any $j \neq k$, either $a_{k,j} \neq 0$ or else there exist $i_1, i_2, \ldots, i_s$ such that (5.4) holds. Hence, $b_{k,j} \neq 0$ or else $b_{k,i_1} b_{i_1,i_2} \cdots b_{i_s,j} \neq 0$. But this implies $|u_j| = M$ or else $|u_{i_1}| = |u_{i_2}| = \cdots = |u_j| = M$. In either case, then, $|u_j| = M$. Therefore we have

$$|u_i| = M, \qquad i = 1, 2, \ldots, N \tag{5.10}$$

Let i^* be any value of i such that (5.8) holds. By (5.5) and (5.10) we have

$$M = |u_{i^*}| \leq \sum_{j=1}^{N} |b_{i^*,j}| \, M$$

or

$$1 \leq \sum_{j=1}^{M} |b_{i^*,j}|$$

which contradicts (5.8). This completes the proof of Theorem 5.3.

If (5.2) holds for all i, we say that A has *strong diagonal dominance.* From the proof of Theorem 5.3 we have

Corollary 5.4. If A has strong diagonal dominance, then $\det A \neq 0$.

We now give a sufficient condition for an Hermitian matrix to be positive definite.

Theorem 5.5. If A is an Hermitian matrix with nonnegative diagonal elements and with weak diagonal dominance, then A is nonnegative definite. If A is also irreducible or nonsingular, then A is positive definite.

Proof. All eigenvalues of the Hermitian matrix A are real by Theorem 1.13. If $\lambda < 0$, then $A - \lambda I$ has strong diagonal dominance and, by Corollary 5.4, $A - \lambda I$ is nonsingular. Thus all eigenvalues of A are nonnegative and by Theorem 2.3, A is nonnegative definite. If A is irreducible, then by Theorem 5.3, A is nonsingular. If A is nonsingular, $\lambda = 0$ is not an eigenvalue of A. Hence all eigenvalues of A are positive and, by Theorem 2.3, A is positive definite.

2.6. PROPERTY A

We shall frequently be concerned with matrices which are sparse and whose nonzero elements form a certain pattern. One class of such matrices have "Property A" defined as follows.

Definition 6.1. A matrix A of order N has *Property A* if there exist two disjoint subsets S_1 and S_2 of W, the set of the first N positive integers, such that $S_1 + S_2 = W$ and such that if $i \neq j$ and if either $a_{i,j} \neq 0$ or $a_{j,i} \neq 0$, then $i \in S_1$ and $j \in S_2$ or else $i \in S_2$ and $j \in S_1$.

We remark that either S_1 or S_2 may be empty. If S_1 or S_2 is empty, then A is, of course, diagonal. If neither S_1 nor S_2 is empty, we can, as will be shown in Chapter 5, rearrange the rows and corresponding columns of a matrix A which has Property A to obtain the form

$$\begin{pmatrix} D_1 & H \\ K & D_2 \end{pmatrix} \tag{6.1}$$

where D_1 and D_2 are square diagonal matrices.

There are many classes of matrices which are related to, or are generalizations of, the class of matrices with Property A. We shall discuss these in Chapters 5 and 13.

2.7. *L*-MATRICES AND RELATED MATRICES

Definition 7.1. A real matrix A of order N is an *L-matrix* if

$$a_{i,i} > 0, \qquad i = 1, 2, \ldots, N \tag{7.1}$$

and

$$a_{i,j} \leq 0, \qquad i \neq j, \qquad i, j = 1, 2, \ldots, N \tag{7.2}$$

As special cases we have the following classes of matrices.

Definition 7.2. A real matrix A is a *Stieltjes matrix* if A is positive definite and if (7.2) holds.

Definition 7.3. A real matrix A is an *M-matrix* if (7.2) holds, if A is nonsingular, and if $A^{-1} \geq 0$.

It follows from Theorem 2.5 that (7.1) holds for any Stieltjes matrix. Hence a Stieltjes matrix is a positive definite L-matrix. By Theorem 5.5 we have

Theorem 7.1. If A is an L-matrix which is symmetric, irreducible, and has weak diagonal dominance, then A is a Stieltjes matrix.

To show that any M-matrix A is an L-matrix we need only show that A has positive diagonal elements. If $A^{-1} = (\alpha_{i,j})$, then we have

$$\sum_{k=1}^{N} a_{i,k}\alpha_{k,i} = 1, \qquad i = 1, 2, \ldots, N \tag{7.3}$$

since $AA^{-1} = I$. Hence, since $A^{-1} \geq 0$,

$$a_{i,i}\alpha_{i,i} = 1 - \sum_{\substack{k=1 \\ k \neq i}}^{N} a_{i,k}\alpha_{k,i} \geq 1 \tag{7.4}$$

by (7.2). Since $\alpha_{i,i} \geq 0$, it follows that $a_{i,i} > 0$ for all i.

Theorem 7.2. If A is an L-matrix, then A is an M-matrix if and only if

$$S(B) < 1 \tag{7.5}$$

where

$$B = D^{-1}C \tag{7.6}$$

$$D = \text{diag } A \tag{7.7}$$

$$A = D - C \tag{7.8}$$

Proof. If $S(B) < 1$, then by Theorem 4.4, $I - B$ is nonsingular and the series $I + B + B^2 + \cdots$ converges to $(I - B)^{-1}$. Since $D \geq 0$ and $C \geq 0$, we have $B \geq 0$ and $(I - B)^{-1} \geq 0$. Since D and $I - B$ are nonsingular, it follows that $A = D(I - B)$ is nonsingular and

$$A^{-1} = (I - B)^{-1}D^{-1} \geq 0 \tag{7.9}$$

Hence, A is an M-matrix.

If A is an M-matrix, let us consider $\hat{A}$ defined by

$$\hat{A} = D^{-1/2}AD^{-1/2} = I - D^{-1/2}CD^{-1/2} \tag{7.10}$$

Evidently, $\hat{A}$ is an M-matrix. Moreover, B is similar to $\tilde{B}$ where

$$\tilde{B} = D^{1/2}BD^{-1/2} \tag{7.11}$$

By direct calculation we verify that

$$(I - \tilde{B})(I + \tilde{B} + \tilde{B}^2 + \cdots + \tilde{B}^m) = I - \tilde{B}^{m+1}$$

and hence

$$(\hat{A})^{-1} = (I - \tilde{B})^{-1} = (I + \tilde{B} + \tilde{B}^2 + \cdots + \tilde{B}^m) + (I - \tilde{B})^{-1}\tilde{B}^{m+1} \quad (7.12)$$

Since $(I - \tilde{B})^{-1} \geq 0$ and $\tilde{B} \geq 0$, it follows that each element of the matrix $I + \tilde{B} + \tilde{B}^2 + \cdots + \tilde{B}^m$ is a nondecreasing function of m which is bounded by the corresponding element of $(I - \tilde{B})^{-1}$. Hence the series $I + \tilde{B} + \tilde{B}^2 + \cdots + \tilde{B}^m$ must converge. But by Theorem 4.4 this can happen only if $S(\tilde{B}) = S(B) < 1$.

Theorem 7.3. If A is a Stieltjes matrix, then A is an M-matrix.

Proof. Let $A = D - C$ where $D = \text{diag } A$. Since A is positive definite, A is nonsingular and $A^{-1} = [D(I - B)]^{-1} = (I - B)^{-1}D^{-1}$ where $B = D^{-1}C$. From Theorem 7.2 we need only show that $S(B) < 1$. Suppose $\bar{\mu} = S(B) \geq 1$. Since $B \geq 0$ it follows from Theorem 1.17 that $\bar{\mu}$ is an eigenvalue of B.

Let $\hat{A} = D^{-1/2}AD^{-1/2} = I - D^{-1/2}CD^{-1/2} = I - \tilde{B}$, where $\tilde{B} = D^{-1/2} \times CD^{-1/2} = D^{1/2}BD^{-1/2}$. Since $\tilde{B}$ is similar to B, $\bar{\mu}$ is an eigenvalue of $\tilde{B}$ and $\hat{A}$ is not positive definite. But on the other hand, by Corollary 2.8, $\hat{A}$ must be positive definite. This contradiction proves that $\bar{\mu} < 1$ and shows that A is an M-matrix.

The class of M-matrices is a subclass of the class of *monotone matrices.* We have

Definition 7.3. A matrix A is a *monotone matrix*, if A is nonsingular and $A^{-1} \geq 0$.

Theorem 7.4. A matrix A is monotone if and only if $Ax \geq 0$ implies $x \geq 0$.

Proof. Suppose A is monotone. If $Ax = y \geq 0$, then $x = A^{-1}y \geq 0$. On the other hand, suppose $Ax \geq 0$ implies $x \geq 0$, and let z be any vector such that $Az = 0$. Then $z \geq 0$. On the other hand, $A(-z) = 0$ and $-z \geq 0$. Thus $z = 0$. Since $Az = 0$ implies $z = 0$, it follows that A is nonsingular.

Let $v_1, v_2, \ldots, v_N$ be the columns of A^{-1}. Then the columns of $AA^{-1} = I$ are $Av_1, Av_2, \ldots, Av_N$. Since each $Av_i \geq 0$, it follows that $v_i \geq 0$. Hence $A^{-1} \geq 0$.

From the definitions of monotone matrices and M-matrices we have

Theorem 7.5. If A is an M-matrix, then A is a monotone matrix. On the other hand, if A is a monotone matrix, such that (7.2) holds, then A is an M-matrix.

We now present a somewhat stronger theorem than Theorem 1.17 for the case where A is a real, symmetric, nonnegative matrix.

Theorem 7.6. If A is a real, symmetric, nonnegative matrix of order N, then

(a) An eigenvalue of A is $S(A)$ and there exists an eigenvector v of A associated with $S(A)$ such that $v \geq 0$.

(b) If A is irreducible, then $S(A)$ is a simple eigenvalue of A and there exists a positive eigenvector associated with $S(A)$. Moreover, $S(A) > 0$ unless $N = 1$ and $A = 0$.

(c) If A is irreducible and if $-S(A)$ is an eigenvalue of A, then A has Property A and the diagonal elements of A vanish. Moreover, $-S(A)$ is a simple eigenvalue of A.

Proof. Since A is real and symmetric, it follows from Theorem 1.13 that A has real eigenvalues and there exists a basis of real eigenvectors of A. Consequently, if the eigenvalues of A are $\lambda_1 \leq \lambda_2 \leq \cdots \leq \lambda_N$, then $S(A) = -\lambda_1$ and/or $S(A) = \lambda_N$. Suppose $v^{(1)}$ is a real eigenvector of A associated with λ_1. Since A is Hermitian, it follows from Theorem 2.1 that (v, Av) is real for any v. Moreover, by (2.9) we have

$$\lambda_1 = (v^{(1)}, Av^{(1)})/(v^{(1)}, v^{(1)}) = \sum_{i,j=1}^{N} a_{i,j} v_i^{(1)} v_j^{(1)}/(v^{(1)}, v^{(1)}) \tag{7.13}$$

and

$$|\lambda_1| \leq \sum_{i,j=1}^{N} a_{i,j} w_i w_j/(w, w)$$

where $w_i = |v_i^{(1)}|$, $i = 1, 2, \ldots, N$. On the other hand, by (2.9) we have

$$\lambda_N \geq (w, Aw)/(w, w) \geq 0$$

Hence

$$S(A) = \lambda_N \geq |\lambda_1| \tag{7.14}$$

and thus $S(A) = \lambda_N$ is an eigenvalue of A.

Let $v^{(N)}$ be a real eigenvector of A associated with λ_N. Then by (2.9) we have

$$\lambda_N = (v^{(N)}, Av^{(N)})/(v^{(N)}, v^{(N)}) = \sum_{i,j=1}^{N} a_{i,j} v_i^{(N)} v_j^{(N)}/(v^{(N)}, v^{(N)})$$

and

$$|\lambda_N| \leq \sum_{i,j=1}^{N} a_{i,j} z_i z_j/(z, z) = (z, Az)/(z, z) \tag{7.15}$$

where $z_i = |v_i^{(N)}|$, $i = 1, 2, \ldots, N$. On the other hand, by (2.9) we have $\lambda_N \geq (z, Az)/(z, z)$ so that

$$S(A) = \lambda_N = (z, Az)/(z, z) \tag{7.16}$$

where

$$z \geq 0 \tag{7.17}$$

To prove (a), we need only show that z is an eigenvector of A associated with $S(A)$. We prove

Lemma 7.7. If A is a real symmetric matrix of order N and if $z \neq 0$ is any vector such that

$$(z, Az)/(z, z) = \lambda_N \tag{7.16$'$}$$

where λ_N is the largest eigenvalue of A, then

$$Az = \lambda_N z \tag{7.17$'$}$$

Proof. Let the eigenvalues of A be $\lambda_1 \leq \lambda_2 \leq \cdots \leq \lambda_N$, and let $v^{(1)}, v^{(2)}, \ldots, v^{(N)}$ be the associated eigenvectors. There exist constants $c_1, c_2, \ldots, c_N$ such that

$$z = \sum_{i=1}^{N} c_i v^{(i)} \tag{7.18}$$

and, as in the proof of Theorem 2.2, we have

$$(z, Az)/(z, z) = \sum_{i=1}^{N} \left(|c_i|^2 \Big/ \sum_{j=1}^{N} |c_j|^2\right)\lambda_i \tag{7.19}$$

If $c_i \neq 0$ for any i such that $\lambda_i < \lambda_N$, then $(z, Az)/(z, z) < \lambda_N$. Hence z must be a linear combination of those eigenvectors $v^{(i)}$ associated with λ_N, and therefore must itself be such an eigenvector.

Suppose now that $A \neq 0$ and A is irreducible. Since $A = A^T$ we have $S(A) = \|A\|$ which is positive since $A \neq 0$.

We now show that if z is any real eigenvector associated with the eigenvalue $S(A)$, then $z_i \neq 0$ for all i, and if z has at least one positive component, then $z > 0$. From (7.16) and (2.9) we have

$$S(A) = \sum_{i,j=1}^{N} a_{i,j} z_i z_j \Big/ \sum_{i=1}^{N} z_i^2$$
$$\leq \sum_{i,j=1}^{N} a_{i,j} \mid z_i \mid \mid z_j \mid \Big/ \sum_{i=1}^{N} z_i^2 \leq S(A)$$

Therefore, $(\mid z \mid, A \mid z \mid)/(\mid z \mid, \mid z \mid) = S(A)$, and it follows from Lemma 7.7 that $\mid z \mid$ is an eigenvector of A associated with $S(A)$. By the irreducibility of A it follows that $\mid z_i \mid \neq 0$ for all i and hence $z_i \neq 0$ for all i.

Since $z_i \neq 0$ for all i, it follows that the expression $\sum_{i,j=1}^{N} a_{i,j} z_i z_j$ is strictly less than $\sum_{i,j=1}^{N} a_{i,j} \mid z_i \mid \mid z_j \mid$ unless $z_i z_j > 0$ for all i, j such that $a_{i,j} \neq 0$. Using the irreducibility of A and the fact that $z_i \neq 0$ for all i we can show that since $z_i > 0$ for some i, then $z_i > 0$ for all i and $z > 0$.

Suppose now that $S(A)$ is not a simple eigenvalue of A. Then since A is Hermitian, there exist at least two real orthogonal eigenvectors, say z and z', associated with $S(A)$. Moreover, $z_i \neq 0$ for all i and $z_i' \neq 0$ for all i. Also, from the above discussion it follows that $z_1^{-1} z > 0$ and $(z_1')^{-1} z' > 0$ since $(z_1^{-1} z)_1 = 1$ and $((z_1')^{-1} z')_1 = 1$. But it is not possible for z and z' to be orthogonal since

$$(z, z') = (z_1(z_1^{-1} z), z_1'(z_1')^{-1} z') = z_1 z_1'(z_1^{-1} z, (z_1')^{-1} z') > 0$$

since the inner product of two positive vectors is positive. This completes the proof of (b).

Now suppose that A is irreducible and $-S(A)$ is an eigenvalue of A. Then by (2.9) we have

$$-S(A) = \min_{v \neq 0} [(v, Av)/(v, v)] = (v^{(1)}, Av^{(1)})/(v^{(1)}, v^{(1)})$$

where $Av^{(1)} = -S(A)v^{(1)}$. First, we show that $\mid v^{(1)} \mid$ maximizes $(v, Av)/(v, v)$ and hence, by Lemma 7.7, is an eigenvector of A associated with $S(A)$. Since $S(A)$ is a simple eigenvalue of A it follows from (b) that $\mid v^{(1)} \mid > 0$ and $v_i^{(1)} \neq 0$ for all i. Next, it is easy to show that we must have $v_i^{(1)} v_j^{(1)} < 0$ for all i, j such that $a_{i,j} \neq 0$. (This, of course, implies that all diagonal elements of A vanish.) Because of the irreducibility of A it is necessary that A has Property A. All values $v_i^{(1)}$ are of one sign if $i \in S_1$ and of opposite sign if $i \in S_2$. To show that $-S(A)$ is a simple eigenvalue of A we use a procedure similar to that given in

(b) and show that there cannot exist two orthogonal vectors w and w' associated with the eigenvalue $-S(A)$. The details are left to the reader.

Suppose now that A is irreducible. If $N > 1$, then clearly $A \neq 0$, and $S(A) = \| A \| > 0$. Thus $S(A) > 0$ unless $N = 1$ and $A = 0$. We now seek to show that any nonnegative eigenvector associated with $S(A)$ is *positive*. This is obvious for $N = 1$. If $N > 1$ and if for some i_1 we have $z_{i_1} = 0$, then since

$$(Az)_{i_1} = \sum_{j=1}^{N} a_{i_1,j} z_j = 0$$

it follows that for each j such that $a_{i_1,j} \neq 0$ we must have $z_j = 0$. Since A is irreducible, there exists $i_2 \neq i_1$ such that $a_{i_1,i_2} \neq 0$. Hence $z_{i_2} = 0$. Therefore, for any j such that $a_{i_1,j} \neq 0$ or $a_{i_2,j} \neq 0$ we must have $z_j = 0$. But since A is irreducible, there exists i_3 such that $i_3 \neq i_2$ and $i_3 \neq i_1$, and such that $a_{i_1,i_3} \neq 0$ or $a_{i_2,i_3} \neq 0$. Therefore $z_{i_3} = 0$. Continuing in this way we can show that $z_i = 0$ for all i and hence $z = 0$. This contradiction shows that $z > 0$.

2.8. ILLUSTRATIONS

We now describe some problems involving elliptic partial differential equations and show how their solution by finite difference methods often leads to linear systems whose matrices have some of the properties described above.

Let R be a bounded plane region with boundary S and let $g(x, y)$ be a given function defined and continuous on S. The problem is to find a function $u(x, y)$ defined in $R + S$ and continuous in $R + S$ such that $u(x, y)$ is twice differentiable in R and satisfies the linear second-order partial differential equation†

$$L[u] = Au_{xx} + Cu_{yy} + Du_x + Eu_y + Fu = G \tag{8.1}$$

in R. Here, A, C, D, E, F, and G are analytic functions of the independent variables x and y in R and satisfy the conditions $A \geq m$, $C \geq m$, for some positive constant m, and $F \leq 0$. The function u is also required to satisfy the condition

$$u(x, y) = g(x, y) \tag{8.2}$$

on the boundary S.

† We shall use interchangeably the notation $u_x = \partial u/\partial x$, $u_{xy} = \partial^2 u/(\partial x\, \partial y)$, etc.

We shall refer to the above problem as the *generalized Dirichlet problem.* For the Dirichlet problem one considers Laplace's equation (1-1.3). For the model problem, defined in Chapter 1, we considered Poisson's equation (1-1.1).

It is frequently desirable to consider instead of (8.1) the *self-adjoint* differential equation

$$L[u] = (Au_x)_x + (Cu_y)_y + Fu = G \tag{8.3}$$

where $A \geq m > 0$, $C \geq m > 0$, and $F \leq 0$ in R. We can obtain (8.3) from (8.1) if we have

$$A_x = D, \qquad C_y = E \tag{8.4}$$

Even if the conditions (8.4) are not satisfied one can sometimes obtain the form (8.3) by multiplying both sides of (8.1) by an "integrating factor" $\mu(x, y)$ so that we have

$$(\mu A)_x = \mu D, \qquad (\mu C)_y = \mu E \tag{8.5}$$

The function $\mu(x, y)$ exists if and only if

$$\frac{\partial}{\partial y}\left(\frac{D - A_x}{A}\right) = \frac{\partial}{\partial x}\left(\frac{E - C_y}{C}\right) \tag{8.6}$$

If (8.6) is satisfied we say that the equation (8.1) is *essentially self-adjoint.*

We now construct a discrete generalized Dirichlet problem. First we define the set Ω_h as follows. Given the point (x_0, y_0) and a mesh size $h > 0$, we let Ω_h be the set of all points $(x_0 + ih,\ y_0 + jh)$, $i, j = 0, \pm 1, \pm 2, \ldots$. (More generally, we could use different mesh sizes in each coordinate direction.) We say that two points (x, y) and (x', y') of Ω_h are *adjacent* if $(x - x')^2 + (y - y')^2 = h^2$. Two adjacent points of Ω_h are *properly adjacent* if both are in $R + S$ and if the open segment joining them, not necessarily including the end points, is in R. We let $R_h = R \cap \Omega_h$. A point P of R_h is *regular* if the four adjacent mesh points in Ω_h lie in $R + S$ and are properly adjacent to P. We shall assume here that R and Ω_h are such that all points of R_h are regular points. We let S_h denote the set of all points of Ω_h which are properly adjacent to a point of R_h but which do not belong to R_h. Clearly, we must have $S_h \subseteq S$.

Let us now develop a discrete representation of the differential equation (8.1). For a regular point (x, y) of R_h we represent the partial derivatives at (x, y) by

$$\begin{aligned}
u_x &\sim [u(x+h, y) - u(x-h, y)]/2h \\
u_y &\sim [u(x, y+h) - u(x, y-h)]/2h \\
u_{xx} &\sim [u(x+h, y) + u(x-h, y) - 2u(x, y)]/h^2 \\
u_{yy} &\sim [u(x, y+h) + u(x, y-h) - 2u(x, y)]/h^2
\end{aligned}$$

Substituting in the differential equation (8.1), multiplying by $-h^2$, we obtain

$$a_0 u(x, y) - a_1 u(x+h, y) - a_2 u(x, y+h) - a_3 u(x-h, y) - a_4 u(x, y-h) = t(x, y) \tag{8.7}$$

where

$$\begin{aligned}
a_0 &= a_1 + a_2 + a_3 + a_4 - h^2 F \\
&= 2(A + C - \tfrac{1}{2}h^2 F) \\
a_1 &= A + \tfrac{1}{2}hD, \qquad a_2 = C + \tfrac{1}{2}hE \\
a_3 &= A - \tfrac{1}{2}hD, \qquad a_4 = C - \tfrac{1}{2}hE \\
t &= -h^2 G
\end{aligned} \tag{8.8}$$

Our discrete generalized Dirichlet problem is to find a function u defined on $R_h + S_h$ which satisfies (8.7) on R_h and which satisfies (8.2) on S_h.

If instead of the differential equation (8.1) we are dealing with the self-adjoint equation (8.3), we can replace the differential operators $(Au_x)_x$ and $(Cu_y)_y$ by

$$\begin{aligned}
(Au_x)_x &\sim h^{-2}\{A(x + \tfrac{1}{2}h, y)[u(x+h, y) - u(x, y)] \\
&\qquad - A(x - \tfrac{1}{2}h, y)[u(x, y) - u(x-h, y)]\} \\
(Cu_y)_y &\sim h^{-2}\{C(x, y + \tfrac{1}{2}h)[u(x, y+h) - u(x, y)] \\
&\qquad - C(x, y - \tfrac{1}{2}h)[u(x, y) - u(x, y-h)]\}
\end{aligned} \tag{8.9}$$

Substituting in the differential equation and multiplying by $-h^2$ we obtain the difference equation

$$\hat{a}_0 u(x, y) - \hat{a}_1 u(x+h, y) - \hat{a}_2 u(x, y+h) - \hat{a}_3 u(x-h, y) - \hat{a}_4 u(x, y-h) = t(x, y) \tag{8.10}$$

where

$$\begin{aligned} \hat{a}_0 &= \hat{a}_1 + \hat{a}_2 + \hat{a}_3 + \hat{a}_4 - h^2F \\ \hat{a}_1 &= A(x + \tfrac{1}{2}h, y), \qquad \hat{a}_2 = C(x, y + \tfrac{1}{2}h) \\ \hat{a}_3 &= A(x - \tfrac{1}{2}h, y), \qquad \hat{a}_4 = C(x, y - \tfrac{1}{2}h) \\ t &= -h^2G \end{aligned} \tag{8.11}$$

We shall consider two discrete generalized Dirichlet problems involving the self-adjoint equation (8.3). The discrete problem defined by (8.10) and (8.2) is referred to as "Variant II." If we write (8.3) in the expanded form

$$L[u] = Au_{xx} + Cu_{yy} + A_xu_x + C_yu_y + Fu = G \tag{8.12}$$

and use (8.7) with $D = A_x$, $E = C_y$, then we have "Variant I." We also refer to the use of (8.7) for the differential equation (8.1) as "Variant I."

With either Variant I or Variant II, the problem of solving the discrete generalized Dirichlet problem reduces to the solution of a system of linear algebraic equations of the form

$$Au = b \tag{8.13}$$

where there is one equation and one unknown for each of the N points of R_h. The row of the matrix corresponding to the point (x, y) has $a_0(x, y)$ or $\hat{a}_0(x, y)$ as the diagonal element and $-a_i(x, y)$ or $-\hat{a}_i(x, y)$ in the column corresponding to a point of R_h properly adjacent to (x, y). Terms of (8.7) which involve values of $u(x, y)$ on S_h are brought to the right-hand side of the equation for (x, y). Thus for Variant I the element on the right-hand side of the equation for (x, y) is

$$t(x,y)+a_1'u(x+h,y)+a_2'u(x,y+h)+a_3'u(x-h,y)+a_4'u(x,y-h) \tag{8.14}$$

where $a_1' = a_1$ if $(x + h, y)$ is in S_h; otherwise $a_1' = 0$. Similarly, for a_2', a_3', and a_4'.

From (8.8) and (8.11) it is evident that since $F \leq 0$ we have

$$a_{i,i} > 0, \qquad i = 1, 2, \ldots, N \tag{8.15}$$

Moreover, for Variant I we have $a_{i,j} \leq 0$, for all $i \neq j$ provided h is sufficiently small, i.e., provided

$$h < \min[\min(2A/|D|), \min(2C/|E|)] \tag{8.16}$$

where the minima are taken over $R + S$. In the case of Variant II we always have $a_{i,j} \leq 0$ for $i \neq j$.

Let us now assume that we are either using Variant II or else that we are using Variant I with a mesh size small enough so that not only is $a_{i,j} \leq 0$ for $i \neq j$ but for each point of R_h we have

$$a_i > 0, \qquad i = 1, 2, 3, 4 \tag{8.17}$$

A subset T_h of R_h is *connected* if any two points of T_h can be joined by a path consisting of horizontal and vertical segments joining properly adjacent points of R_h. For each point P of R_h we can consider the set $T_h(P)$ of all mesh points which can be joined to P in this manner. Corresponding to each distinct set of this kind we obtain a system of linear algebraic equations of the form (8.13). The systems thus obtained are independent of each other. We now show that for each such system the matrix is irreducible. Let N be the order of the system. If $N = 1$, the matrix is clearly irreducible. If $N > 1$ and the matrix is not irreducible, there exist two nonempty disjoint subsets S and T of W, the set of the first N integers, such that $S + T = W$ and such that $a_{i,j} = 0$ for all $i \in S$ and $j \in T$. Let P be any point in the set $T_h(P)$ and assume $P \in S$.† Assume that some point Q of $T_h(P)$ belongs to T. Evidently P can be joined to Q by a path consisting of line segments connecting properly adjacent mesh points. Somewhere along this path there must be a segment joining a point P' of S to a point P'' of T. However, the coefficient of $u(P'')$ in the equation for $u(P')$ must be positive by (8.17). Hence $a_{i,j} \neq 0$ where $i \in S$ and $j \in T$. This contradiction shows that T is empty and that A is irreducible.

Next we show that A has weak diagonal dominance. Since $F \leq 0$, (5.1) holds by (8.8) and (8.11). Moreover, since R is bounded, there exists points of R_h which are properly adjacent to one or more points of S_h. For such a point certain of the terms in (8.7), or (8.10), are transferred to the right-hand side to obtain the form (8.13). Thus some of the a_i do not appear in the matrix A. By (8.17), and by (8.8) and (8.11) it follows that (5.2) holds for certain i and hence A has weak diagonal dominance.

The matrix A corresponding to Variant II is symmetric. For by (8.10) and (8.11), the coefficient of $u(x + h, y)$ in the equation for $u(x, y)$, namely $A(x + \frac{1}{2}h, y)$, is the same as the coefficient of $u(x' - h, y)$ in

† Actually we shall number the mesh points 1, 2, ..., N; when we say $P \in S$ we actually mean that the number corresponding to P belongs to S.

the equation for $u(x', y)$, where $x' = x + h$. This latter coefficient is

$$A(x' - \tfrac{1}{2}h, y) = A(x + h - \tfrac{1}{2}h, y) = A(x + \tfrac{1}{2}h, y)$$

In the case of Variant I, the matrix A is not in general symmetric, but, as will be shown in Chapter 6, in many cases A is similar to a symmetric matrix.

We now show that with either Variant I or Variant II we obtain a matrix with Property A. Indeed, let S_1 correspond to those mesh points $(x_0 + ph, y_0 + qh)$ where $p + q$ is even and let S_2 correspond to those mesh points such that $p + q$ is odd. Evidently, from (8.7), if $i \neq j$ and if i and j both belong to S_1 or if i and j both belong to S_2 we have $a_{i,j} = 0$.

In summary we have:

(a) For Variants I and II, $a_{i,i} > 0$, $i = 1, 2, \ldots, N$. Moreover, for Variant I with h sufficiently small, and for Variant II A is an L-matrix.

(b) For Variant II and for Variant I with h sufficiently small, the system (8.13) can be written as one or more independent subsystems each of which has an irreducible matrix.

(c) For Variant II and for Variant I with h sufficiently small, A has weak diagonal dominance.

(d) For Variant II, A is symmetric.

(e) For Variants I and II, A has Property A.

Let us now assume that we are dealing with Variant II or else with Variant I and that h is small enough so that (8.17) holds. We assume that R_h is connected, or if not, we assume that (8.13) corresponds to one of the connected subsets of R_h. Hence A is an irreducible L-matrix with weak diagonal dominance. Moreover, by Theorem 5.5, with Variant II, A is a Stieltjes matrix. In Chapter 4 we will show that if A is an irreducible matrix with weak diagonal dominance, then $S(B) < 1$, where $B = I - (\operatorname{diag} A)^{-1}A$. From this it follows by Theorem 7.2 that A is an M-matrix.

SUPPLEMENTARY DISCUSSION

Section 2.1. A proof of Theorem 1.8 is given by Faddeev and Faddeeva [1963]. Except as otherwise indicated, all other theorems given without proof in Section 2.1 can be found in Birkhoff and MacLane [1953]. An elementary discussion of the determination of the sign of a term in the expansion of a determinant is given in Barnard and Child [1952, pp. 49, 120].

Section 2.2. The meaning of the term "positive definite" as applied to a matrix varies in the literature. Our definition is closely related to that used by Wachspress [1966] except that we require that A be Hermitian while Wachspress requires that (v, Av) be real for all v. By Theorem 2.1 the definitions are equivalent.

Section 2.3. For a thorough discussion of matrix and vector norms see Householder [1964]. A good discussion of the topology of linear spaces can be found in the book of Taylor [1963].

For the determination of a vector norm consistent with a given matrix norm see Householder [1964, p. 42].

Section 2.5. Our definition of irreducibility agrees with that of Geiringer [1949] but differs slightly from Varga's definition [1962] for the case $N = 1$ since we do not require that $a_{1,1} \neq 0$ in this case. Forsythe and Moler [1967] require that for a matrix to be irreducible there must not exist permutation matrices P and Q such that PAQ has the form (5.3).

For a thorough treatment of the use of graph theory for matrix analysis see Varga [1962].

As noted by Taussky [1949], Theorem 5.3 has been proved independently many times.

Section 2.6. The class of matrices with Property A was introduced by Young [1950, 1954]. A matrix with Property A and with nonvanishing diagonal elements is "2-cyclic" in the sense of Varga [1962].

Section 2.7. The introduction of the class of L-matrices is motivated by the work of Stein and Rosenberg [1948]. The class of M-matrices was considered by Ostrowski [1955]. Monotone matrices are discussed by Collatz [1952] and Bramble and Hubbard [1964]. Bramble and Hubbard use the condition of Theorem 7.4 for their definition.

Theorem 7.6(c) is a special case of a result of Frobenius [1912]. (See Varga [1962, Theorem 2.3].)

Section 2.8. If one were given the more general equation $L[u] = Au_{xx} + 2Bu_{xy} + Cu_{yy} + Du_x + Eu_y + Fu = G$, where $B^2 - AC \leq m < 0$ for some constant m, then one could by suitable change of independent variables obtain the simpler equation (8.1).

More sophisticated procedures are normally used to handle cases where S_h does not belong to S. One such procedure is to define R_h as before and to determine S_h as follows. A point P of R_h is said to be ***regular*** if the four adjacent points belong to $R + S$ and if the open line segments joining P to those points belong to R. For any nonregular point P of R_h, if a segment from P to an adjacent point Q does not lie in R, then we let the nearest point on the segment to P be in S_h. Thus for points of R_h near the boundary we may have points of S_h closer than a distance h. For such points we use special difference equations.

EXERCISES

Section 2.1

1. Given the matrix

$$A = \begin{pmatrix} 3+i & 2-i & 1-i \\ 1+2i & 0 & 1 \\ 2 & 0 & 4-2i \end{pmatrix}$$

determine diag A, trace A, A^{T}, A^{H}, and A^*. Verify (1.9). Determine C_L and C_U, which are strictly lower and strictly upper triangular matrices, respectively, such that $A = \operatorname{diag} A - C_L - C_U$.

2. If

$$B = \begin{pmatrix} 3 & 4+i \\ 1 & 3 \\ 4-i & 2 \end{pmatrix}$$

and A is given in the preceding example, verify (1.10).

3. For the matrices given below indicate which of the following properties are satisfied: singular or nonsingular; unitary, orthogonal, normal; Hermitian, symmetric; permutation

$$A = \begin{pmatrix} 1 & 0 \\ 0 & 1 \end{pmatrix}, \qquad B = \begin{pmatrix} 1 & 0 & 1 \\ 0 & 0 & 1 \\ 0 & 1 & 0 \end{pmatrix}$$

$$C = \begin{pmatrix} 0 & 0 & 0 \\ 1 & 0 & 2 \\ 5 & i & 0 \end{pmatrix}, \qquad D = \begin{pmatrix} 0 & 1+i & 0 \\ 1+i & 1 & 2-i \\ 0 & 2-i & i \end{pmatrix}$$

$$E = \begin{pmatrix} 0 & 1+i & 0 \\ 1-i & 1 & 2-i \\ 0 & 2-i & i \end{pmatrix}, \qquad F = \begin{pmatrix} 0 & 1+i & 0 \\ 1-i & 2 & 2-i \\ 0 & 2-i & 0 \end{pmatrix}$$

$$G = \begin{pmatrix} 0 & 1 & 0 \\ 0 & 0 & 1 \\ 1 & 0 & 0 \end{pmatrix}$$

4. Find the most general orthogonal matrix of order 2. Also, find the most general unitary matrix and the most general normal matrix of order two.

5. Verify the statement that $(x, Ay) = (A^{H}x, y)$ for all matrices A and all vectors x and y in the case where A is a 3×3 matrix and where x and y are column matrices of order 3.

6. Construct the permutation matrix P corresponding to the permutation $\sigma(1) = 3$, $\sigma(2) = 1$, $\sigma(3) = 4$, $\sigma(4) = 2$ and verify that $PP^{T} = I$.

7. Determine (v, w) and (w, v) where $v = (1 + i, 2 - i)^{T}$, $w = (3,1)^{T}$ and verify (1.14). Also verify (1.15) if

$$A = \begin{pmatrix} 2-i & 1+i \\ 2 & i \end{pmatrix}$$

Compute $\| v \|_2$ and $\| w \|_2$ and verify (1.17).

8. Compute all eigenvalues and eigenvectors of

$$A = \begin{pmatrix} 2 & -1 & 0 \\ -1 & 3 & -1 \\ 0 & -1 & 2 \end{pmatrix}$$

and show that the eigenvectors are linearly independent and form a pairwise orthogonal set. Normalize the eigenvectors so that each has Euclidean length 1 and show that the matrix P whose columns are the normalized eigenvectors is orthogonal. Also, show that $P^{-1}AP$ is a diagonal matrix. Express the vector $v = (1, 2, 3)^{T}$ as a linear combination of the eigenvectors.

9. Verify Theorem 1.7 and (1.27) for the matrix A of the preceding exercise.

10. Consider the permutation defined by $\sigma(1) = 2$, $\sigma(2) = 4$, $\sigma(3) = 1$, $\sigma(4) = 5$, $\sigma(5) = 3$. Use each of the three methods described in the text to determine whether σ is even or odd (i.e., consider transpositions, inversions, and disjoint cycles).

11. Solve the system $Au = b$ by Cramer's rule where

$$A = \begin{pmatrix} 2 & -1 & 0 \\ -1 & 2 & -1 \\ 0 & -1 & 2 \end{pmatrix}, \qquad b = \begin{pmatrix} 1 \\ 0 \\ 1 \end{pmatrix}$$

Also, find A^{-1} in this way by successively considering $b^{(1)} = (1, 0, 0)^T$, $b^{(2)} = (0, 1, 0)^T$, $b^{(3)} = (0, 0, 1)^T$. Use A^{-1} to solve $Au = (2, 1, 0)^T$.

12. Verify that A is singular, where

$$A = \begin{pmatrix} 1 & -1 \\ 2 & -2 \end{pmatrix}$$

and find $v \neq 0$ such that $Av = 0$. Also verify that $Av = b^{(1)}$ has no solution and $Av = b^{(2)}$ has an infinite number of solutions where

$$b^{(1)} = (-1, 0)^T, \qquad b^{(2)} = (-1, -2)^T.$$

13. Verify Theorem 1.12 if

$$A = (2, 1, 0), \qquad B = (1, 2, 1)^T$$

and if

$$A = \begin{pmatrix} 2 & 3 \\ 1 & 4 \end{pmatrix}, \qquad B = \begin{pmatrix} 1 & 1 \\ 2 & 2 \end{pmatrix}$$

14. Find a nonsingular matrix P such that $P^{-1}AP = J$ where J is a Jordan canonical form of A and

$$A = \begin{pmatrix} -\frac{1}{4} & 1 \\ -\frac{1}{4} & \frac{3}{4} \end{pmatrix}$$

Hint: P satisfies $AP = PJ$. The columns of P can be found one at a time.

15. Verify Theorem 1.17 for A and Theorem 1.16 for A and B where

$$A = \begin{pmatrix} 0 & 1 & 1 \\ 1 & 0 & 0 \\ 1 & 0 & 0 \end{pmatrix}, \qquad B = \begin{pmatrix} 0 & -1 & 0 \\ 1 & 0 & 0 \\ 0 & 0 & 0 \end{pmatrix}$$

16. Let $\mathcal{N}(A)$ and $\mathcal{R}(A)$ denote, respectively, the ***null space*** and ***range*** of the $N \times N$ matrix A. Prove that if A is ***normal***, then $\mathcal{N}(A) = \mathcal{N}(A^H)$. Also prove that if $v \in \mathcal{N}(A)$ and if $w \in \mathcal{R}(A)$, then $(v, w) = 0$. Thus show that $\mathcal{N}(A) \cap \mathcal{R}(A) = 0$ and if $A^p v = 0$ for some positive integer p, then $v \in \mathcal{N}(A)$.

17. Show that the eigenvalues of an Hermitian matrix are real and that if $Av = \lambda_1 v$ and $Aw = \lambda_2 w$, with $\lambda_1 \neq \lambda_2$, then $(v, w) = 0$.

Section 2.2

1. By direct calculation verify that (v, Av) is real for all complex v for any 2×2 Hermitian matrix.

2. Verify (2.9) for the matrix

$$A = \begin{pmatrix} 0 & -1 & 0 \\ -1 & 1 & -1 \\ 0 & -1 & 0 \end{pmatrix}$$

3. Is the matrix

$$A = \begin{pmatrix} 4 & 1+i \\ 1-i & 5 \end{pmatrix}$$

positive definite? What can be said about the values of

$$(v, Av)/(v, v)$$

for all nonzero vectors v? Verify this result for the cases

$$v = \begin{pmatrix} 1 \\ 0 \end{pmatrix}, \quad \begin{pmatrix} 0 \\ 1 \end{pmatrix}, \quad \begin{pmatrix} 1 \\ 1 \end{pmatrix}$$

If A is positive definite, show that Re A is positive definite and $(v, Cv) = 0$ for all $v \neq 0$, where $C = \text{Im } A$.

4. Verify Theorem 2.5 for the matrix

$$A = \begin{pmatrix} 2 & -1 & 0 & 0 \\ -1 & 2 & -1 & 0 \\ 0 & -1 & 2 & -1 \\ 0 & 0 & -1 & 2 \end{pmatrix}$$

5. Verify Theorem 2.6 for the cases

$$L = \begin{pmatrix} 1 & 0 & i \\ 2 & 1 & 1+i \\ 1 & 0 & 2 \end{pmatrix}, \quad L = \begin{pmatrix} 1 & i \\ i & -1 \end{pmatrix}, \quad L = \begin{pmatrix} 1 & 0 & 0 \\ 2 & 1 & 0 \\ 3 & 2 & 1 \end{pmatrix}$$

6. Let A be the matrix

$$A = \begin{pmatrix} 2 & -1 & 0 \\ -1 & 2 & -1 \\ 0 & -1 & 2 \end{pmatrix}$$

(a) Verify that A is positive definite.

(b) Find an orthogonal matrix P such that $P^{-1}AP$ is a diagonal matrix.

(c) Find an orthonormal set of eigenvectors which forms a basis for E^3.

(d) Find $S(A)$.

(e) Find the positive definite square root of A. (Answer may be given as a product of matrices.)

Note: In all cases you may leave radicals in the answers.

7. Verify that the matrix

$$A = \begin{pmatrix} 2 & 0 \\ 1 & 2 \end{pmatrix}$$

is positive real but not positive definite.

Section 2.3

1. Compute $S(A)$ and $\| A \|_2$ for

$$A = \begin{pmatrix} 2 & 3 \\ 1 & 2 \end{pmatrix}$$

2. Verify that (3.6) is the induced matrix norm corresponding to the vector norm (3.3).

3. Given a matrix norm $\| \cdot \|_\beta$ one can define a consistent vector norm $\| \cdot \|_\alpha$ as follows

$$\| v \|_\alpha = \| B \|_\beta$$

where B is a matrix whose first column is v and all other elements vanish. Verify that $\| \cdot \|_\alpha$ is a norm and that $\| \cdot \|_\alpha$ and $\| \cdot \|_\beta$ are consistent.

4. Show that $\| \cdot \|_{\beta,L}$ and $\| \cdot \|_{\alpha,L}$ define a matrix norm and a vector norm, respectively, and that $\| \cdot \|_{\beta,L}$ and $\| \cdot \|_{\alpha,L}$ are compatible if and only if $\| \cdot \|_\beta$ and $\| \cdot \|_\alpha$ are compatible.

5. Show that for any matrix A we have

$$(1/N) \| A \|_M \leq \| A \|_2 \leq \| A \|_M$$

(Faddeev and Faddeeva, 1963).
Hint: Note that the sum of the eigenvalues of $A^{\mathrm{H}}A$, namely trace AA^{H}, is $\| A \|_T^2$.

6. For the matrices

$$A = \begin{pmatrix} 2 & -1 & 0 \\ -1 & 2 & -1 \\ 0 & -1 & 2 \end{pmatrix}, \qquad A = \begin{pmatrix} -\frac{1}{4} & 1 \\ -\frac{1}{4} & \frac{3}{4} \end{pmatrix}$$

find $S(A)$, $\| A \|_2$, $\| A \|_\infty$, $\| A \|_1$, $\| A \|_M$, $\| A \|_T$.

7. Verify directly from (3.4) that $\| \cdot \|_\infty$ defined by (3.5) is a matrix norm.
Hint: First verify the fourth norm property for the case $N = 2$.

Section 2.4

1. Give the details of the proof of Theorem 4.2, first for the case where A is the null matrix and then for the general case.

Section 2.5

1. Test the following matrices for (a) irreducibility, (b) weak diagonal dominance, (c) nonvanishing determinant.

$$\begin{pmatrix} i & 2 & 1 \\ 0 & 2 & 3 \\ 0 & 1 & i \end{pmatrix}, \quad \begin{pmatrix} 2 & -2 \\ -1 & 1 \end{pmatrix}, \quad \begin{pmatrix} 2 & -1 & 0 \\ 0 & 2 & -2 \\ 0 & -2 & 2 \end{pmatrix}$$

$$\begin{pmatrix} 4 & 2 & 1 \\ 0 & 2 & -1 \\ 0 & -1 & 2 \end{pmatrix}, \quad \begin{pmatrix} 2 & -1 & 0 \\ -1 & 2 & -1 \\ 0 & -1 & 2 \end{pmatrix}$$

Wherever possible, answer (c) without actually evaluating the determinant.

2. Show that if A is an irreducible matrix with weak diagonal dominance and positive diagonal elements, then all eigenvalues of A have positive real parts (Taussky, 1949).

3. Give a direct proof that $S(A) \leq \| A \|_\infty$ using Corollary 5.4.

Section 2.7

1. Give the details of the proof of Theorem 7.6, part (c).

2. Show that if A is an irreducible M-matrix, then $A^{-1} > 0$.

Section 2.8

1. Prove that a function $\mu(x, y)$ satisfying (8.5) exists if and only if (8.6) holds.

2. Test whether the following equations are self-adjoint or essentially self-adjoint and where possible transform the equation into the form (8.3).

(a) $u_{xx} + (k/y)u_y + u_{yy} = 0$, k an integer,

(b) $u_{xx} + (2/(x + y))u_x + u_{yy} + (2/(x + y))u_y = 0$,

(c) $u_{xx} + (1/x)u_x + xu_{yy} + (x/y)u_y = 0$,

(d) $u_{xx} + xu_x + yu_{yy} + xyu_y = 0$.

3. Consider the differential equation

$$u_{xx} - (k/y)u_y + u_{yy} = 0$$

in the region $0 < x < 1$, $\frac{1}{2} < y < \frac{3}{2}$. How small must h be chosen so that, if $k = 6$, the matrix A corresponding to Variant I of the discrete Generalized Dirichlet problem is an L-matrix?

4. For the problem of Exercise 3, with $k = 1$ and $h = \frac{1}{3}$ determine A and verify that with Variant I one obtains an M-matrix and with Variant II one obtains a Stieltjes-matrix.

5. Consider the problem of solving the differential equation

$$L(u) = u_{xx} - (1/y)u_y + u_{yy} = 1$$

in the region shown in the accompanying figure with indicated boundary values. With a mesh size $h = \frac{1}{2}$, derive a system $Au = b$ based on the difference representation Variant I. Also, find an "integrating factor" $\mu(x, y)$ such that $\mu(x, y)L(u)$ has the self-adjoint form (8.3). Then derive the difference representation Variant II.

(a) Show that in both cases the matrices are irreducible L-matrices and have weak diagonal dominance and Property A.

(b) Show that the matrix for Variant II is positive definite.

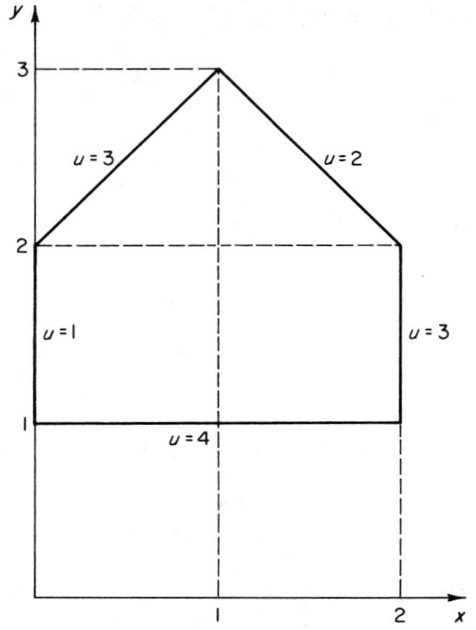

Chapter 3 / LINEAR STATIONARY ITERATIVE METHODS

3.1. INTRODUCTION

We have seen how the use of finite difference methods for solving a boundary value problem involving an elliptic partial differential equation may lead to a linear system

$$Au = b \tag{1.1}$$

where A is a given $N \times N$ matrix and b is a given vector. The order of the matrix equals the number of interior mesh points and may be very large, so large, in fact, that it may be impractical to store the matrix even in a large computer or to solve the system by direct methods. On the other hand, since the matrix is sparse it is usually possible to store all of the nonzero elements and apply various iterative methods.

In general, an iterative method for solving (1.1) is defined by the functions $\phi_0(A, b)$, $\phi_1(u^{(0)}; A, b)$, $\phi_2(u^{(0)}, u^{(1)}; A, b)$, $\ldots$, $\phi_k(u^{(0)}, u^{(1)}, \ldots, u^{(k-1)}; A, b)\ldots$. The sequence $u^{(0)}, u^{(1)}, \ldots$ is defined by

$$\begin{aligned} u^{(0)} &= \phi_0(A, b) \\ u^{(n+1)} &= \phi_{n+1}(u^{(0)}, u^{(1)}, \ldots, u^{(n)}; A, b) \end{aligned} \tag{1.2}$$

Frequently, for some $s \geq 0$ the choice of $u^{(0)}, u^{(1)}, \ldots, u^{(s)}$ is arbitrary, i.e., $\phi_0, \phi_1, \ldots, \phi_s$ are arbitrary functions of A and b.

If for some integer $s > 0$, ϕ_n is independent of n for all $n \geq s$, then the method is said to be *stationary*. Otherwise it is *nonstationary*. In the stationary case we let $\phi = \phi_s = \phi_{s+1} = \cdots$. Evidently $u^{(n+1)}$ depends on at most the s vectors $u^{(n)}, u^{(n-1)}, \ldots, u^{(n-s+1)}$. In the case $s = 1$ we have

$$\begin{aligned} u^{(0)} &= \phi_0(A, b) \\ u^{(n+1)} &= \phi(u^{(n)}; A, b), \qquad n = 0, 1, \ldots \end{aligned} \tag{1.3}$$

If $s = 2$, we have

$$\begin{aligned} u^{(0)} &= \phi_0(A, b) \\ u^{(1)} &= \phi_1(u^{(0)}; A, b) \\ u^{(n+1)} &= \phi(u^{(n-1)}, u^{(n)}; A, b), \qquad n = 1, 2, \ldots \end{aligned} \tag{1.4}$$

The *degree* of a stationary method is $\hat{s}$ for $\hat{s} \leq s$ if, for $n \geq s - 1$, $u^{(n+1)}$ depends on $u^{(n)}, u^{(n-1)}, \ldots, u^{(n-\hat{s}+1)}$ but not on $u^{(k)}$ for $k < n - \hat{s} + 1$. Thus the methods defined by (1.3) and (1.4) are of degree at most *one* and *two*, respectively.

If, for each n, ϕ_n is a linear function of $u^{(0)}, u^{(1)}, \ldots$, and $u^{(n-1)}$, then the method is said to be linear. Otherwise, it is *nonlinear*. In the case of a *linear stationary iterative method of first degree*, $\phi(u^{(n)}; A, b)$ has the form

$$u^{(n+1)} = Gu^{(n)} + k \tag{1.5}$$

for some matrix G and for some vector k. Our object in this chapter will be to study conditions under which such methods will converge to a solution of (1.1) and to determine the rapidity of convergence.

In studying the method (1.5) we shall frequently consider the *related linear system*

$$(I - G)u = k \tag{1.6}$$

We shall study the relation between the set of solutions $\mathscr{S}(A, b)$ of (1.1) and the set of solutions $\mathscr{S}(I - G, k)$ of (1.6). If A is nonsingular, then by Theorem 2-1.2 and Eq. (2-1.23), there exists a unique solution $\bar{u}$ of (1.1) given by

$$\bar{u} = A^{-1}b \tag{1.7}$$

We say that the iterative method defined by (1.5) is *consistent* with the system (1.1) if $\mathscr{S}(A, b) \subseteq \mathscr{S}(I - G, k)$, *reciprocally consistent* if $\mathscr{S}(I - G, k) \subseteq \mathscr{S}(A, b)$, and *completely consistent* if $\mathscr{S}(I - G, k) = \mathscr{S}(A, b)$.

In Section 3.2 we will discuss characterizations and implications of these conditions.

Six "basic" linear stationary iterative methods are defined in Section 3.3 and will be shown to be completely consistent under very general conditions. Methods for generating completely consistent methods will be given in Section 3.4. The balance of the chapter is concerned with convergence and rates of convergence.

Unless otherwise noted, the matrix A and the vector b in (1.1) are real. In this case, if A is nonsingular, then the solution $\bar{u}$ is also real. Moreover, unless otherwise noted, we assume that the matrix G and the vector k of (1.5) are real.

3.2. CONSISTENCY, RECIPROCAL CONSISTENCY, AND COMPLETE CONSISTENCY

Ideally, an iterative method should have the property that for any starting vector $u^{(0)}$ it converges to a solution of (1.1). As a first requirement, we shall insist that if at any stage of the iterative process we obtain a solution of (1.1), then all subsequent iterants remain the same. This requirement is equivalent to *consistency* as shown by the following theorem.

Theorem 2.1. The iterative method (1.5) is consistent with (1.1) if and only if the following holds: if, for some n, $u^{(n)}$ is a solution, say $\bar{u}$, of (1.1), then $u^{(n+1)} = u^{(n+2)} = \cdots = \bar{u}$.

Proof. If consistency holds and if $\bar{u} \in \mathscr{S}(A, b)$, then $(I - G)\bar{u} = k$. If $u^{(n)} = \bar{u}$, then $u^{(n+1)} = Gu^{(n)} + k = G\bar{u} + k = \bar{u}$. Similarly, $\bar{u} = u^{(n+2)} = u^{(n+3)} = \cdots$. On the other hand, suppose that $\bar{u} \in \mathscr{S}(A, b)$ and $u^{(n)} = \bar{u}$ implies $u^{(n+1)} = \bar{u}$. Then $\bar{u} = u^{(n+1)} = Gu^{(n)} + k = G\bar{u} + k$ and $\bar{u} \in \mathscr{S}(I - G, k)$, which implies consistency.

Another characterization of consistency is the following.

Theorem 2.2. If A is nonsingular, then the iterative method (1.5) is consistent with (1.1) if and only if

$$k = (I - G)A^{-1}b \tag{2.1}$$

Proof. If the method is consistent, then since $\mathscr{S}(A, b)$ contains the single element $\bar{u} = A^{-1}b$ which satisfies (1.6) we must have (2.1). On the other hand, if (2.1) is satisfied, and if $\bar{u} = A^{-1}b$, then $k = (I - G)\bar{u}$; hence (1.6) is satisfied and the method is consistent.

As a second requirement we might logically insist that if the sequence defined by (1.5) converges, then it converges to a solution of (1.1). We prove

Theorem 2.3. If the sequence $u^{(0)}, u^{(1)}, \ldots$ defined by (1.5) converges to a limit, $\hat{u}$, then $\hat{u} \in \mathscr{S}(I - G, k)$. The method is reciprocally consistent if and only if whenever the sequence $u^{(0)}, u^{(1)}, \ldots$ converges, it converges to a solution of (1.1). If $I - G$ is nonsingular, then the method is reciprocally consistent if and only if

$$b = A(I - G)^{-1}k \tag{2.2}$$

Proof. If the sequence converges to $\hat{u}$, then taking limits of both sides of (1.5) we have $\hat{u} = G\hat{u} + k$ and hence $\hat{u} \in \mathscr{S}(I - G, k)$. If the method is reciprocally consistent, then $\hat{u} \in \mathscr{S}(A, b)$. On the other hand, if whenever the sequence converges it converges to a limit which satisfies (1.1), then $\mathscr{S}(I - G, k) \subseteq \mathscr{S}(A, b)$, and the method is reciprocally consistent; for, every solution $\hat{u}$ of (1.6) is the limit of a sequence defined by (1.5) with $u^{(0)} = \hat{u}$. The proof of (2.2) is the same as that of Theorem 2.2 with A and b interchanged with $I - G$ and k, respectively.

As an example, let us consider the iterative method defined by

$$u^{(n+1)} = u^{(n)} \tag{2.3}$$

Evidently every vector satisfies the related equation $u = u$. On the other hand, if A is nonsingular, then (1.1) has one and only one solution. Hence the method is consistent but not reciprocally consistent.

As another example, consider the iterative method

$$u^{(n+1)} = u^{(n)} + (1, 0, \ldots, 0)^{\mathrm{T}} \tag{2.4}$$

for solving the system $u = 0$. The related equation has no solution; hence the method is reciprocally consistent, but not consistent. The same iterative method applied to the system $u_1 + u_2 = 1$, $u_1 + u_2 = 2$, however, would be both consistent and reciprocally consistent since neither (1.6) nor (1.1) has a solution.

Our first and second requirements together imply that the method is completely consistent. We now prove

Theorem 2.4. If A is nonsingular, then the iterative method (1.5) is completely consistent with (1.1) if and only if it is consistent and $I - G$ is nonsingular. If $I - G$ is nonsingular, then complete consistency holds if and only if the method (1.5) is reciprocally consistent and A is nonsingular.

Proof. If A is nonsingular, if (1.5) is consistent, and if $I - G$ is nonsingular, then (2.1) holds by Theorem 2.2 and hence (2.2) holds. Therefore, by Theorem 2.3, the method is reciprocally consistent and hence completely consistent. On the other hand, if A is nonsingular and (1.5) is completely consistent, then it is consistent. Since (1.1) has a unique solution, then (1.6) must also. But by Theorem 2-1.4 this implies $I - G$ is nonsingular. This proves the first part of the theorem. The proof of the second part is similar.

We now prove

Theorem 2.5. Let A be a nonsingular matrix and let $\bar{u}$ be the solution of (1.1). If the sequence $u^{(0)}, u^{(1)}, \ldots$ determined by the iterative method (1.5) converges to $\bar{u}$ for all $u^{(0)}$, then (1.5) is completely consistent. On the other hand, if (1.5) is completely consistent and if the sequence $u^{(0)}, u^{(1)}, \ldots$ determined by (1.5) converges, it converges to $\bar{u}$.

Proof. If the sequence $u^{(0)}, u^{(1)}, \ldots$ converges to $\bar{u}$, then, by Theorem 2.3, $\bar{u} \in \mathscr{S}(I - G, k)$; hence the method is consistent. If $u' \neq \bar{u}$ is a solution of (1.6), then if $u^{(0)} = u'$, we have $u^{(1)} = u^{(2)} = \cdots = u'$. Hence u' is a solution of (1.6), a contradiction. Thus $\bar{u}$ is the only solution of (1.6) and complete consistency follows.

If the sequence determined by (1.5) converges, then it converges to a solution w of (1.6) by Theorem 2.3. If complete consistency holds, then $w = \bar{u}$.

While Theorem 2.5 shows that if A is nonsingular, then a completely consistent iterative method will converge to the solution of (1.1) whenever it converges; nevertheless, it may happen that for some starting vectors $u^{(0)}$ the method may not converge. To see this, let us consider the very simple system

$$2Iu = b \tag{2.5}$$

and the iterative method

$$u^{(n+1)} = -u^{(n)} + b \tag{2.6}$$

Since $G = -I$, $I - G$ is nonsingular. Moreover, $k = b = (I - G)A^{-1}b$ so that the method is completely consistent. However, if $u^{(0)} = 0$, then $u^{(1)} = b$, $u^{(2)} = 0$, $u^{(3)} = b$, $u^{(4)} = 0$, etc., and the method does not converge. We note, however, that if the sequence does converge we must have $u^{(0)} = \frac{1}{2}b$ and in this case, as guaranteed by Theorem 2.5, the sequence converges to the solution of (2.5).

Clearly, some additional condition is needed in order to guarantee that the iterative method will indeed converge for all $u^{(0)}$ to a limit independent of $u^{(0)}$. As we shall see in Section 3.5, this condition is simply $S(G) < 1$.

Up to this point we have been concerned with cases where A and/or $I - G$ are nonsingular. We now consider the more general case. For convenience, we introduce the following terminology.

Definition 2.1. The system (1.1) is *solvable* if $\mathscr{S}(A, b)$ is not empty.

Theorem 2.6. If (1.1) is solvable, then the iterative method (1.5) is consistent with (1.1) if and only if there exists a matrix M such that

$$G = I - MA, \qquad k = Mb \tag{2.7}$$

The method (1.5) is completely consistent if and only if a nonsingular matrix M exists such that (2.7) holds. If (1.6) is solvable, then the iterative method (1.5) is reciprocally consistent with (1.1) if and only if there exists a matrix Q such that

$$A = Q(I - G), \qquad b = Qk \tag{2.8}$$

Proof. We remark that the first two parts of the theorem follow at once, for A nonsingular, by Theorems 2.2 and 2.4, where we let $M = (I - G)A^{-1}$. The third part follows, for the case where $I - G$ is nonsingular, by Theorem 2.3 with $Q = A(I - G)^{-1}$.

In the general case the conditions are clearly sufficient. For the proof of the necessity consider first the case where (1.1) is solvable, and let $\bar{u}$ be any solution of (1.1). Let w be any element of $\mathscr{N}(A)$. Since $\bar{u}$ and $\bar{u} + w$ are solutions of (1.1), if the method (1.5) is consistent then $\bar{u}$ and $\bar{u} + w$ must satisfy (1.6). Hence $w \in \mathscr{N}(I - G)$ and $\mathscr{N}(A) \subseteq \mathscr{N}(I - G)$. If, on the other hand, (1.6) is solvable, let $\hat{u}$ be any solution of (1.6) and let w be any element of $\mathscr{N}(I - G)$. If the method (1.5) is reciprocally consistent, then since $\hat{u}$ and $\hat{u} + w$ are solutions of (1.6), they must also be solutions of (1.1) and $w \in \mathscr{N}(A)$. Therefore $\mathcal{N}(I - G) \subseteq \mathcal{N}(A)$. Finally, if (1.6) is completely consistent with (1.1) and if either (1.1) or (1.6), and hence both, is solvable, then we can show that $\mathcal{N}(A) = \mathcal{N}(I - G)$. The theorem now follows as a consequence of the following lemma.

Lemma 2.7. If A and B are any two matrices such that $\mathcal{N}(A) \subseteq \mathcal{N}(B)$, then there exists a matrix M such that

$$MA = B \tag{2.9}$$

If $\mathcal{N}(A) = \mathcal{N}(B)$, then M can be taken to be nonsingular.

Proof. Let V be a nonsingular matrix such that $V^{-1}AV = J_A$ is a Jordan canonical form of A. Since $\mathcal{N}(A) \subseteq \mathcal{N}(B)$, J_A and $V^{-1}BV$ have only zeros in the p columns corresponding to the p linearly independent eigenvectors of A associated with the eigenvalue zero, where p is the dimension of $\mathcal{N}(A)$. Moreover, p of the rows of J_A contain all zeros. Let P_1 and P_2 be permutation matrices such that

$$P_1 J_A P_2 = \begin{pmatrix} K & 0 \\ 0 & 0 \end{pmatrix} \tag{2.10}$$

where K is a square nonsingular matrix. For example, if

$$J_A = \begin{bmatrix} 0 & 0 & 0 & 0 & 0 & 0 & 0 \\ 0 & 0 & 1 & 0 & 0 & 0 & 0 \\ 0 & 0 & 0 & 0 & 0 & 0 & 0 \\ 0 & 0 & 0 & 0 & 1 & 0 & 0 \\ 0 & 0 & 0 & 0 & 0 & 0 & 0 \\ 0 & 0 & 0 & 0 & 0 & 2 & 1 \\ 0 & 0 & 0 & 0 & 0 & 0 & 2 \end{bmatrix} \tag{2.11}$$

then $p = 3$ and we have

$$P_1 J_A P_2 = \begin{bmatrix} 1 & 0 & 0 & 0 & 0 & 0 & 0 \\ 0 & 1 & 0 & 0 & 0 & 0 & 0 \\ 0 & 0 & 2 & 1 & 0 & 0 & 0 \\ 0 & 0 & 0 & 2 & 0 & 0 & 0 \\ 0 & 0 & 0 & 0 & 0 & 0 & 0 \\ 0 & 0 & 0 & 0 & 0 & 0 & 0 \\ 0 & 0 & 0 & 0 & 0 & 0 & 0 \end{bmatrix} \tag{2.12}$$

which has the form (2.10).

Evidently $P_1(V^{-1}BV)P_2$ has the form

$$P_1(V^{-1}BV)P_2 = \begin{pmatrix} R & 0 \\ S & 0 \end{pmatrix} \tag{2.13}$$

Therefore, if

$$Y = \begin{pmatrix} RK^{-1} & 0 \\ SK^{-1} & I \end{pmatrix} \tag{2.14}$$

then $YP_1J_AP_2 = P_1(V^{-1}BV)P_2$. Hence (2.9) holds with $M = VP_1^{-1} \times YP_1V^{-1}$.

If $\mathscr{N}(A) = \mathscr{N}(B)$ then the rank of B, and hence that of $P_1V^{-1}BVP_2$ is $N - p$.† Thus the column rank of

$$W = \begin{pmatrix} RK^{-1} \\ SK^{-1} \end{pmatrix}$$

is $N - p$. Choose the $(N - p) \times p$ matrix E and the $p \times p$ matrix F such that the columns of W together with the columns of

$$\begin{pmatrix} E \\ F \end{pmatrix}$$

form a linearly independent set. This is possible since the columns of W are linearly independent. Thus the matrix

$$Y = \begin{pmatrix} RK^{-1} & E \\ SK^{-1} & F \end{pmatrix}$$

is nonsingular. Moreover $YP_1J_AP_2 = P_1(V^{-1}BV)P_2$ and (2.9) holds with $M = VP_1^{-1}YP_1V^{-1}$. Since M is nonsingular, the lemma follows.

3.3. BASIC LINEAR STATIONARY ITERATIVE METHODS

We now define six basic linear stationary iterative methods. We will illustrate the methods for the system

$$\begin{aligned} a_{1,1}u_1 + a_{1,2}u_2 + a_{1,3}u_3 &= b_1 \\ a_{2,1}u_1 + a_{2,2}u_2 + a_{2,3}u_3 &= b_2 \\ a_{3,1}u_1 + a_{3,2}u_2 + a_{3,3}u_3 &= b_3 \end{aligned} \tag{3.1}$$

† We recall from Section 2.1 that the multiplication of a rectangular matrix on either the right or left by a nonsingular matrix of appropriate size does not change the rank.

or

$$\begin{pmatrix} a_{1,1} & a_{1,2} & a_{1,3} \\ a_{2,1} & a_{2,2} & a_{2,3} \\ a_{3,1} & a_{3,2} & a_{3,3} \end{pmatrix} \begin{pmatrix} u_1 \\ u_2 \\ u_3 \end{pmatrix} = \begin{pmatrix} b_1 \\ b_2 \\ b_3 \end{pmatrix} \tag{3.2}$$

If the diagonal elements of A do not vanish, we can solve each equation for the corresponding unknown obtaining

$$\begin{aligned} u_1 &= \phantom{b_{2,1}u_1 + {}} b_{1,2}u_2 + b_{1,3}u_3 + c_1 \\ u_2 &= b_{2,1}u_1 \phantom{{}+ b_{1,2}u_2} + b_{2,3}u_3 + c_2 \\ u_3 &= b_{3,1}u_1 + b_{3,2}u_2 \phantom{{}+ b_{2,3}u_3} + c_3 \end{aligned} \tag{3.3}$$

where

$$b_{i,j} = \begin{cases} -a_{i,j}/a_{i,i}, & i \neq j \\ 0, & i = j \end{cases} \tag{3.4}$$

and

$$c_i = b_i/a_{i,i}, \qquad i = 1, 2, 3 \tag{3.5}$$

We have thus replaced the system (1.1) by the equivalent system

$$u = Bu + c \tag{3.5'}$$

where

$$B = D^{-1}C \tag{3.6}$$

$$c = D^{-1}b \tag{3.7}$$

and

$$D = \operatorname{diag} A, \qquad C = D - A \tag{3.8}$$

Clearly, D^{-1} exists since the diagonal elements of A do not vanish.

With the *Jacobi method* (J method), we choose arbitrary starting values $u_1^{(0)}, u_2^{(0)}, u_3^{(0)}$ and compute $u_1^{(1)}, u_2^{(1)}, u_3^{(1)}$ from (3.3) using the $u_i^{(0)}$ in the right-hand side of (3.3). One then determines the $u_i^{(2)}$ from the $u_i^{(1)}$, etc. Thus, in general, given $u_i^{(n)}$, one determines $u_i^{(n+1)}$ by

$$\begin{aligned} u_1^{(n+1)} &= \phantom{b_{2,1}u_1^{(n)} + {}} b_{1,2}u_2^{(n)} + b_{1,3}u_3^{(n)} + c_1 \\ u_2^{(n+1)} &= b_{2,1}u_1^{(n)} \phantom{{}+ b_{1,2}u_2^{(n)}} + b_{2,3}u_3^{(n)} + c_2 \\ u_3^{(n+1)} &= b_{3,1}u_1^{(n)} + b_{3,2}u_2^{(n)} \phantom{{}+ b_{2,3}u_3^{(n)}} + c_3 \end{aligned} \tag{3.9}$$

or, equivalently,

$$u^{(n+1)} = Bu^{(n)} + c \tag{3.10}$$

Since $I - B = I - D^{-1}C = D^{-1}(D - C) = D^{-1}A$ we have $(I - B)A^{-1}b = c$. Thus if A is nonsingular, $I - B$ is nonsingular, and by Theorem 2.4 the method is completely consistent. We remark that complete consistency can also be verified from Theorem 2.6 where $M = D^{-1}$. The complete consistency of other methods discussed in this section can also be verified using Theorem 2.6.

Related to the J method is the *simultaneous overrelaxation method* (JOR method). Here we choose a real parameter ω and replace (3.9) by

$$\begin{aligned} u_1^{(n+1)} &= \omega(\phantom{b_{2,1}u_1^{(n)} +{}} b_{1,2}u_2^{(n)} + b_{1,3}u_3^{(n)} + c_1) + (1 - \omega)u_1^{(n)} \\ u_2^{(n+1)} &= \omega(b_{2,1}u_1^{(n)} \phantom{+ b_{1,2}u_2^{(n)}} + b_{2,3}u_3^{(n)} + c_2) + (1 - \omega)u_2^{(n)} \\ u_3^{(n+1)} &= \omega(b_{3,1}u_1^{(n)} + b_{3,2}u_2^{(n)} \phantom{+ b_{1,3}u_3^{(n)}} + c_3) + (1 - \omega)u_3^{(n)} \end{aligned} \tag{3.11}$$

or, equivalently

$$u^{(n+1)} = B_\omega u^{(n)} + \omega c \tag{3.12}$$

where

$$B_\omega = \omega B + (1 - \omega)I \tag{3.13}$$

If $\omega \neq 0$, the method is completely consistent. If $\omega = 1$, we have the J method. If $\omega > 1$, we are in a sense "overcorrecting" since $u^{(n+1)} = u^{(n)} + \omega(u^{(n+1)} - u^{(n)})$ where $u^{(n+1)}$ is obtained from the J method. Similarly, if $\omega < 1$, we are "undercorrecting." Considerations on the choice of ω will be discussed later.

The *Gauss–Seidel method* (GS method) is the same as the J method except that at each stage one uses the values of $u_i^{(n+1)}$ when available. Thus, instead of (3.9) we use

$$\begin{aligned} u_1^{(n+1)} &= \phantom{b_{2,1}u_1^{(n+1)} +{}} b_{1,2}u_2^{(n)} + b_{1,3}u_3^{(n)} + c_1 \\ u_2^{(n+1)} &= b_{2,1}u_1^{(n+1)} \phantom{+ b_{1,2}u_2^{(n)}} + b_{2,3}u_3^{(n)} + c_2 \\ u_3^{(n+1)} &= b_{3,1}u_1^{(n+1)} + b_{3,2}u_2^{(n+1)} \phantom{+ b_{1,3}u_3^{(n)}} + c_3 \end{aligned} \tag{3.14}$$

or, equivalently,

$$u^{(n+1)} = Lu^{(n+1)} + Uu^{(n)} + c \tag{3.15}$$

where L and U are strictly lower and strictly upper triangular matrices, respectively, such that

$$B = L + U \tag{3.16}$$

Thus, we have

$$L = \begin{pmatrix} 0 & 0 & 0 \\ b_{2,1} & 0 & 0 \\ b_{3,1} & b_{3,2} & 0 \end{pmatrix}, \qquad U = \begin{pmatrix} 0 & b_{1,2} & b_{1,3} \\ 0 & 0 & b_{2,3} \\ 0 & 0 & 0 \end{pmatrix} \tag{3.17}$$

Since L is a strictly lower triangular matrix, $\det(I - L) = 1$; hence, $I - L$ is nonsingular and we can solve (3.15) for $u^{(n+1)}$ obtaining

$$u^{(n+1)} = \mathscr{L} u^{(n)} + (I - L)^{-1} c \tag{3.18}$$

where

$$\mathscr{L} = (I - L)^{-1} U \tag{3.19}$$

Evidently (3.18) is in the standard form (1.5) for a linear stationary iterative method. If A is nonsingular, the method is completely consistent by Theorem 2.4 since $I - \mathscr{L} = (I - L)^{-1}(I - L - U) = (I - L)^{-1} D^{-1} A$ which is nonsingular, and since $(I - \mathscr{L}) A^{-1} b = (I - L)^{-1} D^{-1} b = (I - L)^{-1} c$.

The *successive overrelaxation method* (SOR method) is the same as the JOR method except that one uses values of $u_i^{(n+1)}$ whenever possible. Thus we have

$$\begin{aligned} u_1^{(n+1)} &= \omega(\quad\quad\quad\quad\quad\quad b_{1,2} u_2^{(n)} + b_{1,3} u_3^{(n)} + c_1) + (1 - \omega) u_1^{(n)} \\ u_2^{(n+1)} &= \omega(b_{2,1} u_1^{(n+1)} \quad\quad\quad\quad\quad + b_{2,3} u_3^{(n)} + c_2) + (1 - \omega) u_2^{(n)} \\ u_3^{(n+1)} &= \omega(b_{3,1} u_1^{(n+1)} + b_{3,2} u_2^{(n+1)} \quad\quad\quad + c_3) + (1 - \omega) u_3^{(n)} \end{aligned} \tag{3.20}$$

or, equivalently,

$$u^{(n+1)} = \omega(L u^{(n+1)} + U u^{(n)} + c) + (1 - \omega) u^{(n)} \tag{3.21}$$

Since $\det(I - \omega L) = 1$, $I - \omega L$ is nonsingular and we can solve (3.21) for $u^{(n+1)}$ obtaining

$$u^{(n+1)} = \mathscr{L}_\omega u^{(n)} + (I - \omega L)^{-1} \omega c \tag{3.22}$$

where

$$\mathscr{L}_\omega = (I - \omega L)^{-1} (\omega U + (1 - \omega) I) \tag{3.23}$$

If A is nonsingular and $\omega \neq 0$, the SOR method is completely consistent since $I - \mathscr{L}_\omega = \omega(I - \omega L)^{-1} D^{-1} A$ and since $(I - \mathscr{L}_\omega) A^{-1} b = \omega(I - \omega L)^{-1} c$.

If $\omega = 1$, the SOR method reduces to the GS method. If $\omega > 1$, then we are overcorrecting since $u_i^{(n+1)} = u_i^{(n)} + \omega(\tilde{u}_i^{(n+1)} - u_i^{(n)})$ where

$\tilde{u}_i^{(n+1)}$ is the value computed by the GS method. Similarly if $\omega < 1$, we are undercorrecting.

Let us now drop the assumption that the diagonal elements of A do not vanish. We consider the *stationary generalized Richardson's method* (GRF method) defined by

$$u^{(n+1)} = u^{(n)} + P(Au^{(n)} - b) \tag{3.24}$$

where P is any nonsingular diagonal matrix. Evidently (3.24) is equivalent to

$$u^{(n+1)} = R_P u^{(n)} - Pb \tag{3.25}$$

where

$$R_P = I + PA \tag{3.26}$$

The GRF method is completely consistent if A is nonsingular since $I - R_P = -PA$ and $(I - R_P)A^{-1}b = -Pb$. If $P = -D^{-1}$, we have the Jacobi method. If $P = pI$ where $p \neq 0$, we have the *stationary Richardson method* (RF method) defined by

$$u^{(n+1)} = u^{(n)} + p(Au^{(n)} - b) \tag{3.27}$$

or

$$u^{(n+1)} = R_p u^{(n)} - pb \tag{3.28}$$

where

$$R_p = I + pA \tag{3.29}$$

This method is completely consistent if $p \neq 0$ and A is nonsingular. Richardson [1910] first considered the method but allowed p to vary from iteration to iteration. We will study the nonstationary form of Richardson's method in Chapter 11.

Given a system (1.1) one can transform it into an equivalent system by multiplying on the left by a nonsingular diagonal matrix E and letting $v = E^{-1}u$ obtaining

$$(EAE)v = Eb \tag{3.30}$$

This process is known as "prescaling." We now show that if the diagonal elements of P are positive, the GRF method for (1.1) is equivalent to the RF method applied to (3.30) for suitable E in the sense that the eigenvalues of the two methods are the same. Thus by (3.26) R_P is similar to $I + P^{1/2}AP^{1/2}$ which has the same eigenvalues as the RF method with $p = 1$ applied to the scaled system $P^{1/2}AP^{1/2}v = P^{1/2}b$.

3.4. GENERATION OF COMPLETELY CONSISTENT METHODS

Given a linear system (1.1), one can often obtain a completely consistent iterative method by using one or more reversible transformations to obtain an equation of the form (1.6). This procedure was used in Section 3.3 to derive the six basic iterative methods. It is perhaps natural to ask whether one can arbitrarily choose a matrix G, preferably such that $S(G)$ is very small, and then determine a k such that the iterative method (1.1) is completely consistent. That our choice of G cannot be completely arbitrary can be seen by considering the case $G = 0$. If A is nonsingular, then, by (2.1), for complete consistency we must have $k = A^{-1}b$. But it would not in general be practical to construct such an iterative method since the work needed to compute k would be as great as that needed to solve (1.1).

As shown in Section 3.2, if (1.1) is solvable, i.e., if (1.1) has at least one solution, then for any completely consistent iterative method (1.5) there exists a nonsingular matrix M such that

$$G = I - MA, \qquad k = Mb \tag{4.1}$$

Thus (1.5) has the form

$$u^{(n+1)} = (I - MA)u^{(n)} + Mb \tag{4.2}$$

or

$$u^{(n+1)} = u^{(n)} - M(Au^{(n)} - b) \tag{4.3}$$

This is a generalization of the GRF method where we use a nonsingular matrix M instead of a diagonal matrix P.

We now show that any completely consistent iterative method can be obtained from a "splitting" of the matrix A into the difference

$$A = Q - R \tag{4.4}$$

where Q is nonsingular. Given the splitting of (4.4), we can transform the system

$$(Q - R)u = b \tag{4.5}$$

into the equivalent system

$$u = Q^{-1}Ru + Q^{-1}b \tag{4.6}$$

But (4.6) can be obtained as a special case of (4.2) by letting $M = Q^{-1}$, since we have

$$\begin{aligned} G &= I - MA = I - Q^{-1}A = Q^{-1}(Q - A) = Q^{-1}R \\ k &= Mb = Q^{-1}b \end{aligned} \tag{4.7}$$

On the other hand, (4.2) can be obtained as a special case of (4.6) by letting $Q = M^{-1}$.

The practicality of a method derived from a splitting depends on the ease with which one can, for any given vector s, solve

$$Qt = s \tag{4.8}$$

for t. Having chosen Q one could, of course, determine M by successively solving the systems

$$Qt^{(i)} = s^{(i)} \tag{4.9}$$

where $s_j^{(i)} = 0$ for $j \neq i$ and $s_i^{(i)} = 1$. However, in practice it is more convenient to use the formulas

$$u^{(n+1)} = u^{(n)} + \delta^{(n)} = Gu^{(n)} + k \tag{4.10}$$

where $\delta^{(n)}$ is obtained by solving

$$Q\delta^{(n)} = -(Au^{(n)} - b) \tag{4.11}$$

The reader should verify that the choices of Q shown is the accompanying tabulation lead to the iterative methods indicated. Here $A = D - C_L - C_U$ where $D = \text{diag } A$, and C_L and C_U are strictly lower and strictly upper triangular matrices, respectively.

Q	Method
D	J
$\omega^{-1}D$	JOR
$D - C_L$	GS
$\omega^{-1}D - C_L$	SOR
$-p^{-1}I$	RF
$-P^{-1}$ (P diagonal)	GRF

One might expect that the "closer" Q is to A and the "smaller" R is, then the faster the method will converge. Indeed, Varga [1960] proved this in the case of regular splittings, i.e., splittings such that $Q^{-1} \geq 0$ and $R \geq 0$, and under the assumption that $A^{-1} \geq 0$. We shall discuss this further in Chapter 4.

3.5. GENERAL CONVERGENCE THEOREMS

Let us now determine under what conditions the sequence $u^{(0)}, u^{(1)}, \ldots$ defined by (1.5) converges for each starting vector $u^{(0)}$.

Definition 5.1. The iterative method (1.5) is *weakly convergent* if for all $u^{(0)}$ the sequence $u^{(0)}, u^{(1)}, \ldots$ converges. The method is *convergent* if for all $u^{(0)}$ the sequence converges to a limit independent of $u^{(0)}$.

Theorem 5.1. The iterative method (1.5) is convergent if and only if

$$S(G) < 1 \tag{5.1}$$

It is weakly convergent if and only if

$$S(G) \leq 1 \tag{5.2}$$

and the following hold: (1.6) is solvable; if $|\lambda| = 1$ and λ is an eigenvalue of G, then $\lambda = 1$ is the only eigenvalue of G of modulus unity; the dimension of $\mathcal{N}(I - G)$ is equal to the multiplicity of $\lambda = 1$ as an eigenvalue of G.

Proof. If the sequence $u^{(0)}, u^{(1)}, \ldots$ converges to $\hat{u}$ for all $u^{(0)}$, then by Theorem 2.3 $\hat{u}$ satisfies (1.6). If we let $\varepsilon^{(n)} = u^{(n)} - \hat{u}$, we have, from (1.5) and (1.6), $\varepsilon^{(n+1)} = G\varepsilon^{(n)}$ and

$$\varepsilon^{(n)} = G^n \varepsilon^{(0)} \tag{5.3}$$

By Theorem 2-4.5 $\varepsilon^{(n)} \to 0$ as $n \to \infty$ for all $\varepsilon^{(0)}$ if and only if (5.1) holds.

If, on the other hand, $S(G) < 1$, then $I - G$ is nonsingular and (1.6) has a unique solution, say $\hat{u}$. By Theorem 2-4.5 we have $\varepsilon^{(n)} \to 0$ as $n \to \infty$.

If the method (1.5) is weakly convergent, then (1.6) is clearly solvable since the limit of any sequence $u^{(0)}, u^{(1)}, \ldots$ corresponding to a given starting vector $u^{(0)}$ is a solution of (1.6) by Theorem 2.3. If $\hat{u}$ is any solution of (1.6), then (5.3) holds. With weak convergence $\varepsilon^{(n)}$ must converge to a limit which need not be zero. Thus for all $v \in E^N$ the limit of $G^n v$ as $n \to \infty$ must exist. By considering successively the cases where v is a unit vector in E^N we can show that G^n and hence J_G^n, where J_G is the Jordan canonical form of G, must converge as $n \to \infty$. This is possible only if the stated conditions hold.

Conversely, if these conditions hold, then the limit of $G^n v$ as $n \to \infty$ must exist for all v. Since (1.6) is solvable, (5.3) holds and hence $\{u^{(n)}\}$ converges for each $u^{(0)}$.

Theorem 5.2. If (1.1) is solvable and if the iterative method (1.5) is consistent with (1.1), then for $n = 0, 1, 2, \ldots$ we have

$$\varepsilon^{(n)} = G^n \varepsilon^{(0)} \tag{5.4}$$

where

$$\varepsilon^{(n)} = u^{(n)} - \hat{u}, \qquad n = 0, 1, 2, \ldots \tag{5.5}$$

Here $u^{(0)}, u^{(1)}, \ldots$ is the sequence defined by (1.5) for a given $u^{(0)}$ and $\hat{u}$ is any solution of (1.1). If the method is also convergent, then A is nonsingular, the method is completely consistent, and the sequence $u^{(0)}, u^{(1)}, \ldots$ converges to the unique solution of (1.1) for all $u^{(0)}$.

Proof. Let $\hat{u}$ be any solution of (1.1). By consistency, $\hat{u}$ satisfies (1.6), and from (1.5) and (1.6) we have $\varepsilon^{(n+1)} = G\varepsilon^{(n)}$ which implies (5.4). If the method is convergent, then, by Theorem 5.1, (5.1) holds and $I - G$ is nonsingular. Hence $\hat{u}$ is the only solution of (1.6). By consistency, $\hat{u}$ is the only solution of (1.1), and by Theorem 2-1.5, A is nonsingular and $\hat{u}$ is unique. Complete consistency follows from Theorem 2.4, and since the method is convergent, the sequence $u^{(0)}, u^{(1)}, \ldots$ converges to a limit independent of $u^{(0)}$. Since this limit satisfies (1.6), by Theorem 5.3, it must be $\hat{u}$, by the complete consistency.

To summarize, in the normal situation we have the following conditions:

(a) A is nonsingular, and (1.1) has a unique solution.
(b) $S(G) < 1$ and hence $I - G$ is nonsingular.
(c) $k = (I - G)A^{-1}b$.

Under these conditions the iterative method (1.5) is completely consistent with (1.1). Moreover, for any starting vector $u^{(0)}$ the sequence $u^{(0)}$, $u^{(1)}, \ldots$ converges to the unique solution of (1.1).

The following convergence theorem applies to the case where A is positive definite.

Theorem 5.3. If A is a positive definite matrix and if the iterative method (1.5) is completely consistent with (1.1), then the method is convergent if

$$M = Q + Q^{\mathrm{T}} - A \tag{5.6}$$

is positive definite, where

$$Q = A(I - G)^{-1} \tag{5.7}$$

Moreover, we have

$$\| G \|_{A^{1/2}} < 1 \tag{5.8}$$

Conversely, if (5.8) holds, then M is positive definite.

We remark that as noted in Section 3.4 one frequently generates a completely consistent linear stationary iterative method by choosing a nonsingular matrix Q such that $Qs = t$ can be conveniently solved for s for any given t. Thus in order to obtain Q it may not be necessary to find $(I - G)^{-1}$ and then use (5.7).

Proof. We first note that from (5.7) we have

$$G = I - Q^{-1}A \tag{5.9}$$

Moreover, since A is real and positive definite there exists, by Theorem 2-2.7, a real positive definite matrix, say $A^{1/2}$, whose square is A. We let

$$G' = A^{1/2}GA^{-1/2} = I - A^{1/2}Q^{-1}A^{1/2} \tag{5.10}$$

Following Wachspress [1966] we have

$$G'(G')^{\mathrm{T}} = I - A^{1/2}Q^{-1}M(Q^{\mathrm{T}})^{-1}A^{1/2} \tag{5.11}$$

If M is positive definite, then since M is real, by Theorem 2-2.7 the matrix $L = M^{1/2}$ exists. Therefore

$$T = A^{1/2}Q^{-1}M(Q^{\mathrm{T}})^{-1}A^{1/2} = (A^{1/2}Q^{-1}L)(A^{1/2}Q^{-1}L)^{\mathrm{T}} \tag{5.12}$$

Since $A^{1/2}$, Q, and L, and hence $A^{1/2}Q^{-1}L$, are nonsingular it follows by Theorem 2-2.6 that T is positive definite. By (5.11) the eigenvalues of

$G'(G')^T$ are less than unity and we have

$$\| G \|_{A^{1/2}} = \| G' \| = [S(G'(G')^T)]^{1/2} < 1 \tag{5.13}$$

Hence (5.8) follows. From Theorem 2-3.4, it follows that $S(G) < 1$.

On the other hand, if (5.8) holds, then $S(G'(G')^T) < 1$, and hence $A^{1/2}Q^{-1}M(Q^T)^{-1}A^{1/2}$ and M must be positive definite. This completes the proof of Theorem 5.3.

3.6. ALTERNATIVE CONVERGENCE CONDITIONS

In Section 3.5 we showed that in order that the linear stationary iterative method (1.5) should converge for all starting vectors $u^{(0)}$ to a limit independent of $u^{(0)}$ it was necessary and sufficient that $S(G) < 1$. We now give some alternative convergence conditions which hold if and only if $S(G) < 1$.

The following theorem is due to Stein [1952].

Theorem 6.1. *$S(G) < 1$ if and only if there exists a positive definite matrix P such that the matrix M given by*

$$M = P - GPG^H \tag{6.1}$$

is positive definite. If G is real, then P can be taken to be real.

Proof. If $S(G) < 1$, then by Corollary 2-3.6 there exists a nonsingular matrix, say L_1, such that $\| G \|_{L_1} < 1$. We let $P = LL^H$, where $L = L_1^{-1}$. Since L is nonsingular, P is positive definite by Theorem 2-2.6. Evidently, we have

$$\hat{M} = L^{-1}M(L^H)^{-1} = I - (L^{-1}GL)(L^{-1}GL)^H \tag{6.2}$$

The eigenvalues of $\hat{M}$ are $1 - \nu_i$ where the ν_i are the real and nonnegative eigenvalues of the nonnegative definite matrix $(L^{-1}GL)(L^{-1}GL)^H$. But the largest such ν_i does not exceed

$$S((L^{-1}GL)(L^{-1}GL)^H) = \| L^{-1}GL \|^2 = \| L_1GL_1^{-1} \|^2 = \| G \|_{L_1}^2 < 1$$

Since $\hat{M}$ is Hermitian and has positive eigenvalues, it follows from Theorem 2-2.3 that $\hat{M}$ is positive definite. Moreover, M is positive definite by Corollary 2-2.8.

Suppose, on the other hand, that there exists a positive definite matrix P such that M is positive definite. By Theorem 2-2.7 there exists a positive definite matrix, say $P^{1/2}$, whose square is P. Moreover, by Corollary 2-2.8

$$P^{-1/2}MP^{-1/2} = I - P^{-1/2}GPG^{\mathrm{H}}P^{-1/2} = I - (P^{-1/2}GP^{1/2})(P^{-1/2}GP^{1/2})^{\mathrm{H}} \tag{6.3}$$

is positive definite. Hence the eigenvalues of the nonnegative definite matrix $(P^{-1/2}GP^{1/2})(P^{-1/2}GP^{1/2})^{\mathrm{H}}$ are less than unity. Thus $\| G \|_{P^{-1/2}} < 1$, and by Theorem 2-3.4 we have $S(G) < 1$.

It remains to show that if G is real and $S(G) < 1$, then there exists a real positive definite matrix P such that M is positive definite. We already know that there exists a positive definite matrix Q, not necessarily real, such that $\tilde{M} = Q - GQG^{\mathrm{T}}$ is positive definite. Let $Q = Q_1 + iQ_2$ where Q_1 and Q_2 are real. The real part of $\tilde{M}$, namely $Q_1 - GQ_1G^{\mathrm{T}}$, is positive definite, by Theorem 2-2.4. Thus, if we let $P = Q_1$, then M is positive definite, and the proof of Theorem 6.1 is complete.

Suppose now that $S(G) < 1$. If P is a positive definite matrix such that $M = P - GPG^{\mathrm{T}}$ is positive definite, then

$$P^{-1/2}MP^{-1/2} = I - (P^{-1/2}GP^{1/2})(P^{-1/2}GP^{1/2})^{\mathrm{H}} \tag{6.4}$$

is positive definite. Hence $S((P^{-1/2}GP^{1/2})(P^{-1/2}GP^{1/2})^{\mathrm{H}}) < 1$ and

$$\| G \|_{P^{-1/2}} < 1 \tag{6.5}$$

We have thus proved

Corollary 6.2. If G is a matrix such that $S(G) < 1$, then there exists a positive definite matrix R such that

$$\| G \|_R < 1 \tag{6.6}$$

If G is real, then R can be taken to be real.

We note that R can be taken as $P^{-1/2}$ where P is a positive definite matrix satisfying the conditions of Theorem 6.1. Corollary 6.2 is a stronger version of Corollary 2-3.6.

In the study of the convergence properties of the iterative method (1.5) it is sometimes convenient to consider the matrix H defined by

$$H = (I - G)^{-1}(I + G) \tag{6.7}$$

Evidently, the matrix H exists if the method is completely consistent, for in that case, by Theorem 2.4, $I - G$ is nonsingular.† From (6.7) we have $H + I = 2(I - G)^{-1}$, $H - I = 2(I - G)^{-1}G$, and

$$G = (H + I)^{-1}(H - I) \tag{6.8}$$

If ν is an eigenvalue of H, then λ is an eigenvalue of G where

$$\lambda = \frac{\nu - 1}{\nu + 1} = \frac{(\mathrm{Re}\ \nu - 1) + i\ \mathrm{Im}\ \nu}{(\mathrm{Re}\ \nu + 1) + i\ \mathrm{Im}\ \nu} \tag{6.9}$$

and

$$|\lambda|^2 = \frac{(\mathrm{Re}\ \nu - 1)^2 + (\mathrm{Im}\ \nu)^2}{(\mathrm{Re}\ \nu + 1)^2 + (\mathrm{Im}\ \nu)^2} \tag{6.10}$$

Evidently $S(G) < 1$ if and only if the real parts of all of the eigenvalues of H are positive. Matrices such that the real parts of all of their eigenvalues are *negative* are known as *stable* matrices. We are thus led to the following

Definition 6.1. A matrix is *negative-stable* (N-stable) if all of its eigenvalues have positive real parts.

From (6.10) we have

Theorem 6.3. $S(G) < 1$ if and only if $I - G$ is nonsingular and $H = (I - G)^{-1}(I + G)$ is N-stable.

We are now in a position to prove the following theorem which is analogous to Lyapunov's theorem for stable matrices.

Theorem 6.4. A real matrix H is N-stable if and only if there exists a real positive definite matrix Q such that the matrix K given by

$$K = HQ + QH^{\mathrm{T}} \tag{6.11}$$

is positive definite.

Proof. If H is N-stable, then $H + I$ is nonsingular. If $G = (H + I)^{-1} \times (H - I)$, then $I - G = 2(H + I)^{-1}$ and $I - G$ is nonsingular. Moreover, $H = (I - G)^{-1}(I + G)$ and, by Theorem 6.3, $S(G) < 1$. Since G is real, there exists, by Theorem 6.1, a real positive definite matrix Q such that $M = Q - GQG^{\mathrm{T}}$ is positive definite i.e., such that $Q-(H+I)^{-1}$

† We are assuming here that the matrix A corresponding to the original linear system (1.1) is nonsingular.

$(H - I)Q(H^T - I)(H^T + I)^{-1}$ is positive definite. But since

$$M = 2(H + I)^{-1}K(H^T + I)^{-1} \tag{6.12}$$

it follows that

$$K = \tfrac{1}{2}(H + I)M(H^T + I) \tag{6.13}$$

and, by Corollary 2-2.8, K is positive definite.

On the other hand, suppose that K is positive definite for some positive definite matrix Q. We seek to show that $H + I$ is nonsingular. If $H + I$ were singular, then $H^T + I$ would also be singular and for some $w \neq 0$ we would have $H^Tw = -w$. But in that case we would have

$$\begin{aligned}(w, Kw) &= (w, (HQ + QH^T)w) = (QH^Tw, w) + (w, QH^Tw)\\ &= -(Qw, w) - (w, Qw) < 0\end{aligned}$$

since Q is positive definite. This contradicts the assumption that K is positive definite.

Evidently we have

$$(H + I)Q(H^T + I) - (H - I)Q(H^T - I) = 2(HQ + QH^T) \tag{6.14}$$

Hence,

$$2(H + I)^{-1}K(H^T + I)^{-1} = Q - GQG^T \tag{6.15}$$

where $G = (H + I)^{-1}(H - I)$. Thus, $Q - GQG^T$ is positive definite, by Corollary 2-2.8 since K is positive definite. By Theorem 6.1, this implies that $S(G) < 1$, and by Theorem 6.3 it follows that $H = (I - G)^{-1}(I + G)$ is N-stable.

To summarize, we have

Theorem 6.5. Let G be a real matrix. Then $S(G) < 1$ if and only if any one (and hence all) of the following conditions holds:

(a) There exists a real positive definite matrix R such that

$$\| G \|_R = \| RGR^{-1} \| < 1 \tag{6.16}$$

(b) There exists a real positive definite matrix P such that

$$M = P - GPG^T \tag{6.17}$$

is positive definite.

(c) $I - G$ is nonsingular and there exists a real positive definite matrix Q such that

$$K = HQ + QH^{\mathrm{T}} \tag{6.18}$$

is positive definite, where

$$H = (I - G)^{-1}(I + G) \tag{6.19}$$

(d) $I - G$ is nonsingular and H is N-stable.

Moreover, if $S(G) < 1$, then there exist real positive definite matrices P, Q, and R such that

$$P = Q = R^{-2} \tag{6.20}$$

and such that P, Q, and R satisfy conditions (b), (c), and (a), respectively.

As an application of these results let us consider the SOR method for the case where A is positive definite. By (3.23) we have

$$\begin{aligned} I - \mathscr{L}_\omega &= \omega(I - \omega L)^{-1}D^{-1}A \\ I + \mathscr{L}_\omega &= (I - \omega L)^{-1}((2 - \omega)I + \omega(U - L)) \end{aligned} \tag{6.21}$$

Therefore,

$$H = (I - \mathscr{L}_\omega)^{-1}(I + \mathscr{L}_\omega) = A^{-1}D[(2 - \omega/\omega)I + U - L] \tag{6.22}$$

Moreover, since $(DL)^{\mathrm{T}} = DU$ we have

$$HA^{-1} + A^{-1}H^{\mathrm{T}} = [2(2 - \omega)/\omega]A^{-1}DA^{-1} \tag{6.23}$$

which is positive definite for $0 < \omega < 2$. Therefore, by Theorem 6.4, H is N-stable, and, by Theorem 6.5,

$$S(\mathscr{L}_\omega) \le \| \mathscr{L}_\omega \|_{A^{1/2}} < 1 \tag{6.24}$$

3.7. RATES OF CONVERGENCE

We now study the rate of convergence of a convergent linear stationary iterative method. By Theorem 5.1, if (5.1) holds, then for any $u^{(0)}$ the sequence $u^{(0)}, u^{(1)}, \ldots$ defined by (1.5) converges to the unique solution $\hat{u}$ of (1.6). Letting $\varepsilon^{(n)} = u^{(n)} - \hat{u}$ we seek to determine the rate at which $\| \varepsilon^{(n)} \|$ converges to zero as $n \to \infty$.

First, we define two quantities known as *condition numbers* for a given matrix A. The *spectral condition number* of A is defined by

$$\nu(A) = \| A \| \, \| A^{-1} \| \tag{7.1}$$

Evidently we have

$$\nu(A) \geq 1$$

since $I = AA^{-1}$ and $1 = \| I \| \leq \| A \| \, \| A^{-1} \|$. On the other hand, if A is unitary, then $\| A \| = [S(AA^{\mathrm{H}})]^{1/2} = [S(I)]^{1/2} = 1$. Similarly, $\| A^{-1} \| = 1$ and $\nu(A) = 1$ if A is unitary.

The *Jordan condition number* of a matrix is defined by

$$\mathscr{J}(A) = \inf_{V} \nu(V) \tag{7.2}$$

where V is any matrix which reduces A to Jordan canonical form. If A is normal, then $\mathscr{J}(A) = 1$ since A can be reduced to Jordan canonical form by a unitary matrix (Theorem 2-1.13). In some sense, the Jordan condition number gives a measure of the "nonnormality" of A.

The reader should show that if J is any matrix which is a Jordan canonical form of A, then

$$\mathscr{J}(A) = \inf_{V \in S_1} \nu(V) \tag{7.3}$$

where S_1 is the set of all matrices V such that $V^{-1}AV = J$.

We now prove

Theorem 7.1. If G is an arbitrary complex matrix, with $S(G) > 0$, then

$$\frac{1}{\nu(V)} \leq \mathscr{J}(G)^{-1} \leq \varliminf_{n\to\infty} \frac{\| G^n \|}{\binom{n}{p-1} S(G)^{n-p+1}}$$

$$\leq \varlimsup_{n\to\infty} \frac{\| G^n \|}{\binom{n}{p-1} S(G)^{n-p+1}} \leq \mathscr{J}(G) \leq \nu(V) \tag{7.4}$$

where p is the order of the largest block in the Jordan canonical form of G which is associated with an eigenvalue of modulus $S(G)$, $\nu(V)$ is the spectral condition number of any matrix V which reduces G to Jordan canonical form, and $\mathscr{J}(G)$ is the Jordan condition number of G. Moreover, if $S(G) = 0$, then for sufficiently large n we have $\| G^n \| = 0$.

We first prove the following.

Lemma 7.2. If $0 < |\lambda| < 1$, then

$$\lim_{n\to\infty} \frac{\| J_{\lambda p}^n \|}{\binom{n}{p-1} |\lambda|^{n-p+1}} = 1 \tag{7.5}$$

where $J_{\lambda,p}$ is a matrix of order p of the form

$$J_{\lambda,p} = \begin{pmatrix} \lambda & 1 & 0 & \cdots & 0 \\ 0 & \lambda & 1 & \cdots & 0 \\ 0 & 0 & \lambda & \cdots & 0 \\ \vdots & \vdots & \vdots & \cdots & \vdots \\ 0 & 0 & 0 & \cdots & \lambda \end{pmatrix} = \lambda I + E \tag{7.6}$$

Proof. Evidently $E^p = 0$. Hence we have

$$\begin{aligned} J_{\lambda,p}^n &= (\lambda I + E)^n \\ &= \lambda^n I + \binom{n}{1}\lambda^{n-1}E + \binom{n}{2}\lambda^{n-2}E^2 + \cdots + \binom{n}{p-1}\lambda^{n-p+1}E^{p-1} \\ &= \binom{n}{p-1}\lambda^{n-p+1}[\lambda^{p-1}c_0 I + \lambda^{p-2}c_1 E + \cdots + c_{p-2}\lambda E^{p-2} + E^{p-1}] \end{aligned} \tag{7.7}$$

for suitable $c_0, c_1, \ldots, c_{p-2}$. Therefore

$$J_{\lambda,p}^n \Big/ \left[\binom{n}{p-1}\lambda^{n-p+1}\right] = E^{p-1} + [\lambda^{p-1}c_0 I + \lambda^{p-2}c_1\lambda^{p-2}E + \cdots + c_{p-2}\lambda E^{p-2}]$$

Since $\| E \| = \| E^2 \| = \cdots = \| E^{p-1} \| = 1$, since $|\lambda| < 1$, and since each c_i tends to zero as $n \to \infty$, the result (7.5) follows.

Evidently, since

$$J^n = \begin{pmatrix} J_1^n & 0 & \cdots & 0 \\ 0 & J_2^n & \cdots & 0 \\ \vdots & \vdots & \cdots & \vdots \\ 0 & 0 & \cdots & J_p^n \end{pmatrix}$$

it follows that $\| J^n \| = \max_i \| J_i^n \|$, $i = 1, 2, \ldots, p$. Therefore,

$$\lim_{n\to\infty} \left\{ \| J^n \| \Big/ \left[\binom{n}{p-1} S(G)^{n-p+1}\right]\right\} = 1 \tag{7.8}$$

Let V be any matrix such that

$$V^{-1}GV = J$$

Then we have $G = VJV^{-1}$, $G^n = VJ^nV^{-1}$, $J^n = V^{-1}G^nV$ and

$$\| J^n \|/\nu(V) \le \| G^n \| \le \nu(V) \, \| J^n \| \tag{7.9}$$

Since $\mathscr{J}(G) \le \nu(V)$ the result (7.4) follows. The case $S(G) = 0$ follows since $J^n = 0$ for $n \ge N$, the order of G. This completes the proof of Theorem 7.1.

Taking the nth root of the last two members of (7.9) and using (7.7), we have $\overline{\lim}_{n\to\infty} \| G^n \|^{1/n} \le S(G)$. Similarly, $\underline{\lim}_{n\to\infty} \| G^n \|^{1/n} \ge S(G)$ and we have

Corollary 7.3.

$$S(G) = \lim_{n\to\infty} (\| G^n \|)^{1/n} \tag{7.10}$$

From (5.3) and the definition of $\| G^n \|$ we have

$$\| \varepsilon^{(n)} \|/\| \varepsilon^{(0)} \| \le \| G^n \| = \sup_{\varepsilon^{(0)} \ne 0} \| \varepsilon^{(n)} \|/\| \varepsilon^{(0)} \| \tag{7.11}$$

Thus $\| G^n \|$ gives a measure of the amount by which the norm of the error is reduced after n iterations. Usually, one iterates until $\| \varepsilon^{(n)} \|$ is reduced to a fraction, say ϱ, of $\| \varepsilon^{(0)} \|$. Such a reduction can be achieved by choosing n so that

$$\| \varepsilon^{(n)} \| \le \varrho \, \| \varepsilon^{(0)} \| \tag{7.12}$$

By Theorem 5.1 we know that $\| G^n \| \to 0$ as $n \to \infty$ if $S(G) < 1$. Hence we can satisfy (7.12) if we choose n large enough that

$$\| G^n \| \le \varrho \tag{7.13}$$

For all n sufficiently large that $\| G^n \| < 1$ this inequality is equivalent to

$$n \ge -\log \varrho \Big/ \Big(-\frac{1}{n} \log \| G^n \| \Big) \tag{7.14}$$

Thus, the minimum number of iterations is inversely proportional to the quantity $n^{-1} \log \| G^n \|$. We are thus led to define as the *average rate of convergence* the quantity

$$R_n(G) = -(1/n) \log \| G^n \| \tag{7.15}$$

We also define the *asymptotic average rate of convergence*, or *asymptotic rate of convergence* by

$$R(G) = \lim_{n\to\infty} R_n(G) = -\log S(G) \tag{7.16}$$

the latter equality holding because of Corollary 7.3. Frequently we shall refer to $R(G)$ as the *rate of convergence.*

To obtain a crude estimate of the number of iterations needed for convergence we replace the denominator of (7.14) by $R(G)$ and obtain

$$n \doteq -\log \varrho/R(G) \tag{7.17}$$

However, if the integer p appearing in (7.4) is greater than unity, then the value of n obtained from (7.17) will be much too low. Thus, from (7.4) it follows that if $p = 2$, then $\| G^n \|$ will behave like $nS(G)^{n-1}$ rather than like $S(G)^n$ as in the case $p = 1$. In Figure 7.1 we have shown graphs of the functions λ^n and $n\lambda^{n-1}$ for $\lambda = 0.8$. We note that $n\lambda^{n-1}$ increases at first before decreasing. This, of course, slows the convergence

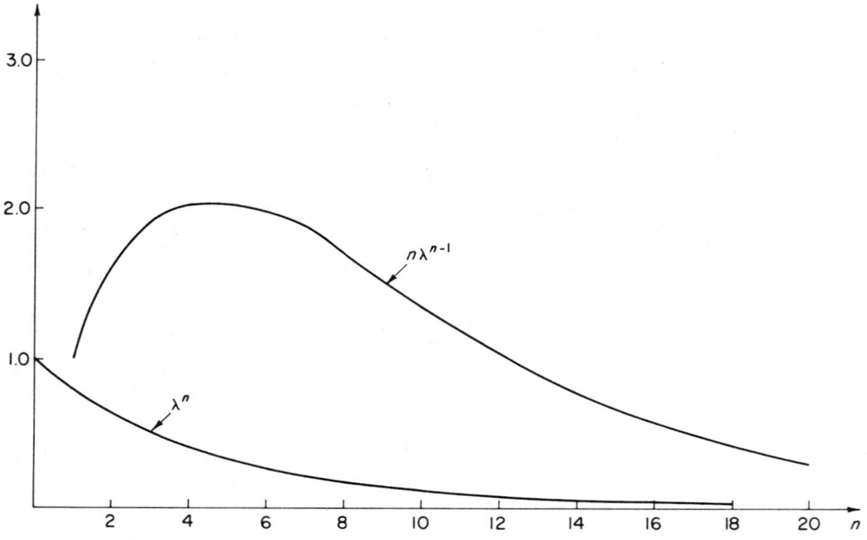

Figure 7.1. Graphs of λ^n and $n\lambda^{n-1}$ for $\lambda = 0.8$.

by a considerable amount. For larger values of p the effect is even worse. Later, however, we shall show how one can often choose other norms so that the effect is mitigated.

3.8. THE JORDAN CONDITION NUMBER OF A 2×2 MATRIX

Theorem 7.1 provides information on $\| G^n \|$ in terms of the Jordan condition number of G. Later we shall show how, in many cases, one can reduce the study of the convergence properties of certain methods to the study of norms of 2×2 matrices. We now derive explicit formulas for the Jordan condition number of the general 2×2 matrix, whose entries may be complex.

Theorem 8.1. Let $A = (a_{i,j})$ be a 2×2 matrix and let λ_1 and λ_2 be the eigenvalues of A.

(i) If $\lambda_1 \neq \lambda_2$, then

$$\mathscr{J}(A) = (\alpha^{1/2} + \beta^{1/2})/| \lambda_1 - \lambda_2 | \tag{8.1}$$

where

$$\begin{aligned} \alpha &= | a_{1,2} |^2 + | a_{2,1} |^2 + \tfrac{1}{2}[| a_{1,1} - a_{2,2} |^2 + | \lambda_2 - \lambda_1 |^2] \\ \beta &= | a_{1,2} |^2 + | a_{2,1} |^2 + \tfrac{1}{2}[| a_{1,1} - a_{2,2} |^2 - | \lambda_2 - \lambda_1 |^2] \end{aligned} \tag{8.2}$$

(ii) If $\lambda_1 = \lambda_2$ and the Jordan canonical form of A is diagonal, then $\mathscr{J}(A) = 1$.

(iii) If $\lambda_1 = \lambda_2$ and the Jordan canonical form of A is not diagonal, then

$$\mathscr{J}(A) = \max[(| a_{1,2} | + | a_{2,1} |), (| a_{1,2} | + | a_{2,1} |)^{-1}] \tag{8.3}$$

Proof. We first state without proof the following.

Lemma 8.2. Let J be a Jordan canonical form of A and let V be any nonsingular matrix such that

$$V^{-1}AV = J \tag{8.4}$$

If W is any nonsingular matrix such that $W^{-1}AW = J$ then the matrix $\tilde{V}$ given by

$$\tilde{V} = V^{-1}W \tag{8.5}$$

satisfies

$$J\tilde{V} = \tilde{V}J \tag{8.6}$$

To prove the Theorem, let us first consider the case $\lambda_1 \neq \lambda_2$. If $a_{1,2} = a_{2,1} = 0$, then A is diagonal, and hence, since $I^{-1}AI = A = J$ and $\| I \| = 1$, we have $\mathscr{J}(A) = 1$. Moreover, the right member of (8.1) is also unity; hence (8.1) holds in this case.

If $a_{1,2} = 0$ but $a_{2,1} \neq 0$, let us consider the matrix $A' = P^{-1}AP$, where

$$P = P^{-1} = \begin{pmatrix} 0 & 1 \\ 1 & 0 \end{pmatrix}$$

If $V^{-1}AV$ is a Jordan canonical form of A, then $(P^{-1}V)^{-1}A'(P^{-1}V)$ is a Jordan canonical form of A'. Since $\| P \| = \| P^{-1} \| = 1$, it follows that $\mathscr{J}(A) = \mathscr{J}(A')$. Hence, unless $a_{1,2} = a_{2,1} = 0$ we can, without loss of generality, assume that $a_{1,2} \neq 0$.

If $a_{1,2} \neq 0$, then we have

$$V^{-1}AV = \begin{pmatrix} \lambda_1 & 0 \\ 0 & \lambda_2 \end{pmatrix} = J \tag{8.7}$$

where

$$V = \begin{pmatrix} 1 & 1 \\ p & q \end{pmatrix} \tag{8.8}$$

$$p = (\lambda_1 - a_{1,1})/a_{1,2}, \qquad q = (\lambda_2 - a_{1,1})/a_{1,2} \tag{8.9}$$

Since the most general matrix $\tilde{V}$ such that (8.6) holds is

$$\tilde{V} = \begin{pmatrix} a & 0 \\ 0 & b \end{pmatrix} \tag{8.10}$$

where a and b are arbitrary, it follows from Lemma 8.2 that the most general matrix W such that $W^{-1}AW = J$ is given by

$$W = V\tilde{V} = \begin{pmatrix} a & b \\ ap & bq \end{pmatrix} \tag{8.11}$$

where $a \neq 0$, $b \neq 0$. We seek to determine a and b so that $\nu(W)$ is minimized.

Evidently, the eigenvalues Γ_1 and Γ_2 of WW^{H} satisfy

$$\Gamma^2 - T\Gamma + \Delta = 0 \tag{8.12}$$

where

$$\begin{aligned} \Delta &= \det WW^{\mathrm{H}} = |a|^4 |c|^2 |q-p|^2 \\ T &= \operatorname{trace} WW^{\mathrm{H}} = |a|^2[1 + |c|^2 |q|^2 + |p|^2 + |c|^2] \end{aligned} \tag{8.13}$$

and

$$c = b/a \tag{8.14}$$

Since the eigenvalues of $W^{-1}(W^{\mathrm{H}})^{-1}$ are Γ_1^{-1} and Γ_2^{-1}, if we let

$$\Gamma_1, \Gamma_2 = \tfrac{1}{2}(T \pm [T^2 - 4\Delta]^{1/2}) \tag{8.15}$$

then we have

$$\nu(W) = \left(\frac{\Gamma_1}{\Gamma_2}\right)^{1/2} = \left(\frac{1 + [1 - (4/M^2)]^{1/2}}{1 - [1 - (4/M^2)]^{1/2}}\right)^{1/2} = \frac{1 + [1 - 4/M^2]^{1/2}}{2/M} \tag{8.16}$$

where

$$M = \frac{T}{\Delta^{1/2}} = \frac{1}{|q-p|}\left[|c|(1 + |q|^2) + \frac{1 + |p|^2}{|c|}\right] \tag{8.17}$$

For given p and q we minimize M and hence $\nu(W)$ by letting

$$|c|^2 = \frac{1 + |p|^2}{1 + |q|^2} \tag{8.18}$$

The corresponding value of M is

$$M_0 = \frac{2[1 + |p|^2]^{1/2}[1 + |q|^2]^{1/2}}{|q-p|} \tag{8.19}$$

and, by (8.16)

$$\mathscr{J}(A) = \frac{1 + [1 - (4/M_0^2)]^{1/2}}{2/M_0} \tag{8.20}$$

Since λ_1 and λ_2 satisfy the quadratic equation

$$\lambda^2 - (a_{1,1} + a_{2,2})\lambda + a_{1,1}a_{2,2} - a_{1,2}a_{2,1} = 0 \tag{8.21}$$

we have

$$\lambda_1, \lambda_2 = \tfrac{1}{2}\{a_{1,1} + a_{2,2} \pm [(a_{1,1} - a_{2,2})^2 + 4a_{1,2}a_{2,1}]^{1/2}\} \tag{8.22}$$

and

$$|\lambda_2 - \lambda_1|^2 = |(a_{1,1} - a_{2,2})^2 + 4a_{1,2}a_{2,1}| \tag{8.23}$$

Moreover, by (8.9) we have

$$\begin{aligned} p + q &= \frac{\lambda_1 + \lambda_2 - 2a_{1,1}}{a_{1,2}} = \frac{a_{2,2} - a_{1,1}}{a_{1,2}} \\ pq &= \frac{(\lambda_1 - a_{1,1})(\lambda_2 - a_{1,1})}{a_{1,2}^2} = -\frac{a_{2,1}}{a_{1,2}} \end{aligned} \tag{8.24}$$

so that p and q satisfy the quadratic equation

$$\mu^2 - \left(\frac{a_{2,2} - a_{1,1}}{a_{1,2}^2}\right)\mu - \frac{a_{2,1}}{a_{1,2}} = 0 \tag{8.25}$$

Therefore, we have (see Exercise 1)

$$\begin{aligned} |p|^2 + |q|^2 &= \frac{1}{2}\left|\frac{a_{2,2} - a_{1,1}}{a_{1,2}}\right|^2 + \frac{1}{2}\left|\left(\frac{a_{2,2} - a_{1,1}}{a_{1,2}}\right)^2 + 4\frac{a_{2,1}}{a_{1,2}}\right| \\ &= \frac{|a_{2,2} - a_{1,1}|^2 + |\lambda_2 - \lambda_1|^2}{2|a_{1,2}|^2} \end{aligned} \tag{8.26}$$

Hence, by (8.19), (8.26), (8.24), and (8.2) we have

$$M_0 = 2\alpha^{1/2}/|\lambda_2 - \lambda_1| \tag{8.27}$$

Substituting in (8.20) we obtain (8.1).

Suppose now that $\lambda_1 = \lambda_2 = \lambda$ and the Jordan canonical form of A is diagonal. By (8.22) we have

$$(a_{1,1} - a_{2,2})^2 + 4a_{1,2}a_{2,1} = 0 \tag{8.28}$$

and

$$\lambda = (a_{1,1} + a_{2,2})/2 \tag{8.29}$$

If $(v_1, v_2)^{\mathrm{T}}$ is an eigenvector associated with λ, then

$$\begin{aligned} \tfrac{1}{2}(a_{1,1} - a_{2,2})v_1 + a_{1,2}v_2 &= 0 \\ a_{2,1}v_1 + \tfrac{1}{2}(a_{2,2} - a_{1,1})v_2 &= 0 \end{aligned} \tag{8.30}$$

If any one of the three quantities $a_{1,1} - a_{2,2}$, $a_{1,2}$, or $a_{2,1}$ does not vanish, then there can be only one linearly independent eigenvector and the Jordan canonical form cannot be diagonal. Hence we have $a_{1,2} = a_{2,1} = a_{1,1} - a_{2,2} = 0$, and $\lambda = a_{1,1} = a_{2,2}$. Thus $A = \lambda I$ and $\mathscr{J}(A) = 1$, and (ii) follows.

Suppose now that $\lambda_1 = \lambda_2 = \lambda$ but the Jordan canonical form of A is not diagonal. Evidently $a_{1,2}$ and $a_{2,1}$ cannot both vanish; otherwise, A and hence its Jordan canonical form would be diagonal. As in the proof of (i) we can assume that $a_{1,2} \neq 0$. One can easily verify that

$$V^{-1}AV = J = \begin{pmatrix} \lambda & 1 \\ 0 & \lambda \end{pmatrix} \tag{8.31}$$

where

$$V = \begin{pmatrix} 1 & 0 \\ p & a_{1,2}^{-1} \end{pmatrix}, \qquad p = (a_{2,2} - a_{1,1})/2a_{1,2} \tag{8.32}$$

Since the most general solution of (8.6) is

$$\tilde{V} = \begin{pmatrix} a & b \\ 0 & a \end{pmatrix} \tag{8.33}$$

then by Lemma 8.2 the most general matrix W such that $W^{-1}AW = J$ is given by

$$W = V\tilde{V} = \begin{pmatrix} a & b \\ ap & bp + aa_{1,2}^{-1} \end{pmatrix} \tag{8.34}$$

As before, the eigenvalues of WW^{H} satisfy (8.12) and

$$\begin{aligned} \Delta &= |a|^4/|a_{1,2}|^2 \\ T &= |a|^2(1 + |c|^2 + |p|^2 + |cp + a_{1,2}^{-1}|^2) \end{aligned} \tag{8.35}$$

where $c = a/b$. We again seek to minimize

$$M = T/\Delta^{1/2} = |a_{1,2}|\,[1 + |c|^2 + |p|^2 + |cp + a_{1,2}^{-1}|^2] \tag{8.36}$$

Let us define real values of r, θ, and ϕ such that

$$c = |c|\,e^{i\phi}, \qquad p/a_{1,2}^{*} = re^{i\theta} \tag{8.37}$$

Evidently,

$$\begin{aligned} M/|a_{1,2}| = 1 &+ |c|^2 + |p|^2 + |c|^2|p|^2 + 2r|c|\cos(\theta + \phi) \\ &+ (1/|a_{1,2}|^2) \end{aligned} \tag{8.38}$$

To minimize M we first choose ϕ so that $\cos(\theta + \phi) = -1$, i.e., so that $\phi = (\pi - \theta)$ and obtain

$$M/|a_{1,2}| = |c|^2(1 + |p|^2) + 1 + |p|^2 + (1/|a_{1,2}|^2) - 2|p/a_{1,2}|\,|c| \tag{8.39}$$

The value of $|c|$ which minimizes the above expression is given by

$$|c| = |p|/(|a_{1,2}|(1 + |p|^2)) \tag{8.40}$$

and the corresponding value of M is

$$M_0 = K + K^{-1} \tag{8.41}$$

where

$$K = |a_{1,2}|(1 + |p|^2) = |a_{1,2}|(1 + [|a_{2,2} - a_{1,1}|^2/(4|a_{1,2}|^2)]) \tag{8.42}$$

But by (8.28), we have

$$K = |a_{1,2}| + |a_{2,1}| \tag{8.43}$$

From (8.41) and (8.20) we have

$$\mathscr{J}(A) = \tfrac{1}{2}(K + K^{-1} + |K - K^{-1}|) = \max(K, K^{-1}) \tag{8.44}$$

and (8.3) follows from (8.43). This completes the proof of Theorem 8.1.

SUPPLEMENTARY DISCUSSION

Section 3.1. Our terminology for the first part of Section 3.1 is based on that of Forsythe [1953]. (See also Forsythe and Wasow [1960].)

Section 3.2. The discussion given in the text appears in abbreviated form in Young [1971b]. Keller [1965] treated the solution of singular linear systems with A positive semidefinite. Such problems frequently arise in the solution of the Neumann problem for an elliptic partial differential equation.

Section 3.3. For a historical discussion of the J and GS methods see Forsythe [1953]. Work on the SOR method was done independently by Frankel [1950] and by Young [1950, 1954]. The RF method is a special case of the method of Richardson [1910]. While Richardson allowed the iteration parameter to vary from iteration to iteration, the parameter is kept fixed with the RF method.

Section 3.5. Theorem 5.1, in a slightly different form, was proved by Geiringer [1949]. Theorem 5.3 is based on a result of Wachspress [1966, p. 17].

Section 3.6. The term "N-stable" was suggested by the fact that an N-stable matrix is the negative of a stable matrix as frequently defined (see, for instance, Taussky [1961]), as having eigenvalues with negative real parts. Ostrowski and Schneider [1962] consider many kinds of stability for matrices, and refer to N-stable matrices as *positive stable*, while stable matrices are *negative stable.*

The fact that Theorem 6.4 fcllows from Stein's theorem was noted by Householder [1964, p. 63, Exercise 55].

The proof of the convergence of the SOR method when A is positive definite also follows from Theorem 5.3 (see Wachspress [1966]).

Section 3.7. Theorem 7.1 is closely related to a result of Varga [1962], but makes use of the Jordan condition number. Further discussion of the Jordan condition number and other condition numbers can be found in papers by Henrici [1962], Smith [1967], and Loizou [1969].

Young [1950, 1954] defined the *rate of convergence* of a stationary iterative method. This definition is equivalent to *asymptotic rate of convergence* as defined by Varga [1962].

EXERCISES

Section 3.1

1. Consider the iterative method

$$\begin{aligned} u^{(0)} &= \text{arbitrary} \\ u^{(1)} &= \text{arbitrary} \\ u^{(2)} &= u^{(1)} + u^{(0)} \\ u^{(3)} &= u^{(2)} + u^{(1)} + u^{(0)} \\ u^{(4)} &= u^{(3)} + u^{(2)} + u^{(1)} + u^{(0)} \\ u^{(n+1)} &= 2u^{(n)} + u^{(n-1)}, \qquad n \geq 4 \end{aligned}$$

Is the method stationary? What is the degree of the method?

2. Classify the following methods according to degree, stationary or nonstationary, linear or nonlinear.

(a) $$u^{(n+1)} = \begin{cases} u^{(n)}, & \text{if } u_1^{(n)} \geq 0 \\ Gu^{(n)} + k, & \text{if } u_1^{(n)} < 0 \end{cases}$$

$u^{(0)}$ arbitrary

(b)
$$u^{(n+1)} = \begin{cases} G_1 u^{(n)} + k, & n \ \text{odd} \\ G_2 u^{(n)} + k', & n \ \text{even} \end{cases}$$
$$u^{(0)} \quad \text{arbitrary}$$

(c)
$$\begin{aligned} u^{(n+1)} &= Gu^{(n)} + Hu^{(n-1)} + k, \qquad n \geq 1 \\ u^{(1)} &= G'u^{(0)} + k' \\ u^{(0)} & \quad \text{arbitrary} \end{aligned}$$

(d)
$$\begin{aligned} u^{(n+1)} &= Gu^{(n)} + Hu^{(n-1)} + k, \qquad n \geq 1 \\ u^{(1)}, u^{(0)} & \quad \text{arbitrary} \end{aligned}$$

(e)
$$\begin{aligned} u^{(n+1)} &= G_{n+1} u^{(n)} + k^{(n+1)}, \qquad n \geq 1 \\ u^{(0)} & \quad \text{arbitrary} \end{aligned}$$

(f)
$$\begin{aligned} u^{(n+1)} &= 2u^{(n)} - u^{(n-1)}, \qquad n \geq 1 \\ u^{(1)} &= \begin{pmatrix} 1 \\ 1 \\ 1 \end{pmatrix} \\ u^{(0)} &= 0 \end{aligned}$$

3. In each of the following cases test the iterative method $u^{(n+1)} = Gu^{(n)} + k$ for consistency, complete consistency, and reciprocal consistency with the linear system $Au = b$.

	A	b	G	k
(a)	$\begin{pmatrix} 2 & -1 \\ -1 & 2 \end{pmatrix}$	$\begin{pmatrix} 3 \\ -3 \end{pmatrix}$	$\begin{pmatrix} 0 & 0 \\ 0 & 0 \end{pmatrix}$	$\begin{pmatrix} 1 \\ -1 \end{pmatrix}$
(b)	same	same	same	$\begin{pmatrix} 1 \\ 0 \end{pmatrix}$
(c)	same	same	$\begin{pmatrix} 1 & 0 \\ 0 & 1 \end{pmatrix}$	$\begin{pmatrix} 0 \\ 0 \end{pmatrix}$
(d)	same	same	$\begin{pmatrix} 0 & \frac{1}{2} \\ \frac{1}{2} & 0 \end{pmatrix}$	$\begin{pmatrix} \frac{3}{2} \\ -\frac{3}{2} \end{pmatrix}$
(e)	same	same	$\begin{pmatrix} 0 & \frac{1}{2} \\ 0 & \frac{1}{4} \end{pmatrix}$	$\begin{pmatrix} \frac{3}{2} \\ -\frac{3}{4} \end{pmatrix}$
(f)	same	same	$\begin{pmatrix} 1 & 0 \\ 0 & 1 \end{pmatrix}$	$\begin{pmatrix} 1 \\ 0 \end{pmatrix}$
(g)	same	$\begin{pmatrix} 1 \\ 1 \end{pmatrix}$	same	same

	A	b	G	k
(h)	same	same	$\begin{pmatrix} 0 & 0 \\ 0 & 0 \end{pmatrix}$	same
(i)	same	$\begin{pmatrix} 2 \\ -1 \end{pmatrix}$	same	$\begin{pmatrix} 2 \\ 1 \end{pmatrix}$
(j)	same	same	same	$\begin{pmatrix} 1 \\ 1 \end{pmatrix}$
(k)	same	same	same	$\begin{pmatrix} 1 \\ -1 \end{pmatrix}$
(l)	same	$\begin{pmatrix} 1 \\ 0 \end{pmatrix}$	$\begin{pmatrix} 0 & \frac{1}{2} \\ \frac{1}{2} & 0 \end{pmatrix}$	$\begin{pmatrix} \frac{1}{2} \\ 0 \end{pmatrix}$
(m)	same	same	$\begin{pmatrix} 1 & 0 \\ 0 & 1 \end{pmatrix}$	$\begin{pmatrix} 0 \\ 0 \end{pmatrix}$
(n)	same	same	same	$\begin{pmatrix} 1 \\ 0 \end{pmatrix}$
(o)	same	$\begin{pmatrix} 2 \\ -1 \end{pmatrix}$	same	$\begin{pmatrix} 0 \\ 0 \end{pmatrix}$
(p)	same	same	$\begin{pmatrix} 0 & -1 \\ 0 & 1 \end{pmatrix}$	$\begin{pmatrix} 1 \\ 0 \end{pmatrix}$

Section 3.2

1. For each of the methods in Exercise 3 of Section 3.1 involving a nonsingular matrix A, determine $(I - G)A^{-1}b$ and verify Theorem 2.2. For each method such that $I - G$ is nonsingular verify the last statement of Theorem 2.3.

2. Verify Theorem 2.6 for each of the examples of Exercise 3 of Section 3.1.

3. Verify Lemma 2.7 for the following pairs of matrices

(a) $$A = \begin{pmatrix} 2 & 1 & 0 \\ 1 & 0 & -1 \\ 0 & 1 & 2 \end{pmatrix}, \qquad B = \begin{pmatrix} 1 & 1 & 1 \\ 1 & 1 & 1 \\ 1 & 1 & 1 \end{pmatrix}$$

(b) $$A = \begin{pmatrix} 2 & 1 & 0 \\ 1 & 0 & -1 \\ 0 & 1 & 2 \end{pmatrix}, \qquad B = \begin{pmatrix} 0 & 1 & 2 \\ 1 & 0 & -1 \\ 2 & 1 & 0 \end{pmatrix}$$

4. Consider the iterative method defined by

$$u^{(n+1)} = \tfrac{1}{2}u^{(n)} + \begin{pmatrix} \frac{1}{2} \\ 1 \end{pmatrix} = Gu^{(n)} + k$$

for solving the linear system $Au = b$, where

$$A = \begin{pmatrix} 1 & -2 \\ -2 & 4 \end{pmatrix}, \qquad b = \begin{pmatrix} -3 \\ 6 \end{pmatrix}$$

Show that the method is reciprocally consistent but not consistent. Also find a matrix Q such that

$$A = Q(I - G), \qquad b = Q\begin{pmatrix} \frac{1}{2} \\ 1 \end{pmatrix}$$

5. Verify Lemma 2.7 for the matrices

$$A = \begin{pmatrix} 1 & 0 \\ 0 & 0 \end{pmatrix}, \qquad B = \begin{pmatrix} 2 & 0 \\ 1 & 0 \end{pmatrix}$$

Show that even though $\mathcal{N}(A) = \mathcal{N}(B)$ there does not exist a matrix X such that $AX = B$.

6. Consider two iterative methods

$$u^{(n+1)} = G_1 u^{(n)} + k_1$$
$$u^{(n+1)} = G_2 u^{(n)} + k_2$$

for solving the solvable system $Au = b$. Prove that if both methods are consistent, then the composite method

$$u^{(n+1)} = G_2 G_1 u^{(n)} + (G_2 k_1 + k_2)$$

is consistent, but that even though both methods are reciprocally consistent the composite method may not be.

7. Prove that both A and $I - G$ are nonsingular, the method (1.5) is completely consistent with (1.1), (1.1) and (1.6) have the same unique solution $\bar{u} = A^{-1}b = (I - G)^{-1}k$, and (2.1) and (2.2) hold under either of the following sets of conditions.

(a) $I - G$ is nonsingular, (1.1) is solvable, and the method (1.5) is consistent with (1.1).

(b) A is nonsingular, (1.6) is solvable, and the method (1.5) is reciprocally consistent with (1.1).

Section 3.3

1. Construct G and k for each of the basic iterative methods for solving

$$\begin{pmatrix} 4 & -1 \\ -1 & 2 \end{pmatrix}\begin{pmatrix} u_1 \\ u_2 \end{pmatrix} = \begin{pmatrix} 3 \\ 1 \end{pmatrix}$$

For the SOR and JOR methods let $\omega = \frac{3}{2}$; for the GRF method let

$$P = \begin{pmatrix} 2 & 0 \\ 0 & 1 \end{pmatrix}$$

and for the RF method let $p = -1$. In each case verify that $k = (I - G)A^{-1}b$.

2. In the previous example carry out three iterations of each method with the starting values $u_1^{(0)} = 1$, $u_2^{(0)} = 0$.

3. Show that the matrix B_E of the Jacobi method for the scaled system (3.30) has the same eigenvalues as the matrix B for the original system.

4. Verify that the related equation corresponding to each of the basic methods can be derived from (1.1) by one or more reversible transformations.

5. If the eigenvalues of the J method are real and lie in the interval $-0.5 \leq \mu \leq 0.8$ find the value of ω which minimizes $S(B_\omega)$. Do the same if $-0.8 \leq \mu \leq 0.8$. In each case find the smallest value of $S(B_\omega)$.

6. Consider the finite difference equation analogue of the Dirichlet problem for the unit square with a $\frac{1}{3} \times \frac{1}{3}$ corner removed, as indicated, and with $h = \frac{1}{3}$.

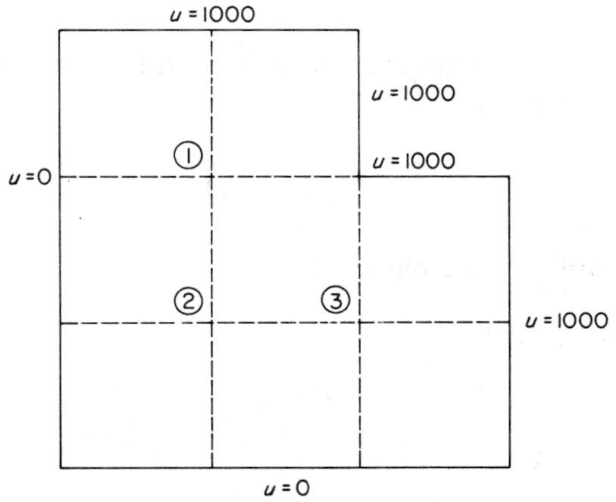

(a) Solve the difference equation directly.

(b) Solve the difference equation by each of the following iterative methods starting with zero initial values

(i) Jacobi,
(ii) Gauss–Seidel.

Iterate in each case until the maximum change from one iteration to the next does not exceed one unit. Use the indicated ordering.

(c) Compute the eigenvalues of the matrix associated with each method. (Note that for the Gauss–Seidel method the eigenvalues satisfy $\det(\lambda L + U - \lambda I) = 0$.)

(d) Compute the spectral radii and interpret in terms of the observations on the iterative procedures.

7. Let A be a matrix whose diagonal elements do not vanish. Consider the symmetric SOR method (SSOR method) which involves two half-iterations using the SOR method. The first half-iteration is the ordinary SOR method and the second half-iteration is the SOR method using the reverse order.

(a) Express the method in the form

$$u^{(n+1)} = Gu^{(n)} + k$$

and determine G and k in terms of L, U, and ω.

(b) Show that unless $\omega = 0$ or $\omega = 2$ the method is completely consistent. (Assume A is nonsingular.)

Section 3.4

1. Verify that the splittings defined at the end of Section 3.4 lead to the indicated iterative methods. If

$$A = \begin{pmatrix} 2 & -1 \\ -1 & 2 \end{pmatrix}$$

which of the splittings are regular?

2. Consider the system

$$\begin{pmatrix} 2 & -1 & 0 \\ -1 & 2 & -1 \\ 0 & -1 & 2 \end{pmatrix} \begin{pmatrix} u_1 \\ u_2 \\ u_3 \end{pmatrix} = \begin{pmatrix} 1 \\ 2 \\ 3 \end{pmatrix}$$

Carry out one iteration of the procedure (4.10) and (4.11) with

$$Q = \begin{pmatrix} 1 & -1 & 0 \\ 0 & 2 & 0 \\ 0 & 0 & 1 \end{pmatrix}$$

and with $u^{(0)} = (1, 1, 1)^T$.

3. Let A be a matrix with nonvanishing diagonal elements. Construct an iterative method from the matrix

$$Q = \frac{\omega}{2-\omega}\left(\frac{1}{\omega} D - C_L\right) D^{-1} \left(\frac{1}{\omega} D - C_U\right)$$

where $\omega \neq 0$ and $\omega \neq 2$. Here $C_L = DL$ and $C_U = DU$. Show that $G = \mathscr{U}_\omega \mathscr{L}_\omega$ where $\mathscr{L}_\omega = (I - \omega L)^{-1}(\omega U + (1-\omega)I)$, $\mathscr{U}_\omega = (I - \omega U)^{-1}$ $(\omega L + (1-\omega)I)$. Show that if $\omega \neq 0$ and $\omega \neq 2$, then the method is completely consistent.

Section 3.5

1. Complete the proof of Theorem 5.1.

2. Apply Theorem 5.1 to the iterative methods

$$u^{(n+1)} = \begin{pmatrix} 1 & 0 \\ 0 & -1 \end{pmatrix} u^{(n)} + \begin{pmatrix} 1 \\ 0 \end{pmatrix}$$

$$u^{(n+1)} = \begin{pmatrix} 1 & 1 \\ 0 & 1 \end{pmatrix} u^{(n)} + \begin{pmatrix} 1 \\ 0 \end{pmatrix}$$

$$u^{(n+1)} = \begin{pmatrix} 1 & 0 \\ 0 & 1 \end{pmatrix} u^{(n)} + \begin{pmatrix} 1 \\ 0 \end{pmatrix}$$

$$u^{(n+1)} = \begin{pmatrix} 1 & 0 \\ 0 & 1 \end{pmatrix} u^{(n)} + \begin{pmatrix} 0 \\ 0 \end{pmatrix}$$

Apply the methods in each case with

$$u^{(0)} = \begin{pmatrix} 2 \\ 1 \end{pmatrix}$$

3. Consider the linear system

$$\begin{pmatrix} 2 & -1 & 0 \\ -1 & 2 & -1 \\ 0 & -1 & 2 \end{pmatrix} \begin{pmatrix} u_1 \\ u_2 \\ u_3 \end{pmatrix} = \begin{pmatrix} 0 \\ 2 \\ 0 \end{pmatrix}$$

First determine the true solution of the system. Starting with the initial vector

$$u^{(0)} = \begin{pmatrix} 0 \\ 0 \\ 0 \end{pmatrix}$$

carry out five iterations with the SOR method with $\omega = 2$. Does the method appear to be converging? Reconcile this with the fact that the method is completely consistent. (Try the method using the true solution as a starting vector.) What is $S(\mathscr{L}_\omega)$ in this case?

4. Verify directly that the Jacobi method for solving

$$\begin{pmatrix} 1 & 1 \\ 1 & 1 \end{pmatrix} \begin{pmatrix} u_1 \\ u_2 \end{pmatrix} = \begin{pmatrix} 3 \\ 3 \end{pmatrix}$$

is completely consistent but not convergent or weakly convergent. Show that the Gauss–Seidel method is completely consistent and weakly convergent.

5. In the preceding problem apply the Gauss–Seidel method both with the starting vector $\begin{pmatrix} 1 \\ 0 \end{pmatrix}$ and also with $\begin{pmatrix} 0 \\ 1 \end{pmatrix}$. Show that in general, with the starting vector

$$u^{(0)} = \begin{pmatrix} u_1^{(0)} \\ u_2^{(0)} \end{pmatrix}$$

the method converges to

$$\begin{pmatrix} 3 - u_2^{(0)} \\ u_2^{(0)} \end{pmatrix}$$

which is a solution of the original system.

6. Consider the system

$$Au = b$$

where

$$A = \begin{pmatrix} 1 & -1 \\ -1 & 1 \end{pmatrix}, \qquad b = \begin{pmatrix} 1 \\ 0 \end{pmatrix}$$

Show that $S(\mathscr{L}) = 1$, $\lambda = 1$ is the only eigenvalue of $\mathscr{L}$ of modulus unity, and the dimension of the null space of $I - \mathscr{L}$ is equal to the multiplicity of $\lambda = 1$ as an eigenvalue of $\mathscr{L}$. Is the GS method weakly convergent? Try the method with the starting vector $(0, 0)^{\mathrm{T}}$. Reconcile this with Theorem 5.1.

7. Give an example to show that even though a linear stationary iterative method is convergent and is reciprocally consistent with a linear system, it need not be completely consistent.

Section 3.6

1. Let $\mathscr{L}$ be the matrix corresponding to the Gauss–Seidel method for a linear system involving the matrix

$$A = \begin{pmatrix} 4 & -1 & -1 & 0 \\ -1 & 4 & 0 & -1 \\ -1 & 0 & 4 & -1 \\ 0 & -1 & -1 & 4 \end{pmatrix}$$

Find a real positive definite matrix P such that $P - \mathscr{L}P\mathscr{L}^{\mathrm{H}}$ is positive definite. Also find a real positive definite matrix R such that $\| \mathscr{L} \|_R < 1$. Finally, verify that $H = (I - \mathscr{L})^{-1}(I + \mathscr{L})$ is N stable. Find Q such that $HQ + QH^{\mathrm{T}}$ is positive definite.

2. Show that if A is an L-matrix with weak diagonal dominance, then the real parts of the eigenvalues of A are nonnegative.

3. If A is an L-matrix with weak diagonal dominance, must A be N-stable?

4. Prove Corollary 6.2 in the case G is real.

Section 3.7

1. Let J be any matrix which is a Jordan canonical form of A. Show that

$$\inf_{V \in S_1} \nu(V) = \mathscr{J}(A)$$

where S_1 is the set of all matrices V such that $V^{-1}AV = J$.

2. Consider the matrix

$$A = \begin{pmatrix} -\frac{1}{4} & 1 \\ -\frac{1}{4} & \frac{3}{4} \end{pmatrix}$$

(a) Show that the matrix

$$P = \begin{pmatrix} 2 & -2 \\ 1 & 1 \end{pmatrix}$$

reduces A to Jordan canonical form, i.e., $P^{-1}AP = J$ where J is a Jordan canonical form of A.

(b) Compute the spectral condition number $\nu(P) = \| P \| \| P^{-1} \|$.

(c) On the basis of this, compute an approximate upper and an approximate lower bound for n such that $\| A^n \| \leq 10^{-2}$.

(d) Estimate the smallest value of n such that $S(A^n) \leq 10^{-2}$.

(e) Compute the Jordan condition number of A.

(f) On the basis of this, obtain an improved estimate for (c).

(g) Starting with the vector

$$x^{(0)} = \begin{pmatrix} 1 \\ 1 \end{pmatrix}$$

compute $A^n x^{(0)} = x^{(n)}$ for $n = 1, 2, \ldots,$ until $\| x^{(n)} \| \leq 10^{-2} \| x^{(0)} \|$.

(h) Compute A^n where n is the smallest integer such that $\| A^n \| \leq 10^{-2}$.

3. Compute the Jordan condition numbers of the matrices

$$\begin{pmatrix} 2 & -1 \\ -1 & 2 \end{pmatrix}, \quad \begin{pmatrix} 2 & i \\ i & 1 \end{pmatrix}, \quad \begin{pmatrix} 0 & 1 \\ 0 & 0 \end{pmatrix}, \quad \begin{pmatrix} 0 & 1 \\ 1 & 0 \end{pmatrix}, \quad \begin{pmatrix} 3 & 1 \\ 2 & 4 \end{pmatrix}$$

4. Let V be any nonsingular matrix and let Q be a nonsingular matrix such that $\| Q \| = \| Q^{-1} \| = 1$. Show that

$$\nu(VQ) = \nu(V)$$

Here $\nu(A) = \| A \| \| A^{-1} \|$ for any matrix A.

5. Let A be a matrix of order 21 whose Jordan condition number is 5 and whose Jordan canonical form is shown in the accompanying tabulation.

Eigenvalue	Size of block in Jordan canonical form
0.64 + 0.48i	2
0.64 − 0.48i	2
0.8	1
0.01 + 0.5i	4
0.01 + 0.5i	4
0.01 − 0.5i	4
0.01 − 0.5i	4

(a) Compute an approximate upper bound on n such that $\| A^n \| \leq 0.01$.

(b) Determine n if $\mathscr{J}(A)$ were unity and all blocks in the Jordan canonical form were of size unity.

Section 3.8

1. Prove that if x_1 and x_2 are the roots of $x^2 - bx + c = 0$, then $| x_1 |^2 + | x_2 |^2 = (| b |^2 + | b^2 - 4c |)/2$.

2. Verify (8.33).

3. Verify Theorem 8.1 for the matrices considered in Exercise 3, Section 3.7.

4. Prove Lemma 8.2.

Chapter 4 / CONVERGENCE OF THE BASIC ITERATIVE METHODS

In this chapter we seek to establish some convergence theorems for the basic iterative methods which were considered in Section 3.3. Some general convergence theorems will be given in Section 4.1. Subsequent sections will be concerned with cases where A has particular properties such as irreducibility, weak diagonal dominance, positive definiteness, etc. In Section 4.6, we will study the convergence of the J and GS methods for the model problem, as defined in Chapter 1, and show that their convergence is unacceptably slow.

More refined convergence theorems can be obtained if the matrix A is "consistently ordered" as defined in Chapter 5. Such theorems and bounds on the convergence rates for the J, GS, and SOR methods will be given in Chapters 5–7.

As in Chapter 3, we assume that the matrices A and b appearing in our linear system (3-1.1) are real. We also assume that diag A is nonsingular.

4.1. GENERAL CONVERGENCE THEOREMS

We now give two theorems on the convergence of the JOR and SOR methods which are applicable for any matrix with nonvanishing diagonal elements.

Theorem 1.1. If the J method converges, then the JOR method converges for $0 < \omega \leq 1$.

Proof. By (3-3.13) the eigenvalues ϱ_i of the JOR method are given in terms of the eigenvalues μ_i of the J method by

$$\varrho_i = \omega\mu_i + 1 - \omega \tag{1.1}$$

If $\mu_i = re^{i\theta}$, $r < 1$, and $0 < \omega \leq 1$, then

$$\begin{aligned} |\varrho_i|^2 &= \omega^2 r^2 + 2\omega r \cos\theta(1-\omega) + (1-\omega)^2 \\ &\leq (\omega r + 1 - \omega)^2 < 1 \end{aligned}$$

and the JOR method converges.

Theorem 1.2. (Kahan, 1958).

$$S(\mathscr{L}_\omega) \geq |\omega - 1| \tag{1.2}$$

Moreover, if the SOR method converges, then

$$0 < \omega < 2 \tag{1.3}$$

Proof. By (2-1.27) the product of the eigenvalues of $\mathscr{L}_\omega$ is equal to $\det \mathscr{L}_\omega$. By (3-3.23) we have

$$\det \mathscr{L}_\omega = \det(\omega U + (1-\omega)I) = (1-\omega)^N \tag{1.4}$$

Therefore

$$S(\mathscr{L}_\omega) \geq (|1-\omega|^N)^{1/N} = |1-\omega| \tag{1.5}$$

and the theorem follows.

4.2. IRREDUCIBLE MATRICES WITH WEAK DIAGONAL DOMINANCE

We have already seen in Section 2.5 that irreducible matrices with weak diagonal dominance are nonsingular. We now prove

Theorem 2.1. Let A be an irreducible matrix with weak diagonal dominance. Then

(a) The J method converges, and the JOR method converges for $0 < \omega \leq 1$.

(b) The GS method converges, and the SOR method converges for $0 < \omega \leq 1$.

Proof. If $S(B) \geq 1$, there exists an eigenvalue μ of B such that $|\mu| \geq 1$. Now $\det(B - \mu I) = 0$ and $\det(I - \mu^{-1}B) = 0$. Clearly, $Q = I - \mu^{-1}B$ is irreducible since A is, and has weak diagonal dominance since $|\mu^{-1}| \leq 1$. Hence $\det Q \neq 0$ by Theorem 2-5.3, and we have a contradiction. Hence $S(B) < 1$ and the J method converges. By Theorem 1.1, the JOR method converges for $0 < \omega \leq 1$.

Suppose now that $0 < \omega \leq 1$ and that $S(\mathscr{L}_\omega) \geq 1$. Then for some eigenvalue λ of $\mathscr{L}_\omega$ we have $|\lambda| \geq 1$ and

$$\det(\mathscr{L}_\omega - \lambda I) = \det Q = 0 \tag{2.1}$$

where

$$Q = I - \frac{\omega}{\lambda + \omega - 1} U - \frac{\lambda\omega}{\lambda + \omega - 1} L \tag{2.2}$$

Let $\lambda^{-1} = qe^{i\theta}$ where q and θ are real. We have

$$\begin{aligned} \left| \frac{\lambda\omega}{\lambda + \omega - 1} \right| &= \frac{\omega}{[1 - 2q(1 - \omega)\cos\theta + q^2(1 - \omega)^2]^{1/2}} \\ &\leq \frac{\omega}{1 - q(1 - \omega)} \end{aligned} \tag{2.3}$$

since $q \leq 1$ and $0 < \omega \leq 1$. But

$$1 - \frac{\omega}{1 - q(1 - \omega)} = \frac{(1 - q)(1 - \omega)}{1 - q(1 - \omega)} \geq 0 \tag{2.4}$$

and hence

$$\left| \frac{\omega}{\lambda + \omega - 1} \right| \leq \left| \frac{\omega\lambda}{\lambda + \omega - 1} \right| \leq 1 \tag{2.5}$$

Since A has weak diagonal dominance, $D^{-1}A = I - L - U$ also has weak diagonal dominance. Hence, by (2.2) and (2.5) it follows that Q has weak diagonal dominance. Since Q is also irreducible, we have $\det Q \neq 0$, by Theorem 2-5.3, and (b) follows. This completes the proof of Theorem 2.1.

4.3. POSITIVE DEFINITE MATRICES

In Theorem 3-5.3 we gave a sufficient condition for the convergence of any iterative method where the matrix A of the original system (3-1.1) is positive definite. We now apply this result to the basic iterative methods of Section 3.3.

Theorem 3.1. Let A be a positive definite matrix and let $D = \text{diag } A$. Then

(a) $\| B \|_{A^{1/2}} < 1$ if $2D - A$ is positive definite.

(b) $\| B_\omega \|_{A^{1/2}} < 1$ if $2\omega^{-1}D - A$ is positive definite or, equivalently, if

$$0 < \omega < 2/(1 - \mu_{\min}) \leq 2 \tag{3.1}$$

where $\mu_{\min} \leq 0$ is the smallest eigenvalue of B.

(c) $\| \mathscr{L} \|_{A^{1/2}} < 1$.

(d) $\| \mathscr{L}_\omega \|_{A^{1/2}} < 1$ if

$$0 < \omega < 2 \tag{3.2}$$

(e) $\| R_p \|_{A^{1/2}} < 1$ if

$$0 > p > -2/S(A) \tag{3.3}$$

(f) $\| R_P \|_{A^{1/2}} < 1$ if the matrix Z given by

$$Z = -2P^{-1} - A \tag{3.4}$$

is positive definite.

Proof. The proof is based on the use of the matrix $Q = A(I - G)^{-1}$ where G is the matrix corresponding to the iteration method in question. (We note that since each method is completely consistent and A is nonsingular, then, by Theorem 3-2.4, $I - G$ is nonsingular.) The matrix Q in each case is given in Section 3.4. By Theorem 3-5.3, we need only show that for each method the matrix M given by

$$M = Q + Q^{\mathrm{T}} - A \tag{3.5}$$

is positive definite.

For the J method, we have $Q = D$ and $M = 2D - A$. For the JOR method, $Q = \omega^{-1}D$ and $M = 2\omega^{-1}D - A$. We now prove

Lemma 3.2. Let A be a real symmetric matrix with positive diagonal elements, and let $\omega \neq 0$ be a real number. The matrix $2\omega^{-1}D - A$, where $D = \text{diag } A$, is positive definite if and only if ω satisfies (3.1).

Proof. By Corollary 2-2.8, $2\omega^{-1}D - A$ is positive definite if and only if

$2\omega^{-1}I - \hat{A} = (2\omega^{-1} - 1)I + \tilde{B} = H$ is positive definite where

$$\hat{A} = D^{-1/2}AD^{-1/2}, \qquad \tilde{B} = D^{1/2}BD^{-1/2} \tag{3.6}$$

The eigenvalues of H are $2\omega^{-1} - 1 + \mu_i$ where the μ_i are the eigenvalues of B. Since trace $B = 0$ and since the eigenvalues of B are real, it follows from (2-1.27) that $\mu_{\min} \le 0$. Evidently H is positive definite if and only if (3.1) holds. The lemma now follows.

For the GS method, $Q = D - C_L$ and $M = D$ which is positive definite since by Theorem 2-2.5 the diagonal elements of A are positive. For the SOR method, $Q = \omega^{-1}D - C_L$ and $Q + Q^T - A = (2\omega^{-1} - 1)D$ which is positive definite if $0 < \omega < 2$.

For the RF method, $Q = -p^{-1}I$ and $M = -2p^{-1}I - A$, which is positive definite if (3.3) holds. Similarly, for the GRF method $Q = -P^{-1}$ and $M = -2P^{-1} - A$.

We now study the convergence properties of the basic methods in more detail.

Theorem 3.3. Let A be a real, symmetric, nonsingular matrix with positive diagonal elements. Then the JOR method converges if and only if A and $2\omega^{-1}D - A$ are positive definite. The condition that $2\omega^{-1}D - A$ is positive definite may be replaced by the condition (3.1).

Proof. Since $\tilde{B} = D^{1/2}BD^{-1/2} = D^{-1/2}CD^{-1/2}$ and since $C = D - A$ is symmetric, it follows that $\tilde{B}$ is symmetric. Hence $\tilde{B}$ and therefore B have real eigenvalues.

Since $B_\omega = \omega B + (1 - \omega)I$, the eigenvalues ϱ_i of B_ω are given in terms of the eigenvalues μ_i by (1.1). In order that the JOR method converge, it is necessary and sufficient that

$$-1 < \omega\mu_i + 1 - \omega < 1$$

for all eigenvalues μ_i of B. But since $\mu_{\min} \le 0$, these conditions are equivalent to the conditions

$$1 - \mu_i > 0, \qquad \mu_i > 1 - 2\omega^{-1}, \qquad \omega > 0 \tag{3.7}$$

But since

$$\hat{A} = D^{-1/2}AD^{-1/2} = I - D^{-1/2}CD^{-1/2} = I - \tilde{B} \tag{3.8}$$

the eigenvalues of $\hat{A}$ must all be positive. Hence $\hat{A}$ and, by Corollary 2-2.8, A must be positive definite.

The second condition of (3.7) implies that the eigenvalues $\hat{\nu}_i$ of $\hat{A}$ satisfy

$$\hat{\nu}_i < 2\omega^{-1} \tag{3.9}$$

since $\hat{\nu}_i = 1 - \mu_i$. Therefore the matrix $2\omega^{-1}I - \hat{A}$ and, by Corollary 2-2.8, $2\omega^{-1}D - A$ are positive definite.

The sufficiency of the conditions of Theorem 3.3 follows from Theorem 3.1. By Lemma 3.2 we can replace the condition on $2\omega^{-1}D - A$ by (3.1). This completes the proof of Theorem 3.3.

Corollary 3.4. Under the hypotheses of Theorem 3.3, the J method converges if and only if A and $2D - A$ are positive definite.

As an example of a positive definite matrix where the J method fails to converge, let us consider the matrix

$$A = \begin{pmatrix} 1 & a & a \\ a & 1 & a \\ a & a & 1 \end{pmatrix} \tag{3.10}$$

whose eigenvalues are $1 - a$, $1 - a$, $1 + 2a$. If a lies in the range $-\frac{1}{2} < a < 1$, then A is positive definite. The eigenvalues of B are $a, a, -2a$. Thus, if $a = 0.8$, the eigenvalues of A are 0.2, 0.2, and 2.6; thus A is positive definite. On the other hand, the eigenvalues of B are 0.8, 0.8, -1.6; hence $S(B) = 1.6 > 1$ and the J method fails to converge. However, the JOR method converges for all ω in the range

$$0 < \omega < 2/(1 - (-1.6)) = 2/2.6 = 0.770$$

In Section 4.4 we shall show that if A is a positive definite L-matrix, i.e., a Stieltjes matrix (see Section 2.7), then the J method converges.

We remark that if A is symmetric, nonsingular, and has positive diagonal elements, but is not necessarily positive definite, then $S(B_\omega B_{-\omega}) < 1$ if ω lies in the range

$$|\,\omega\,| < \min(2^{1/2}/(1 - \mu_{\min}),\ 2^{1/2}/|\,\mu_{\max} - 1\,|) \tag{3.11}$$

This amounts to using the JOR method with ω and $-\omega$ on alternate iterations.

We will now show that if A is a symmetric matrix with positive diagonal elements, one can often, without loss of generality, assume that A has the form

$$A = I - L - U \tag{3.12}$$

where L and U are strictly lower and strictly upper triangular matrices, respectively, and where

$$L = U^{\mathrm{T}} \tag{3.13}$$

For any matrix A with nonvanishing diagonal elements, we define the matrices

$$\begin{aligned} D[A] &= \operatorname{diag} A \\ C[A] &= D[A] - A \\ B[A] &= (D[A])^{-1}C[A] \\ C_L[A] + C_U[A] &= C[A] \end{aligned} \tag{3.14}$$

where $C_L[A]$ and $C_U[A]$ are strictly lower triangular and strictly upper triangular, respectively. We also define the matrices

$$\begin{aligned} L[A] &= (D[A])^{-1}C_L[A] \\ U[A] &= (D[A])^{-1}C_U[A] \\ \mathscr{L}_\omega[A] &= (I - \omega L[A])^{-1}(\omega U[A] + (1 - \omega)I) \end{aligned} \tag{3.15}$$

If the diagonal elements of A are positive, then we define for any matrix H, the matrices $\tilde{H}$ and $\hat{H}$ by

$$\tilde{H} = D^{1/2}HD^{-1/2}, \qquad \hat{H} = D^{-1/2}HD^{-1/2} \tag{3.16}$$

We now prove

Theorem 3.5. If A is a real symmetric matrix with positive diagonal elements, then $\hat{A} = D^{-1/2}AD^{-1/2}$, where $D = \operatorname{diag} A$, is a real symmetric matrix with positive diagonal elements, and $\hat{A}$ is positive definite if and only if A is. In any case, we have

$$\hat{A} = I - L[\hat{A}] - U[\hat{A}] \tag{3.17}$$

where

$$L[\hat{A}] = \tilde{L}[A], \qquad U[\hat{A}] = \tilde{L}[A] \tag{3.18}$$

and

$$(L[\hat{A}])^{\mathrm{T}} = U[\hat{A}] \tag{3.19}$$

Moreover, we have

$$B[\hat{A}] = \tilde{B}[A], \qquad \mathscr{L}_\omega[\hat{A}] = \tilde{\mathscr{L}}_\omega[A] \tag{3.20}$$

Proof. By Corollary 2-2.8, $\hat{A}$ is positive definite if and only if A is. Since $\hat{C}_L[A]$ and $\hat{C}_U[A]$ are strictly lower and strictly upper triangular matrices, respectively, and since

$$\hat{A} = I - C_L[\hat{A}] - C_U[\hat{A}] \tag{3.21}$$

it follows that

$$C_L[\hat{A}] = \hat{C}_L[A], \qquad C_U[\hat{A}] = \hat{C}_U[A] \tag{3.22}$$

Since $D[\hat{A}] = I$ we have (3.17) and

$$L[\hat{A}] = C_L[\hat{A}] = \hat{C}_L[A] = D^{-1/2}C_L[A]D^{-1/2} = D^{1/2}L[A]D^{-1/2} = \tilde{L}[A] \tag{3.23}$$

and, similarly,

$$U[\hat{A}] = \tilde{U}[A] = D^{-1/2}C_U[A]D^{-1/2} = D^{-1/2}(C_L[A])^{\mathrm{T}}D^{-1/2} = L[\hat{A}]^{\mathrm{T}} \tag{3.24}$$

since $(C_L[A])^{\mathrm{T}} = C_U[A]$. Hence (3.19) holds. Moreover

$$B[\hat{A}] = L[\hat{A}] + U[\hat{A}] = \tilde{L}[A] + \tilde{U}[A] = \tilde{B}[A] \tag{3.25}$$

and

$$\begin{aligned} \mathscr{L}_\omega[\hat{A}] &= (I - \omega L[\hat{A}])^{-1}(\omega U[\hat{A}] + (1-\omega)I) \\ &= (I - \omega\tilde{L}[A])^{-1}(\omega\tilde{U}[A] + (1-\omega)I) = \tilde{\mathscr{L}}_\omega[A] \end{aligned} \tag{3.26}$$

This completes the proof of Theorem 3.5.

Theorem 3.6. Let A be a symmetric matrix with positive diagonal elements. Then the SOR method converges if and only if A is positive definite and $0 < \omega < 2$.

Proof. By Theorem 3.5, it is sufficient to assume that A has the form (3.12) where (3.13) holds. The sufficiency of the conditions has already been shown in Theorem 3.1. We now give an alternative proof. If λ is an eigenvalue of $\mathscr{L}_\omega$, then for some $v \neq 0$ we have $\mathscr{L}_\omega v = \lambda v$ and hence

$$(\omega U + (1-\omega)I)v = \lambda(I - \omega L)v \tag{3.27}$$

Multiplying both sides on the left by v^H and solving for λ we obtain

$$\lambda = [\omega(v, Uv) + (1 - \omega)(v, v)]/[(v, v) - \omega(v, Lv)] \tag{3.28}$$

By (3.13) we have $(v, Uv) = (U^T v, v) = (Lv, v) = (v, Lv)^*$, and hence

$$\lambda = [\omega z^* + 1 - \omega]/[1 - \omega z] \tag{3.29}$$

where

$$z = (v, Lv)/(v, v) \tag{3.30}$$

If we let $z = re^{i\theta}$, where r and θ are real, we have

$$|\lambda|^2 = 1 - [\omega(2 - \omega)(1 - 2r\cos\theta)]/[(1 - \omega r\cos\theta)^2 + \omega^2 r^2 \sin^2\theta] \tag{3.31}$$

We can show $|\lambda| < 1$ for $0 < \omega < 2$ if we can show that $1 - 2r\cos\theta > 0$. But by (3.30) and (3.12) we have

$$2r\cos\theta = 2\,\mathrm{Re}(z) = z + z^* = (v, Bv)/(v, v) = 1 - (v, Av)/(v, v) < 1 \tag{3.32}$$

if A is positive definite. Thus $|\lambda| < 1$ and the convergence follows.

Let us now give a third convergence proof which is based on the use of the quadratic form

$$Q(u) = \tfrac{1}{2}(u, Au) - (b, u) \tag{3.33}$$

The use of this quadratic form is motivated by the fact that if A is positive definite, then the minimum value of $Q(u)$ is assumed when $u = \bar{u}$, where $\bar{u}$ is the unique solution of the system $Au = b$.

To see this, let us consider the case $N = 3$ where the quadratic form $Q(u)$ becomes

$$Q(u) = \tfrac{1}{2}[a_{1,1}u_1^2 + a_{2,2}u_2^2 + a_{3,3}u_3^2 + 2a_{1,2}u_1u_2 + 2a_{1,3}u_1u_3 + 2a_{2,3}u_2u_3] - b_1u_1 - b_2u_2 - b_3u_3 \tag{3.34}$$

Taking partial derivatives of Q with respect to u_1, u_2, u_3, respectively, we have

$$\begin{aligned} \partial Q/\partial u_1 &= a_{1,1}u_1 + a_{1,2}u_2 + a_{1,3}u_3 - b_1 \\ \partial Q/\partial u_2 &= a_{2,1}u_1 + a_{2,2}u_2 + a_{2,3}u_3 - b_2 \\ \partial Q/\partial u_3 &= a_{3,1}u_1 + a_{3,2}u_2 + a_{3,3}u_3 - b_3 \end{aligned} \tag{3.35}$$

Moreover, we have

$$\partial^2 Q/(\partial u_i\, \partial u_j) = a_{i,j}$$

Since A is positive definite, a necessary and sufficient condition for a relative minimum for $Q(u)$ is that $\partial Q/\partial u_i = 0$, $i = 1, 2, 3$, i.e., that u_1, u_2, u_3 satisfy $Au - b = 0$.

From (3.33) it follows that if A is symmetric and if $\bar{u}$ is a solution of $Au = b$, then

$$Q(\bar{u} + \varepsilon) - Q(\bar{u}) = \tfrac{1}{2}(\varepsilon, A\varepsilon) \tag{3.36}$$

If A is positive definite, then

$$Q(u) > Q(\bar{u}) \tag{3.37}$$

unless $u = \bar{u}$. Thus in this case $\bar{u}$ corresponds to an absolute minimum for $Q(u)$.

We now prove

Lemma 3.7. If A is a matrix of the form (3.12) where (3.13) holds and if $u^{(n+1)}$ is obtained from $u^{(n)}$ by the SOR method, then for $0 < \omega < 2$ we have

$$Q(u^{(n+1)}) = Q(u^{(n)}) - \tfrac{1}{2}[(2 - \omega)/\omega](\delta^{(n)}, \delta^{(n)}) \tag{3.38}$$

where

$$\delta^{(n)} = u^{(n+1)} - u^{(n)} \tag{3.39}$$

Moreover,

$$Q(u^{(n+1)}) < Q(u^{(n)}) \tag{3.40}$$

unless $u^{(n)}$ satisfies $Au = b$.

Proof. For any vector u let $r = Au - b$. If $\delta = u' - u$, where u' is obtained from u by the SOR method, then by (3.33) we have

$$Q(u + \delta) = Q(u) + \tfrac{1}{2}(\delta, A\delta) + (\delta, r)$$

Since $\delta = -\omega(I - \omega L)^{-1} r$, we have $r = -\omega^{-1}(I - \omega L)\delta$ and

$$\begin{aligned} Q(u + \delta) &= Q(u) + \tfrac{1}{2}(\delta, [A - (2/\omega)(I - \omega L)]\delta) \\ &= Q(u) + \tfrac{1}{2}(\delta, [(1 - (2/\omega))I + L - U]\delta) \end{aligned}$$

But since $L^{\mathrm{T}} = U$ we have $(\delta, (L - U)\delta) = 0$ so that

$$Q(u + \delta) = Q(u) - \tfrac{1}{2}[(2 - \omega)/\omega](\delta, \delta)$$

and (3.38) follows. If $Q(u + \delta) = Q(u)$, then $\delta = 0$ and hence $r = Au - b = 0$. The lemma now follows.

We remark that we do not assume for Lemma 3.7 that A is positive definite or even that A is nonsingular.

If A is positive definite, then the nonincreasing sequence $\{Q(u^{(n)})\}$ is bounded below by $Q(\bar{u})$. Therefore $\{Q(u^{(n)})\}$ converges to a limit, say Q_0, such that $Q_0 \geq Q(\bar{u})$. In any case, we have, by (3.38),

$$\lim_{n\to\infty} (\delta^{(n)}, \delta^{(n)}) = 0 \tag{3.41}$$

As usual, let

$$\varepsilon^{(n)} = u^{(n)} - \bar{u} \tag{3.42}$$

Then we have $\varepsilon^{(n+1)} = \mathscr{L}_\omega \varepsilon^{(n)}$ and

$$\delta^{(n)} = \varepsilon^{(n+1)} - \varepsilon^{(n)} = (\mathscr{L}_\omega - I)\varepsilon^{(n)} = -\omega(I - \omega L)^{-1} A \varepsilon^{(n)} \tag{3.43}$$

or

$$\varepsilon^{(n)} = -\omega^{-1} A^{-1}(I - \omega L)\delta^{(n)} = -\omega^{-1} K \delta^{(n)} \tag{3.44}$$

where $K = A^{-1}(I - \omega L)$. Then we have

$$\| \varepsilon^{(n)} \| = (1/\omega) \| K\delta^{(n)} \| = (1/\omega) \| \delta^{(n)} \|_K \tag{3.45}$$

Since $\| \delta^{(n)} \| \to 0$ as $n \to \infty$ it follows from Theorem 2-4.1 that $\| \delta^{(n)} \|_K \to 0$ and hence $\| \varepsilon^{(n)} \| \to 0$ and $\varepsilon^{(n)} \to 0$. This proves the convergence of the SOR method.

Suppose now that A is a symmetric matrix with positive diagonal elements but is not necessarily positive definite or even nonsingular. By Theorem 3.5, it is sufficient to assume that A has the form (3.12) where (3.13) holds. Suppose also that the SOR method is convergent. Then by Theorem 3-5.1 we must have $S(\mathscr{L}_\omega) < 1$ and, by Theorem 1.2, $0 < \omega < 2$. By (3-3.23) we have $I - \mathscr{L}_\omega = \omega(I - \omega L)^{-1} A$ and hence

$$A = \omega^{-1}(I - \omega L)(I - \mathscr{L}_\omega) \tag{3.46}$$

Since $I - \mathscr{L}_\omega$ is nonsingular, it follows that A is nonsingular and that (3-1.1) has a unique solution, say $\bar{u}$.

If A is not positive definite, then since A is nonsingular, it must have a negative eigenvalue, say α. Let v be an eigenvector of A corresponding to α. If

$$u^{(0)} = \bar{u} + v \tag{3.47}$$

then $\varepsilon^{(0)} = v$ and

$$(\varepsilon^{(0)}, A\varepsilon^{(0)}) = (v, Av) = \alpha(v, v) < 0 \tag{3.48}$$

Therefore by (3.36) we have

$$Q(u^{(0)}) < Q(\bar{u}) \tag{3.49}$$

But since $Q(u^{(n)})$ is a decreasing function of n it is not possible for $Q(u^{(n)})$ to converge to $Q(\bar{u})$. On the other hand, if $u^{(n)} \to \bar{u}$ as $n \to 0$, then $\varepsilon^{(n)} \to 0$. Since A is symmetric, we have, by Theorem 2-2.2,

$$|(\varepsilon^{(n)}, A\varepsilon^{(n)})|/(\varepsilon^{(n)}, \varepsilon^{(n)}) \le S(A) \tag{3.50}$$

Therefore, by (3.36),

$$|Q(u^{(n)}) - Q(\bar{u})| = \tfrac{1}{2}|(\varepsilon^{(n)}, A\varepsilon^{(n)})| \le \tfrac{1}{2}S(A)(\varepsilon^{(n)}, \varepsilon^{(n)}) \tag{3.51}$$

Since $\varepsilon^{(n)} \to 0$ as $n \to \infty$ it follows by Theorem 2-4.1 that $(\varepsilon^{(n)}, \varepsilon^{(n)}) = \| \varepsilon^{(n)} \|^2 \to 0$. Thus, $Q(u^{(n)}) \to Q(\bar{u})$ which is impossible. Hence with the starting value (3.47) the SOR method does not converge, and the proof of Theorem 3.6 is complete.

For the GRF method we have

Theorem 3.8. Let A be a symmetric nonsingular matrix and let P be a nonsingular diagonal matrix such that $P \ge 0$ or $P \le 0$. Let the (real) eigenvalues of PA be $\sigma_1, \sigma_2, \ldots, \sigma_N$. The GRF method converges if and only if

$$-2 < \sigma_i < 0, \qquad i = 1, 2, \ldots, N \tag{3.52}$$

Moreover, if (3.52) holds then either A is positive definite and $P \le 0$ or else A is negative definite and $P \ge 0$.

Proof. If $P \ge 0$, then the eigenvalues σ_i of PA are the same as those of $P^{1/2}AP^{1/2}$ and hence are real. Since $R_P = I + PA$, the eigenvalues $1 + \sigma_i$ of R_P will be less than unity in modulus if and only if (3.52) holds. In particular, this implies, by Corollary 2-2.8, that A is negative definite. A similar argument holds if $P \le 0$, and the theorem follows.

Corollary 3.9. Under the hypotheses of Theorem 3.8, the RF method converges if and only if either A is positive definite and

$$-2/S(A) < p < 0 \tag{3.53}$$

or else A is negative definite and

$$0 < p < 2/S(A) \tag{3.54}$$

We remark that if A is symmetric and nonsingular, but not necessarily either positive definite or negative definite, then $S(R_p R_{-p}) < 1$ if p lies in the range

$$-2^{1/2}/S(A) < p < 2^{1/2}/S(A) \tag{3.55}$$

This amounts to using the RF method with p and $-p$ on alternate iterations. (See also an analogous procedure based on the J method described earlier in this section.)

4.4. THE SOR METHOD WITH VARYING RELAXATION FACTORS

We now study the convergence of the SOR method when the relaxation factor varies from iteration to iteration. Our treatment is based on that given by Ostrowski [1954] (see also Wachspress [1966]).

Theorem 4.1. If A is a positive definite matrix, then the SOR method based on the use of the relaxation factors $\omega_1, \omega_2, \ldots$, converges provided at least one of the following conditions holds:

(a) for some $\varepsilon > 0$ we have

$$\varepsilon \leq \omega_i \leq 2 - \varepsilon, \qquad i = 1, 2, \ldots \tag{4.1}$$

for all i sufficiently large.

(b) $0 \leq \omega_i \leq 2$ for all i sufficiently large and the series

$$\sum_{i=1}^{\infty} \omega_i(2 - \omega_i) \tag{4.2}$$

diverges.

We first prove

Lemma 4.2. If A is a positive definite matrix, then for any H we have

$$(A^{1/2}HA^{-1/2})(A^{1/2}HA^{-1/2})^{\mathrm{T}} = Z((\hat{A})^{1/2}\tilde{H}(\hat{A})^{-1/2})((\hat{A})^{1/2}\tilde{H}(\hat{A})^{-1/2})^{\mathrm{T}}Z^{-1} \tag{4.3}$$

where $\hat{A}$, $\hat{H}$, and $\tilde{H}$ are defined by (3.6) and (3.16), and where

$$Z = A^{-1/2}D^{1/2}(\hat{A})^{1/2} \tag{4.4}$$

Moreover,

$$\| H \|_{A^{1/2}} = \| \hat{H} \|_{(\hat{A})^{1/2}} = [S(\tilde{H}(A)^{-1}\tilde{H}^{\mathrm{T}}A)]^{1/2} \tag{4.5}$$

and

$$\| \mathscr{L}_{\omega}[A] \|_{A^{1/2}} = \| \mathscr{L}_{\omega}[\hat{A}] \|_{(\hat{A})^{1/2}} \tag{4.6}$$

Proof. The right member of (4.3) is

$$\begin{aligned} A^{-1/2}D^{1/2}\hat{A}\tilde{H}(\hat{A})^{-1}\tilde{H}^{\mathrm{T}}D^{-1/2}A^{1/2} &= A^{1/2}HA^{-1}H^{\mathrm{T}}A^{1/2} \\ &= (A^{1/2}HA^{-1/2})(A^{1/2}HA^{-1/2})^{\mathrm{T}} \end{aligned}$$

hence (4.3) holds. Moreover,

$$[(\hat{A})^{1/2}\tilde{H}(\hat{A})^{-1/2}][(\hat{A})^{1/2}\tilde{H}(\hat{A})^{-1/2}]^{\mathrm{T}} = (\hat{A})^{1/2}\tilde{H}(\hat{A})^{-1}\tilde{H}^{\mathrm{T}}(\hat{A})^{1/2}$$

which is similar to $\tilde{H}(\hat{A})^{-1}\tilde{H}^{\mathrm{T}}\hat{A}$ and (4.5) follows. From Theorem 3.5 we have $\tilde{\mathscr{L}}_{\omega}[A] = \mathscr{L}_{\omega}[\hat{A}]$ and (4.6) follows.

If we let $\varepsilon^{(n)} = u^{(n)} - \bar{u}$, where $\bar{u}$ is the exact solution of (3-1.1), then we have $\varepsilon^{(n)} = \mathscr{G}_n\varepsilon^{(0)}$ where

$$\mathscr{G}_n = \mathscr{G}_n[A] = \prod_{i=n}^{1} \mathscr{L}_{\omega_i}[A] \tag{4.7}$$

Evidently $\varepsilon^{(n)} \to 0$ as $n \to 0$ provided $\mathscr{G}_n \to 0$. By Theorem 2-4.2 we can prove convergence if we can show that

$$\lim_{n\to\infty} \| \mathscr{G}_n \|_{A^{1/2}} = 0 \tag{4.8}$$

By Lemma 4.2 and Theorem 3.5 it is sufficient to consider the case where diag $A = I$.

For any positive definite matrix A and any H we define

$$H' = A^{1/2}HA^{-1/2} \tag{4.9}$$

By (3-3.23) we have, assuming diag $A = I$,

$$\mathscr{L}_{\omega}' = A^{1/2}(I - \omega(I - \omega L)^{-1}A)A^{-1/2} = I - \omega A^{1/2}(I - \omega L)^{-1}A^{1/2} \tag{4.10}$$

and hence, after some calculation, we have

$$\mathscr{L}_{\omega}'(\mathscr{L}_{\omega}')^{\mathrm{T}} = I - \omega(2 - \omega)[A^{1/2}(I - \omega L)^{-1}][A^{1/2}(I - \omega L)^{-1}]^{\mathrm{T}} \tag{4.11}$$

If $0 < \omega < 2$, $\mathscr{L}_\omega'(\mathscr{L}_\omega')^T$ is positive definite and $\| \mathscr{L}_\omega \|_{A^{1/2}} < 1$. We have already shown this (Theorem 3.1).

Let $\nu(\omega)$ denote the smallest eigenvalue of the positive definite matrix $P = A^{1/2}(I - \omega L)^{-1}(A^{1/2}(I - \omega L)^{-1})^T$. Since the elements of P are continuous functions of ω it follows from Theorem 2-1.10 that $\nu(\omega)$ is a continuous function of ω. Hence $\nu(\omega)$ must have a minimum value, say α, in the interval $0 \le \omega \le 2$ and $\alpha > 0$. Therefore, if $0 \le \omega \le 2$, we have

$$\| \mathscr{L}_\omega \|_{A^{1/2}}^2 \le 1 - \omega(2 - \omega)\alpha \tag{4.12}$$

For the sake of simplicity, we assume that each ω_i lies in the interval $0 \le \omega_i \le 2$. Evidently, from (4.7), we have

$$\| \mathscr{G}_n \|_{A^{1/2}} \le \prod_{i=1}^{n} \| \mathscr{L}_{\omega_i} \|_{A^{1/2}} \le \prod_{i=1}^{n} (1 - \omega_i(2 - \omega_i)\alpha) \tag{4.13}$$

and

$$-\log \| \mathscr{G}_n \|_{A^{1/2}} \ge - \sum_{i=1}^{\infty} \log(1 - \omega_i(2 - \omega_i)\alpha) \ge \sum_{i=1}^{\infty} \omega_i(2 - \omega_i)\alpha \tag{4.14}$$

Clearly, (4.8) holds if either (a) or (b) holds, and the proof of Theorem 4.1 is complete.

Further discussion of various choices of the ω_i will be given in Chapter 10. One can choose a single ω, say ω_b, such that

$$S(\mathscr{G}_m) \ge S(\mathscr{L}_{\omega_b}^m) \tag{4.15}$$

for any choice $\omega_1, \omega_2, \ldots, \omega_m$. However, other choices of the ω_i are better if one wishes to minimize certain norms of $\mathscr{G}_m$.

4.5. L-MATRICES AND RELATED MATRICES

We now consider the convergence properties of the J, JOR, GS, and SOR methods when A is an L-matrix. We first prove the following theorem which is an extension of the results of Stein and Rosenberg [1948].

Theorem 5.1. If A is an L-matrix and if $0 < \omega \le 1$, then

(a) $S(B) < 1$ if and only if $S(\mathscr{L}_\omega) < 1$.

(b) $S(B) < 1$ (and $S(\mathscr{L}_\omega) < 1$) if and only if A is an M-matrix;

if $S(B) < 1$, then

$$S(\mathscr{L}_\omega) \le 1 - \omega + \omega S(B) \tag{5.1}$$

(c) if $S(B) \ge 1$ and $S(\mathscr{L}_\omega) \ge 1$, then

$$S(\mathscr{L}_\omega) \ge 1 - \omega + \omega S(B) \ge 1 \tag{5.2}$$

Proof. Since L is a strictly lower triangular matrix, then $L^N = 0$ and, since A is an L-matrix and $0 < \omega \le 1$, we have

$$(I - \omega L)^{-1} = I + \omega L + \omega^2 L^2 + \cdots + \omega^{N-1} L^{N-1} \ge 0 \tag{5.3}$$

and

$$\mathscr{L}_\omega = (I - \omega L)^{-1}(\omega U + (1 - \omega)I) \ge 0 \tag{5.4}$$

Let $\bar{\lambda} = S(\mathscr{L}_\omega)$ and $\bar{\mu} = S(B)$. By Theorem 2-1.17, $\bar{\lambda}$ is an eigenvalue of $\mathscr{L}_\omega$ and for some $w \ne 0$ we have $\mathscr{L}_\omega w = \bar{\lambda} w$ and

$$(\bar{\lambda} L + U)w = [(\bar{\lambda} + \omega - 1)/\omega]w \tag{5.5}$$

Since $(\bar{\lambda} + \omega - 1)/\omega$ is an eigenvalue of $\bar{\lambda} L + U$ we have

$$\bar{\lambda} + \omega - 1 \le \omega S(\bar{\lambda} L + U) \tag{5.6}$$

If $\bar{\lambda} \le 1$, then $S(\bar{\lambda} L + U) \le S(L + U) = \bar{\mu}$, by Theorem 2-1.16, and

$$\bar{\lambda} \le \omega \bar{\mu} + 1 - \omega \tag{5.7}$$

On the other hand, if $\bar{\lambda} \ge 1$, then

$$(\bar{\lambda} + \omega - 1)/\omega \le S(\bar{\lambda} L + U) \le S(\bar{\lambda} L + \bar{\lambda} U) = \bar{\lambda}\bar{\mu} \tag{5.8}$$

and

$$\bar{\mu} \ge (\bar{\lambda} + \omega - 1)/(\omega\bar{\lambda}) = 1 + [(1 - \omega)(\bar{\lambda} - 1)/(\omega\bar{\lambda})] \ge 1 \tag{5.9}$$

We have thus shown

(i) if $\bar{\lambda} \le 1$, then $\bar{\lambda} \le \omega\bar{\mu} + 1 - \omega$.
(ii) if $\bar{\lambda} \ge 1$, then $\bar{\mu} \ge 1$.

which implies

(iii) if $\bar{\mu} < 1$, then $\bar{\lambda} < 1$.

Since $B \geq 0$, it follows by Theorem 2-1.17 that $\bar{\mu}$ is an eigenvalue of B. Therefore, for some $v \neq 0$ we have $Bv = \bar{\mu}v$ and

$$Qv = (1 - \omega + \omega\bar{\mu})v \tag{5.10}$$

where

$$\begin{aligned} Q &= (I - \alpha L)^{-1}(\omega U + (1 - \omega)I) \\ \alpha &= \omega/(1 - \omega + \omega\bar{\mu}) \end{aligned} \tag{5.11}$$

Therefore,

$$1 - \omega + \omega\bar{\mu} \leq S(Q) \tag{5.12}$$

But if $\bar{\mu} \geq 1$, then, since $\alpha \leq \omega$, we have

$$\begin{aligned} (I - \alpha L)^{-1} &= I + \alpha L + \cdots + \alpha^{N-1}L^{N-1} \\ &\leq I + \omega L + \cdots + \omega^{N-1}L^{N-1} = (I - \omega L)^{-1} \end{aligned} \tag{5.13}$$

We have $Q \leq \mathscr{L}_\omega$ and hence

$$1 - \omega + \omega\bar{\mu} \leq S(Q) \leq S(\mathscr{L}_\omega) = \bar{\lambda} \tag{5.14}$$

We have thus shown

(iv) if $\bar{\mu} \geq 1$, then $\bar{\lambda} \geq 1 - \omega + \omega\bar{\mu} \geq 1$.

This implies

(v) if $\bar{\lambda} < 1$, then $\bar{\mu} < 1$.

By (iii) and (v) we have (a). By (i) and Theorem 2-7.2 we have (b). By (iv) we have (c), and the proof of Theorem 5.1 is complete.

Corollary 5.2. If A is an L-matrix, then

(a) $S(B) < 1$ if and only if $S(\mathscr{L}) < 1$.

(b) $S(B) < 1$ and $S(\mathscr{L}) < 1$ if and only if A is an M-matrix; if $S(B) < 1$, then

$$S(\mathscr{L}) \leq S(B) \tag{5.15}$$

(c) if $S(B) \geq 1$ and $S(\mathscr{L}) \geq 1$, then

$$S(\mathscr{L}) \geq S(B) \tag{5.16}$$

Corollary 5.3. If A is a Stieltjes matrix, then the J method converges and the JOR method converges for $0 < \omega \leq 1$.

Proof. Since a Stieltjes matrix is a positive definite L-matrix and since the GS method converges for any positive definite matrix, by Theorem 3.1, then, by Theorem 5.1, the J method converges. By Theorem 1.1 the JOR method converges for $0 < \omega \leq 1$.

The following theorem will prove useful in our later work.

Theorem 5.4. If A is an L-matrix, then

$$S(\bar{\lambda}L + U) = \bar{\lambda} \tag{5.17}$$

where

$$\bar{\lambda} = S(\mathscr{L}) \tag{5.18}$$

Proof. Let $\alpha = S(\bar{\lambda}L + U)$. As in the proof of Theorem 5.1, we can show that $\bar{\lambda}$ is an eigenvalue of $\bar{\lambda}L + U$, and hence $\bar{\lambda} \leq \alpha$. By Theorem 2-1.17, α is an eigenvalue of $\bar{\lambda}L + U$ and, for some $w \neq 0$,

$$(\bar{\lambda}L + U)w = \alpha w$$

and

$$(I - (\bar{\lambda}/\alpha)L)^{-1}Uw = \alpha w \tag{5.19}$$

Since $\alpha \geq \bar{\lambda}$ we have

$$\begin{aligned}(I - (\bar{\lambda}/\alpha)L)^{-1} &= I + (\bar{\lambda}/\alpha)L + (\bar{\lambda}/\alpha)^2L^2 + \cdots + (\bar{\lambda}/\alpha)^{N-1}L^{N-1} \\ &\leq I + L + L^2 + \cdots + L^{N-1} = (I - L)^{-1}\end{aligned} \tag{5.20}$$

Therefore, $(I - (\bar{\lambda}/\alpha)L)^{-1}U \leq (I - L)^{-1}U = \mathscr{L}$ and by Theorem 2-1.16 we have

$$\alpha \leq S[(I - (\bar{\lambda}/\alpha)L)^{-1}U] \leq \bar{\lambda} \tag{5.21}$$

Therefore $\alpha = \bar{\lambda}$ and (5.17) follows.

Let us now consider iterative methods based on "regular splittings" (see Section 3.4). We represent A in the form

$$A = Q - R \tag{5.22}$$

where Q is a monotone matrix (see Section 2.7) and $R \geq 0$. If A is an L-matrix, then the J, GS, and SOR methods, the latter for $0 < \omega \leq 1$, correspond to regular splittings. In general, corresponding to the splitting (5.22) we have the completely consistent iterative method

$$u^{(n+1)} = Q^{-1}Ru^{(n)} + Q^{-1}b \tag{5.23}$$

We now prove

Theorem 5.5. Let $A = Q - R$ be a regular splitting of the nonsingular matrix A. Then A is a monotone matrix if and only if $S(Q^{-1}R) < 1$. Moreover, if A is monotone, then

$$S(Q^{-1}R) = S(A^{-1}R)/[1 + S(A^{-1}R)] \tag{5.24}$$

Proof. If $S(Q^{-1}R) < 1$, then we have

$$A^{-1} = (Q - R)^{-1} = (I - Q^{-1}R)^{-1}Q^{-1} = (I + (Q^{-1}R) + (Q^{-1}R)^2 + \cdots)Q^{-1} \geq 0 \tag{5.25}$$

the series converging by Theorem 2-4.4. Hence A is monotone. If, on the other hand, A is monotone, then since $I - Q^{-1}R = Q^{-1}A$ it follows that $I - Q^{-1}R$ is nonsingular. If $S(Q^{-1}R) = 1$, then, by Theorem 2-1.17, unity would be an eigenvalue of $Q^{-1}R$ and $I - Q^{-1}R$ would be singular. Therefore $S(Q^{-1}R) \neq 1$.

Since $Q^{-1}R \geq 0$ it follows from Theorem 2-1.17 that for some vector $v \geq 0$ we have

$$Q^{-1}Rv = S(Q^{-1}R)v \tag{5.26}$$

and hence, since

$$A^{-1}R = (I - Q^{-1}R)^{-1}Q^{-1}R \tag{5.27}$$

it follows that

$$A^{-1}Rv = [S(Q^{-1}R)/[1 - S(Q^{-1}R)]]v \tag{5.28}$$

If $S(Q^{-1}R) > 1$, then $A^{-1}Rv$ has negative components which is impossible since $A^{-1} \geq 0$ and $R \geq 0$. This proves that

$$S(Q^{-1}R) < 1 \tag{5.29}$$

From (5.28) it follows that if $A^{-1} \geq 0$, then

$$S(A^{-1}R) \geq S(Q^{-1}R)/[1 - S(Q^{-1}R)] \tag{5.30}$$

and hence

$$S(Q^{-1}R) \leq S(A^{-1}R)/[1 + S(A^{-1}R)] \tag{5.31}$$

On the other hand, since $A^{-1}R \geq 0$ it follows from Theorem 2-1.17 that for some $w \neq 0$ we have

$$A^{-1}Rw = S(A^{-1}R)w \tag{5.32}$$

But since

$$Q^{-1}R = (A + R)^{-1}R = (I + A^{-1}R)^{-1}A^{-1}R \tag{5.33}$$

we have

$$Q^{-1}Rw = [S(A^{-1}R)/[1 + S(A^{-1}R)]]w \tag{5.34}$$

and, hence,

$$S(Q^{-1}R) \geq S(A^{-1}R)/[1 + S(A^{-1}R)] \tag{5.35}$$

The result (5.24) follows from (5.31) and (5.35) and the proof of Theorem 5.5 is complete.

From Theorem 5.5 and from the fact that $x(1 + x)^{-1}$ is an increasing function of x for $x \geq 0$ we have

Corollary 5.6. Let A be a monotone matrix and let $A = Q_1 - R_1$ and $A = Q_2 - R_2$ be two regular splittings of A. If $R_2 \leq R_1$, then

$$S(Q_2^{-1}R_2) \leq S(Q_1^{-1}R_1) \tag{5.36}$$

The fact that $S(\mathscr{L}) \leq S(B)$ if A is an M-matrix follows directly from Corollary 5.6 if we let $Q_2 = D - C_L$, $Q_1 = D$. Since a monotone L-matrix is an M-matrix, it follows from Theorem 5.5 that $S(B) < 1$ if and only if A is an M-matrix. Thus we have an alternative proof of (b) of Corollary 5.2.

We now prove for the case where A is a monotone matrix

Theorem 5.7. If A is an L-matrix such that $S(B) < 1$ and if $0 < \omega_1 \leq \omega_2 \leq 1$, then

$$S(\mathscr{L}_{\omega_2}) \leq S(\mathscr{L}_{\omega_1}) < 1 \tag{5.37}$$

Proof. By Theorem 2-7.2, since $S(B) < 1$, and A is an L-matrix, then A is an M-matrix. As in Section 3.3, we have

$$\mathscr{L}_\omega = Q(\omega)^{-1}R(\omega) \tag{5.38}$$

where

$$Q(\omega) = \omega^{-1}D - C_L, \qquad R(\omega) = (\omega^{-1} - 1)D + C_U \tag{5.39}$$

But since $Q(\omega) = \omega^{-1}D(I - D^{-1}C_L)$ it follows that

$$Q(\omega)^{-1} = (I + (D^{-1}C_L) + (D^{-1}C_L)^2 + \cdots + (D^{-1}C_L)^{N-1})\omega D^{-1} \geq 0 \tag{5.40}$$

Hence $Q(\omega)$ is monotone. Moreover, if $0 < \omega \leq 1$, then $R(\omega) \geq 0$. Also, since $\omega^{-1} - 1$ is a decreasing function of ω for $0 \leq \omega \leq 1$, it follows that if $0 < \omega_1 \leq \omega_2 \leq 1$, then

$$R(\omega_2) \leq R(\omega_1) \tag{5.41}$$

The result (5.37) now follows from Corollary 5.6.

As another consequence of Corollary 5.6 we have

Theorem 5.8. Let A be an M-matrix and let $A = Q_1 - R_1$ and $A = Q_2 - R_2$ be two splittings, where Q_1 and Q_2 are formed from A by replacing some of the off-diagonal elements of A by zeros. The splittings are regular and if $R_1 \leq R_2$, then

$$S(Q_1^{-1}R_1) \leq S(Q_2^{-1}R_2) \leq S(B) \tag{5.42}$$

Proof. To show that the splitting $A = Q_1 - R_1$ is regular we need only show that Q_1 is monotone. Let $D = \operatorname{diag} A$ and $C = D - A$. Then $Q_1 = D - C_1$ where $0 \leq C_1 \leq C$. Evidently $D^{-1}C_1 \leq D^{-1}C$ and hence $S(D^{-1}C_1) \leq S(D^{-1}C)$ by Theorem 2-1.16. Moreover, since A is an M-matrix, it follows from Theorem 2-7.2 that $S(D^{-1}C) < 1$. Therefore, $S(D^{-1}C_1) < 1$ and, by Theorem 2-7.2, Q_1 is monotone. Thus, the splitting $A = Q_1 - R_1$ is regular as are the splittings $A = Q_2 - R_2$ and $A = D - C$. Since $R_1 \leq R_2 \leq C$, the result (5.42) follows from Corollary 5.6.

If A is an M-matrix, then, by Theorem 5.1, $S(B) < 1$ and, by Theorem 5.7, $S(\mathscr{L}_\omega)$ is a nonincreasing function in the range $0 \leq \omega \leq 1$. Moreover, $S(\mathscr{L}) < 1$. By the continuity of $S(\mathscr{L}_\omega)$ as a function of ω, we must have $S(\mathscr{L}_\omega) < 1$ for $0 < \omega \leq \bar{\omega}$ for some $\bar{\omega} > 1$. A lower bound for $\bar{\omega}$ is given by the following theorem of Kahan [1958].

Theorem 5.9. If A is an M-matrix and if

$$0 < \omega < 2/[1 + S(B)] \tag{5.43}$$

then $S(\mathscr{L}_\omega) < 1$.

Proof. For $0 < \omega \leq 1$, the convergence follows from Theorem 5.1. If $\omega \geq 1$, then by (5.3) the matrix

$$T_\omega = (I - \omega L)^{-1}[\omega U + (\omega - 1)I] \tag{5.44}$$

is nonnegative, and

$$|\mathscr{L}_\omega| \le T_\omega \tag{5.45}$$

Let $\bar{\gamma} = S(T_\omega)$. Since $T_\omega \ge 0$, then, by Theorem 2-1.17, for some $v \ne 0$ we have $T_\omega v = \bar{\gamma} v$ and

$$(\omega U + \omega\bar{\gamma}L)v = (\bar{\gamma} + 1 - \omega)v \tag{5.46}$$

Hence,

$$\bar{\gamma} + 1 - \omega \le S(\omega U + \omega\bar{\gamma}L) \tag{5.47}$$

and if $\bar{\gamma} \ge 1$, then by Theorem 2-1.16

$$\bar{\gamma} + 1 - \omega \le \omega\bar{\gamma}S(B) \tag{5.48}$$

or

$$\omega \ge (1 + \bar{\gamma})/[1 + \bar{\gamma}S(B)] \ge 2/[1 + S(B)] \tag{5.49}$$

Therefore, if (5.43) holds, then we must have $\bar{\gamma} < 1$. By (5.45) and Theorem 2-1.16, it follows that $S(\mathscr{L}_\omega) \le \bar{\gamma} < 1$ and the theorem is proved.

4.6. RATES OF CONVERGENCE OF THE J AND GS METHODS FOR THE MODEL PROBLEM

We now consider the application of the J and GS methods to the model problem. As shown in Section 2.8, the matrix A is a Stieltjes matrix and hence the convergence of the J and GS methods follows at once from Corollary 5.2. By Corollary 5.2 we also know that the GS method converges faster than the J method; more precisely, we know that $S(\mathscr{L}) \le S(B)$. As we now show, the rate of convergence of each method is very slow and approaches zero very rapidly as the mesh size $h \to 0$. We shall also show, in Chapter 6, that this behavior of the convergence rates holds for nonrectangular as well as rectangular regions.

Instead of the model problem, we consider the case where R is the rectangle $0 < x < a$, $0 < y < b$ where for some $h > 0$ the quantities a/h and b/h are integers. (The numbers a and b are said to be *commensurable*.) We also assume that $G(x, y) \equiv 0$. For any value of h such that a/h and b/h are integers, say, I and J, respectively, we have the difference equation

$$4u(x, y) - u(x + h, y) - u(x, y + h) - u(x - h, y) - u(x, y - h) = 0 \tag{6.1}$$

for each point (x, y) of R_h. On the boundary S_h the function $u(x, y)$ is required to satisfy

$$u(x, y) = g(x, y) \tag{6.2}$$

where $g(x, y)$ is a given function defined on S.

For the J method, given $u^{(n)}(x, y)$ for all $(x, y) \in R_h + S_h$ where on S_h

$$u^{(n)}(x, y) = g(x, y) \tag{6.3}$$

we determine $u^{(n+1)}(x, y)$ by

$$\begin{aligned} u^{(n+1)}(x, y) = {} & \tfrac{1}{4}u^{(n)}(x + h, y) + \tfrac{1}{4}u^{(n)}(x, y + h) + \tfrac{1}{4}u^{(n)}(x - h, y) \\ & + \tfrac{1}{4}u^{(n)}(x, y - h) \end{aligned} \tag{6.4}$$

if $(x, y) \in R_h$ and

$$u^{(n+1)}(x, y) = g(x, y) \tag{6.5}$$

if $(x, y) \in S_h$. Let $\bar{u}(x, y)$ be the exact solution of (6.1) and (6.2) and let

$$\varepsilon^{(n)}(x, y) = u^{(n)}(x, y) - \bar{u}(x, y) \tag{6.6}$$

Then, by (6.3)–(6.5) we have

$$\begin{aligned} \varepsilon^{(n+1)}(x, y) = {} & \tfrac{1}{4}\varepsilon^{(n)}(x + h, y) + \tfrac{1}{4}\varepsilon^{(n)}(x, y + h) + \tfrac{1}{4}\varepsilon^{(n)}(x - h, y) \\ & + \tfrac{1}{4}\varepsilon^{(n)}(x, y - h) \end{aligned} \tag{6.7}$$

for $(x, y) \in R_h$ and

$$\varepsilon^{(n+1)}(x, y) = 0 \tag{6.8}$$

for $(x, y) \in S_h$. For any function $u(x, y)$ defined on $R_h + S_h$ and satisfying (6.2) the error

$$\varepsilon(x, y) = u(x, y) - \bar{u}(x, y) \tag{6.9}$$

corresponds to a vector in E^N, where $N = (I - 1)(J - 1)$ is the number of points in R_h. Moreover, if the function $\varepsilon^{(n)}(x, y)$ corresponds to the vector $\varepsilon^{(n)}$, then the function $\varepsilon^{(n+1)}(x, y)$ defined by (6.7) and (6.8) corresponds to the vector $\varepsilon^{(n+1)} = B\varepsilon^{(n)}$ where B is the matrix corresponding to the Jacobi method.

To determine an eigenvalue μ of B we seek a function $v(x, y)$ defined on $R_h + S_h$, such that

$$\mu v(x, y) = \tfrac{1}{4}v(x + h, y) + \tfrac{1}{4}v(x, y + h) + \tfrac{1}{4}v(x - h, y) + \tfrac{1}{4}v(x, y - h) \tag{6.10}$$

on R_h and

$$v(x, y) = 0 \tag{6.11}$$

on S_h. One can verify directly that for any nonzero integers p and q the conditions (6.10) and (6.11) are satisfied if we let

$$v(x, y) = v_{p,q}(x, y) = \sin \frac{p\pi x}{a} \sin \frac{q\pi y}{b} \tag{6.12}$$

Moreover,

$$\mu = \mu_{p,q} = \frac{1}{2}\left(\cos \frac{p\pi h}{a} + \cos \frac{q\pi h}{b}\right) = \frac{1}{2}\left(\cos \frac{p\pi}{I} + \cos \frac{q\pi}{J}\right) \tag{6.13}$$

It is easy to show that the $(I - 1)(J - 1)$ vectors $v^{(p,q)}$ which correspond to the functions $v_{p,q}(x, y)$ are linearly independent for $p = 1, 2, \ldots, I - 1$, $q = 1, 2, \ldots, J - 1$. Indeed, the $v^{(p,q)}$ are nonvanishing, pairwise orthogonal vectors since for $1 \le p \le I - 1$, $1 \le p' \le I - 1$, $1 \le q \le J - 1$, $1 \le q' \le J - 1$ we have

$$\sum_{(x,y)\in R_h} v_{p,q}(x, y)v_{p',q'}(x, y) = 0 \tag{6.14}$$

unless $p = p'$ and $q = q'$ in which case the equality does not hold. To see this, we need only show that if $1 \le p \le I - 1$, $1 \le p' \le I - 1$, we have

$$\sum_{k=1}^{I-1} \sin \frac{p\pi k}{I} \sin \frac{p'\pi k}{I} = 0 \tag{6.15}$$

if $p \ne p'$ and the equality does not hold if $p = p'$. Thus the vectors $v^{(p,q)}$ being pairwise orthogonal are linearly independent and hence span E^N. We have, then, determined all of the eigenvectors and all of the eigenvalues of B. Since $|\mu_{p,q}|$ is maximized for $p = q = 1$, we have

$$\bar{\mu} = S(B) = \mu_{1,1} = \tfrac{1}{2}[\cos(\pi h/a) + \cos(\pi h/b)] = \tfrac{1}{2}[\cos(\pi/I) + \cos(\pi/J)] \tag{6.16}$$

Evidently,

$$S(B) = \bar{\mu} = 1 - \tfrac{1}{4}[(\pi^2/a^2) + (\pi^2/b^2)]h^2 + O(h^4) \tag{6.17}$$

so that the asymptotic rate of convergence is given by

$$R(B) = -\log \bar{\mu} = \tfrac{1}{4}[(\pi^2/a^2) + (\pi^2/b^2)]h^2 + O(h^4) \tag{6.18}$$

Thus, from Section 3.7 we would expect that the number of iterations required to achieve a specified degree of convergence increases with h^{-2} as $h \to 0$.

Let us now carry out a similar analysis for the GS method. We use the "natural ordering" of the points of R_h, where (x', y') follows (x, y) if $y' > y$ or else $y' = y$ and $x' > x$. Thus the points of R_h are treated row by row from left to right starting at the bottom row.

With this ordering, the iteration formula is given by

$$u^{(n+1)}(x, y) = \tfrac{1}{4}u^{(n)}(x + h, y) + \tfrac{1}{4}u^{(n)}(x, y + h) + \tfrac{1}{4}u^{(n+1)}(x - h, y) + \tfrac{1}{4}u^{(n+1)}(x, y - h) \tag{6.19}$$

for $(x, y) \in R_h$, and

$$u^{(n+1)}(x, y) = g(x, y) \tag{6.20}$$

for $(x, y) \in S_h$. Again defining $\varepsilon^{(n)}(x, y)$ by (6.6) we have

$$\varepsilon^{(n+1)}(x, y) = \tfrac{1}{4}\varepsilon^{(n)}(x + h, y) + \tfrac{1}{4}\varepsilon^{(n)}(x, y + h) + \tfrac{1}{4}\varepsilon^{(n+1)}(x - h, y) + \tfrac{1}{4}\varepsilon^{(n+1)}(x, y - h) \tag{6.21}$$

for $(x, y) \in R_h$ and

$$\varepsilon^{(n+1)}(x, y) = 0 \tag{6.22}$$

on S_h. As in the case of the J method, if the function $\varepsilon^{(n)}(x, y)$ corresponds to the vector $\varepsilon^{(n)}$, then the function $\varepsilon^{(n+1)}(x, y)$ corresponds to the vector $\varepsilon^{(n+1)}$. Moreover, (6.21) corresponds to the matrix equation

$$\varepsilon^{(n+1)} = L\varepsilon^{(n+1)} + U\varepsilon^{(n)} \tag{6.23}$$

where L and U are strictly lower and strictly upper triangular matrices, respectively, such that $L + U = B$. We seek to determine an eigenvalue λ and an eigenvector w of $\mathscr{L} = (I - L)^{-1}U$. But if $\mathscr{L}w = \lambda w$, then

$$\lambda w = \lambda Lw + Uw \tag{6.24}$$

This corresponds to the equation

$$\lambda w(x, y) = \tfrac{1}{4}w(x + h, y) + \tfrac{1}{4}w(x, y + h) + \tfrac{1}{4}\lambda w(x - h, y) + \tfrac{1}{4}\lambda w(x, y - h) \tag{6.25}$$

for $(x, y) \in R_h$, where

$$w(x, y) = 0 \tag{6.26}$$

on S_h. It can be verified† that if $v(x, y)$ and μ satisfy (6.10) and (6.11), then

$$w(x, y) = \lambda^{(x+y)/(2h)} v(x, y) \tag{6.27}$$

satisfies (6.25) and (6.26), where

$$\lambda = \mu^2 \tag{6.28}$$

Conversely, we can show that if $\lambda \neq 0$ and if λ and $w(x, y)$ satisfy (6.25) and (6.26), then

$$v(x, y) = \lambda^{-(x+y)/(2h)} w(x, y) \tag{6.29}$$

satisfies (6.10) and (6.11) where

$$\mu = \lambda^{1/2} \tag{6.30}$$

As a matter of fact, $\mu = \lambda^{1/2}$ and

$$\hat{v}(x, y) = (-\lambda)^{-(x+y)/2h} w(x, y) \tag{6.31}$$

also satisfy (6.10) and (6.11). In any case, the nonzero eigenvalues of $\mathscr{L}$ are given by

$$\lambda = \lambda_{p,q} = \mu_{p,q}^2 = \tfrac{1}{4}[\cos(p\pi/I) + \cos(q\pi/J)]^2$$
$$p = 1, 2, \ldots, I-1, \quad q = 1, 2, \ldots, J-1 \tag{6.32}$$

Moreover, we have

$$S(\mathscr{L}) = S(B)^2 = \tfrac{1}{4}[\cos(\pi/I) + \cos(\pi/J)]^2 \tag{6.33}$$

In Chapters 5 and 6 we shall show that $S(\mathscr{L}) = S(B)^2$ for a much wider class of problems.

For the model problem, we have $a = b = 1$ and $I = J = h^{-1}$. Therefore,

$$\begin{aligned} \bar{\mu} &= S(B) = \cos \pi h \\ \bar{\lambda} &= S(\mathscr{L}) = \cos^2 \pi h \end{aligned} \tag{6.34}$$

† The eigenvectors of B and $\mathscr{L}$ can be derived directly by the method of separation of variables.

Table 6.1 gives the results of some numerical experiments involving the model problem. The function $g(x, y)$ was taken to be identically zero so that $\bar{u}(x, y) \equiv 0$. The starting values were taken as unity for all $(x, y) \in R_h$. The observed number of iterations n_0 needed to reduce the maximum error to $\varrho = 10^{-6}$ is given for each method together with the theoretical value given by

$$n_T = (\log(1/\varrho))/R = (\log 10^6)/R \doteq 13.82/R \tag{6.35}$$

where R is the asymptotic rate of convergence of the method, i.e., $R = -\log \cos \pi h$ and $R = -2 \log \cos \pi h$ for the J and GS methods, respectively.

TABLE 6.1 Numerical Experiments with the Model Problem[a]

h	$\bar{\mu}$	$R(B)$	$R(\mathscr{L})$	n_O		n_T	
				J	GS	J	GS
$\frac{1}{5}$	0.8090	0.212	0.424	67	35	66	33
$\frac{1}{10}$	0.9511	0.0502	0.100	285	143	276	138
$\frac{1}{20}$	0.9877	0.0124	0.0248	1154	578	1116	558
$\frac{1}{40}$	0.9969	0.0031	0.0062	4631	2317	4475	2238

[a] n_O is the observed number of iterations; n_T is the theoretical number of iterations.

It can be seen that the observed and theoretical numbers of iterations increase rapidly as h decreases and behave like h^{-2}. The GS method is about twice as fast as the J method but is still very slow indeed. In Chapters 5 and 6 we shall show that by using the SOR method with a suitable ω one can greatly increase the rate of convergence.

SUPPLEMENTARY DISCUSSION

Section 4.2. Theorem 2.1 is a slight generalization of results of Geiringer [1949].

Section 4.3. A proof of Corollary 3.4 is given by Faddeev and Faddeeva [1963, p. 191]. Theorem 3.6 was proved by Reich [1949] for the case $\omega = 1$ and by Ostrowski [1954] for the more general case.

Section 4.5. Other extensions of the Stein–Rosenberg theory for L-matrices are given by Kahan [1958]. For example, Kahan gives a sharper result than Corollary 5.2. He shows that if $0 < S(B) < 1$, then $0 < S(B) < S(\mathscr{L}) < 1$, and if $1 < S(B)$, then $1 < S(B) < S(\mathscr{L})$. Theorem 5.1 with the statement (c) replaced by "If $S(B) \geq 1$, then $S(\mathscr{L}) \geq S(B)$ and $S(\mathscr{L}_\omega) \geq 1$" follows directly from Kahan's work.

Theorem 5.5 is a slight generalization of a result of Varga [1962, p. 89]. Theorem 5.7 was proved by Kahan [1958, p. 3.10]. Kahan also showed that there exists $\bar{\omega} \geq 1$ such that $S(\mathscr{L}_\omega)$ is decreasing for $0 < \omega \leq \bar{\omega}$.

Section 4.6. The determination of the eigenvectors of the GS method in terms of those of the J method is given by Young [1950]. A similar representation of the eigenvalues of $\mathscr{L}_\omega$ is also given.

EXERCISES

Section 4.1

1. Show that $\lambda = 0$ is an eigenvalue of $\mathscr{L}$.

2. If the eigenvalues of the J method are real and lie in the interval $-0.5 \leq \mu \leq 0.8$, find $S(B_{1/2})$.

3. If the JOR method converges, must ω lie in the interval $0 < \omega < 2$? Consider the special case where all eigenvalues of B are real and also the general case.

Section 4.2

1. Show that the JOR and SOR methods converge for $0 < \omega \leq 1$ when applied to a linear system with the matrix

$$\begin{pmatrix} 4i & -2 & -i & -1 \\ -3 & 4 & -i & 0 \\ 0 & -2i & 4 & -2 \\ 1+i & 0 & 1 & 4i \end{pmatrix}$$

Section 4.3

1. Prove that if A is symmetric, nonsingular, and has positive diagonal elements, and if ω satisfies (3.11) then $S(B_\omega B_{-\omega}) < 1$.

2. Consider the matrix

$$A = \begin{pmatrix} 4 & a \\ a & 1 \end{pmatrix}$$

where a is a real number. Let $D = \operatorname{diag} A$ and $\hat{A} = D^{-1/2} A D^{-1/2}$.

(a) Find the range of values of a such that A is positive definite and compare with the range of values of a such that $\hat{A}$ is positive definite.

(b) For the case $a = -1$, compute $B[A]$ and $B[\hat{A}]$, and show that

$$\tilde{B}[A] = D^{1/2} B[A] D^{-1/2} = B[\hat{A}]$$

(c) For the case $a = -1$, compute $\mathscr{L}_\omega[A]$ and $\mathscr{L}_\omega[\hat{A}]$ and show that

$$\tilde{\mathscr{L}}_\omega[A] = D^{1/2} \mathscr{L}_\omega[A] D^{-1/2} = \mathscr{L}_\omega[\hat{A}]$$

Hint: Note that

$$\begin{pmatrix} 1 & 0 \\ x & 1 \end{pmatrix}^{-1} = \begin{pmatrix} 1 & 0 \\ -x & 1 \end{pmatrix}$$

3. Consider the linear system $Au = b$ where b is given and where

$$A = \begin{pmatrix} 1 & -0.8 \\ -0.8 & 1 \end{pmatrix}$$

(a) Show that if $\omega = 1.25$, then

$$\mathscr{L}_\omega = \begin{pmatrix} -\frac{1}{4} & 1 \\ -\frac{1}{4} & \frac{3}{4} \end{pmatrix}$$

(b) Find the eigenvalues of $\mathscr{L}_{1.25}$ and the eigenvectors and principal vectors.

(c) If L and U are strictly lower and strictly upper triangular matrices, respectively, such that $L + U = I - A$, show that the eigenvalue λ and the eigenvector v found in (b) satisfy

$$\lambda = [\omega(v, Uv) + (1 - \omega)(v, v)]/[(v, v) - \omega(v, Lv)]$$

4. Consider the solution of the linear system $Au = b$ where

$$A = \begin{pmatrix} 1 & -2 \\ -2 & 1 \end{pmatrix}, \qquad b = \begin{pmatrix} -1 \\ -4 \end{pmatrix}$$

(a) Find the solution $\bar{u}$ and compute $Q(\bar{u})$ where $A(\bar{u}) = \frac{1}{2}(u, Au) - (u, b)$.

(b) Starting with the vector

$$u^{(0)} = \begin{pmatrix} 4 \\ 1 \end{pmatrix}$$

carry out one iteration with the Gauss–Seidel method. Compute $Q(u^{(1)})$ and $Q(u^{(0)})$ and verify that

$$Q(u^{(1)}) - Q(u^{(0)}) = -\tfrac{1}{2}\omega(2-\omega)(\varepsilon^{(0)}, P\varepsilon^{(0)})$$

where $\varepsilon^{(0)} = u^{(0)} - \bar{u}$ and

$$P = A(I - \omega U)^{-1}(I - \omega L)^{-1}A$$

and L and U are lower and upper triangular matrices, respectively, such that $L + U = I - A$.

(c) Will the Gauss–Seidel method converge for the given starting value? Give reasons.

(d) Find a starting vector, other than $\bar{u}$, such that the Gauss–Seidel method converges.

5. Consider the linear system $Au = b$ where

$$A = \begin{pmatrix} 1 & -\frac{1}{2} \\ -\frac{1}{2} & 1 \end{pmatrix}, \qquad b = \begin{pmatrix} 3 \\ 0 \end{pmatrix}$$

and the related quadratic form

$$A(u) = \tfrac{1}{2}(u, Au) - (u, b)$$

(a) Compute $Q(\bar{u})$ where $\bar{u}$ is the exact solution of the system.

(b) Letting

$$u^{(0)} = \begin{pmatrix} 0 \\ 0 \end{pmatrix}$$

compute $u^{(1)}$ by the SOR method with $\omega = 1.5$.

(c) Compute $Q(u^{(0)})$ and $Q(u^{(1)})$.

(d) Show that

$$Q(u^{(1)}) - Q(\bar{u}) = \tfrac{1}{2}(\varepsilon^{(1)}, A\varepsilon^{(1)})$$

where

$$\varepsilon^{(1)} = u^{(1)} - \bar{u}$$

6. Show that if $u^{(0)}, u^{(1)}, \ldots$ is determined by the SOR method for the system $Au = b$, where A is a positive definite matrix, then

$$(\varepsilon^{(n+1)}, A\varepsilon^{(n+1)}) - (\varepsilon^{(n)}, A\varepsilon^{(n)}) = -\| \varepsilon^{(n)} \|_{P^{1/2}}^2$$

where

$$P = A - \mathscr{L}_\omega{}^{\mathrm{T}} A \mathscr{L}_\omega = \omega(2-\omega)[A(I-\omega U)^{-1}][A(I-\omega U)^{-1}]^{\mathrm{T}}$$

and

$$\varepsilon^{(n)} = u^{(n)} - \bar{u}$$

Here $\bar{u} = A^{-1}b$. Use this result to give an alternative convergence proof of the SOR method for the case where A is positive definite and $0 < \omega < 2$ and also to show that $\| \varepsilon^{(n)} \|_{A^{1/2}} \to 0$ as $n \to \infty$. Assume diag $A = I$.

7. Consider the linear system $Au = b$ where

$$A = \begin{pmatrix} 1 & a & a \\ a & 1 & a \\ a & a & 1 \end{pmatrix}$$

(a) For what values of a is A positive definite?

(b) For what real values of a do the following methods converge:

J method,
JOR method (give values of ω for which convergence holds and draw a graph of ω versus a),
SOR method (give values of ω for which convergence holds and draw a graph of ω versus a),
RF method (give values of p for which convergence holds and draw a graph of ω versus a).

(c) For what values of a (real or complex) does the J method converge? The GS method?

8. For the SSOR method (see Exercise 7, Section 3.3) show that if A is positive definite and $0 < \omega < 2$, then $S(G) \leq \| G \|_{A^{1/2}} < 1$ and all eigenvalues of G are real and positive.

Section 4.4

1. Carry out the details of the proof of Theorem 4.1 for the case where not all ω_i lie in the interval $0 \leq \omega_i \leq 2$.

2. Determine whether or not the SOR method converges, when A is positive definite, with the parameter choices

(a) $\omega_i = 2^{-i}$,
(b) $\omega_i = (-1)^i 2^{-i}$,
(c) $\omega_i = 1/i$,
(d) $\omega_1 = 5,\ \omega_i = 2^{-i},\ i \geq 2$,
(e) $1, 0, \frac{1}{2}, 2, \frac{1}{4}, 0, \frac{1}{8}, 2, \frac{1}{16}, 0, \ldots$.

Section 4.5

1. Work out the details of the derivation of (5.10).

2. Compute $S(B)$ and $S(\mathscr{L})$ as a function of a for the matrix

$$A = \begin{pmatrix} 1 & -a \\ -4a & 1 \end{pmatrix}$$

Verify Theorem 5.1 for the case $a \geq 0$. Also verify that $S(\mathscr{L}_\omega) < 1$ for the case $a = \frac{1}{4}$, $\omega = \frac{1}{2}$, and show that

$$S(\mathscr{L}_\omega) < 1 - \omega + \omega S(B)$$

In addition, verify Theorem 5.7 for the case $\omega_1 = \frac{1}{2}$, $\omega_2 = 1$.

3. Show that if A is a symmetric L-matrix, then the J and GS methods converge if and only if A is a Stieltjes matrix.

4. If A is an L-matrix such that $S(B) = 0.8$, find bounds on ω such that $S(\mathscr{L}_\omega) < 1$. Also find bounds on $S(\mathscr{L}_{0.5})$ and $S(\mathscr{L})$. If $S(B) = 1.2$, find bounds on $S(\mathscr{L}_{0.5})$ and $S(\mathscr{L})$.

5. Let A be a positive definite matrix and let the eigenvalues of the matrix B corresponding to the Jacobi method lie in the interval $-0.7 \leq \mu \leq 0.8$. For what values of ω will the SOR method converge? If $B \geq 0$, give an upper bound for $S(\mathscr{L}_{1/2})$.

Section 4.6

1. Prove that

$$\sum_{k=1}^{I-1} \sin \frac{p\pi k}{I} \sin \frac{p'\pi k}{I} = 0$$

if $p \neq p'$ and $1 \leq p \leq I-1$, $1 \leq p' \leq I-1$.

2. Consider the five-point discrete analog of the Dirichlet problem for a 2×1 rectangle with $h = \frac{1}{20}$.

(a) Determine $S(B)$ and $S(\mathscr{L})$ where B and $\mathscr{L}$ correspond to the J and GS methods, respectively.

(b) Estimate the number of iterations needed to reduce $\| u^{(n)} - \bar{u} \|$ to 10^{-6} of $\| u^{(0)} - \bar{u} \|$. Here $\bar{u}$ is the exact solution.

3. For the matrix

$$\begin{pmatrix} 4 & -1 & -1 & 0 \\ -1 & 4 & 0 & -1 \\ -1 & 0 & 4 & -1 \\ 0 & -1 & -1 & 4 \end{pmatrix}$$

find the eigenvalues, eigenvectors, and principal vectors of $\mathscr{L}$. Also find the eigenvalues of B and verify that if $\lambda \neq 0$ is an eigenvalue of $\mathscr{L}$, then $\mu = \lambda^{1/2}$ is an eigenvalue of B.

4. Verify the statements following (6.26).

5. Solve the model problem for Laplace's equation with $h = \frac{1}{20}$ using the J, GS, and SOR methods, the latter with $\omega = 1.75$. Let the boundary values be zero and let the starting values at interior mesh points be unity. Iterate until the maximum error does not exceed 10^{-6} in absolute value.

Also compute the theoretical number of iterations for each method by $(\log 10^6)/R(G)$ where $R(G)$ is the rate of convergence corresponding to the method in question.

6. Consider the model problem for Laplace's equation. For a given mesh size h the nine-point formula

$$\begin{aligned} 20u(x, y) &- 4[u(x + h, y) + u(x - h, y) + u(x, y + h) + u(x, y - h)] \\ &-[u(x + h, y + h) + u(x + h, y - h) + u(x - h, y + h) \\ &+u(x - h, y - h)] = 0 \end{aligned}$$

is used.

(a) Show that the matrix A of the corresponding linear system is a Stieltjes matrix.

(b) Derive an expression for $\bar{\mu}$, the spectral radius of the Jacobi method, and evaluate $\bar{\mu}$ for $h^{-1} = 20$.

7. Show that the relation (6.28) between the nonzero eigenvalues of $\mathscr{L}$ and the eigenvalues of B holds even for nonrectangular regions. Show that it also holds for the more general difference equations considered in Section 2.8.

8. Show that, for the difference equations considered in Section 2.8, if μ is an eigenvalue of B and if λ is a root of $(\lambda + \omega - 1)^2 = \omega^2\mu^2\lambda$, then λ is an eigenvalue of $\mathscr{L}_\omega$ with an eigenvector given by (6.27), where $v(x, y)$ is an eigenvector of B. Conversely, if $\lambda \neq 0$ is an eigenvalue of $\mathscr{L}_\omega$, there exists an eigenvalue μ of B such that $(\lambda + \omega - 1)^2 = \omega^2\mu^2\lambda$, and the eigenvector is given by (6.29).

9. Using the results of the previous exercise, if $\mu = 0.8$ is an eigenvalue of B find two eigenvalues of $\mathscr{L}_{1.25}$ and one eigenvalue of $\mathscr{L}$.

Chapter 5 / EIGENVALUES OF THE SOR METHOD FOR CONSISTENTLY ORDERED MATRICES

5.1. INTRODUCTION

In this chapter we develop a relation between the eigenvalues of the matrix $\mathscr{L}_\omega$, associated with the SOR method and the eigenvalues of the matrix B associated with the Jacobi method, which holds for a certain class of matrices. In Section 5.2 we shall derive this relation in the case where A is a block tri-diagonal matrix whose diagonal blocks are diagonal matrices. It will then be shown in Section 5.3 that the relation holds for a wider class of matrices, namely, "consistently ordered matrices." Moreover, it will be shown in Section 5.4 that if A belongs to a still more general class of matrices, those having "Property A" as defined in Section 2.6, then by a suitable permutation of the rows and corresponding columns of A we can obtain a consistently ordered matrix. Methods for choosing such a permutation will be described in Section 5.4. In Section 5.5 we shall give an alternate proof of the eigenvalue relation, based on the use of nonmigratory permutations.

It has been shown in Section 2.8, that matrices arising from the solution by finite difference methods of a wide class of elliptic partial differential equations have Property A. In Section 5.6, methods will be given for reordering the mesh points, which is equivalent to permuting the rows and corresponding columns of A, in order to obtain a consistently ordered matrix.

In Section 5.7, a computer program will be described for testing whether or not a given matrix has Property A and, if so, whether it is consistently ordered. Other developments of the theory will be described in Section 5.8.

In Chapter 6, the relation between the eigenvalues of $\mathscr{L}_\omega$ and those of B will be exploited and methods presented for choosing ω in order to optimize the rate of convergence of the SOR method.

5.2. BLOCK TRI-DIAGONAL MATRICES

We now study the convergence properties of the SOR method when the matrix A has the block tri-diagonal form

$$A = \begin{pmatrix} D_1 & H_1 & 0 & \cdots & 0 & 0 \\ K_1 & D_2 & H_2 & \cdots & 0 & 0 \\ 0 & K_2 & D_3 & \cdots & 0 & 0 \\ \vdots & \vdots & \vdots & \cdots & \vdots & \vdots \\ 0 & 0 & 0 & \cdots & D_{s-1} & H_{s-1} \\ 0 & 0 & 0 & \cdots & K_{s-1} & D_s \end{pmatrix} \tag{2.1}$$

where the D_i are square diagonal matrices. We shall refer to such a matrix as a *T-matrix*. If A is a T-matrix with nonvanishing diagonal elements, then the matrix B corresponding to the J method has the form

$$B = \begin{pmatrix} O_1 & F_1 & 0 & 0 & \cdots & 0 & 0 \\ G_1 & O_2 & F_2 & 0 & \cdots & 0 & 0 \\ 0 & G_2 & O_3 & F_3 & \cdots & 0 & 0 \\ \vdots & \vdots & \vdots & \vdots & \cdots & \vdots & \vdots \\ 0 & 0 & 0 & 0 & \cdots & O_{s-1} & F_{s-1} \\ 0 & 0 & 0 & 0 & \cdots & G_{s-1} & O_s \end{pmatrix} \tag{2.2}$$

where the diagonal blocks O_i are square null matrices. In this section we shall derive a relation between the eigenvalues of the matrix $\mathscr{L}_\omega$ and the eigenvalues μ of B. This relation depends on the following theorem.

Theorem 2.1. Let A be a T-matrix. Then

$$\det(\alpha C_L + \alpha^{-1} C_U - kD)$$

is independent of α for all $\alpha \neq 0$ and for all k. Here $A = D - C_L - C_U$ where $D = \text{diag } A$ and where C_L and C_U are strictly lower and strictly upper triangular matrices, respectively.

Proof. It is easy to verify that

$$Q^{-1}(\alpha C_L + \alpha^{-1} C_U - kD)Q = (C_L + C_U - kD) \tag{2.3}$$

where

$$Q = \begin{pmatrix} I_1 & 0 & 0 & \cdots & 0 & 0 \\ 0 & \alpha I_2 & 0 & \cdots & 0 & 0 \\ 0 & 0 & \alpha^2 I_3 & \cdots & 0 & 0 \\ \vdots & \vdots & \vdots & \cdots & \vdots & \vdots \\ 0 & 0 & 0 & \cdots & 0 & \alpha^{s-1} I_s \end{pmatrix} \tag{2.4}$$

where the I_i are identity matrices of the same sizes as the D_i. For instance, if $s = 3$, we have

$$\begin{aligned} &Q^{-1}(\alpha C_L + \alpha^{-1} C_U - kD)Q \\ &= -\begin{pmatrix} I_1 & 0 & 0 \\ 0 & \alpha^{-1} I_2 & 0 \\ 0 & 0 & \alpha^{-2} I_3 \end{pmatrix} \begin{pmatrix} kD_1 & \alpha^{-1} H_1 & 0 \\ \alpha K_1 & kD_2 & \alpha^{-1} H_2 \\ 0 & \alpha K_2 & kD_3 \end{pmatrix} \begin{pmatrix} I_1 & 0 & 0 \\ 0 & \alpha I_2 & 0 \\ 0 & 0 & \alpha^2 I_3 \end{pmatrix} \\ &= -\begin{pmatrix} kD_1 & H_1 & 0 \\ K_1 & kD_2 & H_2 \\ 0 & K_2 & kD_3 \end{pmatrix} = C_L + C_U - kD \end{aligned}$$

Clearly, the determinant of $C_L + C_U - kD$ is independent of α for all k and the theorem follows.

Theorem 2.2. Let A be a T-matrix with nonvanishing diagonal elements and let $B = I - (\text{diag } A)^{-1} A = L + U$ where L and U are strictly lower and strictly upper triangular matrices, respectively. Then

(a) If μ is any eigenvalue of B of multiplicity p, then $-\mu$ is also an eigenvalue of B of multiplicity p.

(b) λ satisfies

$$(\lambda + \omega - 1)^2 = \omega^2 \mu^2 \lambda \tag{2.5}$$

for some eigenvalue μ of B if and only if λ satisfies

$$\lambda + \omega - 1 = \omega\mu\lambda^{1/2} \tag{2.6}$$

for some eigenvalue μ of B.

(c) If λ satisfies either, and hence both of the relations (2.5) and (2.6), then λ is an eigenvalue of $\mathscr{L}_\omega$.

(d) If λ is an eigenvalue of $\mathscr{L}_\omega$, then there exists an eigenvalue μ of B such that (2.5) and (2.6) hold.

Proof. By Theorem 2.1, with $\alpha = -1$ we have $\det(C_L + C_U + \mu D) = \det(-C_L - C_U + \mu D) = \pm\det(C_L + C_U - \mu D)$. Hence $\det(L + U - \mu I)$ is some power of μ times a polynomial in μ^2. Therefore (a) follows.

Suppose now that μ is an eigenvalue of B and λ satisfies (2.6). Then clearly λ satisfies (2.5). On the other hand, if λ satisfies (2.5), then λ satisfies either (2.6) or the equation $\lambda + \omega - 1 = -\omega\mu\lambda^{1/2}$. But since $-\mu$ is an eigenvalue of B it follows that (2.6) holds for some eigenvalue of B and (b) follows.

Evidently λ is an eigenvalue of $\mathscr{L}_\omega$ if and only if $\det(\mathscr{L}_\omega - \lambda I) = 0$. But by (3-3.23), we have

$$\begin{aligned}\det(\mathscr{L}_\omega - \lambda I) &= \det(I - \omega L)^{-1} \det(\omega U + (1 - \omega)I - \lambda(I - \omega L)) \\ &= \det(\omega U + \omega\lambda L - (\lambda + \omega - 1)I)\end{aligned} \tag{2.7}$$

If $\lambda = 0$ satisfies (2.5) and (2.6), then $\omega = 1$. Since $\det \mathscr{L}_\omega = (1 - \omega)^N$ it follows that $\lambda = 0$ is an eigenvalue of $\mathscr{L}_\omega$ if $\lambda = 0$ satisfies (2.5) and (2.6). On the other hand, if $\lambda \neq 0$, then

$$\begin{aligned}\det(\mathscr{L}_\omega - \lambda I) &= \omega^N\lambda^{N/2} \det[\lambda^{1/2}L + \lambda^{-1/2}U - (\lambda + \omega - 1)/(\omega\lambda^{1/2})I] \\ &= \omega^N\lambda^{N/2} \det[L + U - (\lambda + \omega - 1)/(\omega\lambda^{1/2})I]\end{aligned} \tag{2.8}$$

by Theorem 2.1. Therefore, if $\lambda \neq 0$ satisfies (2.5) and (2.6) for some eigenvalue μ of B, then $(\lambda + \omega - 1)/(\omega\lambda^{1/2})$ is an eigenvalue of B and hence $\det(B - [(\lambda + \omega - 1)/(\omega\lambda^{1/2})]I)$ and $\det(\mathscr{L}_\omega - \lambda I)$ vanish. Thus λ is an eigenvalue of $\mathscr{L}_\omega$ and (c) follows.

If λ is an eigenvalue of $\mathscr{L}_\omega$ and if $\lambda = 0$, then $\omega = 1$ as above. In this case, since any eigenvalue μ of B satisfies (2.5) and (2.6) it follows that (d) holds for the case $\lambda = 0$. If $\lambda \neq 0$, then by (2.8) we have

$$0 = \det(\mathscr{L}_\omega - \lambda I) = \omega^N\lambda^{N/2} \det[B - (\lambda + \omega - 1)/(\omega\lambda^{1/2})I] = 0 \tag{2.9}$$

Hence $(\lambda + \omega - 1)/(\omega\lambda^{1/2})$ must equal an eigenvalue μ of B and (2.6) is satisfied. This completes the proof of Theorem 2.2.

Corollary 2.3. Under the hypotheses of Theorem 2.2, the set of eigenvalues of $\mathscr{L}$ includes the number zero together with the numbers μ_1^2, $\mu_2^2, \ldots, \mu_q^2$ where $\pm\mu_1, \pm\mu_2, \ldots, \pm\mu_q$ are the nonzero eigenvalues of B. Moreover,

$$S(\mathscr{L}) = (S(B))^2, \qquad R(\mathscr{L}) = 2R(B) \tag{2.10}$$

where $R(B) = -\log S(B)$ and $R(\mathscr{L}) = -\log S(\mathscr{L})$ are the rates of convergence of the J and GS methods, respectively.

5.3. CONSISTENTLY ORDERED MATRICES AND ORDERING VECTORS

In this section we shall show that the main results of Section 5.2 hold for a wider class of matrices, namely the class of consistently ordered matrices. We shall develop the theory of consistently ordered matrices first in terms of ordering vectors. Later in Section 5.5 we shall present a longer, though perhaps more intuitive, approach based on the use of nonmigratory permutations which relate the matrix A to a certain T-matrix.

Definition 3.1. Given a matrix $A = (a_{i,j})$, the integers i and j are *associated* with respect to A if $a_{i,j} \neq 0$ or $a_{j,i} \neq 0$.

Definition 3.2. The matrix A of order N is *consistently ordered* if for some t there exist disjoint subsets $S_1, S_2, \ldots, S_t$ of $W = \{1, 2, \ldots, N\}$ such that $\sum_{k=1}^{t} S_k = W$ and such that if i and j are associated, then $j \in S_{k+1}$ if $j > i$ and $j \in S_{k-1}$ if $j < i$, where S_k is the subset containing i.

We remark that the sets $S_1, S_2, \ldots, S_t$ are not necessarily all nonempty. However, if we discard the empty sets, the remaining sets, say $S_1', S_2', \ldots, S_{t'}'$, where $t' \leq t$ still have the properties required in Definition 3.2. For suppose i and j are associated. Then $i \in S_k'$ for some k. Then $i \in S_{k_1}$ for some $k_1 \geq k$. If $j > i$, then, since i and j are associated, we have $j \in S_{k_1+1}$; hence S_{k_1+1} is not empty and $S_{k+1}' = S_{k_1+1}$. Therefore, $j \in S_{k+1}'$. Similarly, if $j < i$ we have $j \in S_{k-1}'$. Thus the sets $S_1', S_2', \ldots, S_{t'}'$ satisfy the conditions of Definition 3.2.

As an example, consider the matrix

$$A = \begin{pmatrix} 1 & 0 & 0 \\ 0 & 1 & 1 \\ 0 & 1 & 1 \end{pmatrix} \tag{3.1}$$

If we let $S_1 = \{1\}$, S_2 be empty, $S_3 = \{2\}$, and $S_4 = \{3\}$, then the conditions of Definition 3.2 are satisfied. However, we can also discard the empty set S_2 and let $S_1' = \{1\}$, $S_2' = \{2\}$, $S_3' = \{3\}$.

In the previous example, the matrix A was a T-matrix. However, the following matrix is not a T-matrix.

$$A = \begin{pmatrix} 4 & 0 & 0 & -1 \\ -1 & 4 & -1 & 0 \\ 0 & -1 & 4 & 0 \\ -1 & 0 & 0 & 4 \end{pmatrix} \tag{3.2}$$

However, A is consistently ordered with the sets $S_1 = \{1\}$, $S_2 = \{2, 4\}$, $S_3 = \{3\}$.

We now prove

Theorem 3.1. If A is a T-matrix, then A is consistently ordered.

Proof. If A has the form (2.1), let the orders of $D_1, D_2, \ldots, D_s$ be $r_1, r_2, \ldots, r_s$, respectively, and let $S_1 = \{1, 2, \ldots, r_1\}$, $S_2 = \{r_1 + 1.$ $r_1+2, \cdots, r_1+r_2\}, \ldots, S_s = \{q_{s-1}+1, q_{s-1}+2, \ldots, q_{s-1}+r_s\}$ where $q_{s-1} = r_1 + \cdots + r_{s-1}$. Clearly, A is consistently ordered with the sets $S_1, S_2, \ldots, S_s$. Thus, if $i \in S_k$, $a_{i,j} \neq 0$, and $i < j$, we have $j \in S_{k+1}$, while if $i \in S_k$, $a_{i,j} \neq 0$, and $i > j$, then $j \in S_{k-1}$.

Definition 3.3. The vector $\gamma = (\gamma_1, \gamma_2, \ldots, \gamma_N)^{\mathrm{T}}$, where $\gamma_1, \gamma_2, \ldots, \gamma_N$ are integers, is an *ordering vector* for the matrix A of order N if for any pair of associated integers i and j with $i \neq j$ we have $|\gamma_i - \gamma_j| = 1$.

Definition 3.4. An ordering vector $\gamma = (\gamma_1, \gamma_2, \ldots, \gamma_N)^{\mathrm{T}}$ for the matrix A of order N is a *compatible ordering vector* for A if

(a) $\gamma_i - \gamma_j = 1$ if i and j are associated and $i > j$,

(b) $\gamma_i - \gamma_j = -1$ if i and j are associated and $i < j$.

As an example, the vector $\gamma = (1, 2, 1, 2)^{\mathrm{T}}$ is an ordering vector for the matrix A of (3.2) but is not a compatible ordering vector since 2 and 3 are associated and $\gamma_2 - \gamma_3 = 1$. Another ordering vector is $\gamma' = (1, 2, 3, 2)^T$, which is compatible.

We now prove

Theorem 3.2. A matrix A of order N is consistently ordered if and only if there exists a compatible ordering vector for A.

Proof. If the matrix is consistently ordered, we can construct a compatible ordering vector simply by letting $\gamma_i = k$ if $i \in S_k$. For, if $i \in S_k$ and if i and j are associated, we have $j \in S_{k+1}$, i.e., $\gamma_j = k + 1 = \gamma_i + 1$ if $j > i$ and $j \in S_{k-1}$, i.e., $\gamma_j = k - 1 = \gamma_i - 1$, if $j < i$.

If a compatible ordering vector exists, let $\alpha = \min\{\gamma_1, \gamma_2, \ldots, \gamma_N\}$, and $\beta = \max\{\gamma_1, \gamma_2, \ldots, \gamma_N\}$. For $k = 1, 2, \ldots, \beta - \alpha + 1$, let S_k be the set of all i such that $\gamma_i = \alpha + k - 1$. Some of the S_k may be empty, but in any case they are disjoint and $S_1 + S_2 + \cdots + S_{\beta-\alpha-1} = W$. Moreover, if $i \in S_k$ and i and j are associated, then $\gamma_j = \gamma_i + 1$, and hence $j \in S_{k+1}$ if $j > i$; or, $\gamma_j = \gamma_i - 1$, and hence $j \in S_{k-1}$ if $j < i$. Thus the sets $S_1, S_2, \ldots, S_{\beta-\alpha+1}$ have the required properties for the matrix to be consistently ordered.

In the consistently ordered matrix (3.2), the correspondence between the sets $S_1 = \{1\}$, $S_2 = \{2, 4\}$, $S_3 = \{3\}$ and the ordering vector $\gamma' = (1, 2, 3, 2)^T$, which is suggested by the proof of Theorem 3.2, is easily seen.

It is easy to show that the matrix

$$A = \begin{pmatrix} 4 & -1 & 0 & -1 \\ -1 & 4 & -1 & 0 \\ 0 & -1 & 4 & -1 \\ -1 & 0 & -1 & 4 \end{pmatrix} \tag{3.3}$$

has several ordering vectors but no compatible ordering vectors. Thus it is not consistently ordered. Obviously, if a matrix does not have any ordering vectors, it certainly has no compatible ordering vectors and is therefore not consistently ordered. For example, the matrix

$$A = \begin{pmatrix} 4 & -1 & -1 \\ -1 & 4 & -1 \\ -1 & -1 & 4 \end{pmatrix} \tag{3.4}$$

has no ordering vectors. In the next section we shall show that in order that a matrix have at least one ordering vector, it is necessary and sufficient that it have Property A.

We now prove

Theorem 3.3. If A is a consistently ordered matrix, then

$$\Delta = \det(\alpha C_L + \alpha^{-1} C_U - kD) \tag{3.5}$$

is independent of α for $\alpha \neq 0$ and for all k, where D, C_L and C_U are defined in Theorem 2.1.

Proof. The general term of Δ is

$$t(\sigma) = \pm \prod_{i=1}^{N} a_{i,\sigma(i)} \alpha^{n_L - n_U} k^{N-(n_L+n_U)}$$

where σ is a permutation defined on the integers $1, 2, \ldots, N$. Here n_L and n_U are, respectively, the number of values of i such that $i > \sigma(i)$ and such that $i < \sigma(i)$. Let γ be a compatible ordering vector for A. Evidently we have

$$n_L = \sum_{\substack{i=1 \\ \sigma(i)<i}}^{N} (\gamma_i - \gamma_{\sigma(i)}), \qquad n_U = \sum_{\substack{i=1 \\ \sigma(i)>i}}^{N} (\gamma_{\sigma(i)} - \gamma_i)$$

since $\sigma(i) > i$ and $a_{i,\sigma(i)} \neq 0$ if and only if $\gamma_{\sigma(i)} = \gamma_i + 1$ and since $\sigma(i) < i$ and $a_{i,\sigma(i)} \neq 0$ if and only if $\gamma_{\sigma(i)} - \gamma_i = -1$. However, since σ is a permutation we have

$$n_L - n_U = \sum_{\substack{i=1 \\ \sigma(i)\neq i}}^{N} (\gamma_i - \gamma_{\sigma(i)}) = \sum_{\substack{i=1 \\ \sigma(i)\neq i}}^{N} \gamma_i - \sum_{\substack{i=1 \\ \sigma(i)\neq i}}^{N} \gamma_{\sigma(i)} = 0$$

Hence $t(\sigma)$ is independent of α and the result follows.

From the proof of Theorem 2.2 we have

Theorem 3.4. Let A be a consistently ordered matrix with nonvanishing diagonal elements and let $B = I - (\text{diag } A)^{-1}A$. Then the conclusions of Theorem 2.2 and Corollary 2.3 are valid.

As an immediate corollary we have

Theorem 3.5. Let A be a symmetric consistently ordered matrix with positive diagonal elements. Then $\bar{\mu} = S(B) < 1$ if and only if A is positive definite.

For, by Theorem 4-5.6, the GS method converges, when A is symmetric and has positive diagonal elements, if and only if A is positive definite. But by Theorem 3.4 the GS method converges if and only if $\bar{\mu} < 1$.

5.4. PROPERTY A

We have seen that in order for a matrix to be consistently ordered, it is necessary, though not sufficient, that an ordering vector exist. We now show that an ordering vector exists if and only if A has Property A, as defined in Section 2.6, and that if A has Property A, one can permute the rows and columns of A and obtain a consistently ordered matrix.

Theorem 4.1. There exists an ordering vector for a matrix A if and only if A has Property A. Moreover, if A is consistently ordered, then A has Property A.

Proof. If A has Property A, we let $\gamma_i = 1$ if $i \in S_1$, and $\gamma_i = 2$ if $i \in S_2$, where S_1 and S_2 are defined in Definition 2-6.1. Since $|\gamma_i - \gamma_j| = 1$, if $i \neq j$ and if i and j are associated, then it follows that $\gamma = (\gamma_1, \gamma_2, \ldots, \gamma_N)^T$ is an ordering vector.

On the other hand, if an ordering vector exists, let S_1 be the set of all i such that γ_i is odd and let S_2 be the set of all i such that γ_i is even. If i and j are associated and if $i \neq j$, then $|\gamma_i - \gamma_j| = 1$. If $i \in S_1$, then $j \in S_2$, for if $j \in S_1$, then $\gamma_i - \gamma_j$ would be even and hence $|\gamma_i - \gamma_j|$ could not be unity. A similar argument holds if $i \in S_2, j \in S_1$. Obviously S_1 and S_2 are disjoint and $S_1 + S_2 = W$. Therefore, S_1 and S_2 satisfy the requirements of Definition 2-6.1 and hence A has Property A.

If A is consistently ordered, then there exists an ordering vector by Theorem 3.2. Hence A has Property A. This completes the proof of Theorem 4.1.

From the above proof we have

Corollary 4.2. If A has Property A, then there exists an ordering vector γ for A whose components have at most two different values.

We now show that for any matrix A and any permutation matrix P, if $A' = (a'_{i,j}) = P^{-1}AP$, then

$$a'_{i,j} = a_{\sigma^{-1}(i),\sigma^{-1}(j)} \tag{4.1}$$

where σ is the permutation corresponding to P. If we refer to the definition of a permutation matrix given in Section 2.6, we see that $A_1 = AP$ can be obtained from A by permuting the columns of A so that the ith column of A is moved to the $\sigma(i)$th column of A_1. Similarly, the matrix $A' = P^{-1}AP = P^{-1}A_1$ can be obtained from A_1 by permuting the rows of A_1 so that the ith row of A_1 is moved to the $\sigma(i)$th row of A'. The result is that A' is obtained from A by moving the element $a_{i,j}$ to the $\sigma(i)$th row and the $\sigma(j)$th column of A'. Thus (4.1) follows.

Theorem 4.3. If A has Property A, then for any permutation matrix P the matrix $A' = P^{-1}AP$ has Property A.

Proof. Since A has Property A, there exist sets S_1 and S_2 satisfying the conditions of Definition 2-6.1. Let $\sigma(i)$ be the permutation corresponding to P and let S_1' and S_2' denote, respectively, the set of all i such that $i = \sigma(j)$ for $j \in S_1$ and such that $i = \sigma(j)$ for $j \in S_2$, respectively. Evidently, by (4.1), the sets S_1' and S_2' satisfy the requirements of Definition 2-6.1, for A', and hence A' has Property A.

Theorem 4.4. A matrix A has Property A if and only if A is a diagonal matrix or else there exists a permutation matrix P such that $P^{-1}AP$ has the form

$$A' = P^{-1}AP = \begin{pmatrix} D_1 & H \\ K & D_2 \end{pmatrix} \tag{4.2}$$

where D_1 and D_2 are square diagonal matrices.

Proof. If A has Property A, let S_1 and S_2 be the sets specified in Definition 2-6.1. If S_1 or S_2 are empty, then A is a diagonal matrix. Otherwise, let s_1 and s_2 be the number of elements of S_1 and S_2, respectively, and let the elements of S_k be $i_1^{(k)}, i_2^{(k)}, \ldots, i_{s_k}^{(k)}$ where $i_1^{(k)} < i_2^{(k)} < \cdots < i_{s_k}^{(k)}$, $k = 1, 2$. Let $\sigma(i)$ be defined as follows:

$$\begin{aligned} &\sigma(i_1^{(1)}) = 1, && \sigma(i_2^{(1)}) = 2, \ldots, \sigma(i_{s_1}^{(1)}) = s_1 \\ &\sigma(i_1^{(2)}) = s_1 + 1, && \sigma(i_2^{(2)}) = s_1 + 2, \ldots, \sigma(i_{s_2}^{(2)}) = s_1 + s_2 = N \end{aligned} \tag{4.3}$$

We show that if P is the permutation matrix corresponding to the permutation σ, then $A' = P^{-1}AP$ has the form (4.2). Let $T_1 = \{1, 2, \ldots, s_1\}$ and $T_2 = \{s_1 + 1, s_1 + 2, \ldots, s_1 + s_2\}$. It suffices to show that if $a'_{i,j} \neq 0$ and $i \neq j$, then $j \in T_2$ if $i \in T_1$, and $j \in T_1$ if $i \in T_2$. If $a'_{i,j} \neq 0$ and $i \neq j$, then $a_{\sigma^{-1}(i),\sigma^{-1}(j)} \neq 0$ and hence $\sigma^{-1}(i)$ and $\sigma^{-1}(j)$ are associated. Since A has Property A and since $\sigma^{-1}(i) \neq \sigma^{-1}(j)$, it follows that either

$\sigma^{-1}(i) \in S_1$ and $\sigma^{-1}(j) \in S_2$ or $\sigma^{-1}(i) \in S_2$ and $\sigma^{-1}(j) \in S_1$. By the construction of σ, it follows that either $i = \sigma(\sigma^{-1}(i)) \in T_1$ and $j = \sigma(\sigma^{-1}(j)) \in T_2$ or else $i \in T_2$ and $j \in T_1$. Hence A' has the form (4.2) if both T_1 and T_2 are nonempty or else A' is a diagonal matrix.

Conversely, if for some permutation matrix P, the matrix $A' = P^{-1}AP$ has the form (4.2), then A' has Property A since A' is a T-matrix and hence is consistently ordered. Therefore, by Theorem 4.3, $A = PA'P^{-1}$ has Property A. This completes the proof of Theorem 4.4.

Theorem 4.5. A matrix A has Property A if and only if there exists a permutation matrix P such that $A' = P^{-1}AP$ is consistently ordered.

Proof. If A has Property A, then, by Theorem 4.4, for some permutation matrix P, $P^{-1}AP$ has the form (4.2) and hence is consistently ordered. On the other hand, if $A' = P^{-1}AP$ is consistently ordered for some permutation matrix P, then A' has Property A by Theorem 4.1 and hence $A = PA'P^{-1}$ has Property A by Theorem 4.3.

The following theorem shows how a consistently ordered matrix can be constructed from a matrix which has Property A and from a given ordering vector.

Theorem 4.6. Let A be a matrix with Property A and let γ be any ordering vector for A. There exists a permutation matrix P such that $A' = P^{-1}AP$ is consistently ordered and such that $\gamma' = (\gamma_{\sigma^{-1}(i)})$ is a compatible ordering vector for A', where σ is the permutation corresponding to P.

Proof. We first construct the sets $S_1, S_2, \ldots, S_t$ as in the proof of Theorem 3.2, where $t = \beta - \alpha + 1$ and $\alpha = \min(\gamma_1, \gamma_2, \ldots, \gamma_N)$, $\beta = \max(\gamma_1, \gamma_2, \ldots, \gamma_N)$. We label the elements of each set S_k, taken in any order, by $i_1^{(k)}, i_2^{(k)}, \ldots, i_{s_k}^{(k)}$ where s_k is the number of elements in S_k.

We define the permutation

$$\begin{array}{ll}
\sigma(i_1^{(1)}) = 1, & \sigma(i_2^{(1)}) = 2, \ldots, \sigma(i_{s_1}^{(1)}) = s_1 \\
\sigma(i_1^{(2)}) = s_1 + 1, & \sigma(i_2^{(2)}) = s_1 + 2, \ldots, \sigma(i_{s_2}^{(2)}) = s_1 + s_2 \\
\vdots & \\
\sigma(i_1^{(t)}) = q_{t-1} + 1, & \sigma(i_2^{(t)}) = q_{t-1} + 2, \ldots, \sigma(i_{s_t}^{(t)}) = q_t = N
\end{array}$$

where $q_{t-1} = s_1 + s_2 + \cdots + s_{t-1}$, $q_t = q_{t-1} + s_t$. The reader should complete the proof of Theorem 4.6 by showing that $P^{-1}AP$ is a con-

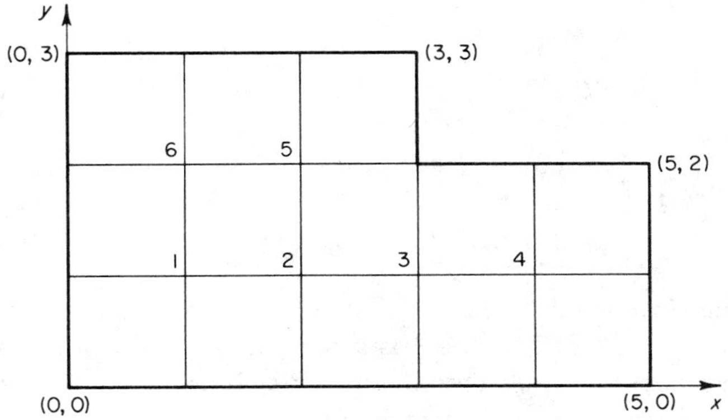

Figure 4.1.

sistently ordered matrix and that $\gamma' = (\gamma_{\sigma^{-1}(i)})$ is a compatible ordering vector.

To illustrate, let us consider the discrete analogue of the Dirichlet problem for the region shown in Figure 4.1 with the mesh size $h = 1$. The matrix is

$$A = \begin{pmatrix} 4 & -1 & 0 & 0 & 0 & -1 \\ -1 & 4 & -1 & 0 & -1 & 0 \\ 0 & -1 & 4 & -1 & 0 & 0 \\ 0 & 0 & -1 & 4 & 0 & 0 \\ 0 & -1 & 0 & 0 & 4 & -1 \\ -1 & 0 & 0 & 0 & -1 & 4 \end{pmatrix} \tag{4.4}$$

It is easy to show that there does not exist a compatible ordering vector for A. However, the vector $\gamma = (1, 2, 3, 4, 3, 2)^{\mathrm{T}}$ is an ordering vector for A. We construct the sets $S_1 = \{1\}$, $S_2 = \{2, 6\}$, $S_3 = \{3,5\}$, $S_4 = \{4\}$, and the permutation

$$\sigma(1) = 1; \quad \sigma(2) = 2, \quad \sigma(6) = 3; \quad \sigma(3) = 4, \quad \sigma(5) = 5; \quad \sigma(4) = 6 \tag{4.5}$$

The corresponding permutation matrix is

$$P = \begin{pmatrix} 1 & 0 & 0 & 0 & 0 & 0 \\ 0 & 1 & 0 & 0 & 0 & 0 \\ 0 & 0 & 0 & 1 & 0 & 0 \\ 0 & 0 & 0 & 0 & 0 & 1 \\ 0 & 0 & 0 & 0 & 1 & 0 \\ 0 & 0 & 1 & 0 & 0 & 0 \end{pmatrix} \tag{4.6}$$

and

$$A' = P^{-1}AP = \begin{pmatrix} (4) & (-1 \quad -1) & (0 \quad 0) & (0) \\ \begin{pmatrix} -1 \\ -1 \end{pmatrix} & \begin{pmatrix} 4 & 0 \\ 0 & 4 \end{pmatrix} & \begin{pmatrix} -1 & -1 \\ 0 & -1 \end{pmatrix} & \begin{pmatrix} 0 \\ 0 \end{pmatrix} \\ \begin{pmatrix} 0 \\ 0 \end{pmatrix} & \begin{pmatrix} -1 & 0 \\ -1 & -1 \end{pmatrix} & \begin{pmatrix} 4 & 0 \\ 0 & 4 \end{pmatrix} & \begin{pmatrix} -1 \\ 0 \end{pmatrix} \\ (0) & (0 \quad 0) & (-1 \quad 0) & (4) \end{pmatrix} \quad (4.7)$$

is a T-matrix and is therefore consistently ordered. Also, the vector $\gamma' = (\gamma_{\sigma^{-1}(i)}) = (1, 2, 2, 3, 3, 4)^{T}$ is a compatible ordering vector for A'.

The permutation thus described corresponds to labeling the mesh points by diagonals, as indicated in Figure 4.2.

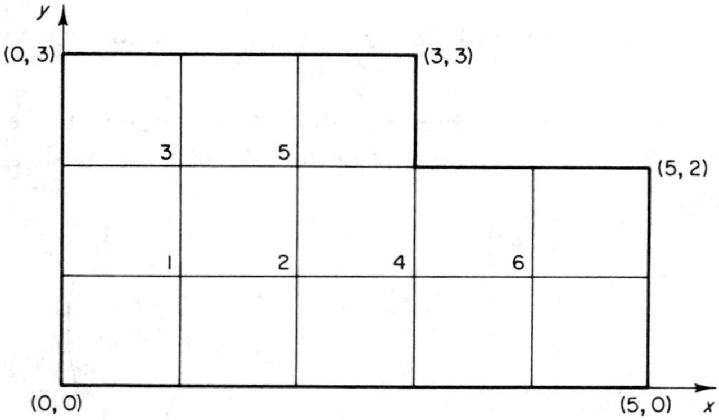

Figure 4.2.

Theorem 4.7. Let A be a matrix with Property A which has non-vanishing diagonal elements and let $B = I - (\text{diag } A)^{-1}A$. Then conclusion (a) of Theorem 2.2 follows.

Proof. Since A has Property A, by Theorem 4.5 there exists a permutation matrix P so that $A' = P^{-1}AP$ is consistently ordered. But since A' is consistently ordered, the conclusion (a) of Theorem 2.2 is valid for $B[A']$, by Theorem 3.4. But $B[A] = D[A]^{-1}C[A]$, where $A = D[A] - C[A]$. Since $P^{-1}D[A]P$ is a diagonal matrix and since the diagonal elements of $P^{-1}C[A]P$ vanish, it follows that $D[P^{-1}AP] = P^{-1}D[A]P$ and $C[P^{-1}AP] = P^{-1}C[A]P$. Hence

$$B[A'] = D[A']^{-1}C[A'] = P^{-1}D[A]^{-1}C[A]P = P^{-1}B[A]P \quad (4.8)$$

and the eigenvalues of $B[A']$ are the same as those of $B[A]$. Hence the conclusion (a) of Theorem 2.2 is also valid for $B = B[A]$.

5.5. NONMIGRATORY PERMUTATIONS

We now present an alternative proof of Theorem 3.4 involving the use of "nonmigratory" permutations.

Definition 5.1. Given a matrix A, a permutation matrix P is *nonmigratory* if

$$P^{-1}C_L[A]P = C_L[P^{-1}AP], \qquad P^{-1}C_U[A]P = C_U[P^{-1}AP] \tag{5.1}$$

where $C_L[A]$ and $C_U[A]$ are defined by (4-3.14).

Theorem 5.1. A matrix A is consistently ordered if and only if there exists a nonmigratory permutation matrix P such that $P^{-1}AP$ is a T-matrix.

Proof. We construct the permutation function $\sigma(i)$ as follows. Let $S_1, S_2, \ldots, S_t$ be any sets having the properties of Definition 3.2. Assume that all empty sets have been discarded so that the S_k are nonempty. Let $s_1, s_2, \ldots, s_t$ be the number of elements of $S_1, S_2, \ldots, S_t$, respectively. Let the elements of S_k be $i_1^{(k)}, i_2^{(k)}, \ldots, i_{s_k}^{(k)}$. where $i_1^{(k)} < i_2^{(k)} < \cdots < i_{s_k}^{(k)}$. Then we define $\sigma(i)$ as follows:

$$\begin{array}{ll} \sigma(i_1^{(1)}) = 1, & \sigma(i_2^{(1)}) = 2, \ldots, \sigma(i_{s_1}^{(1)}) = s_1 \\ \sigma(i_1^{(2)}) = s_1 + 1, & \sigma(i_2^{(2)}) = s_1 + 2, \ldots, \sigma(i_{s_2}^{(2)}) = s_1+s_2 \\ \cdots & \cdots \\ \sigma(i_1^{(t)}) = q_{t-1} + 1, & \sigma(i_2^{(t)}) = q_{t-1} + 1, \ldots, \sigma(i_{s_t}^{(t)}) = q_t = N \end{array} \tag{5.2}$$

where $q_{t-1} = s_1 + s_2 + \cdots + s_{t-1}$, $q_t = q_{t-1} + s_t$.

If i and j are associated and if $i > j$, then since A is consistently ordered $j \in S_{k-1}$, where S_k is the subset containing i. By the construction of the permutation σ it follows that $\sigma(i) > \sigma(j)$. Similarly, we can show that if i and j are associated and $i < j$, then $\sigma(i) < \sigma(j)$. Since $\sigma(i) = \sigma(j)$ if $i = j$, it follows that if i and j are associated, then $\sigma(i) > \sigma(j)$ if and only if $i > j$; similarly, $\sigma(i) < \sigma(j)$ if and only if $i < j$.

We now seek to show that the permutation P is nonmigratory. First, we seek to show that if $A' = P^{-1}AP$ then

$$C_L[A'] = P^{-1}C_L[A]P \tag{5.3}$$

Let $A'' = (a''_{i,j}) = P^{-1}C_L[A]P$ and $A^{(3)} = (a^{(3)}_{i,j}) = C_L[A']$. Evidently, by (4.1) we have

$$a''_{i,j} = \begin{cases} a_{\sigma^{-1}(i),\sigma^{-1}(j)} & \text{if } \sigma^{-1}(i) > \sigma^{-1}(j) \\ 0 & \text{if } \sigma^{-1}(i) \leq \sigma^{-1}(j) \end{cases} \tag{5.4}$$

since $C_L[A]$ is a strictly lower triangular matrix. If $a^{(3)}_{i,j} \neq 0$, then $i > j$ and by (4.1) we have $a^{(3)}_{i,j} = a_{\sigma^{-1}(i),\sigma^{-1}(j)} \neq 0$. Hence $\sigma^{-1}(i)$ and $\sigma^{-1}(j)$ are associated. Therefore, since $\sigma(\sigma^{-1}(i)) = i > j = \sigma(\sigma^{-1}(j))$ we have $\sigma^{-1}(i) > \sigma^{-1}(j)$, and, by (5.4), $a^{(3)}_{i,j} = a''_{i,j}$. Suppose, on the other hand, that $a''_{i,j} \neq 0$. Then from (5.4) it follows that $\sigma^{-1}(i) > \sigma^{-1}(j)$ and, moreover, $\sigma^{-1}(i)$ and $\sigma^{-1}(j)$ are associated. Hence, $\sigma(\sigma^{-1}(i)) > \sigma(\sigma^{-1}(j))$ and $i > j$. Therefore, $a''_{i,j} = a_{\sigma^{-1}(i),\sigma^{-1}(j)} = a^{(3)}_{i,j}$ since $i > j$. From this, it follows that (5.3) holds. Similarly, we can show that $C_U[A'] = P^{-1}C_U[A]P$ and $D[A'] = P^{-1}D[A]P$. Hence (5.1) follows.

We seek to show that the matrix

$$A' = P^{-1}AP = (a'_{i,j}) = (a_{\sigma^{-1}(i),\sigma^{-1}(j)}) \tag{5.5}$$

is a T-matrix where the sizes of the square diagonal blocks are $s_1, s_2, \ldots, s_t$. Let us define the sets $T_1, T_2, \ldots, T_t$ of W as follows:

$$\begin{aligned} T_1 &= \{1, 2, \ldots, s_1\} \\ T_2 &= \{s_1 + 1, s_2 + 1, \ldots, s_1 + s_2\} \\ &\vdots \\ T_t &= \{q_{t-1} + 1, q_{t-1} + 2, \ldots, N\} \end{aligned}$$

We show that if $a'_{i,j} \neq 0$ and $i \neq j$, then $j \in T_{k-1} + T_{k+1}$, where T_k is the set containing i. If $a'_{i,j} \neq 0$, then by (5.5) we have $a_{\sigma^{-1}(i),\sigma^{-1}(j)} \neq 0$, and hence $\sigma^{-1}(i)$ and $\sigma^{-1}(j)$ are associated. Since A is consistently ordered and $\sigma^{-1}(i) \neq \sigma^{-1}(j)$, it follows that $\sigma^{-1}(j) \in S_{k+1} + S_{k-1}$ where S_k is the subset containing $\sigma^{-1}(i)$. But by (5.2) it follows that $\sigma(i) \in T_k$ if and only if $i \in S_k$; hence $i \in T_k$ if and only if $\sigma^{-1}(i) \in S_k$. Therefore, $i \in T_k$ and $j \in T_{k-1} + T_{k+1}$. Hence A' is a T-matrix.

To prove the converse, suppose that a nonmigratory permutation matrix P exists such that $A' = P^{-1}AP$ is a T-matrix. Then by Theorem 3.1, A' is consistently ordered. Let $S_1, S_2, \ldots, S_t$ be sets satisfying the requirements of Definition 3.2. To show that $A = PA'P^{-1}$ is consistently ordered, for each $k = 1, 2, \ldots, t$ let T_k be the set of all i such that $\sigma(i) \in S_k$, where $\sigma(i)$ is the permutation corresponding to P. By (4.1) we have $a'_{i,j} = a_{\sigma^{-1}(i),\sigma^{-1}(j)}$ or

$$a_{i,j} = a'_{\sigma(i),\sigma(j)} \tag{5.6}$$

If $i \in T_k$ and if $a_{i,j} \neq 0$ and $i > j$, then since the permutation is nonmigratory, $a'_{\sigma(i),\sigma(j)}$ must be below the main diagonal in A'; hence $\sigma(i) > \sigma(j)$. Since $\sigma(i) \in S_k$, then $\sigma(j) \in S_{k-1}$ and $j \in T_{k-1}$. Similarly, if $i < j$ we have $j \in T_{k+1}$. Therefore, the sets $T_1, T_2, \ldots, T_t$ satisfy the requirements of Definition 3.2 and hence A is consistently ordered. This completes the proof of Theorem 5.1.

We now show that given a consistently ordered matrix A with nonvanishing diagonal elements, one can find a T-matrix A' such that the matrices associated with the SOR and Jacobi methods have the same eigenvalues for A' as for A.

Theorem 5.2. Let A be a consistently ordered matrix with nonvanishing diagonal elements. There exists a T-matrix A' with nonvanishing diagonal elements such that $\mathscr{L}_\omega[A]$ is similar to $\mathscr{L}_\omega[A']$ and such that $B[A]$ is similar to $B[A']$.

Proof. The proof follows at once from:

Lemma 5.3. If A is a consistently ordered matrix with nonvanishing diagonal elements, then there exists a nonmigratory permutation matrix P such that $A' = P^{-1}AP$ is a T-matrix with nonvanishing diagonal elements and such that

$$\mathscr{L}_\omega[P^{-1}AP] = P^{-1}\mathscr{L}_\omega[A]P, \qquad B[P^{-1}AP] = P^{-1}B[A]P \tag{5.7}$$

Proof. By (4-3.15), we have

$$\mathscr{L}_\omega[A'] = (I - \omega L[A'])^{-1}(\omega U[A'] + (1 - \omega)I) \tag{5.8}$$

and

$$\begin{aligned} L[A'] &= D[A']^{-1}C_L[A'] \\ U[A'] &= D[A']^{-1}C_U[A'] \end{aligned} \tag{5.9}$$

By (5.1), it follows that

$$L[A'] = P^{-1}L[A]P, \qquad U[A'] = P^{-1}U[A]P \tag{5.10}$$

and

$$\mathscr{L}_\omega[A'] = P^{-1}\mathscr{L}_\omega[A]P \tag{5.11}$$

Moreover, by (4-3.14) and (4-3.15), we have

$$\begin{aligned} B[A'] &= L[A'] + U[A'] \\ &= P^{-1}L[A]P + P^{-1}U[A]P \\ &= P^{-1}(L[A] + U[A])P \\ &= P^{-1}(B[A])P \end{aligned} \tag{5.12}$$

This completes the proof of the lemma.

To illustrate Lemma 5.3, let us consider the discrete analog of the Dirichlet problem for the region shown in Figure 5.1 with $h = 1$. The matrix is given by

$$A = \begin{pmatrix} 4 & -1 & 0 & 0 & -1 & 0 \\ -1 & 4 & -1 & 0 & 0 & -1 \\ 0 & -1 & 4 & -1 & 0 & 0 \\ 0 & 0 & -1 & 4 & 0 & 0 \\ -1 & 0 & 0 & 0 & 4 & -1 \\ 0 & -1 & 0 & 0 & -1 & 4 \end{pmatrix} \tag{5.13}$$

One can verify that A is consistently ordered with the sets $S_1 = \{1\}$, $S_2 = \{2, 5\}$, $S_3 = \{3, 6\}$, $S_4 = \{4\}$. Thus, following (5.2) we let

$$\begin{array}{lll} \sigma(1) = 1, & \sigma(2) = 2, & \sigma(3) = 4 \\ \sigma(4) = 6, & \sigma(5) = 3, & \sigma(6) = 5 \end{array} \tag{5.14}$$

The corresponding permutation matrix is

$$P = \begin{pmatrix} 1 & 0 & 0 & 0 & 0 & 0 \\ 0 & 1 & 0 & 0 & 0 & 0 \\ 0 & 0 & 0 & 1 & 0 & 0 \\ 0 & 0 & 0 & 0 & 0 & 1 \\ 0 & 0 & 1 & 0 & 0 & 0 \\ 0 & 0 & 0 & 0 & 1 & 0 \end{pmatrix} \tag{5.15}$$

The matrix $A' = P^{-1}AP$ is given by (4.7), which has the form (2.1). Moreover, one can verify that $P^{-1}C_L[A]P$ is a strictly lower triangular matrix and $P^{-1}C_U[A]P$ is a strictly upper triangular matrix.

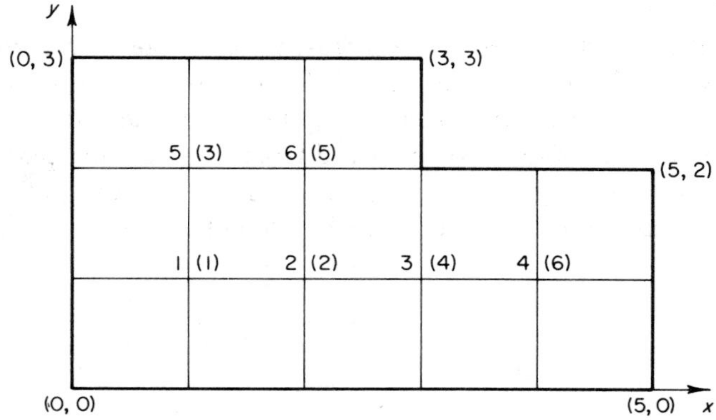

Figure 5.1.

The matrix A' could be obtained alternatively by relabeling the points as indicated in parentheses in Figure 5.1. Thus the point originally labeled i is relabeled $\sigma(i)$. The reader should verify this. One can easily see that the relabeling does not effect the ordering relation between any pair of adjacent mesh points.

5.6. CONSISTENTLY ORDERED MATRICES ARISING FROM DIFFERENCE EQUATIONS

We have already seen (in Section 2.8) that matrices arising in the five-point finite difference solution of elliptic partial differential equations frequently have Property A. If such is the case, given a configuration of N mesh points R_h with the ordering indicated by numbers $1, 2, \ldots, N$ we can determine whether or not the matrix arising from a five-point difference equation is consistently ordered in the following way. Draw arrows between each pair of adjacent mesh points in the direction of increasing labels. For each closed path consisting of mesh segments, if the number of arrows in clockwise direction equals the number of arrows in the counterclockwise direction, then the corresponding matrix is consistently ordered. For a simply connected mesh configuration (such that for each closed path all mesh points contained inside the path are in R_h), it is sufficient that for each set of four mesh points arranged in a square there should be two counterclockwise-oriented arrows and two clockwise arrows.

If the situation described above holds, we can construct a compatible ordering vector by first letting $\gamma_i = 1$ for one point in each connected subset of R_h. Let j be any mesh point in the connected subset of R_h which contains the point i. We proceed from i to j by any path of mesh segments, increasing γ_i by one for each segment of the path with the arrow and decreasing γ_i by one for each segment against the arrow. Since

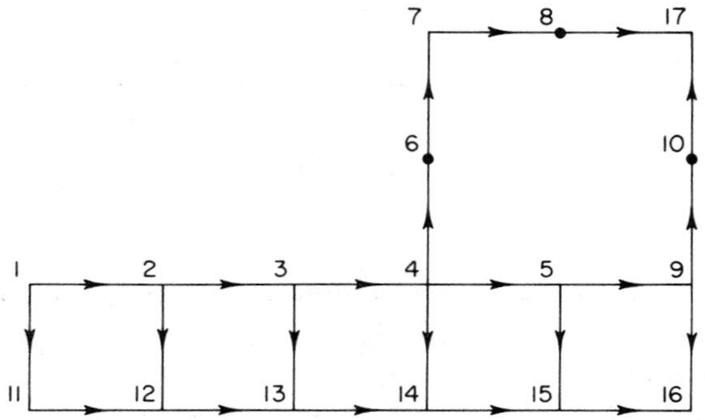

Figure 6.1.

the net change in γ in going around any closed path is zero, it follows that the value of γ_j thus obtained will be independent of the path. Thus we are able to define a compatible ordering vector. In the example shown in Figure 6.1 the "circulation" around each closed path is zero, and we can define $\gamma_1 = 1$, $\gamma_2 = \gamma_{11} = 2$, $\gamma_3 = \gamma_{12} = 3$, $\gamma_4 = \gamma_{13} = 4$, $\gamma_5 = \gamma_{14} = \gamma_6 = 5$, $\gamma_{15} = \gamma_9 = \gamma_7 = 6$, $\gamma_{16} = \gamma_{10} = \gamma_8 = 7$, $\gamma_{17} = 8$. The ordering vector is easily seen to be compatible.

In general, given a set of mesh points $(x_0 + p_i h, y_0 + q_i h)$, $i = 1, 2, \ldots, N$ we first seek to determine an ordering vector. Having done so, we know by Theorem 4.5 that for a suitable relabeling of the mesh points one obtains a matrix A such that the ordering vector is compatible. The existence of at least one ordering vector is guaranteed by Theorem 4.1, since with any labeling of the points we have Property A.

Two ordering vectors $\gamma^{(0)}$ and $\gamma^{(1)}$ are the following:

$$\gamma^{(0)} = \begin{cases} 1 & \text{if} \quad p_i + q_i \quad \text{is even} \\ 2 & \text{if} \quad p_i + q_i \quad \text{is odd} \end{cases} \tag{6.1}$$

$$\gamma_i^{(1)} = p_i + q_i \tag{6.2}$$

Methods of labeling the mesh points such that the resulting matrices have $\gamma^{(0)}$ or $\gamma^{(1)}$ as a compatible ordering vector are as follows:

(1) All points $(x_0 + ph, y_0 + qh)$ with $p + q$ even ("red" points) occur before those with $p + q$ odd ("black" points).

This ordering is known as a "red–black" ordering. Thus $\gamma^{(0)}$ is a compatible ordering vector. Moreover, A is a T-matrix.

(2) A point $(x_0 + ph, y_0 + qh)$ occurs before $(x_0 + p'h, y_0 + q'h)$ if $q < q'$ or if $q = q'$ and $p < p'$.

This ordering is known as the "natural" ordering. Thus $\gamma^{(1)}$ is a compatible ordering vector. In general, A is not a T-matrix but there exists a nonmigratory permutation P such that $P^{-1}AP$ is a T-matrix.

(3) A point $(x_0 + ph, y_0 + qh)$ occurs before $(x_0 + p'h, y_0 + q'h)$ if $p + q < p' + q'$.

This ordering is known as "ordering by diagonals." Thus $\gamma^{(1)}$ is a compatible ordering vector. A is a T-matrix and can be obtained from the matrix corresponding to the natural ordering by a nonmigratory permutation.

Thus, for a problem arising from the solution of a five-point difference equation on a square mesh, the red–black ordering, the natural ordering, and ordering by diagonals all lead to consistently ordered matrices. The same is true for many other orderings as well. It is important to note that by Theorem 3.4 the eigenvalues of $\mathscr{L}_\omega$ are the same for all such orderings.

5.7. A COMPUTER PROGRAM FOR TESTING FOR PROPERTY A AND CONSISTENT ORDERING

We now describe a systematic procedure for testing whether or not a matrix has Property A and, if so, whether or not it is consistently ordered.

The procedure involves an attempt to construct two ordering vectors $\bar{\gamma}$ and γ. The ordering vector $\bar{\gamma}$ has at most two different components and, by Corollary 4.2, $\bar{\gamma}$ must exist if the matrix has Property A. The ordering vector γ is a compatible ordering vector which must exist if the matrix is consistently ordered.

The first step in the procedure is to construct a matrix $R = (r_{i,j})$ such that $r_{i,j} = 1$ if $i \neq j$, and if either $a_{i,j} \neq 0$ or $a_{j,i} \neq 0$, i.e., if i and j are associated. For other values of i and j, we let $r_{i,j} = 0$.

To begin with, we let $\bar{\gamma}_1 = \gamma_1 = 1$ and seek all j such that $r_{1,j} \neq 0$. For all such j we assign $\bar{\gamma}_j = \gamma_j = 2$. When a value of $\bar{\gamma}$ and γ has been assigned to an integer i we set $a_i = 1$. (Initially $a_1 = 1$, $a_2 = a_3 = \cdots = a_N = 0$.) The vector $d = (d_1, d_2, \ldots, d_N)$ is determined as follows: Initially, $d_1 = 1$, $d_2 = d_3 = \cdots = d_N = 0$. For all j such that $r_{1,j} \neq 0$ we let $d_j = 1$. When we have considered all integers j in relation to 1 (i.e., assigned $\bar{\gamma}_j$ and γ_j for all j with $r_{1,j} \neq 0$), we then set $d_1 = 0$ and choose i as the smallest integer such that $d_i \neq 0$. If no such i exists, then we choose i as the smallest integer such that $a_i = 0$, i.e., such that $\bar{\gamma}_i$ and γ_i have not been assigned. In the former case $\bar{\gamma}_i$ and γ_i have already been assigned; in the latter case we arbitrarily let $\bar{\gamma}_i = \gamma_i = 1$.

In the general case, suppose we are working with the integer i and are trying to assign $\bar{\gamma}_j$ and γ_j for all j with $r_{i,j} \neq 0$. If $a_j = 0$, we set $d_j = 1$. We can also set $\bar{\gamma}_j = 1 - \bar{\gamma}_i$ and $\gamma_j = \gamma_i + 1$ if $j > i$ or $\gamma_j = \gamma_i - 1$ if $j < i$. However, if $a_j = 1$, then $\bar{\gamma}_j$ and γ_j have already been assigned. We must test to see if the proposed new assigned values agree with the old. If the proposed value of γ_j does not agree with the already assigned value, then no compatible ordering vector can be found, and hence A is not consistently ordered. The switch α is set to α_2 so that no further attempts will be made to construct γ. If the proposed value of $\bar{\gamma}_j$ does not agree with the already assigned value, then the matrix does not have Property A and the program is terminated.

Having completed the "processing" of i we see if $d_j \neq 0$ for any j and, if so, we process the smallest such integer. If no such j exists, we see if there are any other integers j remaining to be processed, i.e., if $a_j = 0$ for some j. If so, we process the smallest such integer. Otherwise, $\bar{\gamma}$ and, possibly, γ have been assigned without inconsistency and the matrix has Property A and, possibly, is consistently ordered. The vector $\bar{\gamma}$ and possibly γ are then printed and the program is terminated. (See Figure 7.1.)

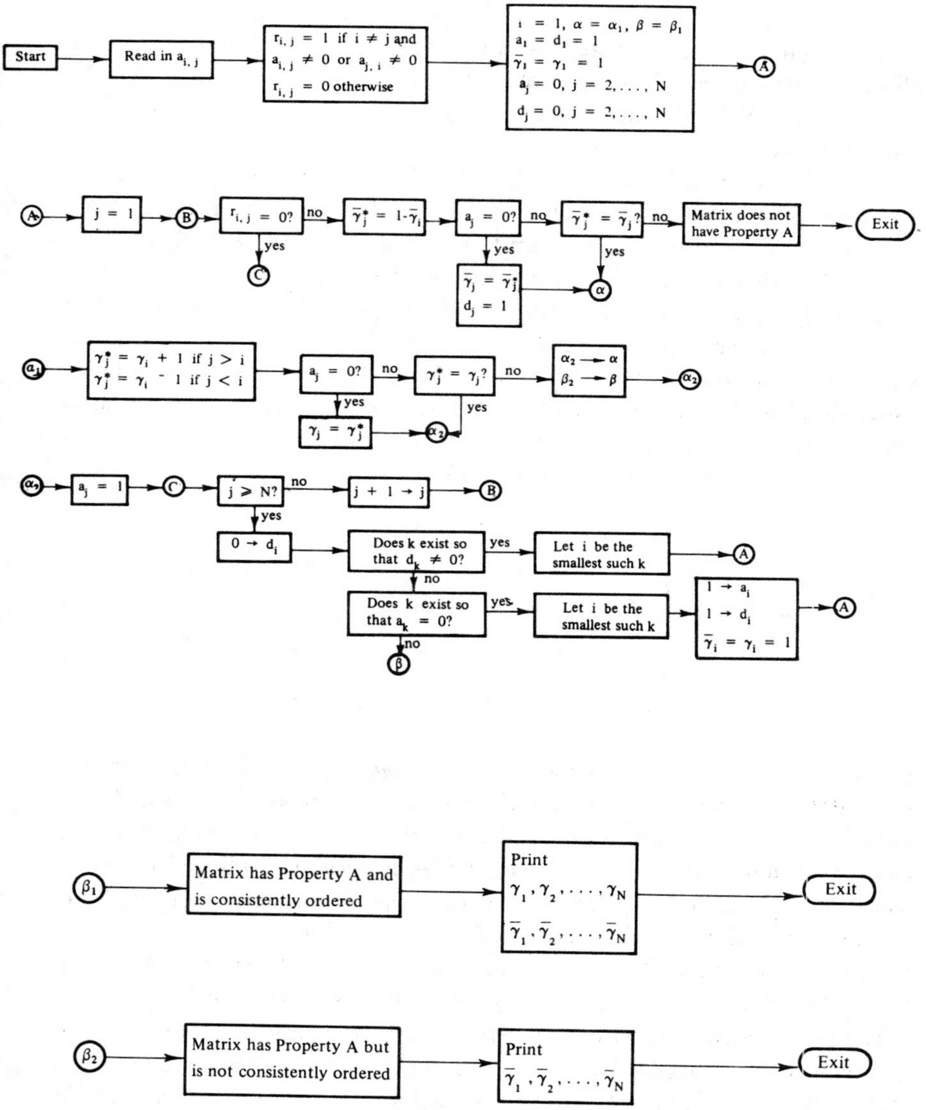

Figure 7.1. Test for consistently ordered matrix and Property A.

5.8. OTHER DEVELOPMENTS OF THE SOR THEORY

Consistently ordered matrices and matrices with Property A are closely related to certain classes of matrices considered by Varga [1962]. We give a very brief summary of Varga's development and relate it to our discussion.

From Theorem 2-1.17, we know that if $B \geq 0$, then $S(B)$ is an eigenvalue of B. We showed in Theorem 2-7.6 that if B is an irreducible symmetric nonnegative matrix and if $-S(B)$ is an eigenvalue of B, then B has Property A and has vanishing diagonal elements. Moreover, both $S(B)$ and $-S(B)$ are simple. More generally, Varga defined a matrix B to be *cyclic of index* $p \geq 2$ as an irreducible nonnegative matrix with p (simple) eigenvalues of modulus $S(B)$. If there is only one eigenvalue of B with modulus $S(B)$, then the matrix B is *primitive*. Varga showed that if B is cyclic of index p, then there exists a permutation matrix P such that $P^{-1}BP$ has the block form

$$P^{-1}BP = \begin{pmatrix} 0 & 0 & \cdots & 0 & \cdots & B_{1,p} \\ B_{2,1} & 0 & \cdots & 0 & \cdots & 0 \\ 0 & B_{3,2} & \cdots & 0 & \cdots & 0 \\ \vdots & \vdots & \cdots & \vdots & \cdots & \vdots \\ 0 & 0 & \cdots & B_{p,p-1} & \cdots & 0 \end{pmatrix} \tag{8.1}$$

A matrix B is said to be *weakly cyclic of index* $p \geq 2$ if there exists a permutation matrix P such that $P^{-1}BP$ has the form (8.1). Finally, a *p-cyclic* matrix A, where $p \geq 2$, is a matrix with nonvanishing diagonal elements such that $B = I - (\text{diag } A)^{-1}A$ is weakly cyclic of index p. Varga [1962] defines a p-cyclic matrix A as "consistently ordered" if all the eigenvalues of the matrix $\alpha L + \alpha^{-(p-1)}U$ are independent of α for all $\alpha \neq 0$, where $B = L + U$ and where L and U are strictly lower triangular and strictly upper triangular matrices, respectively.

Evidently, a matrix with Property A and nonvanishing diagonal elements is either a diagonal matrix or a 2-cyclic matrix. The corresponding matrix $B = I - (\text{diag } A)^{-1}A$ is either a null matrix or is weakly cyclic of index 2. If A is also an irreducible L-matrix, then B is cyclic of index 2. If A is consistently ordered in our sense and has nonvanishing diagonal elements, then, by Theorem 3.3, it is consistently ordered in the sense of Varga. However, as we shall see in Chapter 13, there exist matrices which are consistently ordered in Varga's sense which are not in our sense. On the other hand, it will also be shown in Chapter 13 that if A is an irreducible L-matrix such that $\det(\alpha C_L + \alpha^{-1} C_U - kD)$ is independent of α for all $\alpha \neq 0$ and for all k, then A is consistently ordered in our sense as well as in Varga's sense. This result is originally due to Broyden [1968].

These matters will be discussed in more detail in Chapter 13 where we will consider the class of generalized consistently ordered matrices as well as generalizations of Property A.

SUPPLEMENTARY DISCUSSION

Theorem 3.3 is equivalent to a result of Young [1950, 1954]. Theorem 3.4 was also proved by Young [1950, 1954]. Friedman [1957] developed the theory for the case of T-matrices. While, at first glance, it might appear that the development of the theory along the lines of T-matrices and nonmigratory permutations should be simpler than the alternative development based on ordering vectors, such is not necessarily the case, at least in the opinion of the author.

EXERCISES

Section 5.2

1. Verify directly that Theorem 2.1 holds for the T-matrix

$$A = \begin{pmatrix} 4 & -1 & -1 & 0 \\ -1 & 4 & 0 & -1 \\ -1 & 0 & 4 & -1 \\ 0 & -1 & -1 & 4 \end{pmatrix}$$

Compute the eigenvalues of B, $\mathscr{L}$, and $\mathscr{L}_{1/2}$; verify Theorem 2.2.

Section 5.3

1. Determine all of the ordering vectors of the matrix (3.3) and show that none is compatible.

2. Show that the matrix (3.4) has no ordering vectors.

3. Show that the matrix

$$A = \begin{pmatrix} 4 & -1 & 0 & -1 & 0 \\ -1 & 4 & -1 & 0 & -1 \\ 0 & -1 & 4 & 0 & 0 \\ -1 & 0 & 0 & 4 & -1 \\ 0 & -1 & 0 & -1 & 4 \end{pmatrix}$$

is not a T-matrix but is consistently ordered. Verify Theorem 3.3.

4. Let A be a consistently ordered matrix with nonvanishing diagonal elements.

(a) What can be said about ω and about the eigenvalues of B if it is known that $\lambda = 0$ is an eigenvalue of $\mathscr{L}_\omega$?

(b) If $N = 7$ and if the eigenvalues of B are 0, 0.4, -0.4, 0.8, -0.8, 0.9, -0.9, what are the eigenvalues of $\mathscr{L}_1$ and $\mathscr{L}_{1.25}$?

(c) If $N = 4$ and if the eigenvalues of $\mathscr{L}_{1.64}$ are $0.32(2)^{1/2} \pm 0.32(2)^{1/2}i$, 0.8, 0.512, what are the eigenvalues of B?

Section 5.4

1. Which of the following matrices have Property A?

$$(0), \quad \begin{pmatrix} 2 & 1 \\ 1 & 7 \end{pmatrix}, \quad \begin{pmatrix} 2 & 0 & 0 \\ 1 & 8 & 0 \\ 5 & 1 & 2^{1/2} \end{pmatrix}, \quad \begin{pmatrix} 2 & 1 & 0 \\ 3 & 5 & 8 \\ 2 & 0 & 1 \end{pmatrix}, \quad \begin{pmatrix} 4 & -1 & -1 & 0 \\ -1 & 4 & 0 & -1 \\ -1 & 0 & 4 & -1 \\ 0 & -1 & -1 & 4 \end{pmatrix}$$

2. Show that the matrix

$$A = \begin{pmatrix} 4 & -1 & 0 & 0 & 0 & -1 \\ -1 & 4 & -1 & 0 & -1 & 0 \\ 0 & -1 & 4 & -1 & 0 & 0 \\ 0 & 0 & -1 & 4 & -1 & 0 \\ 0 & -1 & 0 & -1 & 4 & -1 \\ -1 & 0 & 0 & 0 & -1 & 4 \end{pmatrix}$$

has Property A but is not consistently ordered. Construct an ordering vector and from this find a permutation matrix P such that $P^{-1}AP$ is consistently ordered. Verify Theorem 4.4.

3. Show that the matrix (4.4) does not have a compatible ordering vector.

4. For the model problem for Laplace's equation with $h = \frac{1}{3}$, consider the six matrices arising from all orderings of the four interior mesh points such that the point $(\frac{1}{3}, \frac{1}{3})$ is labeled 1. Indicate which of the matrices have Property A, which are consistently ordered, and which are T-matrices. Compute the eigenvalues of B for each ordering and the eigenvalues of $\mathscr{L}$ for each ordering which corresponds to a consistently ordered matrix.

5. For the matrix (3.3) find a permutation matrix P such that $A' = P^{-1}AP$ is a consistently ordered matrix. Compare $C_L[A']$ with $P^{-1}C_L[A]P$.

6. Complete the proof of Theorem 4.6.

Section 5.5

1. Show that if A is a tri-diagonal matrix, then $P^{-1}AP$ is consistently ordered for all permutation matrices P.

2. Consider the matrix

$$A = \begin{pmatrix} 4 & 0 & 0 & -1 & 0 & -1 & 0 & 0 & 0 \\ 0 & 4 & 0 & 0 & -1 & 0 & -1 & 0 & -1 \\ 0 & 0 & 4 & 0 & 0 & -1 & 0 & -1 & 0 \\ -1 & 0 & 0 & 4 & 0 & 0 & -1 & 0 & -1 \\ 0 & -1 & 0 & 0 & 4 & 0 & 0 & -1 & 0 \\ -1 & 0 & -1 & 0 & 0 & 4 & 0 & 0 & -1 \\ 0 & -1 & 0 & -1 & 0 & 0 & 4 & 0 & 0 \\ 0 & 0 & -1 & 0 & -1 & 0 & 0 & 4 & -1 \\ 0 & -1 & 0 & -1 & 0 & -1 & 0 & -1 & 4 \end{pmatrix}$$

(a) Show that A has Property A.

(b) Is A a consistently ordered matrix? If not, find a permutation matrix P such that $P^{-1}AP$ is consistently ordered.

(c) If A is consistently ordered, find a nonmigratory permutation matrix Q such that $Q^{-1}AQ$ is a T-matrix. Otherwise, find a migratory permutation matrix Q such that $Q^{-1}AQ$ is a T-matrix.

3. Consider the discrete analog of the Dirichlet problem for the region shown in the accompanying figure with $h = 1$.

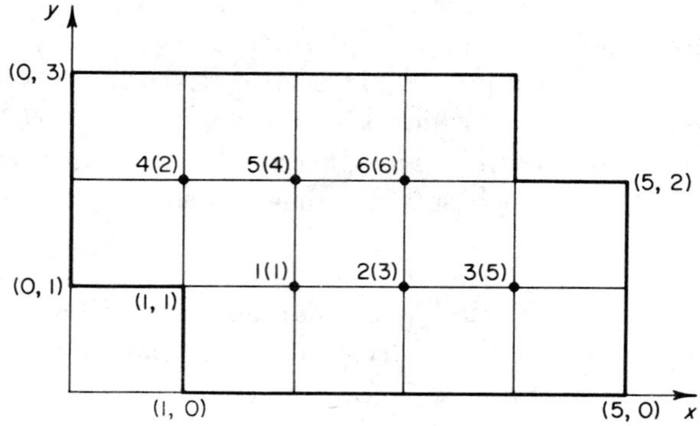

(a) Show that replacing derivatives by difference quotients and multiplying by $-h^2$ one obtains a linear system $Au = b$ where

$$A = \begin{pmatrix} 4 & -1 & 0 & 0 & -1 & 0 \\ -1 & 4 & -1 & 0 & 0 & -1 \\ 0 & -1 & 4 & 0 & 0 & 0 \\ 0 & 0 & 0 & 4 & -1 & 0 \\ -1 & 0 & 0 & -1 & 4 & -1 \\ 0 & -1 & 0 & 0 & -1 & 4 \end{pmatrix}$$

Here the interior mesh points are labeled as indicated by the numbers without parentheses.

(b) Show that the matrix has Property A and is, in fact, consistently ordered.

(c) Consider the permutation $\sigma(i)$ where $\sigma(1) = 1$, $\sigma(4) = 2$, $\sigma(2) = 3$, $\sigma(5) = 4$, $\sigma(3) = 5$, $\sigma(6) = 6$. Construct the corresponding permutation matrix P and show that $A' = P^{-1}AP$ is a T-matrix. Verify that $C_L[A'] = P^{-1}C_L[A]P$ and $C_U[A'] = P^{-1}C_U[A]P$.

(d) With the points relabeled as indicated by the numbers in the parentheses, obtain the matrix A'' as in (a) and show that $A'' = A'$.

4. In the previous example, if the points are labeled as in this accompanying diagram show that A has Property A but is not consistently ordered. Find an ordering vector for A and construct a permutation matrix P such that $A' = P^{-1}AP$ is consistently ordered.

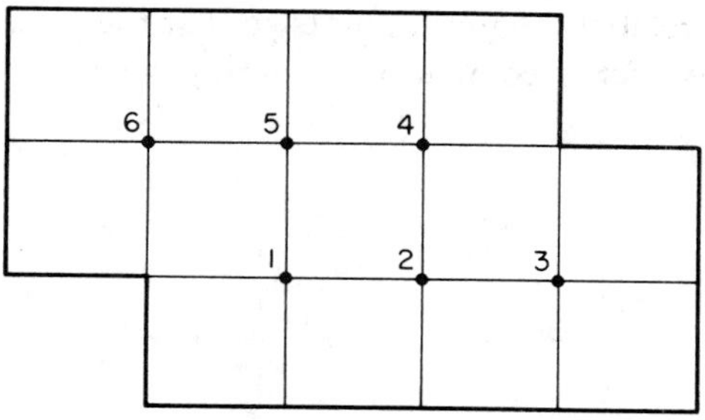

Section 5.6

1. Show that the ordering vector given for the problem of Figure 6.1 is compatible.

2. Verify that the red–black ordering, the natural ordering, and the ordering by diagonals all lead to consistently ordered matrices. Specifically consider the model problem in the unit square with $h = \frac{1}{4}$.

3. Consider the model problem for Laplace's equation with $h = \frac{1}{4}$ with the ordering of the interior mesh points given in the accompanying figure. Is the corresponding matrix consistently ordered? If so, find a compatible ordering vector and a nonmigratory permutation matrix P such that $P^{-1}AP$ is a T-matrix.

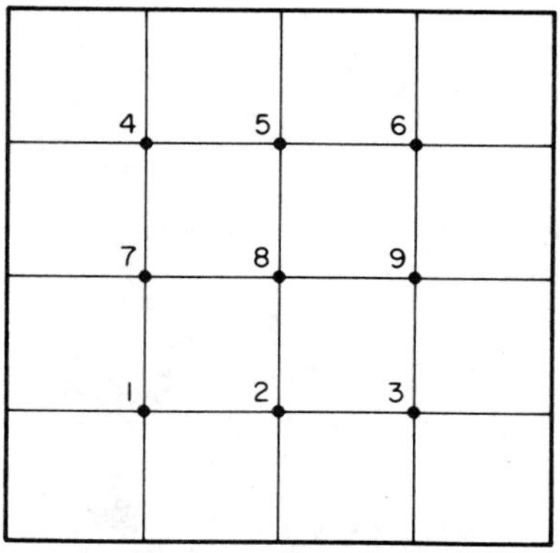

4. Consider the discrete analog of the Dirichlet problem for the region shown below with mesh size $h = 1$. With the interior mesh points labeled as indicated in the accompanying figure, find an ordering vector and, if possible, a compatible ordering vector. If the matrix is not consistently ordered relabel the mesh points so that the matrix is consistently ordered and construct a compatible ordering vector.

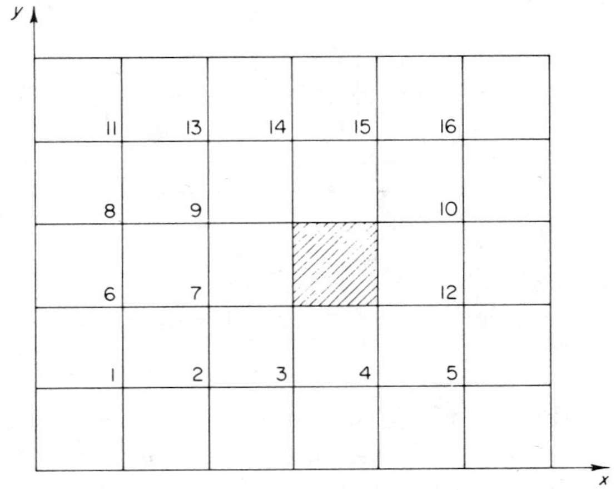

Section 5.7

1. Verify the validity of the procedure described for determining whether or not a matrix A has Property A and is consistently ordered. Specifically, consider the case of the matrix corresponding to the problem of Figure 6.1 with the labeling indicated and also with the points 12 and 4 interchanged.

Chapter 6 / DETERMINATION OF THE OPTIMUM RELAXATION FACTOR

In Chapter 5 we derived a relation between the eigenvalues of B and those of $\mathscr{L}_\omega$ which holds whenever A is a consistently ordered matrix with nonvanishing diagonal elements. In this chapter we use this relation to study the convergence and rate of convergence of $\mathscr{L}_\omega$ under various assumptions on the eigenvalues of B. The case where all eigenvalues of B are real will be treated in Section 6.2. It will be shown that if $\bar{\mu} < 1$ then the value of ω which is optimum in the sense of minimizing $S(\mathscr{L}_\omega)$ is given by

$$\omega_b = \frac{2}{1 + (1 - \bar{\mu}^2)^{1/2}} = 1 + \left(\frac{\bar{\mu}}{1 + (1 - \bar{\mu}^2)^{1/2}}\right)^2 \tag{1}$$

where $\bar{\mu} = S(B)$. It will be shown in Section 6.3 that with this choice of ω the SOR method converges faster than the GS method by an order of magnitude.

In Section 6.4 convergence conditions for the SOR method will be derived under the assumption that the eigenvalues μ of B lie in certain regions in the complex plane. It will be shown that the SOR method converges for sufficiently small ω, if for all μ we have $|\operatorname{Re} \mu| < 1$. The optimum value of ω is determined under various assumptions on the location of the eigenvalues of B.

The remainder of the chapter will be devoted to methods for estimating a value of $\bar{\mu}$ to be used in (1) to determine ω_b. General methods will be given as well as methods applicable to the discrete generalized Dirichlet problem. Numerical results will be given in Section 6.10.

6.1. VIRTUAL SPECTRAL RADIUS

Let G be a matrix which is a function of the matrix A and suppose that for some quadratic equation

$$P(\lambda) = \lambda^2 - b(\mu)\lambda + c(\mu) = 0 \tag{1.1}$$

with continuous coefficients $b(\mu)$ and $c(\mu)$, we have

(a) If λ is an eigenvalue of G, then, for some eigenvalue μ of B, λ satisfies (1.1).

(b) If μ is any eigenvalue of B, then at least one of the roots of (1.1) is an eigenvalue of G and, except for isolated values of μ, both roots of (1.1) are eigenvalues of G.

For each complex number μ we let $\psi(\mu)$ be the root radius of (1.1), i.e., the maximum of the moduli of the roots of (1.1).

We define the *virtual spectral radius* of G by

$$\bar{S}(G) = \max_{\mu \in \bar{S}_B} \psi(\mu) \tag{1.2}$$

where $\bar{S}_B$ is the smallest convex set containing S_B, the set of all eigenvalues of B. It is evident that

$$S(G) \leq \bar{S}(G) \tag{1.3}$$

It is possible that $S(G)$ may be less than $\bar{S}(G)$ for two reasons. First, it may happen that for certain values of μ only one of the roots of (1.1) is an eigenvalue of G. Second, we are allowing μ to vary over $\bar{S}_B$ rather than S_B. On the other hand, there are usually so many eigenvalues of B that it would be impractical to try to consider them individually. We are more likely to be able to estimate bounds on the eigenvalues, and such bounds frequently lead to a good choice of $\bar{S}_B$. For example, if we know that all eigenvalues of B are real, we need only estimate upper and lower bounds. If A has Property A, then the upper bound will be the negative of the lower bound so that only one number need be estimated.

We shall seek to show that the SOR method and certain other methods are strongly convergent in the following sense:

Definition 1.1. Let (3-1.5) be an iterative method such that a relation of the form (1.1) exists between the eigenvalues of G and the eigenvalues of B. The iterative method is *strongly convergent* if

$$\bar{S}(G) < 1 \tag{1.4}$$

As an example, let us consider the case of $\mathscr{L}_\omega$ where A is a consistently ordered matrix with nonvanishing diagonal elements. Here, by Theorem 5-3.4 the equation (1.1) becomes

$$\lambda^2 - (\omega^2\mu^2 - 2(\omega - 1))\lambda + (\omega - 1)^2 = 0 \tag{1.5}$$

In this case if λ is an eigenvalue of $\mathscr{L}_\omega$, then (1.5) is satisfied for some eigenvalue μ of B while, on the other hand, if μ is any eigenvalue of B and if λ satisfies (1.5), then λ is an eigenvalue of $\mathscr{L}_\omega$. Hence in this case we have

$$\bar{S}(\mathscr{L}_\omega) = \max_{\mu \in \bar{S}_B} \psi(\mu) \tag{1.6}$$

6.2. ANALYSIS OF THE CASE WHERE ALL EIGENVALUES OF B ARE REAL

In this section we study the implications of the correspondence given in Chapter 5 between the eigenvalues of $\mathscr{L}_\omega$ and those of B under the assumption that the eigenvalues of B are real. We shall first show that the SOR method converges if and only if $|\omega - 1| < 1$ and $S(B) < 1$. For the case $S(B) < 1$ we shall determine the optimum value ω_b of ω which minimizes $S(\mathscr{L}_\omega)$ and we shall also give the spectral radius $S(\mathscr{L}_{\omega_b})$. Finally, we shall show that in order for the eigenvalues of B to be real and for $S(B)$ to be less than unity it is sufficient that A be positive definite and consistently ordered.

Convergence

To establish the convergence of $\mathscr{L}_\omega$ it is clearly necessary and sufficient to show that for any eigenvalue μ of B both roots of (5-2.6) are less than one in modulus. We first prove the following lemma relating to the roots of a quadratic equation.

Lemma 2.1. If b and c are real, then both roots of the quadratic equation

$$x^2 - bx + c = 0 \tag{2.1}$$

are less than one in modulus if and only if

$$|c| < 1, \qquad |b| < 1 + c \tag{2.2}$$

Proof. Let r_1 and r_2 be the roots of (2.1). If $|r_1| < 1$ and $|r_2| < 1$, then $|c| < 1$ since $c = r_1 r_2$. Since $b = r_1 + r_2$ we have

$$1 + c - |b| = 1 + r_1 r_2 - (r_1 + r_2) = (1 - r_1)(1 - r_2) > 0 \tag{2.3}$$

if $r_1 + r_2 \geq 0$ and

$$1 + c - |b| = 1 + r_1 r_2 + r_1 + r_2 = (1 + r_1)(1 + r_2) > 0 \tag{2.4}$$

if $r_1 + r_2 < 0$. In either case $|b| < 1 + c$.

On the other hand, if (2.2) holds, then (2.3) or (2.4) holds. If (2.3) holds, then either r_1 and r_2 are both less than one or both greater than one. The latter case is impossible since $c = r_1 r_2 < 1$. If either $r_1 \leq -1$ or $r_2 \leq -1$, then, since $r_1 + r_2 \geq 0$, we would have $r_1 \geq 1$ or $r_2 \geq 1$ which is impossible. Thus in this case $|r_1| < 1$ and $|r_2| < 1$. A similar argument holds if $r_1 + r_2 < 0$, and the lemma follows.

By (5-2.6) the eigenvalues λ of $\mathscr{L}_\omega$ satisfy

$$\lambda - \omega\mu\lambda^{1/2} + \omega - 1 = 0 \tag{2.5}$$

Considering this as a quadratic in $\lambda^{1/2}$ and applying Lemma 2.1 with $b = \omega\mu$, $c = \omega - 1$, we have

Theorem 2.2. If A is a consistently ordered matrix with nonvanishing diagonal elements such that $B = I - (\text{diag } A)^{-1}A$ has real eigenvalues, then $S(\mathscr{L}_\omega) = \bar{S}(\mathscr{L}_\omega) < 1$ if and only if

$$0 < \omega < 2 \tag{2.6}$$

and

$$S(B) < 1 \tag{2.7}$$

Determination of ω_b

Theorem 2.3. Let A be a consistently ordered matrix with nonvanishing diagonal elements such that the matrix $B = I - (\text{diag } A)^{-1}A$ has real eigenvalues and such that $\bar{\mu} = S(B) < 1$. If ω_b is defined by (1), then

$$\bar{S}(\mathscr{L}_{\omega_b}) = S(\mathscr{L}_{\omega_b}) = \omega_b - 1 \tag{2.8}$$

and if $\omega \neq \omega_b$, then

$$\bar{S}(\mathscr{L}_\omega) = S(\mathscr{L}_\omega) > S(\mathscr{L}_{\omega_b}) \tag{2.9}$$

Moreover, for any ω in the range $0 < \omega < 2$, we have

$$\bar{S}(\mathscr{L}_\omega) = S(\mathscr{L}_\omega) = \begin{cases} \left[\dfrac{\omega\bar{\mu} + (\omega^2\bar{\mu}^2 - 4(\omega - 1))^{1/2}}{2}\right]^2 & \text{if } 0 < \omega \leq \omega_b \\ \omega - 1 & \text{if } \omega_b \leq \omega < 2 \end{cases} \tag{2.10}$$

Finally, if $0 < \omega < \omega_b$, then $S(\mathscr{L}_\omega)$ is a strictly decreasing function of ω.

Proof. We shall first give a purely analytic proof. An alternative proof, based on geometrical considerations, will be given later.

We remark that ω_b is uniquely determined by the conditions

$$\begin{gathered} \omega_b{}^2\bar{\mu}^2 = 4(\omega_b - 1) \\ 1 \leq \omega_b < 2 \end{gathered} \tag{2.11}$$

To see this, we note that ω_b given by (1) is one of the roots of the quadratic $\omega^2\bar{\mu}^2 = 4(\omega - 1)$, and clearly, ω_b lies in the range $1 \leq \omega < 2$. Since the product of the roots of the quadratic is $4/\bar{\mu}^2$, it follows that the other root lies outside of that range. Hence ω_b is uniquely determined by (2.11).

Since $0 \leq \omega_b - 1 < 1$, then by Theorem 2.2 we have $\bar{S}(\mathscr{L}_{\omega_b}) < 1$. Moreover, since $\bar{S}(\mathscr{L}_\omega) \geq 1$, if $|\omega - 1| \geq 1$ by Theorem 4-1.2, it follows that (2.9) holds if $|\omega - 1| \geq 1$. Thus we need consider only the case $|\omega - 1| < 1$.

Let us define the function $\Gamma(\omega, \mu)$ by

$$\Gamma(\omega, \mu) = \left|\frac{\omega|\mu| + [\omega^2\mu^2 - 4(\omega - 1)]^{1/2}}{2}\right|^2 \tag{2.12}$$

Evidently, $\Gamma(\omega, \mu)$ is the maximum of the moduli of the values of λ satisfying (2.5). Hence by Theorem 5-3.4 we have

$$S(\mathscr{L}_\omega) = \max_{\mu \in S_B} \Gamma(\omega, \mu) \leq \bar{S}(\mathscr{L}_\omega) = \max_{\mu \in \bar{S}_B} \Gamma(\omega, \mu) \tag{2.13}$$

where S_B is the set of all eigenvalues μ of B and $\bar{S}_B$ is the smallest convex set containing S_B. We first prove

Lemma 2.4. Under the hypotheses of Theorem 2.3,

$$\bar{S}(\mathscr{L}_\omega) = \Gamma(\omega, \bar{\mu}) = S(\mathscr{L}_\omega) \tag{2.14}$$

and (2.10) holds.

Proof. We show that

$$\max_{-\bar{\mu} \le \mu \le \bar{\mu}} \Gamma(\omega, \mu) = \Gamma(\omega, \bar{\mu}) \tag{2.15}$$

First, if $\omega^2\bar{\mu}^2 - 4(\omega - 1) < 0$, then $\omega > 1$ and $\omega^2\mu^2 - 4(\omega - 1) < 0$ for all μ such that $|\mu| \le \bar{\mu}$. Moreover, for all μ such that $|\mu| \le \bar{\mu}$ we have

$$\left| \frac{\omega|\mu| + [\omega^2\mu^2 - 4(\omega - 1)]^{1/2}}{2} \right|^2 = \omega - 1 = \Gamma(\omega, \bar{\mu}) \tag{2.16}$$

Next, if $\omega^2\bar{\mu}^2 - 4(\omega - 1) \ge 0$, let us define μ_c by

$$\mu_c{}^2 = \begin{cases} 4(\omega - 1)/\omega^2 & \text{if} \quad \omega \ge 1 \\ 0 & \text{if} \quad \omega < 1 \end{cases} \tag{2.17}$$

If $\mu^2 \le \mu_c{}^2$, then we have $\Gamma(\omega, \mu) = \omega - 1$. Moreover, for $\mu_c{}^2 \le \mu^2 \le \bar{\mu}^2$ the function $\Gamma(\omega, \mu)$ is clearly an increasing function of $|\mu|$. Consequently, we have $\Gamma(\omega, \mu) \le \Gamma(\omega, \bar{\mu})$ for $|\mu| \le \bar{\mu}$. Therefore, (2.15) holds. Since $\bar{\mu} = S(B)$ and since the eigenvalues of B are real, it follows that $\bar{\mu}$ or $-\bar{\mu}$ must be an eigenvalue of B. But by Theorem 5-3.4, both $\bar{\mu}$ and $-\bar{\mu}$ are eigenvalues of B. Hence (2.14) follows.

The function $4(\omega - 1)/\omega^2$ is an increasing function of ω in the range $0 < \omega < 2$ since

$$\frac{d}{d\omega}\left[\frac{4(\omega - 1)}{\omega^2}\right] = \frac{4}{\omega^3}(2 - \omega) > 0$$

Since $4(\omega_b - 1)/\omega_b{}^2 = \bar{\mu}^2$ it follows that if $\omega \ge \omega_b$, then $\bar{\mu}^2 \le 4(\omega - 1)/\omega^2$. Therefore, by (2.12) and (2.14) it follows that $S(\mathscr{L}_\omega) = \omega - 1$ if $2 > \omega \ge \omega_b$.

If $0 < \omega \le \omega_b$, then $\bar{\mu}^2 \ge 4(\omega - 1)/\omega^2$. Consequently, by (2.12) and (2.14) we have

$$\bar{S}(\mathscr{L}_\omega) = S(\mathscr{L}_\omega) = \left[\frac{\omega\bar{\mu} + [\omega^2\bar{\mu}^2 - 4(\omega - 1)]^{1/2}}{2}\right]^2$$

Hence (2.10) holds and the proof of Lemma 2.4 is complete.

From (2.10) it follows that $S(\mathscr{L}_\omega) > S(\mathscr{L}_{\omega_b})$ if $2 > \omega > \omega_b$. We now seek to show that if $0 < \omega < \omega_b$, then $S(\mathscr{L}_\omega)$ is a decreasing function of ω. But if $\omega^2\bar{\mu}^2 - 4(\omega - 1) \geq 0$, we have

$$\frac{d}{d\omega}[\omega\bar{\mu} + [\omega^2\bar{\mu}^2 - 4(\omega - 1)]^{1/2}]$$
$$= \frac{\bar{\mu}(\omega^2\bar{\mu}^2 - 4(\omega - 1))^{1/2} + \omega\bar{\mu}^2 - 2}{[\omega^2\bar{\mu}^2 - 4(\omega - 1)]^{1/2}} \tag{2.18}$$

Moreover,

$$(\omega\bar{\mu}^2 - 2)^2 = \omega^2\bar{\mu}^4 - 4\omega\bar{\mu}^2 + 4$$

and

$$(\bar{\mu}[\omega^2\bar{\mu}^2 - 4(\omega - 1)]^{1/2})^2 = \omega^2\bar{\mu}^4 - 4\omega\bar{\mu}^2 + 4\bar{\mu}^2$$

Therefore, since $\bar{\mu}^2 < 1$ and since $\omega\bar{\mu}^2 - 2 < 0$, it follows that the numerator in the right member of (2.18) is negative. Therefore, by (2.10) it follows that $S(\mathscr{L}_\omega)$ is a decreasing function of ω for $0 < \omega < \omega_b$. Hence (2.9) holds for $0 < \omega < \omega_b$ and the proof of Theorem 2.3 is complete.

Sufficient Conditions for the Eigenvalues of B to Be Real

We now give *sufficient* conditions for the hypotheses of Theorems 2.2 and 2.3 to hold.

Lemma 2.5. If A is a positive definite consistently ordered matrix, then A has positive diagonal elements and the matrix $B = I - (\text{diag } A)^{-1}A$ has real eigenvalues. Moreover, $\bar{\mu} = S(B) < 1$.

Proof. Since A is positive definite, the diagonal elements of A are positive by Theorem 2-2.5. Evidently B is similar to $\tilde{B} = D^{1/2}BD^{-1/2}$ where $D = \text{diag } A$. Moreover, $\tilde{B}$ is symmetric since $\tilde{B} = I - D^{-1/2}AD^{-1/2}$ and since A is symmetric. Therefore B has real eigenvalues. By Theorem 5-3.5, $S(B) < 1$.

Theorem 2.6. Let A be a matrix which has Property A and whose diagonal elements are positive. If, for some diagonal matrix E the matrix $A_1 = EAE^{-1}$ is positive definite, then $B = I - (\text{diag } A)^{-1}A$ has real eigenvalues and $S(B) < 1$.

Proof. Let $B_1 = I - (\text{diag } A_1)^{-1} A_1$. Since $A_1 = EAE^{-1} = E(D - DB)E^{-1} = EDE^{-1} - EDBE^{-1}$ we have diag $A_1 = EDE^{-1}$ and hence

$$B_1 = I - ED^{-1}E^{-1}EAE^{-1} = I - ED^{-1}AE^{-1} = EBE^{-1}$$

But since A_1 is positive definite and has Property A it follows from Theorem 5-3.5 that $S(B_1) < 1$. Hence $S(B) < 1$. Since the eigenvalues of B_1 are real by Lemma 2.5, the eigenvalues of B are real.

Tables of ω_b and $\bar{\mu}$

Table 2.1 gives ω_b, as defined by (1), as a function of $\bar{\mu}$ for $\bar{\mu} = 0(0.001)0.9$ and $\bar{\mu} = 0.9(0.0001)1.0$. In Table 2.2, values of $\bar{\mu} = \cos \pi h$ and $\omega_b = 2(1 + \sin \pi h)^{-1}$ are given for the model problem for $h^{-1} = 2(1)200$.

Alternative Proof of Theorem 2.3

We now give an alternative proof of Theorem 2.3, based on the use of three lemmas relating to the quadratic equation. For convenience we define the *root radius* of a quadratic equation as the maximum of the moduli of its roots. We now study the behavior of $\varrho(b, c)$, the root radius of (2.1), for real values of b and c.

Lemma 2.7. $\varrho(b, c) = 1$ if and only if the point (b, c) lies on the boundary Γ of the triangle bounded by the three lines.

$$c = 1, \qquad b = 1 + c, \qquad b = -(1 + c) \tag{2.19}$$

Proof. Let r_1 and r_2 be the roots of (2.1). If $|b| = 1 + c$, then by (2.3) and (2.4) at least one root of (2.1) must have modulus unity. If $|c| \leq 1$, then the root of largest modulus must have modulus unity. Thus $\varrho(b, c) = 1$ if $|b| = 1 + c$, $|c| \leq 1$. Also, if $c = 1$ and $|b| < 2$, then the roots of (2.1) are complex and have modulus unity. If $c = 1$ and $|b| = 2$, then $r_1 = r_2 = 1$ or $r_1 = r_2 = -1$. Thus for every point of Γ we have $\varrho(b, c) = 1$.

On the other hand, if $\varrho(b, c) = 1$ and if the roots of (2.1) are complex, then $r_1 = e^{i\theta}$, $r_2 = e^{-i\theta}$, and $c = 1$, $b = 2 \cos \theta$. If the roots are real, then one of them equals unity, and by (2.3) and (2.4), $|b| = 1 + c$. For the largest root to equal unity we must have $|c| \leq 1$. This proves the lemma.

TABLE 2.1. $\omega_b = 2(1 + [1 - \bar{\mu}^2]^{1/2})^{-1}$ As a Function of $\bar{\mu} = S(B)$ for $\bar{\mu} = 0(0.001)0.9$ and $\bar{\mu} = 0.9(0.0001)1.0$

$\bar{\mu}$	000	001	002	003	004	005	006	007	008	009
.00	1.00000	1.00000	1.00000	1.00000	1.00000	1.00001	1.00001	1.00001	1.00002	1.00002
.01	1.00003	1.00003	1.00004	1.00004	1.00005	1.00006	1.00006	1.00007	1.00008	1.00009
.02	1.00010	1.00011	1.00012	1.00013	1.00014	1.00016	1.00017	1.00018	1.00020	1.00021
.03	1.00023	1.00024	1.00026	1.00027	1.00029	1.00031	1.00032	1.00034	1.00036	1.00038
.04	1.00040	1.00042	1.00044	1.00046	1.00048	1.00051	1.00053	1.00055	1.00058	1.00060
.05	1.00063	1.00065	1.00068	1.00070	1.00073	1.00076	1.00079	1.00081	1.00084	1.00087
.06	1.00090	1.00093	1.00096	1.00099	1.00103	1.00106	1.00109	1.00112	1.00116	1.00119
.07	1.00123	1.00126	1.00130	1.00134	1.00137	1.00141	1.00145	1.00149	1.00153	1.00157
.08	1.00161	1.00165	1.00169	1.00173	1.00177	1.00181	1.00186	1.00190	1.00194	1.00199
.09	1.00203	1.00208	1.00213	1.00217	1.00222	1.00227	1.00231	1.00236	1.00241	1.00246
.10	1.00251	1.00256	1.00261	1.00267	1.00272	1.00277	1.00282	1.00288	1.00293	1.00299
.11	1.00304	1.00310	1.00316	1.00321	1.00327	1.00333	1.00339	1.00345	1.00351	1.00357
.12	1.00363	1.00369	1.00375	1.00381	1.00387	1.00394	1.00400	1.00407	1.00413	1.00420
.13	1.00426	1.00433	1.00439	1.00446	1.00453	1.00460	1.00467	1.00474	1.00481	1.00488
.14	1.00495	1.00502	1.00509	1.00517	1.00524	1.00531	1.00539	1.00546	1.00554	1.00561
.15	1.00569	1.00577	1.00584	1.00592	1.00600	1.00608	1.00616	1.00624	1.00632	1.00640
.16	1.00648	1.00657	1.00665	1.00673	1.00682	1.00690	1.00699	1.00707	1.00716	1.00724
.17	1.00733	1.00742	1.00751	1.00760	1.00769	1.00778	1.00787	1.00796	1.00805	1.00814
.18	1.00823	1.00833	1.00842	1.00852	1.00861	1.00871	1.00880	1.00890	1.00900	1.00909
.19	1.00919	1.00929	1.00939	1.00949	1.00959	1.00969	1.00979	1.00990	1.01000	1.01010
.20	1.01021	1.01031	1.01041	1.01052	1.01063	1.01073	1.01084	1.01095	1.01106	1.01117
.21	1.01128	1.01139	1.01150	1.01161	1.01172	1.01183	1.01194	1.01206	1.01217	1.01229
.22	1.01240	1.01252	1.01263	1.01275	1.01287	1.01299	1.01311	1.01323	1.01335	1.01347
.23	1.01359	1.01371	1.01383	1.01395	1.01408	1.01420	1.01433	1.01445	1.01458	1.01470
.24	1.01483	1.01496	1.01509	1.01521	1.01534	1.01547	1.01560	1.01574	1.01587	1.01600

TABLE 2.1. *(continued)*

$\bar{\mu}$	000	001	002	003	004	005	006	007	008	009
.25	1.01613	1.01627	1.01640	1.01654	1.01667	1.01681	1.01694	1.01708	1.01722	1.01736
.26	1.01750	1.01764	1.01778	1.01792	1.01806	1.01820	1.01834	1.01849	1.01863	1.01878
.27	1.01892	1.01907	1.01921	1.01936	1.01951	1.01966	1.01981	1.01996	1.02011	1.02026
.28	1.02041	1.02056	1.02071	1.02087	1.02102	1.02118	1.02133	1.02149	1.02164	1.02180
.29	1.02196	1.02212	1.02228	1.02244	1.02260	1.02276	1.02292	1.02308	1.02325	1.02341
.30	1.02357	1.02374	1.02390	1.02407	1.02424	1.02441	1.02457	1.02474	1.02491	1.02508
.31	1.02525	1.02543	1.02560	1.02577	1.02594	1.02612	1.02629	1.02647	1.02665	1.02682
.32	1.02700	1.02718	1.02736	1.02754	1.02772	1.02790	1.02808	1.02826	1.02845	1.02863
.33	1.02882	1.02900	1.02919	1.02937	1.02956	1.02975	1.02994	1.03013	1.03032	1.03051
.34	1.03070	1.03089	1.03109	1.03128	1.03148	1.03167	1.03187	1.03206	1.03226	1.03246
.35	1.03266	1.03286	1.03306	1.03326	1.03346	1.03366	1.03387	1.03407	1.03428	1.03448
.36	1.03469	1.03489	1.03510	1.03531	1.03552	1.03573	1.03594	1.03615	1.03636	1.03658
.37	1.03679	1.03700	1.03722	1.03744	1.03765	1.03787	1.03809	1.03831	1.03853	1.03875
.38	1.03897	1.03919	1.03941	1.03964	1.03986	1.04009	1.04031	1.04054	1.04077	1.04100
.39	1.04122	1.04145	1.04169	1.04192	1.04215	1.04238	1.04262	1.04285	1.04309	1.04332
.40	1.04356	1.04380	1.04404	1.04428	1.04452	1.04476	1.04500	1.04524	1.04549	1.04573
.41	1.04598	1.04622	1.04647	1.04672	1.04697	1.04722	1.04747	1.04772	1.04797	1.04823
.42	1.04848	1.04873	1.04899	1.04925	1.04950	1.04976	1.05002	1.05028	1.05054	1.05080
.43	1.05107	1.05133	1.05159	1.05186	1.05213	1.05239	1.05266	1.05293	1.05320	1.05347
.44	1.05374	1.05401	1.05429	1.05456	1.05484	1.05511	1.05539	1.05567	1.05595	1.05623
.45	1.05651	1.05679	1.05707	1.05736	1.05764	1.05793	1.05821	1.05850	1.05879	1.05908
.46	1.05937	1.05966	1.05995	1.06024	1.06054	1.06083	1.06113	1.06143	1.06172	1.06202
.47	1.06232	1.06262	1.06293	1.06323	1.06353	1.06384	1.06414	1.06445	1.06476	1.06507
.48	1.06538	1.06569	1.06600	1.06631	1.06663	1.06694	1.06726	1.06758	1.06789	1.06821
.49	1.06853	1.06886	1.06918	1.06950	1.06983	1.07015	1.07048	1.07081	1.07114	1.07147
.50	1.07180	1.07213	1.07246	1.07280	1.07313	1.07347	1.07381	1.07414	1.07448	1.07483
.51	1.07517	1.07551	1.07586	1.07620	1.07655	1.07690	1.07724	1.07759	1.07795	1.07830
.52	1.07865	1.07901	1.07936	1.07972	1.08008	1.08044	1.08080	1.08116	1.08152	1.08189
.53	1.08225	1.08262	1.08299	1.08336	1.08373	1.08410	1.08447	1.08484	1.08522	1.08560
.54	1.08597	1.08635	1.08673	1.08711	1.08750	1.08788	1.08827	1.08865	1.08904	1.08943

TABLE 2.1. *(continued)*

$\bar{\mu}$	000	001	002	003	004	005	006	007	008	009
.55	1.08982	1.09021	1.09061	1.09100	1.09140	1.09179	1.09219	1.09259	1.09299	1.09339
.56	1.09380	1.09420	1.09461	1.09502	1.09543	1.09584	1.09625	1.09666	1.09708	1.09749
.57	1.09791	1.09833	1.09875	1.09917	1.09959	1.10002	1.10044	1.10087	1.10130	1.10173
.58	1.10216	1.10259	1.10303	1.10347	1.10390	1.10434	1.10478	1.10522	1.10567	1.10611
.59	1.10656	1.10701	1.10746	1.10791	1.10836	1.10882	1.10927	1.10973	1.11019	1.11065
.60	1.11111	1.11157	1.11204	1.11251	1.11298	1.11345	1.11392	1.11439	1.11487	1.11534
.61	1.11582	1.11630	1.11678	1.11727	1.11775	1.11824	1.11873	1.11922	1.11971	1.12020
.62	1.12070	1.12120	1.12169	1.12219	1.12270	1.12320	1.12371	1.12421	1.12472	1.12524
.63	1.12575	1.12626	1.12678	1.12730	1.12782	1.12834	1.12887	1.12939	1.12992	1.13045
.64	1.13098	1.13152	1.13205	1.13259	1.13313	1.13367	1.13421	1.13476	1.13531	1.13585
.65	1.13641	1.13696	1.13751	1.13807	1.13863	1.13919	1.13976	1.14032	1.14089	1.14146
.66	1.14203	1.14261	1.14318	1.14376	1.14434	1.14492	1.14551	1.14609	1.14668	1.14727
.67	1.14787	1.14846	1.14906	1.14966	1.15026	1.15087	1.15148	1.15209	1.15270	1.15331
.68	1.15393	1.15455	1.15517	1.15579	1.15642	1.15704	1.15767	1.15831	1.15894	1.15958
.69	1.16022	1.16086	1.16151	1.16216	1.16281	1.16346	1.16412	1.16477	1.16543	1.16610
.70	1.16676	1.16743	1.16810	1.16878	1.16945	1.17013	1.17081	1.17150	1.17219	1.17288
.71	1.17357	1.17427	1.17496	1.17567	1.17637	1.17708	1.17779	1.17850	1.17922	1.17993
.72	1.18066	1.18138	1.18211	1.18284	1.18357	1.18431	1.18505	1.18579	1.18654	1.18729
.73	1.18804	1.18879	1.18955	1.19031	1.19108	1.19185	1.19262	1.19339	1.19417	1.19495
.74	1.19574	1.19653	1.19732	1.19811	1.19891	1.19971	1.20052	1.20133	1.20214	1.20296
.75	1.20378	1.20460	1.20543	1.2C626	1.20709	1.20793	1.20877	1.20962	1.21047	1.21132
.76	1.21218	1.21304	1.21390	1.21477	1.21565	1.21652	1.21740	1.21829	1.21918	1.22007
.77	1.22097	1.22187	1.22278	1.22369	1.22460	1.22552	1.22644	1.22737	1.22830	1.22924
.78	1.23018	1.23112	1.23207	1.23303	1.23399	1.23495	1.23592	1.23689	1.23787	1.23886
.79	1.23984	1.24084	1.24183	1.24284	1.24384	1.24486	1.24588	1.24690	1.24793	1.24896
.80	1.25000	1.25104	1.25209	1.25315	1.25421	1.25528	1.25635	1.25743	1.25851	1.25960
.81	1.26069	1.26179	1.26290	1.26401	1.26513	1.26625	1.26739	1.26852	1.26967	1.27081
.82	1.27197	1.27313	1.27430	1.27548	1.27666	1.27785	1.27904	1.28024	1.28145	1.28267
.83	1.28389	1.28512	1.28636	1.28760	1.28886	1.29011	1.29138	1.29266	1.29394	1.29523
.84	1.29652	1.29783	1.29914	1.30046	1.30179	1.30313	1.30448	1.30583	1.30719	1.30856
.85	1.30994	1.31133	1.31273	1.31414	1.31555	1.31698	1.31841	1.31986	1.32131	1.32277
.86	1.32425	1.32573	1.32722	1.32872	1.33024	1.33176	1.33329	1.33484	1.33639	1.33796
.87	1.33954	1.34113	1.34273	1.34434	1.34596	1.34760	1.34924	1.35090	1.35258	1.35426
.88	1.35596	1.35767	1.35939	1.36112	1.36287	1.36464	1.36641	1.36820	1.37001	1.37183
.89	1.37366	1.37551	1.37738	1.37926	1.38115	1.38306	1.38499	1.38693	1.38889	1.39087

TABLE 2.1. *(continued)*

$\bar{\mu}$	0000	0001	0002	0003	0004	0005	0006	0007	0008	0009
.900	1.39286	1.39306	1.39327	1.39347	1.39367	1.39387	1.39407	1.39427	1.39447	1.39467
.901	1.39488	1.39508	1.39528	1.39548	1.39569	1.39589	1.39609	1.39629	1.39650	1.39670
.902	1.39691	1.39711	1.39731	1.39752	1.39772	1.39793	1.39813	1.39834	1.39854	1.39875
.903	1.39895	1.39916	1.39936	1.39957	1.39978	1.39998	1.40019	1.40040	1.40060	1.40081
.904	1.40102	1.40123	1.40143	1.40164	1.40185	1.40206	1.40227	1.40248	1.40269	1.40289
.905	1.40310	1.40331	1.40352	1.40373	1.40394	1.40415	1.40436	1.40457	1.40478	1.40500
.906	1.40521	1.40542	1.40563	1.40584	1.40605	1.40627	1.40648	1.40669	1.40690	1.40712
.907	1.40733	1.40754	1.40776	1.40797	1.40818	1.40840	1.40861	1.40883	1.40904	1.40926
.908	1.40947	1.40969	1.40990	1.41012	1.41034	1.41055	1.41077	1.41098	1.41120	1.41142
.909	1.41164	1.41185	1.41207	1.41229	1.41251	1.41272	1.41294	1.41316	1.41338	1.41360
.910	1.41382	1.41404	1.41426	1.41448	1.41470	1.41492	1.41514	1.41536	1.41558	1.41580
.911	1.41602	1.41624	1.41647	1.41669	1.41691	1.41713	1.41736	1.41758	1.41780	1.41802
.912	1.41825	1.41847	1.41870	1.41892	1.41914	1.41937	1.41959	1.41982	1.42004	1.42027
.913	1.42050	1.42072	1.42095	1.42117	1.42140	1.42163	1.42185	1.42208	1.42231	1.42254
.914	1.42276	1.42299	1.42322	1.42345	1.42368	1.42391	1.42414	1.42437	1.42460	1.42483
.915	1.42506	1.42529	1.42552	1.42575	1.42598	1.42621	1.42644	1.42667	1.42691	1.42714
.916	1.42737	1.42760	1.42784	1.42807	1.42830	1.42854	1.42877	1.42900	1.42924	1.42947
.917	1.42971	1.42994	1.43018	1.43041	1.43065	1.43089	1.43112	1.43136	1.43160	1.43183
.918	1.43207	1.43231	1.43254	1.43278	1.43302	1.43326	1.43350	1.43374	1.43398	1.43422
.919	1.43446	1.43470	1.43494	1.43518	1.43542	1.43566	1.43590	1.43614	1.43638	1.43662
.920	1.43687	1.43711	1.43735	1.43759	1.43784	1.43808	1.43832	1.43857	1.43881	1.43906
.921	1.43930	1.43955	1.43979	1.44004	1.44028	1.44053	1.44078	1.44102	1.44127	1.44152
.922	1.44176	1.44201	1.44226	1.44251	1.44276	1.44300	1.44325	1.44350	1.44375	1.44400
.923	1.44425	1.44450	1.44475	1.44500	1.44525	1.44551	1.44576	1.44601	1.44626	1.44651
.924	1.44677	1.44702	1.44727	1.44753	1.44778	1.44804	1.44829	1.44854	1.44880	1.44905
.925	1.44931	1.44957	1.44982	1.45008	1.45033	1.45059	1.45085	1.45111	1.45136	1.45162
.926	1.45188	1.45214	1.45240	1.45266	1.45292	1.45318	1.45344	1.45370	1.45396	1.45422
.927	1.45448	1.45474	1.45500	1.45527	1.45553	1.45579	1.45605	1.45632	1.45658	1.45685
.928	1.45711	1.45737	1.45764	1.45790	1.45817	1.45844	1.45870	1.45897	1.45923	1.45950
.929	1.45977	1.46004	1.46030	1.46057	1.46084	1.46111	1.46138	1.46165	1.46192	1.46219

TABLE 2.1. *(continued)*

$\bar{\mu}$	0000	0001	0002	0003	0004	0005	0006	0007	0008	0009
.930	1.46246	1.46273	1.46300	1.46327	1.46354	1.46382	1.46409	1.46436	1.46463	1.46491
.931	1.46518	1.46545	1.46573	1.46600	1.46628	1.46655	1.46683	1.46711	1.46738	1.46766
.932	1.46793	1.46821	1.46849	1.46877	1.46905	1.46932	1.46960	1.46988	1.47016	1.47044
.933	1.47072	1.47100	1.47128	1.47156	1.47185	1.47213	1.47241	1.47269	1.47298	1.47326
.934	1.47354	1.47383	1.47411	1.47440	1.47468	1.47497	1.47525	1.47554	1.47582	1.47611
.935	1.47640	1.47669	1.47697	1.47726	1.47755	1.47784	1.47813	1.47842	1.47871	1.47900
.936	1.47929	1.47958	1.47987	1.48016	1.48046	1.48075	1.48104	1.48134	1.48163	1.48192
.937	1.48222	1.48251	1.48281	1.48310	1.48340	1.48370	1.48399	1.48429	1.48459	1.48488
.938	1.48518	1.48548	1.48578	1.48608	1.48638	1.48668	1.48698	1.48728	1.48758	1.48788
.939	1.48819	1.48849	1.48879	1.48910	1.48940	1.48970	1.49001	1.49031	1.49062	1.49092
.940	1.49123	1.49154	1.49184	1.49215	1.49246	1.49277	1.49308	1.49338	1.49369	1.49400
.941	1.49431	1.49462	1.49494	1.49525	1.49556	1.49587	1.49618	1.49650	1.49681	1.49713
.942	1.49744	1.49775	1.49807	1.49839	1.49870	1.49902	1.49934	1.49965	1.49997	1.50029
.943	1.50061	1.50093	1.50125	1.50157	1.50189	1.50221	1.50253	1.50285	1.50317	1.50350
.944	1.50382	1.50414	1.50447	1.50479	1.50512	1.50544	1.50577	1.50610	1.50642	1.50675
.945	1.50708	1.50741	1.50774	1.50807	1.50840	1.50873	1.50906	1.50939	1.50972	1.51005
.946	1.51038	1.51072	1.51105	1.51138	1.51172	1.51205	1.51239	1.51273	1.51306	1.51340
.947	1.51374	1.51407	1.51441	1.51475	1.51509	1.51543	1.51577	1.51611	1.51645	1.51680
.948	1.51714	1.51748	1.51783	1.51817	1.51851	1.51886	1.51921	1.51955	1.51990	1.52025
.949	1.52059	1.52094	1.52129	1.52164	1.52199	1.52234	1.52269	1.52304	1.52339	1.52375
.950	1.52410	1.52445	1.52481	1.52516	1.52552	1.52587	1.52623	1.52659	1.52694	1.52730
.951	1.52766	1.52802	1.52838	1.52874	1.52910	1.52946	1.52982	1.53019	1.53055	1.53091
.952	1.53128	1.53164	1.53201	1.53238	1.53274	1.53311	1.53348	1.53385	1.53421	1.53458
.953	1.53495	1.53533	1.53570	1.53607	1.53644	1.53681	1.53719	1.53756	1.53794	1.53831
.954	1.53869	1.53907	1.53945	1.53982	1.54020	1.54058	1.54096	1.54134	1.54172	1.54211
.955	1.54249	1.54287	1.54326	1.54364	1.54403	1.54441	1.54480	1.54519	1.54557	1.54596
.956	1.54635	1.54674	1.54713	1.54752	1.54792	1.54831	1.54870	1.54910	1.54949	1.54989
.957	1.55028	1.55068	1.55108	1.55147	1.55187	1.55227	1.55267	1.55307	1.55348	1.55388
.958	1.55428	1.55469	1.55509	1.55550	1.55590	1.55631	1.55672	1.55712	1.55753	1.55794
.959	1.55835	1.55876	1.55918	1.55959	1.56000	1.56042	1.56083	1.56125	1.56166	1.56208

TABLE 2.1. *(continued)*

$\bar{\mu}$	0000	0001	0002	0003	0004	0005	0006	0007	0008	0009
.960	1.56250	1.56292	1.56334	1.56376	1.56418	1.56460	1.56503	1.56545	1.56587	1.56630
.961	1.56672	1.56715	1.56758	1.56801	1.56844	1.56887	1.56930	1.56973	1.57016	1.57060
.962	1.57103	1.57147	1.57190	1.57234	1.57278	1.57322	1.57365	1.57410	1.57454	1.57498
.963	1.57542	1.57587	1.57631	1.57676	1.57720	1.57765	1.57810	1.57855	1.57900	1.57945
.964	1.57990	1.58035	1.58081	1.58126	1.58172	1.58217	1.58263	1.58309	1.58355	1.58401
.965	1.58447	1.58493	1.58540	1.58586	1.58633	1.58679	1.58726	1.58773	1.58820	1.58867
.966	1.58914	1.58961	1.59009	1.59056	1.59104	1.59151	1.59199	1.59247	1.59295	1.59343
.967	1.59391	1.59439	1.59488	1.59536	1.59585	1.59633	1.59682	1.59731	1.59780	1.59829
.968	1.59878	1.59928	1.59977	1.60027	1.60077	1.60126	1.60176	1.60226	1.60276	1.60327
.969	1.60377	1.60428	1.60478	1.60529	1.60580	1.60631	1.60682	1.60733	1.60784	1.60836
.970	1.60887	1.60939	1.60991	1.61043	1.61095	1.61147	1.61200	1.61252	1.61305	1.61357
.971	1.61410	1.61463	1.61516	1.61569	1.61623	1.61676	1.61730	1.61784	1.61838	1.61892
.972	1.61946	1.62000	1.62055	1.62109	1.62164	1.62219	1.62274	1.62329	1.62384	1.62440
.973	1.62495	1.62551	1.62607	1.62663	1.62719	1.62775	1.62832	1.62888	1.62945	1.63002
.974	1.63059	1.63116	1.63174	1.63231	1.63289	1.63347	1.63405	1.63463	1.63522	1.63580
.975	1.63639	1.63698	1.63757	1.63816	1.63875	1.63935	1.63994	1.64054	1.64114	1.64174
.976	1.64235	1.64295	1.64356	1.64417	1.64478	1.64539	1.64600	1.64662	1.64724	1.64786
.977	1.64848	1.64910	1.64973	1.65036	1.65098	1.65162	1.65225	1.65288	1.65352	1.65416
.978	1.65480	1.65544	1.65609	1.65674	1.65738	1.65804	1.65869	1.65934	1.66000	1.66066
.979	1.66132	1.66199	1.66265	1.66332	1.66399	1.66466	1.66534	1.66602	1.66669	1.66738
.980	1.66806	1.66875	1.66944	1.67013	1.67082	1.67152	1.67221	1.67291	1.67362	1.67432
.981	1.67503	1.67574	1.67645	1.67717	1.67789	1.67861	1.67933	1.68006	1.68079	1.68152
.982	1.68225	1.68299	1.68373	1.68447	1.68522	1.68597	1.68672	1.68747	1.68823	1.68899
.983	1.68975	1.69052	1.69129	1.69206	1.69283	1.69361	1.69439	1.69518	1.69597	1.69676
.984	1.69755	1.69835	1.69915	1.69995	1.70076	1.70157	1.70239	1.70320	1.70402	1.70485
.985	1.70568	1.70651	1.70735	1.70819	1.70903	1.70988	1.71073	1.71158	1.71244	1.71330
.986	1.71417	1.71504	1.71592	1.71679	1.71768	1.71857	1.71946	1.72035	1.72125	1.72216
.987	1.72307	1.72398	1.72490	1.72582	1.72675	1.72768	1.72862	1.72956	1.73051	1.73146
.988	1.73242	1.73338	1.73435	1.73532	1.73630	1.73729	1.73828	1.73927	1.74027	1.74128
.989	1.74229	1.74331	1.74433	1.74536	1.74640	1.74744	1.74849	1.74954	1.75060	1.75167

TABLE 2.1. *(continued)*

$\bar{\mu}$	0000	0001	0002	0003	0004	0005	0006	0007	0008	0009
.990	1.75274	1.75383	1.75491	1.75601	1.75711	1.75822	1.75934	1.76046	1.76160	1.76274
.991	1.76388	1.76504	1.76620	1.76737	1.76856	1.76974	1.77094	1.77215	1.77336	1.77459
.992	1.77582	1.77707	1.77832	1.77958	1.78086	1.78214	1.78344	1.78474	1.78606	1.78739
.993	1.78873	1.79008	1.79144	1.79281	1.79420	1.79560	1.79702	1.79844	1.79988	1.80134
.994	1.80281	1.80429	1.80579	1.80731	1.80884	1.81039	1.81195	1.81353	1.81513	1.81675
.995	1.81839	1.82005	1.82172	1.82342	1.82514	1.82688	1.82865	1.83044	1.83225	1.83409
.996	1.83595	1.83784	1.83977	1.84172	1.84370	1.84571	1.84776	1.84984	1.85196	1.85412
.997	1.85632	1.85856	1.86084	1.86318	1.86556	1.86800	1.87049	1.87304	1.87565	1.87833
.998	1.88109	1.88392	1.88684	1.88985	1.89296	1.89618	1.89952	1.90300	1.90662	1.91042
.999	1.91441	1.91862	1.92309	1.92788	1.93305	1.93870	1.94499	1.95219	1.96079	1.97211
1.000	2.00000									

TABLE 2.2. Values of $\bar{\mu} = \cos(\pi/M)$ and $\omega_b = 2(1 + [1 - \bar{\mu}^2]^{1/2})^{-1} = 2[1 + \sin(\pi/M)]^{-1}$ for the Model Problem for $M = h^{-1} = 2(1)200$

Values of $\bar{\mu}$

M	0	1	2	3	4	5	6	7	8	9
0			.000000	.500000	.707107	.809017	.866025	.900969	.923880	.939693
10	.951057	.959493	.965926	.970942	.974928	.978148	.980785	.982973	.984808	.986361
20	.987688	.988831	.989821	.990686	.991445	.992115	.992709	.993238	.993712	.994138
30	.994522	.994869	.995125	.995472	.995734	.995974	.996195	.996397	.996584	.996757
40	.996917	.997066	.997204	.997332	.997452	.997564	.997669	.997767	.997859	.997945
50	.998027	.998103	.998176	.998244	.998308	.998369	.998427	.998482	.998533	.998583
60	.998630	.998674	.998717	.998757	.998795	.998832	.998867	.998901	.998933	.998964
70	.998993	.999021	.999048	.999074	.999099	.999123	.999146	.999168	.999189	.999209
80	.999229	.999248	.999266	.999284	.999301	.999317	.999333	.999348	.999363	.999377
90	.999391	.999404	.999417	.999429	.999442	.999453	.999465	.999476	.999486	.999497
100	.999507	.999516	.999526	.999535	.999544	.999552	.999561	.999569	.999577	.999585
110	.999592	.999600	.999607	.999614	.999620	.999627	.999633	.999640	.999646	.999652
120	.999657	.999663	.999668	.999674	.999679	.999684	.999689	.999694	.999699	.999703
130	.999708	.999712	.999717	.999721	.999725	.999729	.999733	.999737	.999741	.999745
140	.999748	.999752	.999755	.999759	.999762	.999765	.999769	.999772	.999775	.999778
150	.999781	.999784	.999786	.999789	.999792	.999795	.999797	.999800	.999802	.999805
160	.999807	.999810	.999812	.999814	.999817	.999819	.999821	.999823	.999825	.999827
170	.999829	.999831	.999833	.999835	.999837	.999839	.999841	.999842	.999844	.999846
180	.999848	.999849	.999851	.999853	.999854	.999856	.999857	.999859	.999860	.999862
190	.999863	.999865	.999866	.999868	.999869	.999870	.999872	.999873	.999874	.999875
200	.999877									

TABLE 2.2. *(continued)*

Values of ω_b

M	0	1	2	3	4	5	6	7	8	9
0			1.000000	1.071797	1.171573	1.259616	1.333333	1.394813	1.446463	1.490291
10	1.527864	1.560388	1.588791	1.613794	1.635964	1.655750	1.673514	1.689547	1.704088	1.717336
20	1.729454	1.740580	1.750831	1.760305	1.769088	1.777251	1.784859	1.791966	1.798619	1.804860
30	1.810727	1.816253	1.821465	1.826391	1.831052	1.835470	1.839663	1.843648	1.847440	1.851052
40	1.854498	1.857788	1.860932	1.863941	1.866822	1.869584	1.872234	1.874779	1.877224	1.879575
50	1.881838	1.884018	1.886119	1.888145	1.890100	1.891989	1.893813	1.895577	1.897283	1.898935
60	1.900534	1.902083	1.903585	1.905042	1.906455	1.907826	1.909159	1.910453	1.911711	1.912934
70	1.914123	1.915281	1.916407	1.917505	1.918573	1.919615	1.920630	1.921620	1.922585	1.923527
80	1.924447	1.925344	1.926220	1.927077	1.927913	1.928731	1.929530	1.930311	1.931075	1.931823
90	1.932555	1.933271	1.933972	1.934659	1.935331	1.935990	1.936635	1.937268	1.937888	1.938496
100	1.939092	1.939676	1.940250	1.940813	1.941365	1.941907	1.942439	1.942962	1.943475	1.943979
110	1.944474	1.944960	1.945438	1.945907	1.946369	1.946823	1.947269	1.947708	1.948140	1.948564
120	1.948982	1.949392	1.949797	1.950195	1.950586	1.950972	1.951351	1.951725	1.952093	1.952456
130	1.952813	1.953164	1.953511	1.953852	1.954189	1.954520	1.954847	1.955169	1.955487	1.955800
140	1.956109	1.956413	1.956713	1.957010	1.957302	1.957590	1.957874	1.958155	1.958432	1.958705
150	1.958974	1.959240	1.959503	1.959762	1.960018	1.960271	1.960521	1.960767	1.961011	1.961251
160	1.961489	1.961723	1.961955	1.962184	1.962410	1.962634	1.962855	1.963073	1.963289	1.963502
170	1.963713	1.963921	1.964127	1.964331	1.964532	1.964731	1.964928	1.965123	1.965315	1.965506
180	1.965694	1.965880	1.966064	1.966247	1.966427	1.966606	1.966782	1.966957	1.967130	1.967301
190	1.967470	1.967638	1.967803	1.967967	1.968130	1.968291	1.968450	1.968608	1.968764	1.968918
200	1.969071									

Lemma 2.8. For any $k > 0$, $\varrho(b, c) = k$ if and only if the point (b, c) lies on the boundary T_k of the triangle bounded by the lines

$$c = k^2, \qquad b = k + ck^{-1}, \qquad b = -(k + ck^{-1}) \tag{2.20}$$

Moreover, $\varrho(b, c) = 0$ if and only if $b = c = 0$, and $\varrho(b, c) > k$ or $\varrho(b, c) < k$ depending upon whether (b, c) is outside or inside the triangle bounded by T_k, respectively.

Proof. If $k > 0$, let $b' = b/k$, $c' = c/k^2$, and $y = x/k$. The result now follows from Lemmas 2.1 and 2.7.

Lemma 2.9. $\varrho(b, c) \geq \varrho(b_1, c)$ if and only if $|b| \geq |b_1|$. If $b_1^2 - 4c \geq 0$ and if $|b| > |b_1|$, then $\varrho(b, c) > \varrho(b_1, c)$.

Proof. If $b_1^2 < 4c$, then $c > 0$ and $\varrho(b_1, c) = c^{1/2}$. Evidently $\varrho(b, c) \geq c^{1/2}$ since the product of the roots of (2.1) is c. On the other hand, if $b_1^2 - 4c \geq 0$, then

$$\varrho(b, c) = \tfrac{1}{2}[|b| + (b^2 - 4c)^{1/2}] \geq \tfrac{1}{2}[|b_1| + (b_1^2 - 4c)^{1/2}] = \varrho(b_1, c)$$

with the strict inequality holding if $|b| > |b_1|$.

From Lemma 2.9 it follows that, for a given ω, $\bar{S}(\mathscr{L}_\omega)$ is the square of the root radius of

$$\lambda - \omega\bar{\mu}\lambda^{1/2} + \omega - 1 = 0 \tag{2.21}$$

considered as a quadratic in $\lambda^{1/2}$. Given ω and $\bar{\mu}$ we can find $S(\mathscr{L}_\omega)$ by finding which of the boundaries T_k contains the point $(\omega\bar{\mu}, \omega - 1)$. If $b = \omega\bar{\mu}$ and $c = \omega - 1$, then (b, c) lies on the line L given by

$$b = \bar{\mu}(1 + c) \tag{2.22}$$

Thus, the problem of minimizing $S(\mathscr{L}_\omega)$ is simply that of finding the smallest value of k such that T_k contains a point of L. A typical situation is illustrated in Figure 2.1 for the case $\bar{\mu} = 2(2)^{1/2}/3$. Since for $k < 1$ the slope of the slanting side of any T_k in the region $b \geq 0$ is k and since the slope of L is $\bar{\mu}^{-1} > 1$, it is clear that there is a unique k such that L intersects T_k in just one point. L intersects T_{k_1} in two points, one on the line $c = k_1^2$ and the other on the side $b = k_1 + ck_1^{-1}$ if and only if $k_1 > k$ since the triangles are nested. Thus in Figure 2.1 the line L intersects $T_{(0.75)^{1/2}}$ at two points and $T_{(0.50)^{1/2}}$ at a single point. It does not intersect any T_k with $k < (0.50)^{1/2}$ at any point.

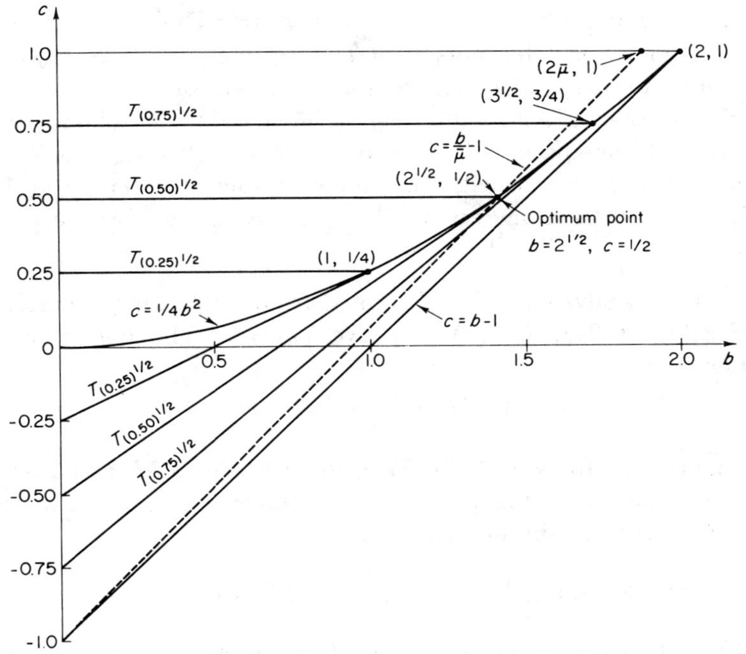

Figure 2.1. Case $\bar{\mu} = 2(2)^{1/2}/3 \doteq 0.943$ with $\omega_b = 1.5$.

We are thus led to seek k such that the corner point $(2k, k^2)$ of T_k lies on L. By (2.22) we require that k satisfy

$$2k = \bar{\mu}(1 + k^2) \tag{2.23}$$

This equation has two real roots, one of which lies in the interval $0 < k < 1$ and is given by

$$k_b = \frac{\bar{\mu}}{1 + (1 - \bar{\mu}^2)^{1/2}} = (\omega_b - 1)^{1/2} = [S(\mathscr{L}_{\omega_b})]^{1/2} \tag{2.24}$$

For the corresponding values of b and c we have

$$b = 2k_b, \qquad c = k_b^2 \tag{2.25}$$

or

$$\begin{aligned} \omega\mu &= \frac{2\bar{\mu}}{1 + (1 - \bar{\mu}^2)^{1/2}} = \omega_b\bar{\mu} \\ \omega - 1 &= \left(\frac{\bar{\mu}}{1 + (1 - \bar{\mu}^2)^{1/2}}\right)^2 = \omega_b - 1 \end{aligned} \tag{2.26}$$

both of which imply that $\omega = \omega_b$. For any other choice of ω, the point $(\omega\bar{\mu}, \omega - 1)$ lies on L but outside of T_{k_b}; hence $S(\mathscr{L}_\omega) > S(\mathscr{L}_{\omega_b})$.

To show that $S(\mathscr{L}_\omega)$ is a decreasing function of ω for $\omega < \omega_b$ we note that if $\omega < \omega_b$, then the point $(\omega\bar{\mu}, \omega - 1)$ lies below the point $(\omega_b\bar{\mu}, \omega_b - 1)$ on the line L. It also lies on the slanting side of some T_k. Since the T_k are nested, the value of k such that $(\omega\bar{\mu}, \omega - 1)$ lies on T_k decreases as ω increases. Thus we have an alternative proof of Theorem 2.3.

The above analysis can be illustrated by the following example. Let $\bar{\mu} = \frac{2}{3}(2)^{1/2} \doteq 0.943$. By (1) we have $\omega_b = \frac{3}{2}$. The line L has the equation

$$c = (1/\bar{\mu})b - 1 = 3(2^{-3/2})b - 1$$

The triangles T_k for $k = 0$, $(0.25)^{1/2}$, $(0.50)^{1/2}$, $(0.75)^{1/2}$, and 1 and the line L are shown in Figure 2.1. The triangle $T_{(0.50)^{1/2}}$ has its vertex $(2(0.50)^{1/2}, 0.50)$ on the line $L(\bar{\mu})$ since

$$c = 0.50 = (1/\bar{\mu})b - 1 = [3/(2)^{1/2}](0.50)^{1/2} - 1 = 0.50$$

Thus $k_b = (0.50)^{1/2}$. As already noted, L intersects $T_{(0.75)^{1/2}}$ at two points and does not intersect $T_{(0.25)^{1/2}}$ at all.

6.3. RATES OF CONVERGENCE: COMPARISON WITH THE GAUSS–SEIDEL METHOD

In Section 3.7 we defined the rate of convergence of the iterative method (3-1.5) to be

$$R(G) = R_\infty(G) = -\log S(G) \tag{3.1}$$

In this section we seek to show that if the eigenvalues of B are real, then $R(\mathscr{L}_{\omega_b})$, the rate of convergence of the SOR method using the optimum relaxation factor ω_b, is greater by an order of magnitude than $R(\mathscr{L})$, when $R(\mathscr{L})$ is small. We prove

Theorem 3.1. Let A be a consistently ordered matrix with non-vanishing diagonal elements such that all eigenvalues of $B = I - (\operatorname{diag} A)^{-1}A$ are real and such that $\bar{\mu} = S(B) < 1$. Let $R(\mathscr{L})$ and $R(\mathscr{L}_{\omega_b})$ denote, respectively, the rates of convergence of the GS and

SOR methods, where the relaxation factor ω_b is given by (1). We have

$$2\bar{\mu}[R(\mathscr{L})]^{1/2} \leq R(\mathscr{L}_{\omega_b}) \leq R(\mathscr{L}) + 2[R(\mathscr{L})]^{1/2} \tag{3.2}$$

the second inequality holding if $R(\mathscr{L}) \leq 3$. Moreover,

$$\lim_{\bar{\mu}\to 1-} [R(\mathscr{L}_{\omega_b})/(2(R(\mathscr{L}))^{1/2})] = 1 \tag{3.3}$$

We remark that by Lemma 2.5 the hypotheses of the theorem are satisfied if A is consistently ordered and is positive definite. We also note that the case $R(\mathscr{L}) \leq 3$ more than covers the range of practical interest since if $R(\mathscr{L})$ were greater than 3, then the GS method itself would be very rapidly convergent and there would be little need to seek a faster method.

For convenience, we let $\alpha = R(\mathscr{L}_{\omega_b})$ and $\beta = -\log \bar{\mu}$. Evidently, by (2.8) and (1) we have

$$\begin{aligned}\alpha &= -\log\left[\frac{2}{1 + (1 - \bar{\mu}^2)^{1/2}} - 1\right] \\ &= -\log \frac{1 - (1 - \bar{\mu}^2)^{1/2}}{1 + (1 - \bar{\mu}^2)^{1/2}} \geq 2(1 - \bar{\mu}^2)^{1/2} = 2(1 - e^{-2\beta})^{1/2} \\ &= 2e^{-\beta}(e^{2\beta} - 1)^{1/2} \geq 2\bar{\mu}(2\beta)^{1/2}\end{aligned}$$

since $-\log[(1 - y)/(1 + y)] \geq 2y$ for $0 \leq y < 1$ and $e^x - 1 \geq x$ for $x \geq 0$. Hence, the first inequality of (3.2) follows since $R(\mathscr{L}) = 2\beta$. On the other hand, we also have

$$\begin{aligned}\alpha &= -2 \log \frac{e^{-\beta}}{1+(1 - e^{-2\beta})^{1/2}} = 2\beta + 2 \log[1 + (1 - e^{-2\beta})^{1/2}] \\ &\leq 2\beta + 2(1 - e^{-2\beta})^{1/2} \leq 2\beta + 2(2\beta)^{1/2}\end{aligned}$$

for $2\beta < 3$, since $\log(1 + x) \leq x$ for $0 \leq x < 1$ and $e^{-y} \geq 1 - y + y^2/2 - y^3/3! \cdots \geq 1 - y$ for $0 < y \leq 3$. This completes the proof of (3.2).

From (3.2) we have

$$\bar{\mu} \leq \alpha/(2(2\beta)^{1/2}) \leq 1 + (\beta/2)^{1/2}$$

if $\beta \leq \frac{3}{2}$. Hence, since $\beta \to 0$ as $\bar{\mu} \to 1-$ the result (3.3) follows. This completes the proof of Theorem 3.1.

Let us illustrate the above result for the model problem. By (4-6.16) we have, if $I = J = h^{-1}$,

$$\bar{\mu} = \cos \pi h \tag{3.4}$$

Therefore,

$$R(B) = -\log \bar{\mu} = -\log \cos \pi h = -\log(1 - \tfrac{1}{2}\pi^2 h^2 + O(h^4))$$
$$= \tfrac{1}{2}\pi^2 h^2 + O(h^4) \tag{3.5}$$

Similarly,

$$R(\mathscr{L}) = 2R(B) = \pi^2 h^2 + O(h^4) \tag{3.6}$$

By Theorem 3.1 we have, for small h, $R(\mathscr{L}_{\omega_b}) \sim 2[R(\mathscr{L})]^{1/2} \sim 2\pi h$ and

$$R(\mathscr{L}_{\omega_b})/R(\mathscr{L}) \sim 2/(\pi h) \tag{3.7}$$

Thus, for example, if $h = \frac{1}{20}$, then

$$R(\mathscr{L}_{\omega_b})/R(\mathscr{L}) \sim 40/\pi \doteq 12.75$$

Thus there is a factor of improvement of over 12 for the SOR method as compared with the GS method. This factor of improvement increases as h decreases and, in fact, is proportional to h^{-1}. If $h = \frac{1}{100}$, then the factor is 63.8.

The approximate formula given by (3.7) for the factor of improvement of $R(\mathscr{L}_{\omega_b})$ over $R(\mathscr{L})$ is fairly accurate. Thus from (1) we have

$$\omega_b = \frac{2}{1 + (1 - \bar{\mu}^2)^{1/2}} = \frac{2}{1 + \sin \pi h} \doteq \frac{2}{1.1565} = 1.730 \tag{3.8}$$

and

$$R(\mathscr{L}_{\omega_b}) = -\log(\omega_b - 1) \doteq 0.315 \tag{3.9}$$

On the other hand,

$$R(\mathscr{L}) = -2 \log \bar{\mu} = -2 \log \cos \pi h \doteq 0.0247 \tag{3.10}$$

Hence

$$R(\mathscr{L}_{\omega_b})/R(\mathscr{L}) \doteq 12.75 \tag{3.11}$$

From (3.7) we are led to state that there is an order-of-magnitude improvement of the SOR method over the GS method. Theorem 3.1 shows that this is true in general. On the other hand, as we shall see in Chapter 7, the actual gain in convergence rate, while still large, is somewhat less than expected. This is due to the fact that the Jordan canonical form of $\mathscr{L}_{\omega_b}$ is not diagonal and hence certain norms of $\mathscr{L}_{\omega_b}^n$ converge to zero less rapidly than would otherwise be the case.

6.4. ANALYSIS OF THE CASE WHERE SOME EIGENVALUES OF B ARE COMPLEX

In order to study the convergence properties of the SOR method when some of the eigenvalues of B may be complex we shall consider a geometric interpretation of the relation (5-2.6) between the eigenvalues λ of $\mathscr{L}_\omega$ and the eigenvalues μ of B.

From (5-2.6) we have

$$\mu = \frac{1}{\omega}\left(\lambda^{1/2} + \frac{\omega - 1}{\lambda^{1/2}}\right) \tag{4.1}$$

If we let

$$\lambda^{1/2} = \varrho e^{i\theta}, \qquad \mu = \alpha + i\beta \tag{4.2}$$

where ϱ, θ, α, and β are real and $\varrho \geq 0$, then we have

$$\alpha = \frac{1}{\omega}\left(\varrho + \frac{\omega - 1}{\varrho}\right)\cos\theta, \qquad \beta = \frac{1}{\omega}\left(\varrho - \frac{\omega - 1}{\varrho}\right)\sin\theta \tag{4.3}$$

Thus, if $\varrho^2 \neq |\omega - 1|$, then (α, β) lies on the ellipse E_ϱ defined by

$$(\alpha^2/a^2) + (\beta^2/b^2) = 1 \tag{4.4}$$

where

$$\begin{aligned} a = a(\omega, \varrho) &= \left|\frac{1}{\omega}\left(\varrho + \frac{\omega - 1}{\varrho}\right)\right| \\ b = b(\omega, \varrho) &= \left|\frac{1}{\omega}\left(\varrho - \frac{\omega - 1}{\varrho}\right)\right| \end{aligned} \tag{4.5}$$

If $\varrho^2 = |\omega - 1|$, then (α, β) lies on the segment $|\alpha| \leq 2(\omega-1)^{1/2}/\omega$, $\beta = 0$, if $\omega \geq 1$ and on the segment $\alpha = 0$, $|\beta| \leq 2(1 - \omega)^{1/2}/\omega$ if $\omega < 1$. We shall refer to these segments as "ellipses" even though the equation (4.4) cannot be used.

Evidently E_ϱ is the image of the circle C_ϱ defined by

$$|\lambda^{1/2}| = \varrho \tag{4.6}$$

and also of $C_{|\omega-1|/\varrho}$. The two circles coincide if $\varrho^2 = |\omega - 1|$ in which case E_ϱ is the segment described above. Thus if $\varrho^2 \geq |\omega - 1|$, then as ϱ increases the ellipses E_ϱ get larger and larger. On the other hand, if $\varrho^2 \leq |\omega - 1|$ and ϱ decreases, the ellipses E_ϱ also expand. Thus, for $\varrho^2 \geq |\omega - 1|$ there is a one-to-one correspondence between the

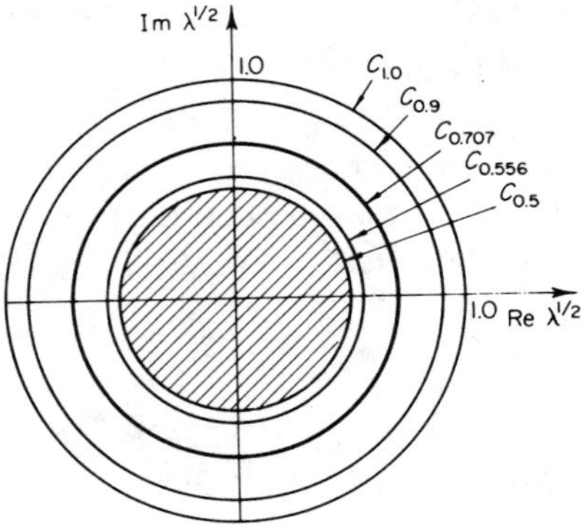

Figure 4.1.(a) $\lambda^{1/2}$-plane ($\omega = 1.5$ and 0.5).

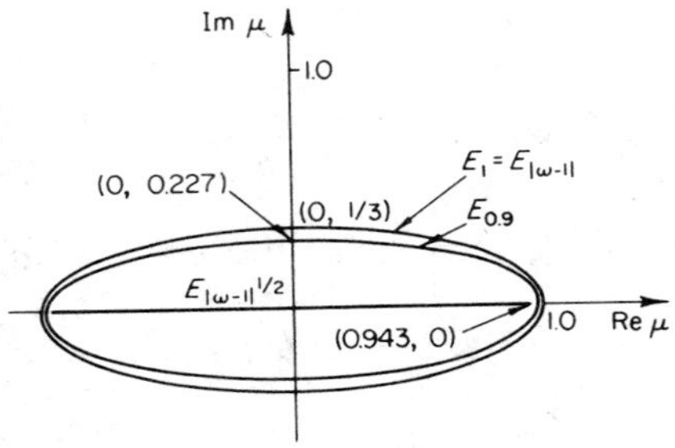

Figure 4.1.(b) μ-plane ($\omega = 1.5$).

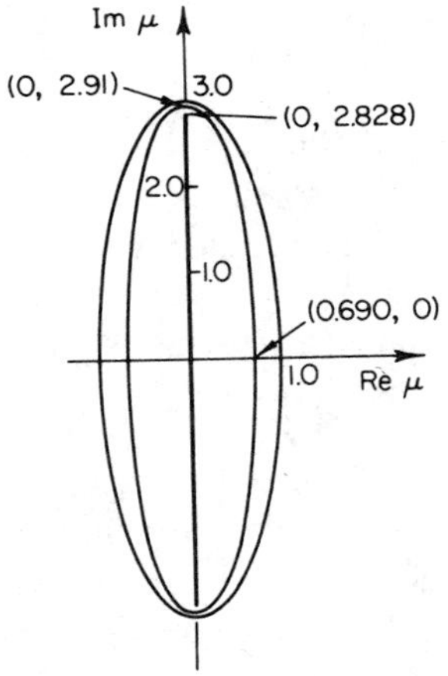

Figure 4.1.(c) μ-plane ($\omega = 0.5$).

exterior of C_ϱ and the exterior of E_ϱ. There is also a one-to-one correspondence between the exterior of E_ϱ and the interior of C_ϱ if $\varrho^2 < |\omega - 1|$. If we can show that no eigenvalue of B lies outside of E_ϱ, then no eigenvalue of $\mathscr{L}_\omega$ can lie outside of the circle $\lambda = \varrho^2$ provided $\varrho^2 > |\omega - 1|$.

The situation is illustrated in Figure 4.1. In the $\lambda^{1/2}$-plane the circles $C_{|\omega-1|^{1/2}}$, C_1, $C_{|\omega-1|}$, C_ϱ, and $C_{|\omega-1|/\varrho}$ are shown where $\varrho = 0.9$. For the case $\omega > 1$, we have the corresponding ellipse in the μ-plane. The slit $E_{|\omega-1|^{1/2}}$ corresponds to $C_{|\omega-1|^{1/2}}$; the ellipse E_1 corresponds to both C_1 and $C_{|\omega-1|}$; and the ellipse E_ϱ corresponds to both C_ϱ and $C_{|\omega-1|/\varrho}$. The same situation occurs in the case $\omega < 1$ except that the slit and the ellipses are oriented in the vertical direction. As ω approaches unity all the circles which are interior to $C_{|\omega-1|^{1/2}}$ shrink to the origin. This corresponds to the fact that $\lambda = 0$ is an eigenvalue of $\mathscr{L}$ of high multiplicity.

Convergence

Theorem 4.1. Let A be a consistently ordered matrix with nonvanishing diagonal elements. The SOR method is strongly convergent if for some positive number D and each eigenvalue $\mu = \alpha + i\beta$ of B, the point (α, β) lies in the interior of the ellipse $E(1, D)$

$$\alpha^2 + (\beta^2/D^2) = 1 \tag{4.7}$$

and ω satisfies

$$0 < \omega < 2/(1 + D) \tag{4.8}$$

Conversely, if the SOR method converges, then all eigenvalues of B lie inside $E(1, D)$ for some $D > 0$. Moreover, if some μ lies on $E(1, D)$ and if the SOR method converges, then (4.8) holds.

We note that Theorem 2.2 follows from the above theorem since all eigenvalues of B are inside $E(1, D)$ for any $D > 0$. To prove convergence for any ω in the range $0 < \omega < 2$ we choose D so small that $D < (2 - \omega)/\omega$.

Evidently, convergence holds if $|\omega - 1| < 1$ and if no eigenvalue $\mu = \alpha + i\beta$ of B lies in the closed exterior of the ellipse E_1 given by

$$\alpha^2 + \beta^2/[(2 - \omega)/\omega]^2 = 1 \tag{4.9}$$

But if (4.8) holds, then $D < (2 - \omega)/\omega$ and the interior of $E(1, D)$ clearly lies in the interior of E_1.

On the other hand, if for every $D > 0$, (α, β) lies outside of the ellipse $E(1, D)$, then we must have $|\alpha| \geq 1$. But this implies that some eigenvalue of B lies outside of E_1 and hence the method cannot converge. If any eigenvalue of B lies on $E(1, D)$ and if $\omega \geq 2(1 + D)^{-1}$, then there is an eigenvalue of B in the closed exterior of $E(1, D)$ and the SOR method does not converge.

Corollary 4.2. Under the hypotheses of Theorem 4.1, the SOR method converges for some ω if and only if for all eigenvalues μ of B we have

$$|\operatorname{Re} \mu| < 1 \tag{4.10}$$

If (4.10) holds, then for some $\bar{\omega}$ such that $0 < \bar{\omega} < 2$ we have $S(\mathscr{L}_{\bar{\omega}}) < 1$ for all ω with $0 < \omega \leq \bar{\omega}$.

To find $\bar{\omega}$ we simply choose D so large that all eigenvalues of B lie in $E(1, D)$. This can be done, for example, by letting

$$D = \bar{\beta}/(1 - \bar{\alpha}^2)^{1/2} \tag{4.11}$$

where

$$\bar{\alpha} = \max_{\mu \in S_B}(\operatorname{Re} \mu), \qquad \bar{\beta} = \max_{\mu \in S_B}(\operatorname{Im} \mu) \tag{4.12}$$

Determination of the Optimum Relaxation Factor

We now consider the choice of ω which minimizes $S(\mathscr{L}_\omega)$ under various assumptions on the eigenvalues of B. First let us assume that $\bar{\alpha}$ and $i\bar{\beta}$ are eigenvalues of B. We seek the smallest value of ϱ such that there exists ω with $\varrho^2 \geq |\omega - 1|$ and such that $\bar{\alpha}$ and $i\bar{\beta}$ lie in the closed interior of the ellipse (4.4). Since $\varrho^2 \geq |\omega - 1|$ we have by (4.5)

$$a = \frac{1}{\omega}\left(\varrho + \frac{\omega - 1}{\varrho}\right), \qquad b = \frac{1}{\omega}\left(\varrho - \frac{\omega - 1}{\varrho}\right) \tag{4.13}$$

Solving (4.13) for ϱ and ω in terms of a and b we have

$$4(\omega - 1)/\omega^2 = a^2 - b^2, \qquad \omega = 2/[1 + (1 + b^2 - a^2)^{1/2}] \tag{4.14}$$

and

$$\varrho = \tfrac{1}{2}\omega(a + b) = (a + b)/[1 + (1 + b^2 - a^2)^{1/2}] \tag{4.15}$$

We remark that with the above choice of ω and ϱ we have $0 < \omega < 2$ and $\varrho^2 \geq | \omega - 1 |$ since

$$\varrho - | \omega - 1 |^{1/2} = \frac{a + b - | b^2 - a^2 |^{1/2}}{1 + (1 + b^2 - a^2)^{1/2}} \geq 0 \tag{4.16}$$

By direct calculation we have

$$\frac{\partial \varrho}{\partial a} = \frac{(1 + b^2 - a^2)^{1/2} + 1 + b^2 + ab}{(1 + b^2 - a^2)^{1/2}[1 + (1 + b^2 - a^2)^{1/2}]^2} > 0$$

$$\frac{\partial \varrho}{\partial b} = \frac{(1 + b^2 - a^2)^{1/2} + 1 - a^2 - ab}{(1 + b^2 - a^2)^{1/2}[1 + (1 + b^2 - a^2)^{1/2}]^2} > 0$$

since

$$1 + b^2 - a^2 - (1 - a^2 - ab)^2 = (a + b)^2(1 - a^2) > 0$$

Evidently, the smallest value of ϱ is obtained when a and b assume their smallest values, namely, $a = \bar{\alpha}$, $b = \bar{\beta}$. Thus we have

$$\begin{aligned} \omega_b &= 2/[1 + (1 + \bar{\beta}^2 - \bar{\alpha}^2)^{1/2}] \\ S(\mathscr{L}_{\omega_b}) &= [(\bar{\alpha} + \bar{\beta})/(1 + (1 + \bar{\beta}^2 - \bar{\alpha}^2))^{1/2}]^2 \end{aligned} \tag{4.17}$$

Clearly, this choice is also optimum if all eigenvalues $\mu = \alpha + i\beta$ of B lie in the closed interior of the ellipse

$$(\alpha^2/\bar{\alpha}^2) + (\beta^2/\bar{\beta}^2) = 1 \tag{4.18}$$

Next, let us consider the case where B has the eigenvalue $\mu = \alpha + i\beta$, where $\alpha \geq 0$, $\beta \geq 0$ and all other eigenvalues of B lie in the rectangle

$$| \operatorname{Re} \mu | \leq \alpha, \qquad | \operatorname{Im} \mu | \leq \beta \tag{4.19}$$

From (4.13) we have

$$(a - b)\varrho^2 - 2\varrho + a + b = 0 \tag{4.20}$$

Evidently, given a and b, we are interested in the smaller root of (4.20) since the larger root of (4.20) is greater than unity. We seek to minimize ϱ as a function of a subject to the restriction that (4.4) holds. Treating ϱ and b as functions of a, differentiating (4.20) with respect to a and letting $\varrho'(a) = 0$ we obtain

$$b'(a) = -(1 + \varrho^2)/(1 - \varrho^2) \tag{4.21}$$

From (4.4) we have

$$b'(a) = -\alpha^2 b^3/(\beta^2 a^3) \tag{4.22}$$

and hence

$$a = (\alpha/\beta)^{2/3}[(1-\varrho^2)/(1+\varrho^2)]^{1/3} b \tag{4.23}$$

From (4.20) we obtain

$$\begin{aligned} b &= \frac{2\varrho(1-\varrho^2)^{-1/3}\beta^{2/3}}{\beta^{2/3}(1-\varrho^2)^{2/3} + \alpha^{2/3}(1+\varrho^2)^{2/3}} \\ a &= \frac{2\varrho(1+\varrho^2)^{-1/3}\alpha^{2/3}}{\beta^{2/3}(1-\varrho^2)^{2/3} + \alpha^{2/3}(1+\varrho^2)^{2/3}} \end{aligned} \tag{4.24}$$

Substituting in (4.4) we have

$$[(1+\varrho^2)/(2\varrho)]^{2/3}\alpha^{2/3} + [(1-\varrho^2)/(2\varrho)]^{2/3}\beta^{2/3} = 1 \tag{4.25}$$

Evidently the coefficients of $\alpha^{2/3}$ and $\beta^{2/3}$ are decreasing functions of ϱ as ϱ increases from 0 to 1. Since they are infinite for $\varrho = 0$ and are one and zero, respectively, for $\varrho = 1$, it follows that if $\alpha < 1$, then there is a unique solution of (4.25). Having solved (4.25) for ϱ (which can be reduced to the solution of a cubic equation) we can determine ω from

$$\omega = \frac{(1+\varrho^2)^{2/3}\alpha^{2/3} + (1-\varrho^2)^{2/3}\beta^{2/3}}{(1+\varrho^2)^{-1/3}\alpha^{2/3} + (1-\varrho^2)^{-1/3}\beta^{2/3}} \tag{4.26}$$

We remark that in the special case $\beta = 0$, we have by (4.25) and (4.26) $\omega = 1 + \varrho^2$, $(1+\varrho^2)^2\alpha^2 = 4\varrho^2$ so that $\omega^2\bar{\mu}^2 = 4(\omega - 1)$ and hence $\omega = \omega_b$.

Table 4.1 gives values of ω, $\varrho^2 = S(\mathscr{L}_\omega)$, and a corresponding to various values of α and β.

Let us now consider the case where, in the first quadrant, B has the eigenvalues $\alpha_1 + i\beta_1$ and $\alpha_2 + i\beta_2$ where $\alpha_2 > \alpha_1 \geq 0$ and where each eigenvalue μ of B lies in one of the two rectangles

$$|\operatorname{Re}\mu| \leq \alpha_k, \qquad |\operatorname{Im}\mu| \leq \beta_k, \qquad k = 1, 2$$

We assume that $\beta_1 > \beta_2$; otherwise, we need consider only the case of the rectangle $|\operatorname{Re}\mu| \leq \alpha_2$, $|\operatorname{Im}\mu| \leq \beta_2$.

We first show that for given α and β the function $\varrho(a)$ defined by (4.15) and (4.4), i.e.,

$$\varrho = \frac{a + \dfrac{a\beta}{(a^2-\alpha^2)^{1/2}}}{1 + \left(1 + \dfrac{a^2\beta^2}{a^2-\alpha^2} - a^2\right)^{1/2}} \tag{4.27}$$

decreases in the range $\alpha < a \leq \hat{a}$ and then increases in the range $\hat{a} \leq a < 1$. Here $\hat{a}$ is the value of a given by (4.24) where ϱ is the unique solution of (4.25). Evidently from (4.20)

$$\varrho'(a) = [1 + b' + (1 - b')\varrho^2]/[2 - 2\varrho(a - b)] \tag{4.28}$$

Moreover, since $\varrho < 1$ for $\alpha < a < 1$ it follows that $2 - 2\varrho(a - b)$ is positive in that interval. Hence $\varrho'(a)$ is continuous. Since $\varrho'(a)$ vanishes for only the single point $\hat{a}$ in the interval, and since $\varrho(\hat{a}) < 1$ the assertion follows.

Let $\varrho_1(a)$ and $\varrho_2(a)$ be given by (4.27) for (α_1, β_1) and (α_2, β_2), respectively, and let $\bar{\varrho}(a) = \max(\varrho_1(a), \varrho_2(a))$. Let $\varrho_1(\hat{a}_1)$ and $\varrho_2(\hat{a}_2)$ be the corresponding minimum values. Also let $\tilde{a}$ be the value of a such that (α_1, β_1) and (α_2, β_2) lie on the same ellipse. Thus we have

$$\tilde{a}^2 = (\alpha_1^2\beta_2^2 - \alpha_2^2\beta_1^2)/(\beta_2^2 - \beta_1^2) \tag{4.29}$$

For $\alpha_2 < a < \tilde{a}$ we have $\varrho_1(a) < \varrho_2(a)$, and for $\tilde{a} < a < 1$ we have $\varrho_1(a) > \varrho_2(a)$. Evidently three cases can occur:

(a) $\hat{a}_2 < \tilde{a}$. In this case $\hat{a}_1 < \tilde{a}$ and

$$\min_a \bar{\varrho}(a) = \varrho_2(\hat{a}_2) \tag{4.30}$$

(b) $\hat{a}_1 > \tilde{a}$. In this case $\hat{a}_2 > \tilde{a}$ and

$$\min_a \bar{\varrho}(a) = \varrho_1(\hat{a}_1) \tag{4.31}$$

(c) $\hat{a}_2 \geq \tilde{a}$ *and* $\hat{a}_1 \leq \tilde{a}$. In this case we have

$$\min_a \bar{\varrho}(a) = \varrho_1(\tilde{a}) \tag{4.32}$$

Thus in cases (a) and (b) we solve the one-point problem for (α_2, β_2) and (α_1, β_1), respectiveiy. For case (c) we compute b using (4.4) with $a = \tilde{a}$ and then obtain $\varrho(a)$ and $\omega(a)$ for $a = \tilde{a}$ by (4.15) and (4.14), respectively.

We remark that the value of ω is optimum not only if all the eigenvalues of B lie in the two rectangles but also if they lie in the two rectangles or within the triangle whose vertices are (α_1, β_1), (α_2, β_2), and (α_1, β_2). The special case $\alpha_1 = \frac{1}{4}$, $\beta_1 = \frac{3}{4}$, $\alpha_2 = \frac{3}{4}$, $\beta_2 = \frac{1}{4}$ is illustrated in Figure 4.2. (See Exercise 6.)

Let us now consider the case in the first quadrant where B has three eigenvalues $\mu_k = \alpha_k + i\beta_k$, $k = 1, 2, 3$, where $\alpha_3 \geq \alpha_2 \geq \alpha_1$. If any eigenvalue μ_j is in the rectangle $0 \leq \alpha \leq \alpha_k$, $0 \leq \beta \leq \beta_k$ for some k,

TABLE 4.1. Values of ω, $S(\mathscr{L}_\omega)$,

α \ β	0.00	.02	.04	.06	.08	.10	.15	.20	.25	.30
	1.11111	1.12104	1.12392	1.12417	1.12248	1.11917	1.10515	1.08473	1.05977	1.03180
.60	.11111	.15241	.18043	.20592	.23004	.25318	.30765	.35770	.40344	.44497
	.60000	.62481	.63750	.64707	.65479	.66122	.67343	.68197	.68814	.69270
	1.13641	1.14800	1.15110	1.15097	1.14844	1.14393	1.12569	1.10005	1.06953	1.03615
.65	.13641	.18505	.21773	.24720	.27484	.30112	.36186	.41619	.46452	.50728
	.65000	.67441	.68668	.69580	.70304	.70898	.72000	.72743	.73264	.73637
	1.16676	1.18025	1.18336	1.18243	1.17847	1.17204	1.14756	1.11464	1.07680	1.03665
.70	.16676	.22426	.26249	.29662	.32826	.35797	.42509	.48309	.53297	.57575
	.70000	.72373	.73539	.74388	.75050	.75582	.76538	.77154	.77568	.77855
	1.20378	1.21939	1.22210	1.21957	1.21315	1.20360	1.16959	1.12640	1.07894	1.03045
.75	.20378	.27226	.31722	.35682	.39296	.42632	.49935	.55960	.60923	.65022
	.75000	.77272	.78354	.79120	.79701	.80156	.80938	.81413	.81716	.81918
	1.25000	1.26796	1.26930	1.26359	1.25272	1.23785	1.18888	1.13113	1.07123	1.01283
.80	.25000	.33263	.38595	.43194	.47293	.50980	.58687	.64651	.69299	.72969
	.80000	.82128	.83095	.83753	.84232	.84592	.85175	.85502	.85698	.85823
	1.30994	1.33021	1.32777	1.31532	1.29582	1.27135	1.19826	1.12016	1.04492	.97552
.85	.30994	.41178	.47562	.52872	.57413	.61323	.68938	.74319	.78233	.81169
	.85000	.86925	.87735	.88251	.88602	.88851	.89219	.89404	.89507	.89569
	1.39286	1.41389	1.40041	1.37160	1.33367	1.29087	1.17898	1.07430	.98216	.90242
.90	.39286	.52292	.59944	.65828	.70457	.74137	.80539	.84538	.87223	.89133
	.90000	.91626	.92216	.92545	.92744	.92871	.93036	.93108	.93144	.93166
	1.43687	1.45633	1.43276	1.39098	1.34012	1.28584	1.15329	1.03700	.93868	.85589
.92	.43687	.58249	.66382	.72267	.76633	.79937	.85354	.88549	.90627	.92076
	.92000	.93462	.93938	.94180	.94315	.94396	.94494	.94535	.94555	.94566
	1.49123	1.50553	1.46359	1.40121	1.33212	1.26324	1.10760	.98028	.87713	.79273
.94	.49123	.65617	.74014	.79518	.83278	.85950	.90054	.92340	.93783	.94773
	.94000	.95257	.95599	.95750	.95825	.95867	.95915	.95934	.95942	.95947
	1.56250	1.56078	1.48186	1.38599	1.29195	1.20551	1.02638	.89061	.78559	.70232
.96	.56250	.75110	.83020	.87402	.90075	.91843	.94387	.95733	.96561	.97121
	.96000	.96985	.97175	.97242	.97271	.97286	.97302	.97308	.97310	.97312
	1.66806	1.60288	1.44068	1.29211	1.16705	1.06252	.86627	.73044	.63120	.55560
.98	.66806	.87733	.92911	.95088	.96256	.96979	.97966	.98468	.98771	.98975
	.98000	.98592	.98640	.98652	.98656	.98658	.98660	.98661	.98662	.98662
	2.00000	2.00000	2.00000	2.00000	2.00000	2.00000	2.00000	2.00000	2.00000	2.00000
1.00	1.00000	1.00000	1.00000	1.00000	1.00000	1.00000	1.00000	1.00000	1.00000	1.00000
	1.00000	1.00000	1.00000	1.00000	1.00000	1.00000	1.00000	1.00000	1.00000	1.00000

we discard that eigenvalue and reduce to the one-eigenvalue or two-eigenvalue case. Otherwise we compute

$$a_j = [(\alpha_3^2\beta_j^2 - \alpha_j^2\beta_3^2)/(\beta_j^2 - \beta_3^2)]^{1/2}, \qquad j = 1, 2 \tag{4.33}$$

If $a_1 \le a_2$, then we discard μ_2 and have the two-eigenvalue problem. If $a_2 < a_1$, then we solve the one-parameter problem in each of the intervals

$$\alpha_3 \le a \le a^{(1)}, \qquad a^{(1)} \le a \le a^{(2)}, \qquad a^{(2)} \le a \le 1$$

with μ_3, μ_2, and μ_1, respectively, where $a^{(1)} = a_2$, and

$$a^{(2)} = [(\alpha_2^2\beta_1^2 - \alpha_1^2\beta_2^2)/(\beta_1^2 - \beta_2^2)]^{1/2} \tag{4.34}$$

and a Corresponding to $\mu = \alpha + i\beta$.

.40	.50	.60	.80	1.00	1.20	1.40	1.60	1.80	2.00
.97150	.91059	.85252	.74973	.66512	.59589	.53883	.49125	.45112	.41688
.51626	.57400	.62090	.69118	.74052	.77671	.80425	.82585	.84321	.85746
.69876	.70242	.70474	.70737	.70872	.70949	.70998	.71030	.71052	.71068
.96644	.89846	.83543	.72715	.64047	.57083	.51416	.46735	.42814	.39488
.57824	.63357	.67722	.74068	.78402	.81525	.83872	.85697	.87154	.88345
.74116	.74394	.74567	.74757	.74853	.74907	.74940	.74963	.74978	.74989
.95597	.88041	.81246	.69934	.61133	.54193	.48617	.44054	.40258	.37056
.64404	.69512	.73421	.78937	.82604	.85201	.87132	.88622	.89804	.90765
.78209	.78406	.78526	.78654	.78718	.78753	.78775	.78789	.78799	.78807
.93735	.85406	.78158	.66477	.57650	.50817	.45397	.41004	.37375	.34331
.71284	.75764	.79088	.83644	.86597	.88657	.90172	.91333	.92250	.92991
.82157	.82284	.82360	.82439	.82477	.82498	.82511	.82520	.82526	.82530
.90641	.81588	.73978	.62116	.53413	.46802	.41625	.37468	.34060	.31216
.78312	.81961	.84588	.88092	.90311	.91838	.92951	.93798	.94464	.95001
.85964	.86036	.86077	.86120	.86140	.86151	.86158	.86163	.86166	.86168
.85625	.76013	.68218	.56476	.48116	.41886	.37073	.33247	.30133	.27550
.85233	.87885	.89742	.92160	.93661	.94681	.95419	.95978	.96415	.96767
.89636	.89669	.89688	.89707	.89716	.89721	.89723	.89725	.89727	.89728
.77380	.67589	.59942	.48824	.41157	.35562	.31301	.27949	.25245	.23017
.91654	.93234	.94315	.95694	.96536	.97103	.97511	.97818	.98058	.98250
.93187	.93198	.93204	.93209	.93212	.93214	.93215	.93215	.93215	.93216
.72585	.62923	.55496	.44862	.37632	.32402	.28446	.25350	.22860	.20816
.93958	.95122	.95911	.96912	.97520	.97928	.98221	.98441	.98613	.98751
.94578	.94583	.94586	.94589	.94590	.94591	.94592	.94592	.94592	.94592
.66383	.57050	.49999	.40071	.33424	.28665	.25091	.22309	.20082	.18259
.96039	.96813	.97334	.97992	.98390	.98656	.98847	.98990	.99102	.99191
.95952	.95955	.95956	.95957	.95958	.95958	.95958	.95958	.95959	.95959
.57907	.49241	.42824	.33963	.28137	.24016	.20947	.18573	.16683	.15142
.97829	.98258	.98546	.98907	.99124	.99270	.99374	.99452	.99512	.99561
.97313	.97314	.97314	.97315	.97315	.97315	.97315	.97315	.97315	.97315
.44815	.37548	.32308	.25256	.20731	.17580	.15261	.13482	.12075	.10934
.99230	.99383	.99486	.99614	.99691	.99742	.99779	.99807	.99828	.99845
.98662	.98662	.98662	.98662	.98662	.98662	.98662	.98662	.98662	.98662
2.00000	2.00000	2.00000	2.00000	2.00000	2.00000	2.00000	2.00000	2.00000	2.00000
1.00000	1.00000	1.00000	1.00000	1.00000	1.00000	1.00000	1.00000	1.00000	1.00000
1.00000	1.00000	1.00000	1.00000	1.00000	1.00000	1.00000	1.00000	1.00000	1.00000

If the optimum a for the one-eigenvalue problem for μ_3 occurs in the interval $\alpha_3 \leq a \leq a^{(1)}$, then that value of a leads to the optimum value of ω. Otherwise we see whether the function $\varrho(a)$ corresponding to the one-parameter problem for μ_2 decreases in the interval $a^{(1)} \leq a \leq a^{(2)}$. If so, then the optimum value of a lies in the interval $a^{(2)} \leq a \leq 1$ and is found by solving the one-parameter problem for μ_1 in that interval. If not, then the optimum a lies in the interval $a^{(1)} \leq a \leq a^{(2)}$ and corresponds to the solution of the one-parameter problem in that interval.

The justification for the procedure lies in the fact that if (α, β) lies on the ellipse E

$$(x/a)^2 + (y/b)^2 = 1$$

and if $(\hat{\alpha}, \hat{\beta})$, where $\hat{\alpha} \geq \alpha$, lies inside or on E, then $(\hat{\alpha}, \hat{\beta})$ lies on or inside any ellipse

$$(x/\hat{a})^2 + (y/\hat{b})^2 = 1$$

which contains (α, β) provided $\hat{a} \geq a$.

Clearly, this procedure can be generalized to an arbitrary number of complex eigenvalues of B. Given a region in the μ-plane known to contain all eigenvalues of B we consider a convex polygon, symmetric with respect to both the real and the imaginary axes containing the region. We then let the (α_k, β_k) be the vertices of the polygon in the first quadrant. For more details see Young and Eidson [1970] who give a description of a computer program for finding the optimum ω.

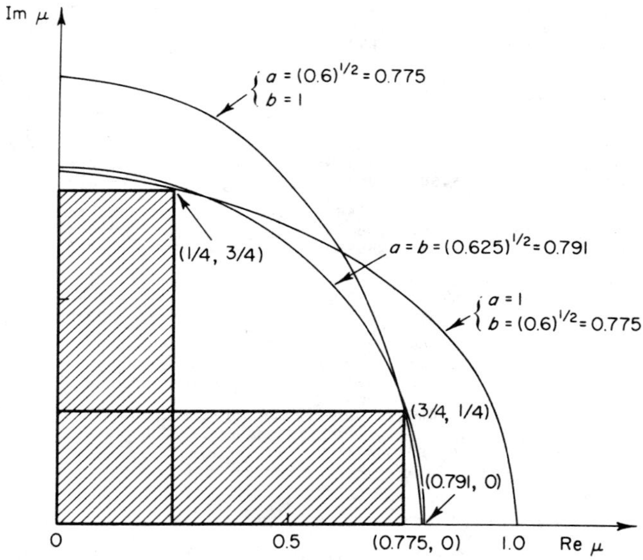

Figure 4.2. Two-eigenvalue case.

6.5. PRACTICAL DETERMINATION OF ω_b: GENERAL CONSIDERATIONS

For the rest of Chapter 6 we shall assume that A is a consistently ordered matrix with nonvanishing diagonal elements and that $B = I - (\text{diag } A)^{-1}A$ has real eigenvalues all of which are less than unity in absolute value.

If we let $\bar{\lambda} = S(\mathscr{L})$ and $\bar{\mu} = S(B)$, we have by (1) and Theorem 2.3,

$$\omega_b = 2/[1 + (1 - \bar{\mu}^2)^{1/2}] = 2/[1 + (1 - \bar{\lambda})^{1/2}], \qquad \bar{\lambda} = \bar{\mu}^2 \tag{5.1}$$

Hence, given either $\bar{\lambda}$ or $\bar{\mu}$ we can obtain all three of the quantities $\bar{\mu}$, $\bar{\lambda}$, ω_b. We shall consider two classes of methods for finding one and hence all three of these quantities: "iterative" and "a priori." In the first class of methods we seek to determine $\bar{\mu}$ or $\bar{\lambda}$ by an iterative process. In some cases the iterations performed may also be considered as iterations with the SOR method itself and hence be "useful," while in others the sole purpose is to obtain an estimate of $\bar{\mu}$ and $\bar{\lambda}$ from which to compute ω_b. Subsequent iterations are performed with the SOR method and $\omega = \omega_b$. The a priori methods involve the approximate determination of $\bar{\mu}$ by the solution of an eigenvalue problem involving either a partial difference operator or a partial differential operator. By the use of monotonicity theorems we are able to study these operators in simple regions such as rectangles.

Behavior of $S(\mathscr{L}_\omega)$

Of fundamental importance in choosing ω is the behavior of $S(\mathscr{L}_\omega)$ as a function of ω. In Figures 5.1 and 5.2 we give a graph of $S(\mathscr{L}_\omega)$ as a function of ω for the case $\bar{\mu} = 0.987688$ which corresponds to the model problem with $h = \frac{1}{20}$ As ω increases from 0 to 1, $S(\mathscr{L}_\omega)$ decreases

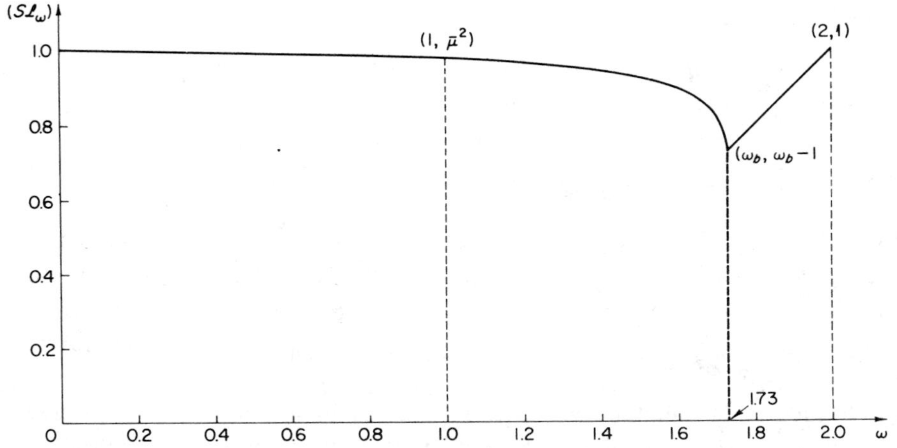

Figure 5.1. $S(\mathscr{L}_\omega)$ vs ω for $\bar{\mu} = \cos(\pi/20) = 0.987688$, $0 \le \omega \le 2$.

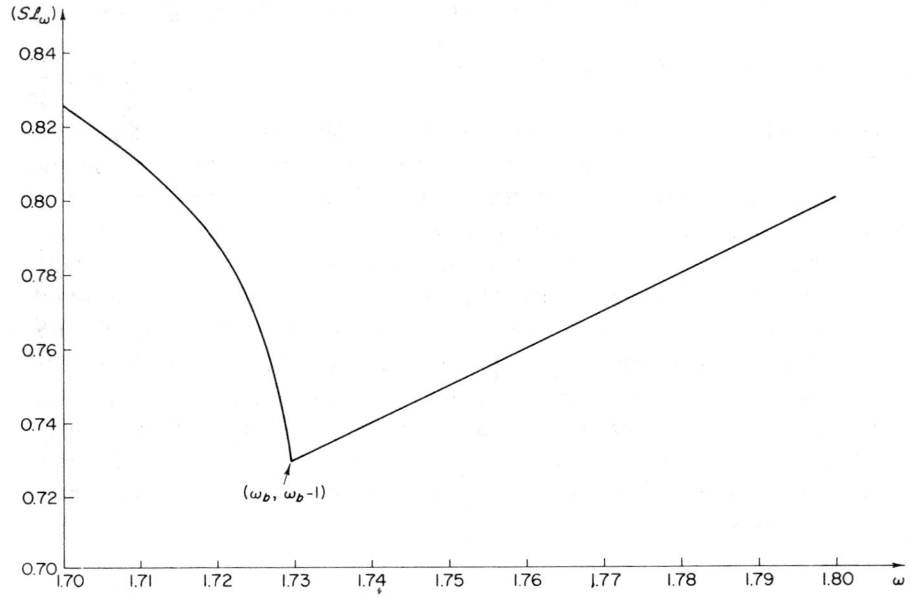

Figure 5.2. $S(\mathscr{L}_\omega)$ vs ω for $\bar{\mu} = 0.987688$, $1.70 \leq \omega \leq 1.80$.

very slowly from 1 to $\bar{\mu}^2 = 0.975528$. As ω increases further $S(\mathscr{L}_\omega)$ decreases slightly more rapidly until ω gets close to $\omega_b = 1.72945$ at which point the decrease is very rapid. As a matter of fact, the left-hand derivative of $S(\mathscr{L}_\omega)$ is infinite at $\omega = \omega_b$ since from (2.10) and (2.18) we have

$$\frac{1}{[S(\mathscr{L}_\omega)]^{1/2}} \frac{d}{d\omega} S(\mathscr{L}_\omega) = \frac{d}{d\omega} [\omega\bar{\mu} + (\omega^2\bar{\mu}^2 - 4(\omega - 1))^{1/2}]$$

$$= \frac{-4(1 - \bar{\mu}^2)}{[\omega^2\bar{\mu}^2 - 4(\omega - 1)]^{1/2}\{\bar{\mu}[\omega^2\bar{\mu}^2 - 4(\omega - 1)]^{1/2} + (2 - \omega\bar{\mu}^2)\}} \tag{5.2}$$

for $0 < \omega < \omega_b$. Therefore, as $\omega \to \omega_b -$ the slope of $S(\mathscr{L}_\omega)$ approaches $-\infty$. The value of $S(\mathscr{L}_\omega)$ for $\omega = \omega_b$ is $\omega_b - 1 = 0.72945$. As ω increases further, $S(\mathscr{L}_\omega)$ increases linearly to a value of unity when $\omega = 2$. (As a matter of fact, $S(\mathscr{L}_\omega) = \omega - 1$ for all ω in the range $\omega_b \leq \omega \leq 2$.)

Evidently, a small decrease in ω_b results in a much larger relative decrease in the rate of convergence than a corresponding increase in ω_b. This is illustrated in Table 5.1. Thus, for example, using $\omega = 1.72$

TABLE 5.1. Convergence of the SOR Method for $\bar{\mu} = 0.987688$

ω	$S(\mathscr{L}_\omega)$	$R(\mathscr{L}_\omega)$	$100\left(\frac{R(\mathscr{L}_{\omega_b}) - R(\mathscr{L}_\omega)}{R(\mathscr{L}_{\omega_b})}\right)$
1.70	0.82620	0.19092	39.48
1.71	0.81085	0.20968	33.53
1.72	0.78881	0.23723	24.80
1.72945	0.72945	0.31546	0
1.73	0.73000	0.31471	0.24
1.74	0.74000	0.30110	4.55
1.75	0.75000	0.28768	8.80
1.76	0.76000	0.27444	13.00
1.77	0.77000	0.26136	17.15
1.78	0.78000	0.24846	21.24
1.79	0.79000	0.23572	25.28
1.80	0.80000	0.22314	29.26

instead of 1.72945 for ω_b results in a relative decrease of 24.80% in the rate of convergence, while using $\omega = 1.74$ (which differs from ω_b by slightly more than 1.72) results in a relative decrease of only 4.55% in the rate of convergence.

Behavior of Eigenvalues of $\mathscr{L}_\omega$

In order to consider methods for finding $S(\mathscr{L}_\omega)$ for various values of ω, as a basis for choosing an accurate value of ω_b, we now study the behavior of the eigenvalues of $\mathscr{L}_\omega$ as functions of ω and μ. Let us first study the behavior of the roots of

$$(\lambda + \omega - 1)^2 = \omega^2\mu^2\lambda \tag{5.3}$$

for given μ as ω varies in the range $0 \leq \omega \leq 2$. If $\mu = 0$, then we have a double root $1 - \omega$ which decreases from 1 to -1 as ω increases from 0 to 2. In general, let

$$\begin{aligned} \lambda^+ &= \tfrac{1}{2}\{\omega^2\mu^2 - 2(\omega - 1) + [\omega^2\mu^2(\omega^2\mu^2 - 4(\omega - 1))]^{1/2}\} \\ \lambda^- &= \tfrac{1}{2}\{\omega^2\mu^2 - 2(\omega - 1) - [\omega^2\mu^2(\omega^2\mu^2 - 4(\omega - 1))]^{1/2}\} \end{aligned} \tag{5.4}$$

When $\omega = 0$, both λ^+ and λ^- equal unity. As ω increases both decrease but λ^- decreases faster. When $\omega = 1$ we have $\lambda^+ = \mu^2$ and $\lambda^- = 0$.

Then further increases of ω will decrease λ^+ and increase λ^- until $\omega = 2[1 + (1 - \mu^2)^{1/2}]^{-1}$ in which case $\lambda^+ = \lambda^- = \omega - 1$. As ω increases still further, λ^+ and λ^- become complex conjugate pairs with increasing modulus moving out to the circle $|\lambda| = 1$ when $\omega = 2$.

For example, if $\mu = 0.8$ we have $\omega_b(\mu) = 2[1 + (1 - \mu^2)^{1/2}]^{-1} = 1.25$ and the values for ω, λ, and λ^- are shown in the following tabulation.

ω	λ^+	λ^-
0	1	1
$\frac{1}{2}$	0.874	0.286
1	0.640	0
1.1	0.556	0.018
$1.25 = \omega_b(\mu)$	0.25	0.25
1.50	$0.220 + 0.449i$	$0.220 - 0.449i$
2.00	$0.280 + 0.960i$	$0.280 - 0.960i$

The behavior of λ^+ and λ^- is shown in Figure 5.3.

The behavior of the eigenvalues of $\mathscr{L}_\omega$ can be seen from Figure 5.3. The curves labeled C_z refer to the value $\mu = z$. Thus, for example, if $z = 0.8$, then the eigenvalues λ^+ and λ^- corresponding to $\mu = 0.8$ are real for $\omega < 1.25$ and behave as indicated in the above table until $\omega = 1.25$. Then they move along the curve $C_{0.8}$ as indicated. The value of the root λ^+ can be found as the intersection of the circle with radius $\omega - 1$ and $C_{0.8}$.

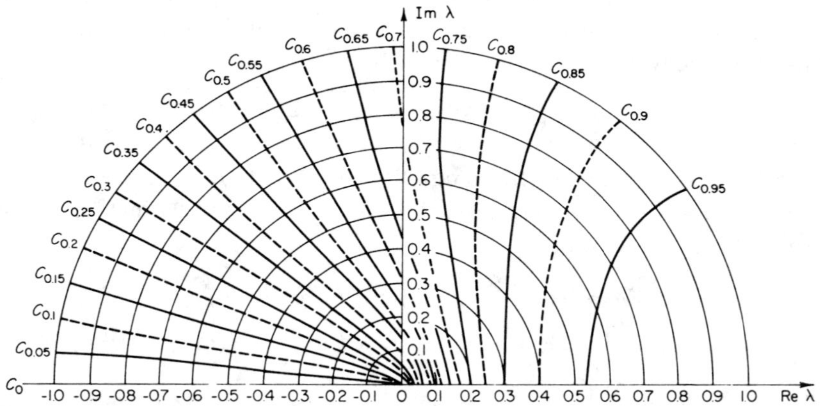

Figure 5.3. Behavior of eigenvalues of $\mathscr{L}_\omega$.

Figure 5.4 shows the behavior of all of the eigenvalues of $\mathscr{L}_\omega$ for the case $\bar{\mu} = S(B) = 0.8$ and $\omega_b = 1.25$. When $\omega = 0$, all eigenvalues of $\mathscr{L}_0$ are unity. As ω increases, the eigenvalues decrease. For $\omega = 0.5$ they lie in the interval $0.286 \leq \lambda \leq 0.874$. When $\omega = 1$, we have many zero eigenvalues and other real eigenvalues in the interval $0 \leq \lambda \leq \bar{\mu}^2 = 0.64$. As ω increases to 1.1 we have a number of complex eigenvalues

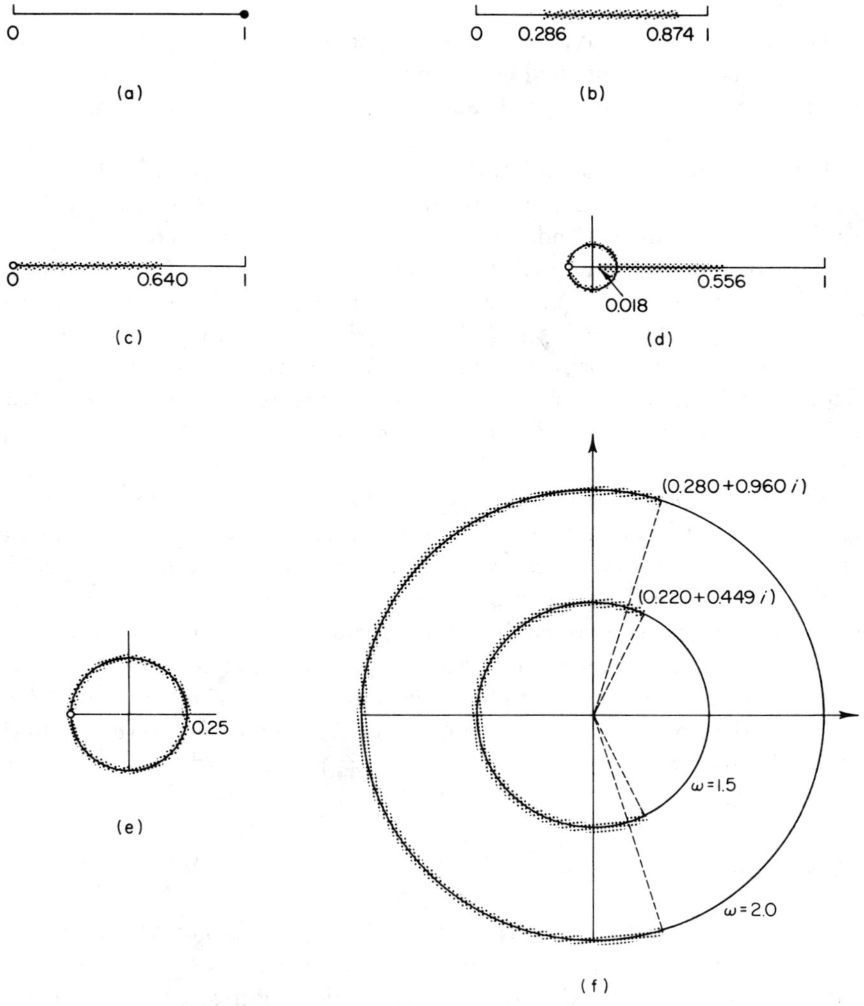

Figure 5.4. Eigenvalues of $\mathscr{L}_\omega$ for $S(B) = 0.8$. (a) $\omega = 0$; (b) $\omega = 0.5$; (c) $\omega = 1$; (d) $\omega = 1.1$; (e) $\omega = \omega_b = 1.25$; (f) $\omega > \omega_b$. (See text for further details.)

on the circle $|\lambda| = \omega - 1 = 0.1$ and real eigenvalues in the interval $0.018 \le \lambda \le 0.556$. As ω increases, the interval of positive real eigenvalues shrinks to the single point $\omega_b - 1 = 0.25$ for $\omega = \omega_b$. If $\bar{\mu}$ is a simple eigenvalue of B, then $\lambda = \omega_b - 1$ is a double eigenvalue of $\mathscr{L}_{\omega_b}$. All other eigenvalues are complex and have modulus 0.25. As ω increases further, the eigenvalues increase in modulus, always having modulus $\omega - 1$. They tend to move away from the positive real axis. For example, when $\omega = 1.5$, there are no eigenvalues on an arc centered about the positive real axis. When $\omega = 2$, all eigenvalues have modulus unity, but again there is none in an arc centered about the positive real axis. The point $0.280 + 0.960i$ is the intersection of circle $|\lambda| = 1$ and the curve $C_{0.8}$ of Figure 5.3.

Let us summarize the situation in the case where $\bar{\mu}$ is a simple eigenvalue of B. This will be the case if A is a Stieltjes matrix since B is similar to a symmetric nonnegative matrix; hence by Theorem 2-7.6 the eigenvalue $\bar{\mu}$ is simple. Evidently, for $0 < \omega < \omega_b$, $S(\mathscr{L}_\omega)$ is a simple eigenvalue of $\mathscr{L}_\omega$ and has a larger modulus than any other eigenvalue of $\mathscr{L}_\omega$. For $\omega = \omega_b$, $S(\mathscr{L}_{\omega_b})$ is a double eigenvalue. Moreover, as we shall see in Chapter 7, there corresponds only one linearly independent eigenvector for $S(\mathscr{L}_{\omega_b})$; hence the Jordan canonical form of $\mathscr{L}_{\omega_b}$ is not diagonal. In addition, all eigenvalues of $\mathscr{L}_{\omega_b}$ have modulus $\omega_b - 1$.

The *power method* (see, for instance, Faddeev and Faddeeva [1963]) for finding the eigenvalue λ_1 of largest modulus of a matrix A involves choosing an arbitrary vector v and successively computing Av, A^2v, $A^3v, \ldots$. If there is only one eigenvalue of A with modulus $S(A)$ then, unless a very unfortunate choice of v is made, the computed vectors will converge to a multiple of the eigenvector v_1 associated with λ_1. From v_1 one can easily obtain λ_1. The power method is not effective if there are more than one eigenvalue of modulus $S(A)$. Thus the power method can be used effectively to determine $S(\mathscr{L}_\omega)$ if $\omega < \omega_b$, but it is not effective if $\omega \ge \omega_b$.

Monotonicity Theorems

The considerations on the behavior of the eigenvalues of $\mathscr{L}_\omega$ are useful if one is planning to estimate $\bar{\mu}$ or $S(\mathscr{L})$ by an iterative process. If one wishes to use a priori methods, however, it is useful to be able to treat a case which is simpler than the given problem. For this purpose the following monotonicity theorems are helpful.

Theorem 5.1. If A is an L-matrix and if A_1 is a matrix formed from A by deleting some of the rows and corresponding columns of A, then $S(B_1) \le S(B)$. Here $B_1 = I - (\text{diag } A_1)^{-1}A_1$ and $B = I - (\text{diag } A)^{-1}A$.

Proof. Evidently $S(B_1) = S(B')$ where B' is obtained from B by replacing the elements corresponding to deleted rows and columns by zeros. By Theorem 2-1.16, we have $S(B_1) \le S(B)$ since $B_1 \le B$.

Theorem 5.2. If A is a symmetric matrix and if A_1 is obtained from A by deleting certain rows and the corresponding columns of A, then

$$\underline{\lambda}(A) \le \underline{\lambda}(A_1) \le \bar{\lambda}(A_1) \le \bar{\lambda}(A) \tag{5.5}$$

where $\underline{\lambda}(A)$ and $\bar{\lambda}(A)$ denote the smallest and largest eigenvalues of A, respectively. If A is positive definite, and if $B[A] = I - (\text{diag } A)^{-1}A$, then

$$S(B[A_1]) \le S(B[A]) \tag{5.6}$$

and

$$K(A) = \bar{\lambda}(A)/\underline{\lambda}(A) \ge \bar{\lambda}(A_1)/\underline{\lambda}(A_1) = K(A_1) \tag{5.7}$$

Proof. For any $N \times N$ matrix A let $\psi(A)$ be the matrix obtained by deleting certain rows and the corresponding columns of A, as in the statement of the theorem. Thus we have $A_1 = \psi(A)$. Without loss of generality we can assume that $\text{diag } A = I$ and $\text{diag } A_1 = I_1$, where $I_1 = \psi(I)$, since otherwise we can consider $\hat{A} = D^{-1/2}AD^{-1/2}$ and $\hat{A}_1 = D_1^{-1/2} \times A_1 D_1^{-1/2}$, where $D = \text{diag } A$, $D_1 = \text{diag } A_1$. For, by Theorem 4-3.5, we have

$$S(B[\hat{A}]) = S(B[A]), \qquad S(B[\hat{A}_1]) = S(B[A]),$$

and

$$\begin{aligned} \hat{A} = D_1^{-1/2}A_1D_1^{-1/2} &= \psi(D)^{-1/2}\psi(A)\psi(D)^{-1/2} \\ &= \psi(D^{-1/2}AD^{-1/2}) = \psi(\hat{A}) \end{aligned}$$

since $D^{-1/2}$ is a diagonal matrix. Moreover, $\text{diag } \hat{A} = I$, $\text{diag } \hat{A}_1 = \psi(I_1) = I_1$.

Let $\bar{\lambda}(A)$ and $\bar{\lambda}(A_1)$ denote the largest eigenvalues of A and A_1, respectively, and let $\underline{\lambda}(A)$ and $\underline{\lambda}(A_1)$ denote the smallest eigenvalues. Then by Theorem 2-2.2 we have

$$\bar{\lambda}(A) = \max_{\substack{v \ne 0 \\ v \in E^N}} (v, Av)/(v, v), \qquad \underline{\lambda}(A) = \min_{\substack{v \ne 0 \\ v \in E^N}} (v, Av)/(v, v) \tag{5.8}$$

and

$$\begin{aligned} \bar{\lambda}(A_1) &= \max_{\substack{v_1 \neq 0, \\ v_1 \in E^{N'}}} (v_1, A_1 v_1)/(v_1, v_1), \\ \underline{\lambda}(A_1) &= \min_{\substack{v_1 \neq 0, \\ v_1 \in E^{N'}}} (v_1, A_1 v_1)/(v_1, v_1) \end{aligned} \tag{5.9}$$

where $E^{N'}$ is the vector space of column vectors of order equal to the order of A_1.

Evidently, we have

$$\bar{\lambda}(A_1) = \max_{\substack{v_1 \neq 0, \\ v_1 \in E^{N'}}} (v_1, A_1 v_1)/(v_1, v_1) = \max_{\substack{v \neq 0 \\ v \in \hat{E}^N}} (v, Av)/(v, v) \tag{5.10}$$

where $\hat{E}^N$ is the set of all column vectors such that those components corresponding to the rows and columns which were deleted in constructing A_1 from A vanish. But since

$$\max_{\substack{v \neq 0 \\ v \in \hat{E}^N}} (v, Av)/(v, v) \leq \max_{\substack{v \neq 0 \\ v \in E^N}} (v, Av)/(v, v) = \bar{\lambda}(A) \tag{5.11}$$

it follows that $\bar{\lambda}(A_1) \leq \bar{\lambda}(A)$. Similarly, we have $\underline{\lambda}(A_1) \geq \underline{\lambda}(A)$ and (5.5) follows.

If A is positive definite, then (5.7) follows from (5.5). Moreover, since $B = I - A$ we have (5.6).

Corollary 5.3. Let A and A' be the matrices corresponding to the discrete generalized Dirichlet problem defined by (2-8.10) for the regions R_h and R_h' where $R_h' \subseteq R_h$. Then

$$S(B') \leq S(B)$$

where $B = I - (\text{diag } A)^{-1}A$, $B' = I - (\text{diag } A')^{-1}A'$.

Proof. As shown in Section 2.8, A and A' are positive definite. Since A' can be obtained from A by deleting certain rows and the corresponding columns, the result follows from Theorem 5.2.

As an example of Theorem 5.1 let us compare $S(B)$ for the Laplace difference equation for the L-shaped region shown in Figure 5.5 and for the unit square with $h = \frac{1}{4}$ in each case. By direct computation we can show that for the square

$$S(B) = \cos(\pi/4) = (2)^{1/2}/2 \doteq 0.7071$$

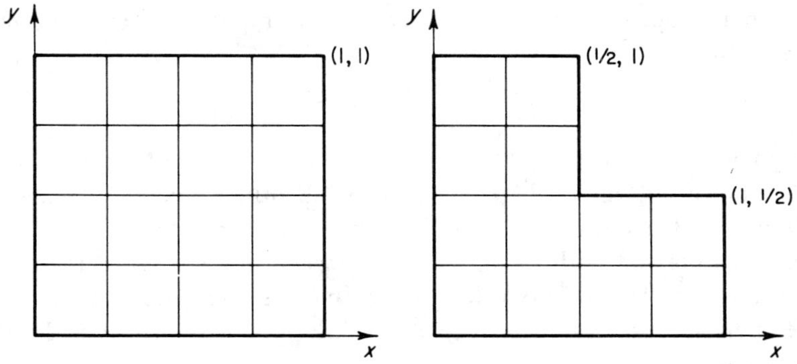

Figure 5.5. The square and the L-shaped region.

while for the L-shaped region

$$S(B) = (3)^{1/2}/4 \doteq 0.432$$

which is less than for the square.

If we estimate ω_b for a given region by determining ω_b for a larger region, the value thus obtained will be at least as large as the true value. The loss in convergence resulting from the inaccurate value of ω_b will be considerably less than if ω_b were inaccurate to the same extent but less than the true value.

6.6. ITERATIVE METHODS FOR CHOOSING ω_b

A very simple scheme for choosing ω is the following "trial and error method" which is feasible if one has to solve many linear systems with the same matrix A. One simply tries various values of ω and sees how many iterations are needed for the method to converge. Various measures of convergence can be used; one of the most common is to require that

$$\| \delta^{(n)} \|_\infty < \varrho$$

where $\varrho \ll 1$ and where $\delta^{(n)} = u^{(n+1)} - u^{(n)}$. In any case, one would choose as ω_b that value of ω which resulted in the fewest iterations. One would expect that the value thus selected would be optimum, or close to optimum, for all linear systems with the same matrix.

A somewhat more sophisticated scheme involves selecting a value of ω and attempting to monitor the rate of convergence, making changes in ω where appropriate. One measure of the rapidity of convergence is the quantity†

$$\theta_n = \| \delta^{(n+1)} \|_\infty / \| \delta^{(n)} \|_\infty$$

It is easy to verify that $\delta^{(n)} = \mathscr{L}_\omega{}^n \delta^{(0)}$; hence, if $S(\mathscr{L}_\omega)$ is a simple eigenvalue of $\mathscr{L}_\omega$ and $|\lambda| < S(\mathscr{L}_\omega)$ for all other eigenvalues λ of $\mathscr{L}_\omega$ it follows that eventually $\delta^{(n)}$ will converge to a multiple of the eigenvector v associated with $S(\mathscr{L}_\omega)$. Thus, if $\omega < \omega_b$, we can carry out the above process. When the values of θ_n have stabilized to a value, say $\bar{\theta}$, we then accept $\bar{\theta}$ as an estimate for $S(\mathscr{L}_\omega)$. We then solve (2.10) for $\bar{\mu}$ obtaining

$$\bar{\mu} \doteq (\bar{\theta} + \omega - 1)/(\omega \bar{\theta}^{1/2})$$

We then can compute an approximate value of ω_b, say $\tilde{\omega}_b$, from (5.1) and resume the SOR iteration process with $\omega = \tilde{\omega}_b$. One can compute the θ_n obtained from this modified process and, when they appear to converge, recompute $\tilde{\omega}_b$ and continue. If the original choice of ω is too large, i.e., if $\omega \geq \omega_b$, then the values of θ_n will oscillate and it will be difficult to estimate $S(\mathscr{L}_\omega)$. One can consider reducing ω successively until one obtains a value such that the θ_n converge.

A number of schemes have been used based on the above idea; see, for instance, Kulsrud [1961], Carré [1961], Rigler [1965], and Reid [1966].

Young [1953] and Young and Shaw [1955] performed a number of iterations with the Gauss–Seidel method in an attempt to get an estimate of $S(\mathscr{L})$. It was found that the number of iterations needed to obtain a good estimate of ω_b was comparable to the number of iterations needed to solve the original problem by the SOR method given the true value of ω_b. Thus the scheme would not be practical unless one wished to solve a number of cases involving the same matrix.

In the schemes described so far an attempt is made to use iterations for estimating $S(\mathscr{L})$, or $S(B)$, which are "useful" in the sense that they contribute to the solution of the original problem. Of course, in the scheme involving iterating with $\mathscr{L}$ the iterations are of very little value as far as the original problem is concerned. It is perhaps natural to consider other iterative procedures for determining $S(\mathscr{L})$, or $S(B)$, where the iterations are not necessarily "useful" as far as the original system is concerned. One such scheme has been proposed by Hageman

† One could, of course, compute the quotient $\| \mathscr{L}_\omega \delta^{(n)} \| / \| \delta^{(n)} \|$ or $\| \mathscr{L}_\omega \delta^{(n)} \|_{A^{1/2}} / \| \delta^{(n)} \|_{A^{1/2}}$ but this would involve additional calculations.

and Kellogg [1966, 1968]. This scheme is designed to yield an accurate value of $\bar{\lambda} = S(\mathscr{L})$ in relatively few iterations if the matrix A is symmetric and has the form

$$A = \begin{pmatrix} D_1 & H \\ K & D_2 \end{pmatrix}$$

where D_1 and D_2 are square diagonal matrices. To carry out the process it is necessary to estimate $\sigma = \bar{\lambda}'/\bar{\lambda}$ where $\lambda \leq \bar{\lambda}' \leq \bar{\lambda}$ for all eigenvalues λ of $\mathscr{L}$ such that $\lambda \neq \bar{\lambda}$. If all the eigenvalues of $\mathscr{L}$ were simple, then $\bar{\lambda}'$ would be the next largest eigenvalue of $\mathscr{L}$. The method consists in starting with an arbitrary vector $v^{(0)}$ and computing for each n

$$v^{(n)} = P_n(\mathscr{L})v^{(0)}$$

where $P_n(\mathscr{L})$ is a polynomial in $\mathscr{L}$ so chosen that $P_n(\bar{\lambda}) = 1$ and $P_n(\lambda)$ is minimized in the range $0 \leq \lambda \leq \bar{\lambda}'$. A special case would be $P_n(\mathscr{L}) = \mathscr{L}^n$ which amounts to repeated iteration with the Gauss–Seidel method. This, as we have seen, is relatively slow. It turns out that the best choice of $P_n(\mathscr{L})$ is related to a certain Chebyshev polynomial. Using $P_n(\mathscr{L})$ one obtains a sequence of vectors $v^{(1)}, v^{(2)}, \ldots$ which converge rapidly to a multiple of an eigenvector associated with $S(\mathscr{L})$. Methods for estimating σ are given by Hageman and Kellogg [1968].

Before studying the papers of Hageman and Kellogg and related work of Varga [1957] and Bilodeau *et al.* [1957], the reader should be familiar with the theory on semi-iterative methods which will be given in Chapter 11.

We remark that it is important to use the red–black ordering for these schemes instead of the natural ordering. With the natural ordering, the eigenvalue zero of $\mathscr{L}$, which has a high multiplicity with all orderings, also has a number of principal eigenvectors. These cause no difficulty when one is iterating with $\mathscr{L}$, but when one uses the Hageman–Kellogg method, the presence of this multiple eigenvalue will greatly impede or even prevent the accurate determination of $\bar{\lambda}$. In Section 7.1, we shall show that with the red–black ordering there are no principal vectors of $\mathscr{L}$ of grade higher than one if A is positive definite.

6.7. AN UPPER BOUND FOR $\bar{\mu}$

In this and subsequent sections we shall consider methods for determining or estimating $\bar{\mu}$ for linear systems arising from the discrete generalized Dirichlet problem considered in Section 2.8. In this section we seek an upper bound for $\bar{\mu}$ corresponding to the difference equation, Variant II, defined by (2-8.10) and (2-8.11), derived from the self-adjoint differential equation (2-8.3).

We seek an upper bound for $\bar{\mu}$ in terms of constants $\underline{A}$, $\bar{A}$, $\underline{C}$, $\bar{C}$, $(-\underline{F})$ and $(-\bar{F})$ where in $R + S$ we have

$$\underline{A} \leq A(x, y) \leq \bar{A}, \quad \underline{C} \leq C(x, y) \leq \bar{C}, \quad (-\underline{F}) \leq -F \leq (-\bar{F}) \tag{7.1}$$

We first show that if B_0 is the matrix corresponding to the Jacobi method with $F \equiv 0$, then

$$S(B) \leq \left[\frac{2(\bar{A} + \bar{C})}{2(\bar{A} + \bar{C}) + h^2(-\underline{F})}\right] S(B_0) \tag{7.2}$$

Let A be the matrix of the linear system

$$Au = b \tag{7.3}$$

corresponding to the difference equation. Then by (2-8.11) we have

$$A = D - C, \qquad D = D_0 + h^2(-F) \tag{7.4}$$

where $D = \operatorname{diag} A$, $D_0 = \operatorname{diag} A_0$, and A_0 corresponds to the system (7.3) with $F \equiv 0$. Moreover,

$$B = D^{-1}C, \qquad B_0 = D_0^{-1}C \tag{7.5}$$

and hence

$$B = (D^{-1}D_0)B_0 \tag{7.6}$$

Since the largest element of $D^{-1}D_0$ is not larger than

$$2(\bar{A} + \bar{C})/[2(\bar{A} + \bar{C}) + h^2(-\underline{F})] \tag{7.7}$$

the result (7.2) follows from Theorem 2-1.16.

Suppose now that the region under consideration is included in a rectangle of sides $a = Ih$ and $b = Jh$ for some integers I and J. By Theorem 5.1, $S(B)$ for the region does not exceed $S(B)$ for the rectangle.

Let us now consider the case where $F \equiv 0$. Evidently we can write (2-8.10) in the form

$$\begin{aligned}[(\hat{a}_1 + \hat{a}_3)u(x, y) - \hat{a}_1 u(x + h, y) - \hat{a}_3 u(x - h, y)] \\ + [(\hat{a}_2 + \hat{a}_4)u(x, y) - \hat{a}_2 u(x, y + h) - \hat{a}_4 u(x, y - h)] = t(x, y)\end{aligned} \tag{7.8}$$

We let

$$A = H + V \tag{7.9}$$

where H corresponds to the first term of (7.8) and V corresponds to the second term. Evidently H is a symmetric L-matrix with weak diagonal dominance. Hence H is nonnegative definite, by Theorem 2-5.5. Similarly if we replace $A(x, y)$ by $\bar{A}$ we do not decrease the eigenvalues of H since the matrix $\bar{H} - H$ where $\bar{H}$ is the matrix corresponding to H with $A(x, y)$ replaced by $\bar{A}$ has weak diagonal dominance and nonnegative diagonal elements. Thus $\bar{H} - H$ is a nonnegative definite matrix. By Theorem 2-2.2 the largest eigenvalue of $\bar{H}$ is at least as large as that of H. Similarly the smallest eigenvalue of $\underline{H}$, corresponding to $\underline{A}$ is not larger than the smallest eigenvalue of H.

Let us now determine a bound on the largest eigenvalue, $\bar{b}(\bar{H})$, of $\bar{H}$. If $M - 1$ is the number of points in the largest row of adjacent points of R_h, then $M \leq I$. Let the points in this row be $(x_0 + h, y)$, $(x_0 + 2h, y)$ $\ldots$, $(x_0 + (M - 1)h, y)$, and for $p = 1, 2, \ldots, M - 1$ let

$$v(x, y) = \sin(p\pi i/M) \tag{7.10}$$

for $x = (x_0 + ih, y)$, $i = 1, 2, \ldots, M - 1$ and let $v(x, y) = 0$ for all other points of R_h. Evidently

$$2\bar{A}v(x, y) - \bar{A}v(x+h, y) - \bar{A}v(x - h, y) = 4\bar{A}\sin^2(p\pi/2M)v(x, y) \tag{7.11}$$

Thus for $p = 1, 2, \ldots, M - 1$

$$(\lambda_x)_p = 4\bar{A}\sin^2(p\pi/2M) \tag{7.12}$$

is an eigenvalue of $\bar{H}$. Therefore

$$\bar{b}(\bar{H}) \leq (\lambda_x)_{M-1} = 4\bar{A}\cos^2(\pi/2M) \leq 4\bar{A}\cos^2(\pi/2I) \tag{7.13}$$

Similarly, we can show that

$$\begin{aligned} &\bar{a}(\bar{H}) \geq 4\underline{A}\sin^2(\pi/2I) \\ &\bar{b}(\bar{V}) \leq 4\bar{C}\cos^2(\pi/2J), \qquad \bar{a}(\bar{V}) \geq 4\underline{C}\sin^2(\pi/2J) \end{aligned} \tag{7.14}$$

Therefore, by Theorem 2-2.2, $K(A)$, the *condition* of A, defined by

$$K(A) = \bar{b}(A)/\bar{a}(A) \tag{7.15}$$

satisfies

$$K(A) \leq \frac{\bar{A}\cos^2(\pi/2I) + \bar{C}\cos^2(\pi/2J)}{\underline{A}\sin^2(\pi/2I) + \underline{C}\sin^2(\pi/2J)} \tag{7.16}$$

We next seek to express $S(B)$ in terms of the condition of A. First we show that

$$S(B) = [K(\hat{A}) - 1]/[K(\hat{A}) + 1] \tag{7.17}$$

where

$$\hat{A} = D^{-1/2}AD^{-1/2} \tag{7.18}$$

and $D = \operatorname{diag} A$. But since $A = D - DB$ it follows that $\hat{A} = I - \tilde{B}$, where $\tilde{B} = D^{1/2}BD^{-1/2}$. Moreover, since A and $\hat{A}$ have Property A, the eigenvalues of B occur in pairs μ and $-\mu$ by Theorem 5-4.7. Since the eigenvalues μ of B and $\hat{\nu}$ of $\hat{A}$ are related by

$$\mu = 1 - \hat{\nu}$$

we have

$$\bar{b}(\hat{A}) = 1 + S(B), \qquad \bar{a}(\hat{A}) = 1 - S(B) \tag{7.19}$$

and (7.17) follows from (7.15).

We now prove the following theorem due to Forsythe and Straus [1955].

Theorem 7.1. If A is a positive definite matrix with Property A, then

$$K(A) \geq K(\hat{A}) \tag{7.20}$$

where

$$\hat{A} = D^{-1/2}AD^{-1/2} \tag{7.21}$$

and $D = \operatorname{diag} A$.

Proof. Since A and $\hat{A}$ have Property A it follows from Theorem 5-4.4 that for some permutation matrix P we have

$$P^{-1}AP = \begin{pmatrix} D_1 & -F_1 \\ -G_1 & D_2 \end{pmatrix}, \qquad P^{-1}\hat{A}P = \begin{pmatrix} I_1 & -F \\ -G & I_2 \end{pmatrix} \tag{7.22}$$

where D_1 and D_2 are square diagonal matrices and where $F_1 = G_1^{\mathrm{T}}$, $F = G^{\mathrm{T}}$. Since the eigenvalues of A and $\hat{A}$ are not affected by the permutation we assume that A and $\hat{A}$ have the form (7.22). We let $\bar{a}(A)$ and $\bar{a}(\hat{A})$ denote the smallest eigenvalues of A and $\hat{A}$, respectively, and let $\bar{b}(A)$ and $\bar{b}(\hat{A})$ be the largest eigenvalues.

Let $v = (v_1^T, v_2^T)^T$ be an eigenvector of $\hat{A}$ associated with $\bar{b}(\hat{A})$. Here v_1 and v_2 are column matrices of sizes compatible with (7.22). Since $\hat{A} = I - \hat{B}$ where $\hat{B} = I - (\text{diag}\,\hat{A})^{-1}\hat{A}$, we have $\bar{b}(\hat{A}) = 1 - \bar{a}(\hat{B})$, $a(\hat{A}) = 1 - \bar{b}(\hat{B})$. But by Theorem 5-4.7, since $\hat{A}$ has Property A, we have $\bar{b}(\hat{B}) = -\bar{a}(\hat{B})$. Moreover, $(v_1^T, v_2^T)^T$ is an eigenvector of $\hat{B}$ associated with $\bar{b}(\hat{B})$ and $\hat{v} = (v_1^T, -v_2^T)^T$ is an eigenvector of $\hat{B}$ associated with $-\bar{b}(\hat{B})$. Therefore, $\hat{v}$ is an eigenvector of $\hat{A}$ associated with $\bar{a}(\hat{A})$.

From Theorem 2-2.2, we have

$$\begin{aligned}\bar{b}(A) &= \max_{w\neq 0} [(w, Aw)/(w, w)] = \max_{w\neq 0} [(w, D^{1/2}\hat{A}D^{1/2}w)/(w, w)] \\ &= \max_{w\neq 0} [(D^{1/2}w, \hat{A}D^{1/2}w)/(w, w)] \geq (v, \hat{A}v)/(D^{-1/2}v, D^{-1/2}v) \qquad (7.23)\end{aligned}$$

where we let $w = D^{-1/2}v$. Therefore since $\hat{A}v = \bar{b}(\hat{A})v$ we have

$$\bar{b}(A) \geq \bar{b}(\hat{A})(v, v)/(D^{-1/2}v, D^{-1/2}v) \qquad (7.24)$$

Evidently $(v, v) = (\hat{v}, \hat{v}) = (v_1, v_1) + (v_2, v_2)$. Moreover, $(D^{-1/2}v, D^{-1/2}v) = (D^{-1/2}\hat{v}, D^{-1/2}\hat{v})$. Therefore,

$$\begin{aligned}\bar{a}(A) &= \min_{w\neq 0} [(w, Aw)/(w, w)] \leq (\hat{v}, \hat{A}\hat{v})/(D^{-1/2}\hat{v}, D^{-1/2}\hat{v}) \\ &= \bar{a}(\hat{A})(v, v)/(D^{-1/2}v, D^{-1/2}v) \qquad (7.25)\end{aligned}$$

Hence,

$$K(A) = \bar{b}(A)/\bar{a}(A) \geq \bar{b}(\hat{A})/\bar{a}(\hat{A}) = K(\hat{A}) \qquad (7.26)$$

and the theorem follows.

From Theorem 7.1 and from the fact that $(x - 1)(x + 1)^{-1}$ is an increasing function of x for $x \geq 1$ it follows that

$$\begin{aligned}S(B) &= [K(\hat{A}) - 1]/[K(\hat{A}) + 1] \leq [K(A) - 1]/[K(A) + 1] \\ &= 1 - 2/[K(A) + 1] \qquad (7.27)\end{aligned}$$

Hence, for the case $F \equiv 0$, we have from (7.16)

$$\begin{aligned}S(B_0) &= \frac{K(\hat{A}) - 1}{K(\hat{A}) + 1} \leq \frac{K(A) - 1}{K(A) + 1} = 1 - \frac{2}{K(A) + 1} \\ &\leq 1 - \frac{2\underline{A}\sin^2(\pi/2I) + 2\underline{C}\sin^2(\pi/2J)}{\frac{1}{2}(\bar{A}+\underline{A}) + \frac{1}{2}(\bar{C}+\underline{C}) + \frac{1}{2}(\bar{A}-\underline{A})\cos(\pi/I) + \frac{1}{2}(\bar{C}-\underline{C})\cos(\pi/J)} \qquad (7.28)\end{aligned}$$

Therefore, by (7.2)

$$S(B) \leq \frac{2(\bar{A} + \bar{C})}{2(\bar{A} + \bar{C}) + h^2(-\underline{F})}$$

$$\times \left\{1 - \frac{2\underline{A} \sin^2(\pi/2I) + 2\underline{C} \sin^2(\pi/2J)}{\frac{1}{2}(\bar{A}+\underline{A}) + \frac{1}{2}(\bar{C}+\underline{C}) + \frac{1}{2}(\bar{A}-\underline{A}) \cos(\pi/I) + \frac{1}{2}(\bar{C}-\underline{C}) \cos(\pi/J)}\right\} \tag{7.29}$$

In the case of the model problem with $I = J = h^{-1}$ and $F \equiv 0$ we have, as in Section 4.6,

$$S(B) \leq 1 - \frac{4 \sin^2(\pi h/2)}{2} = \cos \pi h \tag{7.30}$$

If $h = \frac{1}{20}$ then

$$S(B) \leq 0.9877$$

In the case of the differential equation

$$\frac{\partial}{\partial x}\left(\frac{1}{y} \frac{\partial u}{\partial x}\right) + \frac{\partial}{\partial y}\left(\frac{1}{y} \frac{\partial u}{\partial y}\right) = 0 \tag{7.31}$$

in the square $0 \leq x \leq 1$, $1 \leq y \leq 2$ we have

$$\bar{A} = \bar{C} = 1, \qquad \underline{A} = \underline{C} = \tfrac{1}{2}, \qquad (-\bar{F}) = (-\underline{F}) = 0 \tag{7.32}$$

and, if $I = J = 20$, then

$$S(B) \leq 1 - \frac{2 \sin^2(\pi/40)}{\frac{3}{2} + \frac{1}{2} \cos(\pi/20)} = \frac{1 + 3 \cos(\pi/20)}{3 + \cos(\pi/20)} \doteq 0.9938 \tag{7.33}$$

6.8. A PRIORI DETERMINATION OF $\bar{\mu}$: EXACT METHODS

The determination of an eigenvalue of B for linear systems arising from the five-point generalized Dirichlet problem defined in Section 2.8 is equivalent to the determination of a number μ such that for some function $v(x, y)$ defined on $R_h + S_h$, vanishing on S_h and not vanishing identically on R_h, we have

$$\frac{a_1}{a_0} v(x + h, y) + \frac{a_2}{a_0} v(x, y + h) + \frac{a_3}{a_0} v(x - h, y) + \frac{a_4}{a_0} v(x, y - h) = \mu v(x, y) \tag{8.1}$$

on R_h. Here the $a_i(x, y)$ are given by (2-8.8) for Variant I or by (2-8.11) for the self-adjoint case and Variant II. Multiplying by a_0 we obtain

$$a_1 v(x+h,y)+a_2 v(x,y+h)+a_3 v(x-h,y)+a_4 v(x,y-h)=\mu a_0 v(x,y) \quad (8.2)$$

Let us now consider the "separable" case where there exist positive functions $E_1(x)$, $F_1(y)$, $E_2(x)$, $F_2(y)$ and a constant $\gamma \geq 0$ such that for the functions A, C, and F appearing in (2-8.3), we have

$$A = E_1(x)F_1(y), \qquad C = E_2(x)F_2(y), \qquad F = -\gamma E_2(x)F_1(y) \quad (8.3)$$

while in the case of (2-8.1), we have in addition

$$D = E_3(x)F_1(y), \qquad E = E_2(x)F_3(y) \quad (8.4)$$

for some functions $E_3(x)$, $F_3(y)$. We remark if all of the above conditions are satisfied except for the condition on F, we can neglect F and the corresponding value of $\bar{\mu}$ will not be less than the true value.

We assume that $R + S$ is included in the rectangle $a \leq x \leq b$, $c \leq y \leq d$. By finding $\bar{\mu}$ for the rectangle we shall obtain a value which is not less than the true value. Let us carry out the analysis for the case of the self-adjoint equation (2-8.3), where the $\hat{a}_i$ are given by (2-8.11). If we let

$$v(x, y) = X(x)Y(y) \quad (8.5)$$

we obtain, from (8.2)

$$\frac{1}{E_2(x)}\left\{\frac{\left\{\begin{matrix}\mu[E_1(x+\frac{1}{2}h)+E_1(x-\frac{1}{2}h)]X(x)\\ -E_1(x+\frac{1}{2}h)X(x+h)-E_1(x-\frac{1}{2}h)X(x-h)\end{matrix}\right\}}{X(x)}\right\}+\frac{1}{F_1(y)}$$

$$\times\left\{\frac{\left\{\begin{matrix}\mu[F_2(y+\frac{1}{2}h)+F_2(y-\frac{1}{2}h)]Y(y)\\ -F_2(y+\frac{1}{2}h)Y(y+h)-F_2(h-\frac{1}{2}h)Y(y-h)\end{matrix}\right\}}{Y(y)}\right\}=-\mu\gamma h^2 \quad (8.6)$$

We are thus led to consider the conditions

$$\mu\left[E_1\left(x+\frac{h}{2}\right)+E_1\left(x-\frac{h}{2}\right)\right]X(x)-E_1\left(x+\frac{h}{2}\right)X(x+h)$$
$$-E_1\left(x-\frac{h}{2}\right)X(x-h)=\eta E_2(x)X(x) \quad (8.7)$$

$$\mu\left[F_2\left(y+\frac{h}{2}\right)+F_2\left(y-\frac{h}{2}\right)\right]Y(y)-F_2\left(y+\frac{h}{2}\right)Y(y+h)$$
$$-F_2\left(y-\frac{h}{2}\right)Y(y-h)=[-\mu\gamma h^2-\eta]F_1(y)Y(y) \quad (8.8)$$

where η is a constant and where

$$X(a) = X(b) = Y(c) = Y(d) = 0 \tag{8.9}$$

Given μ, the problem of finding η from (8.7) is equivalent to that of solving the generalized matrix eigenvalue problem $\det(P - \eta Q) = 0$ where P is symmetric and Q is a diagonal matrix with positive diagonal elements. This problem is equivalent to that of finding the eigenvalues of the symmetric matrix $Q^{-1/2}PQ^{-1/2}$. Hence the values of η are real. Similarly, the problem of finding $\eta' = \eta + \mu\gamma h^2$ from (8.8) can be treated in the same way. Thus for given μ, we can find sets of values of η from both (8.7) and (8.8). Evidently μ is an eigenvalue of B if and only if there is a value of η common to both sets.

In some cases we can eliminate η. Let us first consider the Helmholtz equation

$$u_{xx} + u_{yy} - \gamma u = 0, \qquad \gamma \geq 0 \tag{8.10}$$

with $a = c = 0$, $b = d = 1$. Here $E_1 = F_1 = E_2 = F_2 = 1$ and (8.7) becomes

$$2\mu X(x) - X(x + h) - X(x - h) = \eta X(x) \tag{8.11}$$

Letting $X(x) = \sin p\pi x$ we obtain

$$2\mu - 2 \cos p\pi h = \eta \tag{8.12}$$

Similarly, from (8.8), with $Y(y) = \sin q\pi y$ we have

$$2\mu - 2 \cos q\pi h = -\mu\gamma h^2 - \eta \tag{8.13}$$

Eliminating η we have

$$\mu = 2(\cos p\pi h + \cos q\pi h)/(4 + \gamma h^2) \tag{8.14}$$

The largest eigenvalue is given by

$$\bar{\mu} = 4 \cos \pi h/(4 + \gamma h^2) \tag{8.15}$$

which agrees with the result of Section 4.6, for the case $\gamma = 0$.

As another example, let us consider the equation

$$(y^k u_x)_x + (y^k u_y)_y - (\gamma/y)u = 0, \qquad \gamma \geq 0 \tag{8.16}$$

in the rectangle $0 \leq x \leq 1, 0 \leq c \leq y \leq d$. Evidently $E_1(x) = E_2(x) = 1$

and $F_1(y) = F_2(y) = y^k$. From (8.7) we have, letting $X(x) = \sin p\pi x$

$$2\mu - 2\cos p\pi h = \eta \tag{8.17}$$

Substituting in (8.8) we obtain

$$\begin{aligned} F_2\left(y + \frac{h}{2}\right)Y(y+h) &+ F_2\left(y - \frac{h}{2}\right)Y(y-h) \\ &= \left\{\mu\left[F_2\left(y+\frac{h}{2}\right) + F_2\left(y - \frac{h}{2}\right) + (2 + \gamma h^2)F_1(y)\right]\right. \\ &\quad \left. -2\cos(p\pi h)F_1(y)\right\}Y(y) \end{aligned} \tag{8.18}$$

which involves only μ. This problem thus involves the solution of an equation of the form

$$\det(P - \mu Q) = 0 \tag{8.19}$$

where P is a symmetric tri-diagonal matrix and Q is a diagonal matrix with positive diagonal elements. Actually, the solutions of (8.19) are the eigenvalues of the symmetric matrix $Q^{-1/2}PQ^{-1/2}$; therefore the solutions of (8.19) are real.

As an example, let us consider the case $k = -1$, $\gamma = 0$, $c = 1$, $d = 2$, and $h = \frac{1}{3}$. Letting $p = 1$ in (8.18) we obtain (8.19) with

$$P = \begin{pmatrix} \dfrac{2\cos \pi h}{4/3} & \dfrac{1}{3/2} \\ \dfrac{1}{3/2} & \dfrac{2\cos\pi h}{5/3} \end{pmatrix} = \begin{pmatrix} \dfrac{3}{4} & \dfrac{2}{3} \\ \dfrac{2}{3} & \dfrac{3}{5} \end{pmatrix}$$

$$Q = \begin{pmatrix} \dfrac{1}{3/2} + \dfrac{1}{7/6} + \dfrac{2}{4/3} & 0 \\ 0 & \dfrac{1}{11/6} + \dfrac{1}{9/6} + \dfrac{2}{5/3} \end{pmatrix} = \begin{pmatrix} \dfrac{127}{42} & 0 \\ 0 & \dfrac{398}{165} \end{pmatrix}$$

Solving (8.19) we obtain

$$\mu_1 \doteq 0.0015, \qquad \mu_2 \doteq 0.4952$$

and hence

$$\bar{\mu} = 0.4952$$

We remark that for the same problem but with $k = 0$ we have

$$\bar{\mu} = \cos \pi h = 0.5000$$

Other results are given in the accompanying tabulation. (Numerical values of $\bar{\mu}$ for other cases will be given in Section 6.10.)

h^{-1}	$\bar{\mu}\ (k = -1)$	$\bar{\mu}\ (k = 0) = \cos \pi h$
3	0.4952	0.5000
4	0.7033	0.7071
10	0.9502	0.9511
20	0.9875	0.9877

The case of Eq. (2-8.1), which is in general not self-adjoint, can be treated similarly in the separable case. It is of interest to note that even though A is not symmetric, in general, nevertheless if conditions (8.3) and (8.4) hold and if h satisfies (2-8.16), then the eigenvalues of B are real. To show this we first show that there exists a diagonal matrix P with positive diagonal elements such that PA is symmetric.

Let $P(x, y)$ be the function corresponding to the diagonal matrix P. The matrix PA is symmetric provided

$$\begin{aligned} &P(x, y)F_1(y)(E_1(x) + \tfrac{1}{2}hE_3(x)) \\ &\quad = P(x + h, y)F_1(y)(E_1(x + h) - \tfrac{1}{2}hE_3(x + h)) \end{aligned} \tag{8.20}$$

and

$$\begin{aligned} &P(x, y)E_2(x)(F_1(y) + \tfrac{1}{2}hF_3(y)) \\ &\quad = P(x, y + h)E_2(x)(F_2(y + h) - \tfrac{1}{2}hF_3(y + h)) \end{aligned} \tag{8.21}$$

Thus we have

$$\begin{aligned} P(x + h, y) &= \left(\frac{E_1(x) + \tfrac{1}{2}hE_3(x)}{E_1(x + h) - \tfrac{1}{2}hE_3(x + h)}\right)P(x, y) \\ P(x, y + h) &= \left(\frac{F_1(y) + \tfrac{1}{2}hF_3(y)}{F_2(y + h) - \tfrac{1}{2}hF_3(y + h)}\right)P(x, y) \end{aligned} \tag{8.22}$$

For some point $(\bar{x}, \bar{y}) = 1$ we can let $P(\bar{x}, \bar{y}) = 1$. We can then use (8.22) to define $P(x, y)$ for all (x, y) in R_h.

We now show that B has real eigenvalues. Let $A' = PA = PD - PC$ where $A = D - C$ and $D = \text{diag } A$. Then if $D' = PD$ and $C' = PC$ we have

$$B = D^{-1}C = (P^{-1}D')^{-1}P^{-1}C' = (D')^{-1}C'$$

which is similar to the matrix

$$(D')^{-1/2}C'(D')^{-1/2}$$

which is symmetric since C' is symmetric.

Let us now consider the nonsymmetric difference equation corresponding to

$$u_{xx} + (k/y)u_y + u_{yy} = 0 \tag{8.23}$$

which is obtained from (8.16) with $\gamma = 0$ by multiplying by y^{-k}. The nonsymmetric difference equation is

$$\begin{aligned}4u(x, y) - u(x + h, y) - \left(1 + \frac{kh}{2y}\right)u(x, y + h) - u(x - h, y)\\ -\left(1 - \frac{kh}{2y}\right)u(x, y - h) = 0\end{aligned} \tag{8.24}$$

The discrete eigenvalue corresponding to (8.18) is

$$\begin{aligned}\left(1 + \frac{kh}{2y}\right)Y(y + h) + \left(1 - \frac{kh}{2y}\right)Y(y - h)\\ = [\mu(4 + \gamma h^2) - 2 \cos p\pi h]\,Y(y)\end{aligned} \tag{8.25}$$

Let us again consider the case $k = -1$, $\gamma = 0$, $c = 1$, $d = 2$, and $h = \frac{1}{3}$. As in the symmetric case we obtain (8.19) with

$$P = \begin{pmatrix} 2 \cos \pi h & 7/8 \\ 11/10 & 2 \cos \pi h \end{pmatrix} = \begin{pmatrix} 1 & 7/8 \\ 11/10 & 1 \end{pmatrix}$$

$$Q = \begin{pmatrix} 4 & 0 \\ 0 & 4 \end{pmatrix}$$

The solution of (8.19) is

$$\mu = \tfrac{1}{4}[1 \pm (77/80)^{1/2}]$$

and

$$\bar{\mu} = \tfrac{1}{4}[1 + (77/80)^{1/2}] \doteq 0.496$$

which agrees closely with the result $\bar{\mu} = 0.4952$ obtained for the self-adjoint case.

Solutions for other cases have been obtained by Warlick [1955], and by Warlick and Young [1970]. Some of these results will be given in Section 6.10.

We remark that if k is large and c is small, then the matrix A may not be an L-matrix. A study of the convergence properties of the SOR method in such cases is given by Warlick [1955].

6.9. A PRIORI DETERMINATION OF $\bar{\mu}$: APPROXIMATE VALUES

We now seek to determine approximate values of the eigenvalues of B by solving an eigenvalue problem involving a partial differential equation. From (8.2) we have

$$a_0 v(x, y) - a_1 v(x+h, y) - a_2 v(x, y+h) - a_3 v(x - h, y) - a_4 v(x, y - h) \\ = (1 - \mu) a_0 v(x, y) \tag{9.1}$$

Using a slight extension of the method given by Henrici [1960] we divide both sides by $-h^2$ and note that the left side is approximately $L[u]$ while the right side is approximately $-\Lambda a_0(x, y) v(x, y)$ where

$$\Lambda = (1 - \mu)/h^2 \tag{9.2}$$

and

$$\bar{a}_0(x, y) = \lim_{h \to 0} a_0(x, y) \tag{9.3}$$

We are thus led to consider the continuous eigenvalue problem

$$L[v] = -\Lambda \bar{a}_0 v \tag{9.4}$$

From (2-8.8) and (2-8.11), we have

$$\bar{a}_0(x, y) = 2A(x, y) + 2C(x, y) \tag{9.5}$$

If the region is a rectangle and if conditions (8.3) and (8.4) hold, then we can reduce the problem to a pair of eigenvalue problems involving ordinary differential equations. Thus for (2-8.3), if we let $v(x, y) = X(x) Y(y)$, then we have, on substituting in (9.4) and dividing by $E_2(x) F_1(y) X(x) Y(y)$

$$\frac{(E_1 X')'}{E_2 X} + \frac{(F_2 Y')'}{F_1 Y} - \gamma = -2\Lambda\left(\frac{E_1}{E_2} + \frac{F_2}{F_1}\right) \tag{9.6}$$

We thus are led to the two eigenvalue problems

$$\frac{(E_1X')'}{E_2X} + 2\Lambda\frac{E_1}{E_2} = \eta \tag{9.7}$$

$$\frac{(F_2Y')'}{F_1Y} + 2\Lambda\frac{F_2}{F_1} - \gamma = -\eta \tag{9.8}$$

where

$$X(a) = X(b) = Y(c) = Y(d) = 0 \tag{9.9}$$

Given a value of Λ one can determine values of η from each eigenvalue problem. Evidently Λ is an eigenvalue of (9.4) if and only if (9.7) and (9.8) are satisfied for some η.

In some cases we can eliminate η. For example, for Eq. (8.16) in the rectangle $0 \leq x \leq 1$, $0 \leq c \leq y \leq d$, we have, by (9.7)

$$(X''/X) + 2\Lambda = \eta \tag{9.10}$$

Letting $X(x) = \sin p\pi x$ we have

$$\eta = -p^2\pi^2 + 2\Lambda \tag{9.11}$$

From (9.8) we have

$$(y^kY')'/(y^kY) + 2\Lambda - \gamma = -\eta = p^2\pi^2 - 2\Lambda \tag{9.12}$$

or

$$(1/y^k)(y^kY')' + \Gamma Y = 0 \tag{9.13}$$

where

$$\Gamma = 4\Lambda - \gamma - p^2\pi^2 \tag{9.14}$$

For the case $k = 0$, $c = 0$, $d = 1$, we have

$$Y'' + \Gamma Y = 0 \tag{9.15}$$

Letting $Y(y) = \sin q\pi y$ we have

$$\Gamma = q^2\pi^2 \tag{9.16}$$

and

$$\Lambda = \tfrac{1}{4}(\gamma + p^2\pi^2 + q^2\pi^2) \tag{9.17}$$

Therefore, by (9.2) we have

$$\mu \doteq 1 - h^2\Lambda = 1 - \tfrac{1}{4}h^2[\pi^2(p^2 + q^2) + \gamma] \tag{9.18}$$

which agrees with (8.14) to within terms of order h^4.

For the case $k \neq 0$ an analytic solution can be found in terms of Bessel's functions, as shown by Warlick [1955]. The procedure for determining $\bar{\mu}$ is as follows. Let $\underline{\sigma}$ denote the smallest root of

$$J_{|r|}(\sigma)N_{|r|}(\tau\sigma) - J_{|r|}(\tau\sigma)N_{|r|}(\sigma) = 0 \tag{9.19}$$

where $J_{|r|}(x)$ and $N_{|r|}(x)$ are the Bessel's functions of first and second kind, respectively, of order $|r|$ and

$$\tau = d/c, \qquad r = (k-1)/2 \tag{9.20}$$

We then compute

$$\bar{\mu} \sim 1 - \frac{h^2}{4}\left(\frac{\underline{\sigma}^2}{c^2} + \gamma + \pi^2\right) \tag{9.21}$$

A table of the first six roots of (9.19) for various values of τ is given by Jahnke and Emde [1945, pp. 205–206] (see also A. Kalähne, 1907).

As an example, let $c = 1$, $d = 2$, $k = -1$, and $h = \frac{1}{20}$. Then $\tau = 2$, $r = -1$. By the table of roots of (9.19), we have $\underline{\sigma} = 3.197$ and, by (9.21),

$$\bar{\mu} \sim 1 - \tfrac{1}{1600}[(3.197)^2 + (3.142)^2] \doteq 1 - 0.125 \doteq 0.9875$$

For the special case $c = 0$ the limiting value of $(\tau - 1)$ times the root as $\tau \to \infty$ is given in the table. Thus the limiting value

$$\lim_{c\to 0}\,[\underline{\sigma}(\tau - 1)] = \lim_{c\to 0}\left[\underline{\sigma}\left(\frac{d}{c} - 1\right)\right] = \xi \tag{9.22}$$

is given. Hence

$$\lim_{c\to 0}\frac{\underline{\sigma}}{c} = \lim_{c\to 0}\frac{\underline{\sigma}(\tau-1)}{c(\tau-1)} = \lim_{c\to 0}\frac{\underline{\sigma}(\tau-1)}{d-c} = \frac{\xi}{d-c} \tag{9.23}$$

Thus in the above example, if $c = 0$, $d = 1$, then by the table $\xi = 3.832$ and

$$\mu \sim 1 - \frac{h^2}{4}\,[(3.832)^2 + (3.142)^2] \doteq 0.9847$$

6.10. NUMERICAL RESULTS

Table 10.1 gives exact values of $\bar{\mu}$ and ω_b obtained by the methods of Sections 6.8 for the symmetric difference equation corresponding to (8.16) and for the nonsymmetric difference equation corresponding to

TABLE 10.1. Values of $\bar{\mu}$ and ω_b for Difference Equations Derived from (8.16)

k		Exact		Approximate		Bound by (7.29)
		$(c = 0, d = 1)$	$(c = 1, d = 2)$	$(c = 0, d = 1)$	$(c = 1, d = 2)$	$(c = 1, d = 2)$
−1	$\bar{\mu}$	0.98472	0.98747	0.98466	0.98745	0.99383
	ω_b	1.7034	1.7274	1.7028	1.7272	1.8003
0	$\bar{\mu}$	0.98769	0.98769	0.98766	0.98766	0.98766
	ω_b	1.7294	1.7294	1.7292	1.7292	1.7292
1	$\bar{\mu}$	0.98906	0.98776	0.99022	0.98774	0.99383
	ω_b	1.7429	1.7301	1.7551	1.7299	1.8003

(8.23). In each case considered we let $h = \frac{1}{20}$, $\gamma = 0$, $a = 0$, $b = 1$. The values for the symmetric and nonsymmetric cases were identical to the accuracy shown. Also included are approximate values obtained by the methods of Section 6.9, and, for the case $c = 1$, $d = 2$, the bounds on $\bar{\mu}$ given by (7.29).

The agreement between the approximate and the exact values of $\bar{\mu}$ and ω_b is extremely close in all cases except where $k = 1$ and $c = 0$, $d = 1$, where the agreement is reasonably good. In that case the difference between the exact and the theoretical values is about three-fourths as great as the difference between the exact value for $k = 0$ and the exact value for $k = 1$. In every case the approximate value is closer to the exact value than the bound given by (7.29).

TABLE 10.2. Numbers of SOR Iterations Required to Solve Difference Equations Derived from (8.23)

ω	$(c = 0,\ d = 1)$			$(c = 1,\ d = 2)$		
	$k = -1$	$k = 0$	$k = 1$	$k = -1$	$k = 0$	$k = 1$
1.70	55	78	94	80	83	84
1.705	51	75	92	77	80	81
1.710	47	73	89	74	77	78
1.715	49	70	86	71	74	75
1.720	49	66	83	67	70	72
1.725	49	63	80	63	66	68
1.730	52	59	76	58	61	63
1.735	53	55	72	58	57	57
1.740	53	57	68	59	60	60
1.745	53	58	63	59	60	60
1.750	55	58	62	59	60	60
1.755	56	57	63	61	62	63
ω_b (observed)	1.709	1.737	1.747	1.732	1.734	1.735
N	46	54	59	56	57	57
ω_b (exact-theoretical)	1.703	1.729	1.743	1.727	1.729	1.730
N	53	59	65	61	59	63
ω_b (approximate-theoretical)	1.703	1.729	1.755	1.727	1.729	1.730
N	53	59	63	61	59	63

In order to test the effectiveness of the values of ω_b thus found, the SOR method was used to solve both the symmetric and nonsymmetric difference equations corresponding to the solution of (8.23) in each of the square regions $0 \le x \le 1$, $0 \le y \le 1$, and $0 \le x \le 1$, $1 \le y \le 2$. Boundary values of zero were prescribed. Starting with initial values of xy the iteration process was carried out until all values of $u^{(n)}(x, y)$ became less than 10^{-6}. Table 10.2 gives the number of iterations required for ω between 1.70 and 1.755.

It was found that the number of iterations was the same in all cases for the symmetric difference equation as for the nonsymmetric equation. The number of iterations using the approximate-theoretical value of ω_b was at least as small as that using the exact-theoretical value of ω_b. Thus it seems preferable to determine the approximate-theoretical value of ω_b, which is relatively easy, rather than to go through the more laborious procedure of determining the exact-theoretical value.

It should be noted that the actual best value of ω is greater than the theoretical values in each case. This can be explained, at least in part, by the fact that the theoretical value of ω_b minimizes $S(\mathscr{L}_\omega{}^m)$; nevertheless, another value of ω in general minimizes a given norm of $\mathscr{L}_\omega{}^m$.

SUPPLEMENTARY DISCUSSION

Section 6.2. Theorem 2.3 is due to Young [1950, 1954].

Section 6.3. Theorem 3.1 is an extension of a result of Young [1950, 1954].

Section 6.4. The determination of the optimum value of ω corresponding to the two eigenvalues a and ib is given by Wrigley [1962]. The case of the single eigenvalue $\mu = \alpha + i\beta$ is treated by Kjellberg [1958]. See also Russell [1963]. Kjellberg gave a graph which enables one to find ω and $S(\mathscr{L}_\omega)$ given α and β. However, Kjellberg did not obtain the compact result (4.25). The procedure for finding the optimum ω for the case of several complex eigenvalues is believed to be new. Kredell [1962] treated the case where the eigenvalues of B are complex and are not necessarily symmetric with respect to the real axis. In this case the use of a complex relaxation factor is appropriate.

Section 6.5. Early work on the determination of ω_b was done by Young [1955] and by Young and Shaw [1955]. Both iterative and a priori methods were used. Garabedian [1956] (see also Young, 1962) estimated ω_b for the nine-point discrete analog of Laplace's equation by considering a related hyperbolic partial differential equation. Albrecht [1966] obtained an estimate of ω_b for this same problem for an ordering of the mesh points different from the natural ordering.

Section 6.7. The results of Section 6.7 are given in Young [1971].

Section 6.8. Warlick [1955] determined $\bar{\mu}$ for (8.16) for various values of k both by the exact method and also by the a priori method. The determination of the diagonal matrix P such that PA is symmetric is a direct generalization of the method used by Warlick for the differential equation $u_{xx} + (k/y)u_y + u_{yy} = 0$.

Section 6.9. For more general a priori analytic methods for estimating $\bar{\mu}$ see Parter [1965].

EXERCISES

Section 6.1

1. Show that if A has Property A, then $0 \in \bar{S}_B$.

Section 6.2

1. Find the root radius of each of the following quadratics

(a) $x^2 - 2x + 3 = 0$,
(b) $x^2 - 2x + \frac{1}{2} = 0$,
(c) $3x^2 + 2x + 2 = 0$,
(d) $x^2 + x + \frac{1}{2} = 0$,
(e) $x^2 - x - \frac{1}{2} = 0$.

2. Let A be a consistently ordered matrix with nonvanishing diagonal elements such that the eigenvalues of B are real and $S(B) = 0.8$. Find ω_b and $S(\mathscr{L}_{\omega_b})$. Find $S(\mathscr{L}_{1.20})$ and $S(\mathscr{L}_{1.30})$.

3. Prove that if A is a matrix with positive diagonal elements such that for some diagonal matrix E having positive diagonal elements EAE^{-1} is symmetric, then B has real eigenvalues.

4. Carry out an analysis of the alternative proof of Theorem 2.3 for the case $\bar{\mu} = 0.8$.

5. Show that if

$$\omega = \frac{2}{1 + (1 - \bar{\mu}^2)^{1/2}}$$

and

$$\hat{\omega} = \frac{2}{1 + (1 - \alpha^2)^{1/2}}$$

then

$$\hat{\omega} - 1 = (\omega - 1)^2$$

if and only if

$$\alpha^2 = \left(\frac{\bar{\mu}^2}{2 - \bar{\mu}^2}\right)^2$$

Section 6.3

1. Consider the solution by the SOR method of the model problem for Laplace's equation with $h^{-1} = 40$. Assume that the "natural" ordering is used.

(a) Compute $\bar{\mu}$ and ω_b.

(b) Compute $R(\mathscr{L}) = -\log S(\mathscr{L})$ and $R(\mathscr{L}_{\omega_b}) = -\log S(\mathscr{L}_{\omega_b})$. Compare the actual value of $R(\mathscr{L}_{\omega_b})$ with $2[R(\mathscr{L})]^{1/2}$.

(c) Compute ω_b' as determined by using $\bar{\mu}'$ instead of $\bar{\mu}$ in the two cases

(i) $(1 - \bar{\mu}') = 1.10(1 - \bar{\mu})$,
(ii) $(1 - \bar{\mu}') = 0.9(1 - \bar{\mu})$.

In each case compute $S(\mathscr{L}_{\omega_b'})$ and determine the relative decrease in $R(\mathscr{L}_{\omega_b'})$ as compared with $R(\mathscr{L}_{\omega_b})$.

(d) Using ω_b, find an upper bound to the number of iterations which would be required to reduce $\| e^{(0)} \|$, the norm of the initial error, to $10^{-3} \| e^{(0)} \|$. Assume that the Jordan canonical form J of $\mathscr{L}_{\omega_b}$ is a diagonal matrix except for one block of size two of the form

$$\begin{pmatrix} \omega_b - 1 & 1 \\ 0 & \omega_b - 1 \end{pmatrix}$$

Also assume that the Jordan condition number of $\mathscr{L}_{\omega_b}$ is 5.

2. Find ω_b for the model problem for a 2×1 rectangle with $h^{-1} = 20$.

Section 6.4

1. Let A be a consistently ordered matrix with positive diagonal elements. For what values of ω will the SOR method converge if the matrix

$$B = I - (\text{diag } A)^{-1}A$$

has the eigenvalues:

$$0.98, \quad -0.98, \quad 3i, \quad -3i$$

What would be the best value of ω to use and what would the corresponding spectral radius of $\mathscr{L}_\omega$ be?

2. Let A be a consistently ordered matrix with positive diagonal elements such that

$$\mu = \tfrac{3}{4} + \tfrac{1}{4}i$$

is an eigenvalue of B and all eigenvalues of B lie in the rectangle

$$|\operatorname{Re} \mu| \leq \tfrac{3}{4}, \qquad |\operatorname{Im} \mu| \leq \tfrac{1}{4}$$

Find the optimum value of ω for the SOR method and $S(\mathscr{L}_\omega)$. For what values of ω will the SOR method converge?

3. Show that the locus of (α, β) such that the optimum value of ω corresponding to the eigenvalue $\mu = \alpha + i\beta$ is unity is the lemniscate

$$(\alpha^2 + \beta^2)^2 = \alpha^2 - \beta^2$$

(Kjellberg, 1958).

4. Let A be a consistently ordered matrix with positive diagonal elements. Find the optimum value of ω for the SOR method and $S(\mathscr{L}_\omega)$ if all eigenvalues of B lie in the region bounded by 0.9, $0.3 + 0.1i$, $0.3 - 0.1i$, -0.9, $-0.3 - 0.1i$, $-0.3 + 0.1i$. Compare the value of $S(\mathscr{L}_\omega)$ with that obtained if all eigenvalues of B are real and lie in the interval $-0.9 \leq \mu \leq 0.9$.

5. Let A be a consistently ordered matrix with positive diagonal elements such that

$$\mu_1 = \tfrac{1}{2} + \tfrac{3}{4}i$$

and

$$\mu_2 = \tfrac{3}{4} + \tfrac{1}{4}i$$

are eigenvalues of B and all eigenvalues of B lie in one of the rectangles

$$|\operatorname{Re}\mu| \le \tfrac{3}{4}, \qquad |\operatorname{Im}\mu| \le \tfrac{1}{4}$$
$$|\operatorname{Re}\mu| \le \tfrac{1}{2}, \qquad |\operatorname{Im}\mu| \le \tfrac{3}{4}$$

Find the optimum value of ω for the SOR method and $S(\mathscr{L}_\omega)$.

6. Solve the "two-eigenvalue problem" shown in Figure 4.2.

7. Solve the "three-eigenvalue problem" with

$$\mu_1 = \tfrac{1}{4} + \tfrac{7}{8}i, \qquad \mu_2 = \tfrac{1}{2} + \tfrac{3}{4}i, \qquad \text{and} \qquad \mu_3 = \tfrac{3}{4} + \tfrac{1}{4}i$$

Section 6.5

1. Describe the behavior of the eigenvalues of $\mathscr{L}_\omega$ when ω varies from 0 to 2 if A is a positive definite consistently ordered matrix such that $\bar{\mu} = \tfrac{2}{3}(2)^{1/2}$.

Section 6.6

1. Apply the Gauss–Seidel method to the linear system

$$\begin{pmatrix} 4 & -1 & -1 & 0 \\ -1 & 4 & 0 & -1 \\ -1 & 0 & 4 & -1 \\ 0 & -1 & -1 & 4 \end{pmatrix} \begin{pmatrix} u_1 \\ u_2 \\ u_3 \\ u_4 \end{pmatrix} = \begin{pmatrix} 1 \\ 0 \\ 0 \\ 0 \end{pmatrix}$$

with starting values unity and estimate $S(\mathscr{L})$. Then compute ω_b on this basis. Compare with the exact value of ω_b.

2. Consider the model problem for Laplace's equation with $h^{-1} = 20$ and zero boundary values. Starting with initial values unity, iterate with the SOR method where $\omega = 1.5$, and estimate a value of $S(\mathscr{L}_\omega)$; then compute $\bar{\mu}$ and ω_b and compare with the exact values.

Section 6.7

1. Verify Theorem 7.1 for the matrix

$$A = \begin{pmatrix} 2 & -1 \\ -1 & 4 \end{pmatrix}$$

Also verify that

$$S(B) = [K(\hat{A}) - 1]/[K(\hat{A}) + 1] \leq [K(A) - 1]/[K(A) + 1]$$

2. Obtain a bound on $S(B)$ for the case treated in (7.31)–(7.33) with (7.31) replaced by

$$\frac{\partial}{\partial x}\left(\frac{1}{y}\frac{\partial u}{\partial x}\right) + \frac{\partial}{\partial y}\left(\frac{1}{y}\frac{\partial u}{\partial y}\right) - (1 + x + y)u = 0$$

Do the same with (7.31) replaced by

$$\frac{\partial}{\partial x}\left(y\frac{\partial u}{\partial x}\right) + \frac{\partial}{\partial y}\left(y\frac{\partial u}{\partial y}\right) - (1 + x + y)u = 0$$

Section 6.8

1. Derive a formula for $S(B)$ corresponding to the discrete generalized Dirichlet problem for $Au_{xx} + Cu_{yy} + Ku = G$ in a rectangle $Ih \times Jh$, where h is the mesh size. Here A, C, and $-K$ are positive constants. What value of ω would you use if $A = 2$, $C = 3$, $K = -1$, $h = \frac{1}{20}$, $Ih = 3$, $Jh = 4$?

2. Verify the calculations of μ_1 and μ_2 on page 219.

Section 6.9

1. Work out the example at the end of Section 6.9 for the case $k = -1$, $c = 1$, $d = 3$.

2. Derive Eq. (9.19).
Hint: First let $t = y\Gamma^{1/2}$ and $Z(t) = t^{(1-k)/2}Y(t)$; then use (9.13). Show $J_r(c\Gamma^{1/2})N_r(d\Gamma^{1/2}) - J_r(d\Gamma^{1/2})N_r(c\Gamma^{1/2}) = 0$. Note that $J_{-r}(y_1)N_{-r}(y_2) - J_{-r}(y_2)N_{-r}(y_1) = J_r(y_1)N_r(y_2) - J_r(y_2)N_r(y_1)$ for any r, y_1, and y_2.

Section 6.10

1. Verify the approximate values and bounds for $\bar{\mu}$ and ω_b given in Table 10.1.

Chapter 7 / NORMS OF THE SOR METHOD

In this chapter we study certain norms of $\mathscr{L}_\omega^m$ for various values of ω and m under the assumption that A is positive definite and has the form

$$A = \begin{pmatrix} D_1 & H \\ K & D_2 \end{pmatrix} \tag{1}$$

where D_1 and D_2 are square diagonal matrices. Our primary aim is to show why the convergence of the SOR method with the optimum ω is somewhat slower than what one might expect on the basis of $S(\mathscr{L}_\omega)$.

Let G and E be any real matrices which depend on A and are such that E is nonsingular and the eigenvalues of $(EGE^{-1})(EGE^{-1})^{\mathrm{T}}$ are related to those of B. As in Section 6.1, we define the *virtual E-norm* of G by

$$\overline{\| G \|_E} = [\bar{S}((EGE^{-1})(EGE^{-1})^{\mathrm{T}})]^{1/2} \tag{2}$$

Actually, in nearly every case considered in this chapter the virtual norm is equal to the spectral norm. However, in later chapters the difference will be significant.

In Section 7.1 we shall show that the Jordan canonical form of $\mathscr{L}_{\omega_b}$ is not diagonal. Hence based on the discussion of Section 3.7 we expect that $\| \mathscr{L}_{\omega_b}^m \|$ will converge to zero much more slowly than $S(\mathscr{L}_{\omega_b}^m)$ as $m \to \infty$. In Section 7.2 we will derive a relation between the eigenvalues of certain matrices related to A and those of certain 2×2 matrices. This relation will be used not only in this chapter but also in Chapters

11 and 12 for other methods as well. In Section 7.3 we shall study $\overline{\| \mathcal{L}_\omega \|}_{D^{1/2}}$ for various values of ω. In general $\overline{\| \mathcal{L}_{\omega_b}^m \|}_{D^{1/2}} > 1$, but as will be shown in Section 7.4 for m large enough $\overline{\| \mathcal{L}_{\omega_b}^m \|}_{D^{1/2}} < 1$. Eventually $\overline{\| \mathcal{L}_{\omega_b}^m \|}_{D^{1/2}}$ converges to zero, though considerably more slowly than $S(\mathcal{L}_{\omega_b}^m)$. On the other hand, as will be shown in Section 7.5, $\overline{\| \mathcal{L}_\omega \|}_{A^{1/2}} < 1$ for all ω with $0 < \omega < 2$; hence $\overline{\| \mathcal{L}_{\omega_b}^m \|}_{A^{1/2}}$ converges monotonically to zero as $m \to \infty$. A formula for $\overline{\| \mathcal{L}_{\omega_b}^m \|}_{A^{1/2}}$ will be given in Section 7.6. A comparison of the $D^{1/2}$ and $A^{1/2}$ norms of $\mathcal{L}_{\omega_b}^m$ will be given in Section 7.7 along with some numerical results.

7.1. THE JORDAN CANONICAL FORM OF $\mathcal{L}_\omega$

In Section 3.7 we showed that if $S(G) < 1$, then as $m \to \infty$, $\| G^m \|$ tends to zero like

$$\binom{m}{p-1} S(G)^{m-p+1}$$

where p is the order of the largest block in the Jordan canonical form of G associated with an eigenvalue of modulus $S(G)$. Young [1954] showed that if A is a consistently ordered nonsingular matrix with positive diagonal elements and if the eigenvalues of B are real and less than unity in modulus, then the Jordan canonical form of $\mathcal{L}_{\omega_b}$ is not diagonal. We now give a proof of this for the case where A has the form (1). We also show that if A is an irreducible Stieltjes matrix, then $p = 2$.

Eigenvectors of B

Let us first determine the eigenvectors of B assuming that A is a matrix of the form (1) with nonvanishing diagonal elements such that $B = I - D^{-1}A$, where $D = \operatorname{diag} A$, is similar to a diagonal matrix. Evidently, the matrix B has the form

$$B = \begin{pmatrix} O_1 & F \\ G & O_2 \end{pmatrix} \tag{1.1}$$

where O_1 and O_2 are square null matrices of the same sizes as D_1 and D_2, respectively. If A is symmetric and has positive diagonal elements, then B is similar to a diagonal matrix since B is similar to $\tilde{B} = D^{1/2}BD^{-1/2} = I - D^{-1/2}D^{-1/2}$ which is symmetric.

Evidently

$$B^2 = \begin{pmatrix} FG & O \\ O & GF \end{pmatrix} \tag{1.2}$$

is also similar to a diagonal matrix and, in fact, the $r \times r$ matrix FG and the $s \times s$ matrix GF are also similar to diagonal matrices where $r + s = N$. We construct a basis of eigenvectors for B as follows. Let the p eigenvectors of FG associated with the nonzero eigenvalues γ_1, $\gamma_2, \ldots, \gamma_p$ be $\xi^{(1)}, \xi^{(2)}, \ldots, \xi^{(p)}$, i.e.,

$$FG\xi^{(i)} = \gamma_i \xi^{(i)}, \qquad i = 1, 2, \ldots, p \tag{1.3}$$

If we let

$$\eta^{(i)} = G\xi^{(i)}, \qquad i = 1, 2, \ldots, p \tag{1.3'}$$

then $\eta^{(i)} \neq 0$ and $\eta^{(i)}$ is an eigenvector of GF associated with γ_i, i.e.,

$$GF\eta^{(i)} = \gamma_i \eta^{(i)}, \qquad i = 1, 2, \ldots, p \tag{1.3''}$$

Moreover, since the $\xi^{(i)}$ are linearly independent, then so are the $\eta^{(i)}$ since

$$\sum_{i=1}^{p} c_i \eta^{(i)} = 0$$

implies that

$$0 = F\left(\sum_{i=1}^{p} c_i \eta^{(i)}\right) = \sum_{i=1}^{p} \gamma_i c_i \xi^{(i)} = 0$$

and hence the c_i vanish because of the linear independence of the $\xi^{(i)}$. Evidently there can be no more than p eigenvectors of GF associated with nonzero eigenvalues; otherwise there would be more than p linearly independent eigenvectors of FG associated with nonzero eigenvalues. Thus we have

$$p \leq \min(r, s) \tag{1.4}$$

If we let

$$\mu_i = (\gamma_i)^{1/2}, \quad x^{(i)} = \mu_i \xi^{(i)}, \quad y^{(i)} = \eta^{(i)}, \quad v^{(i)} = \begin{pmatrix} x^{(i)} \\ y^{(i)} \end{pmatrix}, \quad i = 1, 2, \ldots, p \tag{1.5}$$

then using (1.5), (1.3), and (1.3″) we have

$$Bv^{(i)} = \begin{pmatrix} Fy^{(i)} \\ Gx^{(i)} \end{pmatrix} = \begin{pmatrix} \gamma_i \xi^{(i)} \\ \mu_i \eta^{(i)} \end{pmatrix} = \mu_i v^{(i)}, \qquad i = 1, 2, \ldots, p \tag{1.6}$$

Here we choose the μ_i to have positive real parts. If A is positive definite, then $\mu_1, \mu_2, \ldots, \mu_p$ are positive.

Let us now define for $i = p+1, p+2, \ldots, 2p$

$$x^{(i)} = x^{(i-p)}, \quad y^{(i)} = -y^{(i-p)}, \quad v^{(i)} = \begin{pmatrix} x^{(i)} \\ y^{(i)} \end{pmatrix}, \quad \mu_i = -\mu_{i-p} \tag{1.7}$$

Evidently, we have

$$Bv^{(i)} = \mu_i v^{(i)}, \qquad i = p+1, p+2, \ldots, 2p \tag{1.8}$$

If $FGx = 0$ where $x \neq 0$, then by (1.2) $B^2(x^{\mathrm{T}}, 0)^{\mathrm{T}} = 0$. Since B is similar to a diagonal matrix it follows from Theorem 2-1.15 that $B(x^{\mathrm{T}}, 0)^{\mathrm{T}} = 0$. Thus, if the eigenvectors of FG associated with the eigenvalue zero are $x^{(2p+1)}, x^{(2p+2)}, \ldots, x^{(p+r)}$, then the vectors

$$v^{(i)} = \begin{pmatrix} x^{(i)} \\ 0 \end{pmatrix}, \qquad i = 2p+1, 2p+2, \ldots, p+r \tag{1.9}$$

are eigenvectors of B associated with the eigenvalue zero. Similarly, if the eigenvectors of GF associated with the eigenvalue zero are $y^{(p+r+1)}$, $y^{(p+r+2)}, \ldots, y^{(N)}$, then the vectors

$$v^{(i)} = \begin{pmatrix} 0 \\ y^{(i)} \end{pmatrix}, \qquad i = p+r+1, p+r+2, \ldots, N \tag{1.9$'$}$$

are eigenvectors of B associated with the eigenvalue zero.

We have thus constructed a basis of eigenvectors for B

$$v^{(i)} = \begin{pmatrix} x^{(i)} \\ y^{(i)} \end{pmatrix}, \qquad i = 1, 2, \ldots, N \tag{1.9$''$}$$

and, moreover, we have by (1.8)

$$Gx^{(i)} = \mu_i y^{(i)}, \qquad Fy^{(i)} = \mu_i x^{(i)} \tag{1.10}$$

We also have

$$\begin{aligned}
&\operatorname{Re}\mu_i \geq 0, && v^{(i)} = \begin{pmatrix} x^{(i)} \\ y^{(i)} \end{pmatrix}, && i = 1, 2, \ldots, p \\
&\operatorname{Re}\mu_i \leq 0, && v^{(i)} = \begin{pmatrix} x^{(i-p)} \\ -y^{(i-p)} \end{pmatrix}, && i = p+1, p+2, \ldots, 2p \\
&\mu_i = 0, && v^{(i)} = \begin{pmatrix} x^{(i)} \\ 0 \end{pmatrix}, && i = 2p+1, 2p+2, \ldots, p+r \\
&\mu_i = 0, && v^{(i)} = \begin{pmatrix} 0 \\ y^{(i)} \end{pmatrix}, && i = p+r+1, p+r+2, \ldots, N
\end{aligned} \tag{1.11}$$

Principal Vectors of $\mathscr{L}_\omega$

We now seek the eigenvectors and principal vectors of $\mathscr{L}_\omega$ for $\omega \neq 0$. By (3-3.23) we have

$$\begin{aligned}\mathscr{L}_\omega &= (I - \omega L)^{-1}(\omega U + (1-\omega)I)\\ &= \begin{pmatrix} I_1 & 0 \\ -\omega\tilde{G} & I_2 \end{pmatrix}^{-1}\begin{pmatrix} (1-\omega)I_1 & \omega F \\ 0 & (1-\omega)I_2 \end{pmatrix}\\ &= \begin{pmatrix} I_1 & 0 \\ \omega G & I_2 \end{pmatrix}\begin{pmatrix} (1-\omega)I_1 & \omega F \\ 0 & (1-\omega)I_2 \end{pmatrix}\\ &= \begin{pmatrix} (1-\omega)I_1 & \omega F \\ \omega(1-\omega)G & \omega^2 GF + (1-\omega)I_2 \end{pmatrix} \qquad (1.12)\end{aligned}$$

where I_1 and I_2 are identity matrices of the same sizes as D_1 and D_2, respectively. For each nonzero eigenvalue μ of B let λ^+ and λ^- be the roots of

$$(\lambda + \omega - 1)^2 = \omega^2\mu^2\lambda \qquad (1.13)$$

Since $B(x^{\mathrm{T}}, y^{\mathrm{T}})^{\mathrm{T}} = \mu(x^{\mathrm{T}}, y^{\mathrm{T}})^{\mathrm{T}}$ then the vectors

$$w = \begin{pmatrix} x \\ (\lambda^+)^{1/2}y \end{pmatrix}, \qquad z = \begin{pmatrix} x \\ (\lambda^-)^{1/2}y \end{pmatrix} \qquad (1.14)$$

are eigenvectors of $\mathscr{L}_\omega$ since by (1.12), (1.10), and (1.13) we have

$$\begin{aligned}\mathscr{L}_\omega\begin{pmatrix} x \\ (\lambda^+)^{1/2}y \end{pmatrix} &= \begin{pmatrix} (1-\omega+\omega\mu(\lambda^+)^{1/2})x \\ [\omega\mu(1-\omega+\omega\mu(\lambda^+)^{1/2}) + (1-\omega)(\lambda^+)^{1/2}]y \end{pmatrix} = \lambda^+ w\\ \mathscr{L}_\omega z &= \lambda^- z\end{aligned} \qquad (1.15)$$

If we let

$$v = \begin{pmatrix} x \\ y \end{pmatrix}, \qquad \hat{v} = \begin{pmatrix} x \\ -y \end{pmatrix} \qquad (1.16)$$

then we have

$$\begin{aligned} w &= \tfrac{1}{2}(v+\hat{v}) + \tfrac{1}{2}(\lambda^+)^{1/2}(v-\hat{v}) = \tfrac{1}{2}(1+(\lambda^+)^{1/2})v + \tfrac{1}{2}(1-(\lambda^+)^{1/2})\hat{v}\\ z &= \tfrac{1}{2}(v+\hat{v}) + \tfrac{1}{2}(\lambda^-)^{1/2}(v-\hat{v}) = \tfrac{1}{2}(1+(\lambda^-)^{1/2})v + \tfrac{1}{2}(1-(\lambda^-)^{1/2})v \end{aligned} \qquad (1.17)$$

If $(\lambda^+)^{1/2} \neq (\lambda^-)^{1/2}$, then w and z are linearly independent. But the discriminant, $\omega^2\mu^2(\omega^2\mu^2 - 4(\omega - 1))$ of (1.13) does not vanish unless

$\omega = 0$ or $\mu = 0$, or

$$\omega^2\mu^2 - 4(\omega - 1) = 0 \tag{1.18}$$

On the other hand, if (1.18) holds and if $\mu \neq 0$, then $(\lambda^+)^{1/2} = (\lambda^-)^{1/2} = \lambda^{1/2} = \omega\mu/2 \neq 0$ and w and z are not linearly independent. We readily verify that if

$$\hat{z} = \tfrac{1}{2}\lambda^{-1/2}\begin{pmatrix} 0 \\ y \end{pmatrix} \tag{1.19}$$

then

$$\mathscr{L}_\omega \hat{z} = w + \lambda \hat{z} \tag{1.20}$$

and hence $\hat{z}$ is a principal vector of grade 2. Moreover, we have

$$\begin{aligned} w &= \tfrac{1}{2}(1 + \lambda^{1/2})v + \tfrac{1}{2}(1 - \lambda^{1/2})\hat{v} \\ \hat{z} &= \tfrac{1}{4}\lambda^{-1/2}v - \tfrac{1}{4}\lambda^{-1/2}\hat{v} \end{aligned} \tag{1.21}$$

Hence w and $\hat{z}$ are linearly independent.

The reader should verify that if $\omega \neq 0$ the vectors

$$\begin{aligned} w^{(i)} &= \begin{pmatrix} x^{(i)} \\ (\lambda_i^+)^{1/2}y^{(i)} \end{pmatrix}, & & i = 1, 2, \ldots, p \\ w^{(i+p)} &= \begin{cases} \begin{pmatrix} x^{(i)} \\ (\lambda_i^-)^{1/2}y^{(i)} \end{pmatrix}, & i = 1, 2, \ldots, p, \quad \text{if } \omega^2\mu_i^2 \neq 4(\omega - 1) \\ \tfrac{1}{2}(\lambda_i^+)^{-1/2}\begin{pmatrix} 0 \\ y^{(i)} \end{pmatrix}, & i = 1, 2, \ldots, p, \quad \text{if } \omega^2\mu_i^2 = 4(\omega - 1) \end{cases} \\ w^{(i)} &= v^{(i)}, & & i = 2p + 1, 2p + 2, \ldots, N \end{aligned} \tag{1.22}$$

are linearly independent and hence form a basis of E^N. The matrix whose columns are the $w^{(i)}$ thus reduces $\mathscr{L}_\omega$ to Jordan canonical form.

Evidently, there are q principal vectors of grade 2 where q is the multiplicity of the positive eigenvalue μ of B such that (1.18) holds. Thus, by (6-2.11) $\mathscr{L}_{\omega_b}$ has q principal vectors of grade 2 where q is the multiplicity of the eigenvalue $\bar{\mu}$ of B. If $\bar{\mu}$ is a simple eigenvalue of B, then $q = 1$. Thus, from Theorem 2-7.6, we have (Young, 1950):

Theorem 1.1. If A is an irreducible Stieltjes matrix of the form (1) or, more generally, if B has the form (1.1) and is such that $S(B) < 1$ for some diagonal matrix E with positive diagonal elements $E^{-1}BE$ is a symmetric, irreducible nonnegative matrix, then the Jordan canonical form of $\mathscr{L}_{\omega_b}$ has one nondiagonal element, which corresponds to the eigenvalue $\omega_b - 1$.

From this result we expect that $\| \mathscr{L}_{\omega_b}^m \|$ behaves like

$$mS(\mathscr{L}_{\omega_b})^{m-1}$$

as $m \to \infty$. In Section 7.4 we shall show that this is the case for $\| \mathscr{L}_{\omega_b}^m \|_{D^{1/2}}$, which in many cases is the same as $\|\mathscr{L}_{\omega_b}^m \|$.

7.2. BASIC EIGENVALUE RELATION

In this section we study the eigenvalues of products of matrices of the form

$$\begin{pmatrix} \alpha_{1,1}(FG) & \alpha_{1,2}(FG)F \\ \alpha_{2,1}(GF)G & \alpha_{2,2}(GF) \end{pmatrix}$$

where $\alpha_{1,1}$ and $\alpha_{1,2}$ are polynomials in FG and $\alpha_{2,1}$ and $\alpha_{2,2}$ are polynomials in GF. Thus, for example, by (1.1) and (1.12) the matrices B and $\mathscr{L}_\omega$ are of this type.

Theorem 2.1. If B is a matrix of the form (1.1), then

(a) The matrix

$$Q = \begin{pmatrix} \alpha_{1,1}(FG) & \alpha_{1,2}(FG)F \\ \alpha_{2,1}(GF)G & \alpha_{2,2}(GF) \end{pmatrix} \tag{2.1}$$

is nonsingular if

$$T(B^2) = \alpha_{1,1}(B^2)\alpha_{2,2}(B^2) - \alpha_{2,1}(B^2)\alpha_{1,2}(B^2)B^2 \tag{2.2}$$

is nonsingular. Moreover $T(B^2)$ is nonsingular if and only if for each eigenvalue μ of B the matrix

$$R(\mu) = \begin{pmatrix} \alpha_{1,1}(\mu^2) & \alpha_{1,2}(\mu^2)\mu \\ \alpha_{2,1}(\mu^2)\mu & \alpha_{2,2}(\mu^2) \end{pmatrix} \tag{2.3}$$

is nonsingular.

(b) Let

$$\mathscr{G}_m = \prod_{k=m}^{1} \begin{pmatrix} \alpha_{1,1}^{(k)}(FG) & \alpha_{1,2}^{(k)}(FG)F \\ \alpha_{2,1}^{(k)}(GF)G & \alpha_{2,2}^{(k)}(GF) \end{pmatrix}^{\nu_k} \tag{2.4}$$

where for each k, $\nu_k = \pm 1$. It is assumed that for any k the matrix

$$T^{(k)}(B^2) = \alpha_{1,1}^{(k)}(B^2)\alpha_{2,2}^{(k)}(B^2) - \alpha_{2,1}^{(k)}(B^2)\alpha_{1,2}^{(k)}(B^2)B^2 \tag{2.5}$$

is nonsingular if $\nu_k = -1$. For each eigenvalue μ of B, let

$$M_m(\mu) = \prod_{k=m}^{1} \begin{pmatrix} \alpha_{1,1}^{(k)}(\mu^2) & \alpha_{1,2}^{(k)}(\mu^2)\mu \\ \alpha_{2,1}^{(k)}(\mu^2)\mu & \alpha_{2,2}^{(k)}(\mu^2) \end{pmatrix}^{\nu_k} \tag{2.6}$$

If μ is a nonzero eigenvalue of B and if λ is an eigenvalue of $M_m(\mu)$, then λ is an eigenvalue of $\mathscr{G}_m$. If $\mu = 0$ is an eigenvalue of B, then at least one of the eigenvalues of $M_m(0)$ is an eigenvalue of $\mathscr{G}_m$.

(c) If λ is an eigenvalue of $\mathscr{G}_m$, then there exists an eigenvalue μ of B such that λ is an eigenvalue of $M_m(\mu)$.

Proof. Let us define the matrix $\hat{Q}$ by

$$\hat{Q} = \begin{pmatrix} \alpha_{1,1}(B^2) & \alpha_{1,2}(B^2)B \\ \alpha_{2,1}(B^2)B & \alpha_{2,2}(B^2) \end{pmatrix} \tag{2.7}$$

Evidently $\hat{Q}$ is singular if and only if there exist vectors v and w not both zero such that

$$\begin{aligned} \alpha_{1,1}(B^2)v + \alpha_{1,2}(B^2)Bw &= 0 \\ \alpha_{2,1}(B^2)Bv + \alpha_{2,2}(B^2)w &= 0 \end{aligned} \tag{2.8}$$

If $w \neq 0$ we multiply the first equation by $\alpha_{2,1}(B^2)B$ and the second by $\alpha_{1,1}(B^2)$. Upon subtraction we have†

$$T(B^2)w = 0 \tag{2.9}$$

But this implies $T(B^2)$ is singular. Similarly, $T(B^2)$ is singular if $v \neq 0$. On the other hand, if $T(B^2)$ is singular, then for some $\tilde{w} \neq 0$, (2.9) holds. If $\alpha_{2,1}(B^2)\tilde{w} = \alpha_{1,2}(B^2)B\tilde{w} = \alpha_{2,1}(B^2)Bv = \alpha_{2,2}(B^2)\tilde{w} = 0$, then $v = \tilde{w}$ and $w = \tilde{w}$ satisfies (2.8) and $\hat{Q}$ is singular. If $\alpha_{1,1}(B^2)\tilde{w} \neq 0$, then (2.8) is satisfied with $v = \alpha_{1,2}(B^2)B\tilde{w}$, $w = -\alpha_{1,1}(B^2)\tilde{w}$, and similarly for other cases. Thus $\hat{Q}$ is nonsingular if and only if $T(B^2)$ is.

By (1.1) we have

$$B^{2m} = \begin{pmatrix} (FG)^m & 0 \\ 0 & (GF)^m \end{pmatrix}, \qquad B^{2m+1} = \begin{pmatrix} 0 & (FG)^m F \\ (GF)^m G & 0 \end{pmatrix}. \tag{2.10}$$

so that

$$\hat{Q} = \begin{pmatrix} \alpha_{1,1}(FG) & 0 & 0 & \alpha_{1,2}(FG)F \\ 0 & \alpha_{1,1}(GF) & \alpha_{1,2}(GF)G & 0 \\ 0 & \alpha_{2,1}(FG)F & \alpha_{2,2}(FG) & 0 \\ \alpha_{2,1}(GF)G & 0 & 0 & \alpha_{2,2}(GF) \end{pmatrix} \tag{2.11}$$

† Here and elsewhere in this chapter we use the fact that any two polynomials in the matrix B commute.

Evidently there exists a permutation matrix P such that

$$P^{-1}\hat{Q}P = \begin{pmatrix} Q & 0 \\ 0 & Q_1 \end{pmatrix} \tag{2.12}$$

where

$$Q_1 = \begin{pmatrix} \alpha_{2,2}(FG) & \alpha_{2,1}(FG)F \\ \alpha_{1,2}(GF)G & \alpha_{1,1}(GF) \end{pmatrix} \tag{2.13}$$

Therefore if $\hat{Q}$ is nonsingular, then Q is nonsingular. Hence if $T(B^2)$ is nonsingular, then Q is nonsingular.

Evidently $R(\mu)$ is nonsingular if and only if

$$T(\mu^2) = \alpha_{1,1}(\mu^2)\alpha_{2,2}(\mu^2) - \alpha_{1,2}(\mu^2)\alpha_{2,1}(\mu^2)\mu^2 \tag{2.14}$$

does not vanish. But the eigenvalues of $T(B^2)$ are simply $T(\mu^2)$ where μ is any eigenvalue of B. Thus $T(B^2)$ is nonsingular if and only if $R(\mu)$ is nonsingular for each eigenvalue μ of B. This completes the proof of (a).

Let μ be any eigenvector of B and let $v = (x^T, y^T)^T$ be an associated eigenvector so that (1.10) holds. For any numbers α and β we have

$$Q^{\nu}\begin{pmatrix} \alpha x \\ \beta y \end{pmatrix} = \begin{pmatrix} \alpha' x \\ \beta' y \end{pmatrix} \tag{2.15}$$

where

$$\begin{pmatrix} \alpha' \\ \beta' \end{pmatrix} = [R(\mu)]^{\nu}\begin{pmatrix} \alpha \\ \beta \end{pmatrix} \tag{2.16}$$

if $\nu = 1$. The result is also true for $\nu = -1$ provided $R(\mu)$ and, hence, $T(B^2)$ are nonsingular. Moreover, by the hypothesis of (b) we have

$$\mathscr{G}_m\begin{pmatrix} \alpha x \\ \beta y \end{pmatrix} = \begin{pmatrix} \alpha'' x \\ \beta'' y \end{pmatrix} \tag{2.17}$$

where

$$\begin{pmatrix} \alpha'' \\ \beta'' \end{pmatrix} = M_m(\mu)\begin{pmatrix} \alpha \\ \beta \end{pmatrix} \tag{2.18}$$

If λ is an eigenvalue of $M_m(\mu)$, then for some $\hat{\alpha}$ and $\hat{\beta}$ not both zero we have

$$M_m(\mu)\begin{pmatrix} \hat{\alpha} \\ \hat{\beta} \end{pmatrix} = \lambda\begin{pmatrix} \hat{\alpha} \\ \hat{\beta} \end{pmatrix} = \begin{pmatrix} \lambda\hat{\alpha} \\ \lambda\hat{\beta} \end{pmatrix} \tag{2.19}$$

By (1.10) it follows that if $\mu \neq 0$, then neither x nor y can vanish. Hence

$(\hat{\alpha}x^{\mathrm{T}}, \hat{\beta}y^{\mathrm{T}})^{\mathrm{T}} \neq 0$ and

$$\mathscr{G}_m\begin{pmatrix}\hat{\alpha}x\\ \hat{\beta}y\end{pmatrix} = \lambda\begin{pmatrix}\hat{\alpha}x\\ \hat{\beta}y\end{pmatrix} \tag{2.20}$$

and λ is an eigenvalue of $\mathscr{G}_m$.

If $\mu = 0$ is an eigenvalue of B, then for some $w = (x^{\mathrm{T}}, y^{\mathrm{T}})^{\mathrm{T}}$ we have $Bw = 0$. Since x and y do not both vanish, then either $(x^{\mathrm{T}}, 0)^{\mathrm{T}}$ or $(0, y^{\mathrm{T}})^{\mathrm{T}}$ are eigenvectors of B associated with the eigenvalue zero. If $x \neq 0$, then

$$\mathscr{G}_m\begin{pmatrix}x\\ 0\end{pmatrix} = \left(\prod_{k=1}^{m} [\alpha_{1,1}^{(k)}(0)]^{\nu_k}\right)\begin{pmatrix}x\\ 0\end{pmatrix} \tag{2.21}$$

Hence $\prod_{k=1}^{m} [\alpha_{1,1}^{(k)}(0)]^{\nu_k}$ is an eigenvalue of $\mathscr{G}_m$. We note that by (a), if $\nu_k = -1$, then $\alpha_{1,1}^{(k)}(0) \neq 0$. Similarly, if $y \neq 0$, then $\prod_{k=1}^{m} [\alpha_{2,2}^{(k)}(0)]^{\nu_k}$ is an eigenvalue of $\mathscr{G}_m$. Thus, at least one of the eigenvalues of $M_m(0)$ is an eigenvalue of $\mathscr{G}_m$. This completes the proof of (b).

We now show that if λ is an eigenvalue of $\mathscr{G}_m$, then λ is an eigenvalue of $\mathscr{J}_m$ where

$$\mathscr{J}_m = \prod_{k=m}^{1} \begin{pmatrix}\alpha_{1,1}^{(k)}(B^2) & \alpha_{1,2}^{(k)}(B^2)B\\ \alpha_{2,1}^{(k)}(B^2)B & \alpha_{2,2}^{(k)}(B^2)\end{pmatrix}^{\nu_k} \tag{2.22}$$

Evidently if $T(B^2)$ is nonsingular, then, as in the proof of (a), $\hat{Q}$ is nonsingular. Hence by (2.12) Q and Q_1 are nonsingular and we have

$$P^{-1}\hat{Q}^{-1}P = \begin{pmatrix}Q^{-1} & 0\\ 0 & Q_1^{-1}\end{pmatrix} \tag{2.23}$$

Hence, by (2.1), (2.4), and (2.18) we have

$$P^{-1}\mathscr{J}_m P = \begin{pmatrix}\mathscr{G}_m & 0\\ 0 & \mathscr{H}_m\end{pmatrix} \tag{2.24}$$

for some matrix $\mathscr{H}_m$. Consequently, if λ is an eigenvalue of $\mathscr{G}_m$, then λ is an eigenvalue of $\mathscr{J}_m$.

It is easily verified that if $T(B^2)$ is nonsingular, then

$$\hat{Q}^{-1} = \begin{pmatrix}T^{-1} & 0\\ 0 & T^{-1}\end{pmatrix}\begin{pmatrix}\alpha_{2,2}(B^2) & -\alpha_{1,2}(B^2)B\\ -\alpha_{2,1}(B^2)B & \alpha_{1,1}(B^2)\end{pmatrix} \tag{2.25}$$

Hence $\mathscr{J}_m$ has the form

$$\mathscr{J}_m = \begin{pmatrix}\psi_{1,1}(B^2) & \psi_{1,2}(B^2)B\\ \psi_{2,1}(B^2)B & \psi_{2,2}(B^2)\end{pmatrix} \tag{2.26}$$

where the $\psi_{i,j}(B^2)$ are rational functions of B^2.

Suppose now that λ is an eigenvalue of $\mathscr{J}_m$. Then for some v and w not both zero we have

$$\begin{aligned}\psi_{1,1}(B^2)v + \psi_{1,2}(B^2)Bw &= \lambda v \\ \psi_{2,1}(B^2)Bv + \psi_{2,2}(B^2)w &= \lambda w\end{aligned} \tag{2.27}$$

If $v \neq 0$, we multiply the first equation by $\psi_{2,1}(B^2)B$ and the second by $\psi_{1,1}(B^2) - \lambda I$. Upon subtraction we obtain

$$[\psi_{1,2}(B^2)\psi_{2,1}(B^2)B^2 - (\psi_{2,2}(B^2) - \lambda I)(\psi_{1,1}(B^2) - \lambda I)]v = 0 \tag{2.28}$$

Hence

$$\Gamma(B^2) = \psi_{1,2}(B^2)\psi_{2,1}(B^2)B^2 - (\psi_{2,2}(B^2) - \lambda I)(\psi_{1,1}(B^2) - \lambda I) \tag{2.29}$$

is singular. (This can also be shown if $w \neq 0$.) Let V be a matrix which reduces B to Jordan canonical form, i.e., such that

$$V^{-1}BV = J_A \tag{2.30}$$

Then

$$V^{-1}\Gamma(B^2)V = \psi_{1,2}(J_A{}^2)\psi_{2,1}(J_A{}^2)J_A{}^2 - (\psi_{2,2}(J_A{}^2) - \lambda I)(\psi_{1,1}(J_A{}^2) - \lambda I) \tag{2.31}$$

is an upper triangular matrix whose diagonal elements are

$$\psi_{1,2}(\mu^2)\psi_{2,1}(\mu^2)\mu^2 - (\psi_{2,2}(\mu^2) - \lambda)(\psi_{1,1}(\mu^2) - \lambda) \tag{2.32}$$

Thus, in order for $\Gamma(B^2)$ to be singular one of the diagonal elements of $V^{-1}\Gamma V$ must vanish and hence λ must be an eigenvalue of $M_m(\mu)$ for some eigenvalue μ of B. This completes the proof of (c).

We now derive a relation between $\mathscr{G}_m$ and B.

Theorem 2.2. Let B be a matrix of the form (1.1) and let $\mathscr{G}_m$ be given by (2.4) where it is assumed that if $\nu_k = -1$ for any k, then $T^{(k)}(B^2)$ is nonsingular. We have

$$(\phi_{1,1}(B) - \mathscr{G}_m)(\phi_{2,2}(B) - \mathscr{G}_m) = \phi_{1,2}(B)\phi_{2,1}(B) \tag{2.33}$$

where

$$\begin{aligned}\phi_{1,1}(B) &= \psi_{1,1}(B^2), & \phi_{1,2}(B) &= \psi_{1,2}(B^2)B \\ \phi_{2,1}(B) &= \psi_{2,1}(B^2)B, & \phi_{2,2}(B) &= \psi_{2,2}(B^2)\end{aligned} \tag{2.34}$$

and where the $\psi_{i,j}(B^2)$ are given by (2.26).

Proof. For any matrix partitioned in accordance with (1.1) let us define the "projection operators"

$$\Delta_{i,j}(E) = E_{i,j}, \qquad i, j = 1, 2 \tag{2.35}$$

where

$$E = \begin{pmatrix} E_{1,1} & E_{1,2} \\ E_{2,1} & E_{2,2} \end{pmatrix} \tag{2.36}$$

Evidently by (2.26) and (2.34) we have

$$\begin{aligned} \mathscr{J}_m &= \begin{pmatrix} \phi_{1,1}(B) & \phi_{1,2}(B) \\ \phi_{2,1}(B) & \phi_{2,2}(B) \end{pmatrix} \\ &= \begin{pmatrix} \Delta_{1,1}(\phi_{1,1}) & \Delta_{1,2}(\phi_{1,1}) & \Delta_{1,1}(\phi_{1,2}) & \Delta_{1,2}(\phi_{1,2}) \\ \Delta_{2,1}(\phi_{1,1}) & \Delta_{2,2}(\phi_{1,1}) & \Delta_{2,1}(\phi_{1,2}) & \Delta_{2,2}(\phi_{1,2}) \\ \Delta_{1,1}(\phi_{2,1}) & \Delta_{1,2}(\phi_{2,1}) & \Delta_{1,1}(\phi_{2,2}) & \Delta_{1,2}(\phi_{2,2}) \\ \Delta_{2,1}(\phi_{2,1}) & \Delta_{2,2}(\phi_{2,1}) & \Delta_{2,1}(\phi_{2,2}) & \Delta_{2,2}(\phi_{2,2}) \end{pmatrix} \end{aligned} \tag{2.37}$$

and by (2.24)

$$P^{-1}\mathscr{J}_m P = \begin{pmatrix} \Delta_{1,1}(\phi_{1,1}) & \Delta_{1,2}(\phi_{1,2}) & 0 & 0 \\ \Delta_{2,1}(\phi_{2,1}) & \Delta_{2,2}(\phi_{2,2}) & 0 & 0 \\ 0 & 0 & \Delta_{1,1}(\phi_{2,2}) & \Delta_{1,2}(\phi_{2,1}) \\ 0 & 0 & \Delta_{2,1}(\phi_{1,2}) & \Delta_{2,2}(\phi_{1,1}) \end{pmatrix} \tag{2.38}$$

Therefore, we have $\Delta_{1,2}(\phi_{k,k}) = \Delta_{2,1}(\phi_{k,k}) = 0$, $k = 1, 2$, and $\Delta_{k,k}(\phi_{1,2}) = \Delta_{k,k}(\phi_{2,1}) = 0$, $k = 1, 2$. Moreover, since by (2.38) and (2.24) we have

$$\mathscr{G}_m = \begin{pmatrix} \Delta_{1,1}(\phi_{1,1}) & \Delta_{1,2}(\phi_{1,2}) \\ \Delta_{2,1}(\phi_{2,1}) & \Delta_{2,2}(\phi_{2,2}) \end{pmatrix} \tag{2.39}$$

and

$$\phi_{1,1} - \mathscr{G}_m = \begin{pmatrix} 0 & -\Delta_{1,2}(\phi_{1,2}) \\ -\Delta_{2,1}(\phi_{2,1}) & \Delta_{2,2}(\phi_{1,1} - \phi_{2,2}) \end{pmatrix} \tag{2.40}$$

$$\phi_{2,2} - \mathscr{G}_m = \begin{pmatrix} \Delta_{1,1}(\phi_{1,1} - \phi_{2,2}) & -\Delta_{1,2}(\phi_{1,2}) \\ -\Delta_{2,1}(\phi_{2,1}) & 0 \end{pmatrix} \tag{2.41}$$

Therefore,

$$\begin{aligned} &(\phi_{1,1} - \mathscr{G}_m)(\phi_{2,2} - \mathscr{G}_m) \\ &\quad = \begin{pmatrix} \Delta_{1,2}(\phi_{1,2})\Delta_{2,1}(\phi_{2,1}) & 0 \\ Z & \Delta_{2,1}(\phi_{2,1})\Delta_{1,2}(\phi_{1,2}) \end{pmatrix} \end{aligned} \tag{2.42}$$

where

$$Z = -\Delta_{2,1}(\phi_{2,1})\Delta_{1,1}(\phi_{1,1} - \phi_{2,2}) - \Delta_{2,2}(\phi_{1,1} - \phi_{2,2})\Delta_{2,1}(\phi_{2,1})$$
$$= -\Delta_{2,1}[\phi_{2,1}(\phi_{2,2} - \phi_{1,1}) + (\phi_{1,1} - \phi_{2,2})\phi_{2,1}] = 0 \tag{2.43}$$

Since we also have

$$\phi_{1,2}\phi_{2,1} = \begin{pmatrix} 0 & \Delta_{1,2}(\phi_{1,2}) \\ \Delta_{2,1}(\phi_{1,2}) & 0 \end{pmatrix}\begin{pmatrix} 0 & \Delta_{1,2}(\phi_{2,1}) \\ \Delta_{2,1}(\phi_{2,1}) & 0 \end{pmatrix}$$
$$= \begin{pmatrix} \Delta_{1,2}(\phi_{1,2})\Delta_{2,1}(\phi_{2,1}) & 0 \\ 0 & \Delta_{2,1}(\phi_{1,2})\Delta_{1,2}(\phi_{2,1}) \end{pmatrix} \tag{2.44}$$

the result (2.33) follows.

As an application of this result we have

Corollary 2.3. If B is a matrix of the form (1.1) and if $\mathscr{L}_\omega$ is given by (1.12), then

$$((1-\omega)I - \mathscr{L}_\omega)^2 = \omega^2\mathscr{L}_\omega B^2 \tag{2.45}$$

Proof. By (1.12) the matrix $\mathscr{J}_m$ corresponding to $\mathscr{L}_\omega$ is given by (2.37) where

$$\begin{aligned} \phi_{1,1}(B) &= (1-\omega)I, & \phi_{1,2}(B) &= \omega B \\ \phi_{2,1}(B) &= \omega(1-\omega)B, & \phi_{2,2}(B) &= \omega^2B^2 + (1-\omega)I \end{aligned} \tag{2.46}$$

Therefore, we have

$$((1-\omega)I - \mathscr{L}_\omega)(\omega^2B^2 + (1-\omega)I - \mathscr{L}_\omega) = \omega^2(1-\omega)B^2 \tag{2.47}$$

which is equivalent to (2.45).

7.3. DETERMINATION OF $\| \mathscr{L}_\omega \|_{D^{1/2}}$

Rather than attempting to determine $\overline{\| \mathscr{L}_\omega \|}$ we shall instead consider $\| \overline{\mathscr{L}_\omega} \|_{D^{1/2}}$. We first show that if A is positive definite it is sufficient to assume that $A = I - L - U$ where $L^{\mathrm{T}} = U$. Otherwise, we consider $\hat{A} = D^{-1/2}AD^{-1/2}$ which is symmetric and has diagonal elements unity. By Theorem 4-3.5 it follows that $B[\hat{A}] = \tilde{B}[A]$; hence $\bar{\mu}$ and ω_b are the same for $\hat{A}$ as for A. Moreover, by Theorem 4-3.5 we also have

$$\| \mathscr{L}_\omega^m[A] \|_{D^{1/2}} = \| \tilde{\mathscr{L}}_\omega^m[A] \| = \| \mathscr{L}_\omega^m[\hat{A}] \| = \| \mathscr{L}_\omega^m[\hat{A}] \|_{D[\hat{A}]^{1/2}} \tag{3.1}$$

with a similar relation holding for the corresponding virtual norms.

From Theorem 2.1 and (1.12) it follows that the eigenvalues of $\mathscr{L}_\omega \mathscr{L}_\omega^{\mathrm{T}}$ are the same as the eigenvalues of $M(\omega, \mu)M(\omega, \mu)^{\mathrm{T}}$ where

$$\begin{aligned} M(\omega, \mu) &= \begin{pmatrix} 1 & 0 \\ \omega\mu & 1 \end{pmatrix} \begin{pmatrix} 1-\omega & \omega\mu \\ 0 & 1-\omega \end{pmatrix} \\ &= \begin{pmatrix} 1-\omega & \omega\mu \\ (1-\omega)\omega\mu & \omega^2\mu^2 + 1 - \omega \end{pmatrix} \end{aligned} \tag{3.2}$$

This follows since by (1.12) we have

$$\mathscr{L}_\omega^{\mathrm{T}} = \begin{pmatrix} (1-\omega)I_1 & 0 \\ \omega G & (1-\omega)I_2 \end{pmatrix} \begin{pmatrix} I_1 & \omega F \\ 0 & I_2 \end{pmatrix} \tag{3.3}$$

and since $L^{\mathrm{T}} = U$ and $F^{\mathrm{T}} = G$. Thus $\mathscr{L}_\omega^{\mathrm{T}}$ has the required form for the applicability of Theorem 2.1.

Evidently the eigenvalues λ of $\mathscr{L}_\omega \mathscr{L}_\omega^{\mathrm{T}}$ satisfy the characteristic equation

$$\lambda^2 - T(\mu^2)\lambda + C = 0 \tag{3.4}$$

where

$$\begin{aligned} T(\mu^2) &= (1-\omega)^2 + \omega^2\mu^2 + (1-\omega)^2\omega^2\mu^2 + (\omega^2\mu^2 + 1 - \omega)^2 \\ C &= (\omega - 1)^4 \end{aligned} \tag{3.5}$$

We prove

Theorem 3.1. If A is a positive definite matrix of the form (1), then $\| \overline{\mathscr{L}_\omega} \|_{D^{1/2}} < 1$ if and only if

$$\omega < 2(1 - \bar{\mu}^2)^{1/2}/[\bar{\mu}^2 + (1 - \bar{\mu}^2)^{1/2}] \tag{3.6}$$

Proof. Evidently

$$T(\mu^2) = 2(1-\omega)^2 + \omega^2\mu^2(2-\omega)^2 + \omega^4\mu^4 \tag{3.7}$$

which is an increasing function of μ. By Lemma 6-2.9, it follows that for given ω, the largest value of the root radius of (3.4) is assumed for

$\mu = \bar{\mu}$. Therefore we have

$$\| \mathscr{L}_\omega \|_{D^{1/2}} = \| \overline{\mathscr{L}_\omega} \|_{D^{1/2}}$$

Moreover, the largest root radius is the largest root of

$$\lambda - (\omega - 1)^2 = \omega\bar{\mu}[\omega^2\bar{\mu}^2 + (2 - \omega)^2]^{1/2}\lambda^{1/2} \tag{3.8}$$

By Lemma 6-2.1, the root radius is less than unity if and only if we have

$$| \omega - 1 | < 1$$

and

$$\omega\bar{\mu}[\omega^2\bar{\mu}^2 + (2 - \omega)^2]^{1/2} < 1 - (\omega - 1)^2 = \omega(2 - \omega)$$

From the second equation we obtain (3.6).

The problem of finding ω which minimizes $\| \mathscr{L}_\omega \|_{D^{1/2}}$ is rather complicated. Values of $\| \mathscr{L}_\omega \|_{D^{1/2}}$ were obtained numerically for various values of $\bar{\mu}$ and for $\omega = 0(0.05)2$ and are shown in the accompanying tabulation. Here ω_0 gave the smallest value of $\| \mathscr{L}_\omega \|_{D^{1/2}}$ for the values computed. We let ω_b' be the tabulated value of ω closest to ω_b. Evidently $\| \mathscr{L}_{\omega_b'} \|_{D^{1/2}}$ is much larger than $S(\mathscr{L}_{\omega_b}) = \omega_b - 1$. In fact $\| \mathscr{L}_{\omega_0} \|_{D^{1/2}}$ is much larger than $S(\mathscr{L}_{\omega_b})$.

$\bar{\mu}$	$\dfrac{2(1 - \bar{\mu}^2)^{1/2}}{\bar{\mu}^2 + (1 - \bar{\mu}^2)^{1/2}}$	ω_0	$\| \mathscr{L}_{\omega_0} \|_{D^{1/2}}$	ω_b	ω_b'	$\| \mathscr{L}_{\omega_b'} \|_{D^{1/2}}$
0	2.00000	1.00	0	1.00000	1.00	0
0.1	1.98010	1.00	0.10050	1.00251	1.00	0.10050
0.2	1.92155	1.00	0.20396	1.01020	1.00	0.20396
0.3	1.82758	1.00	0.31321	1.02357	1.00	0.31321
0.4	1.70275	0.95	0.43014	1.04356	1.05	0.44191
0.5	1.55198	0.95	0.55194	1.07180	1.05	0.57420
0.6	1.37931	0.85	0.67492	1.11111	1.10	0.74994
0.7	1.18614	0.75	0.79081	1.16676	1.15	0.96571
0.8	0.96774	0.65	0.88993	1.25000	1.25	1.29815
0.9	0.69972	0.45	0.96296	1.39286	1.40	1.84513
0.95	0.51410	0.30	0.98748	1.52410	1.50	2.26250
1.0	0	0	1.00000	2.00000	2.00	4.23607

7.4. DETERMINATION OF $\| \mathscr{L}_{\omega_b}^m \|_{D^{1/2}}$

We first prove

Theorem 4.1. Let A be a positive definite matrix of the form (1) and let $B = I - D^{-1}A$ where $D = \operatorname{diag} A$. Then

$$\| \overline{\mathscr{L}_{\omega_b}^m} \|_{D^{1/2}} = \| \mathscr{L}_{\omega_b}^m \|_{D^{1/2}}$$
$$= r^m\{m(r^{-1/2} + r^{1/2}) + [m^2(r^{-1/2} + r^{1/2})^2 + 1]^{1/2}\} = F_1(m) \tag{4.1}$$

where

$$r = \omega_b - 1, \qquad \omega_b = \frac{2}{1 + (1 - \bar{\mu}^2)^{1/2}}, \qquad \bar{\mu} = S(B) \tag{4.2}$$

Proof. By Theorem 5-3.5 we have $\bar{\mu} < 1$, and, by (6-1), $1 \leq \omega_b < 2$. As in Section 7.3 it is sufficient to consider the case where $A = I - L - U$ where $L^{\mathrm{T}} = U$. From Theorem 2.1, it follows that if $\mathscr{L}_\omega$ is given by (1.12), then the eigenvalues of $\mathscr{L}_\omega^m (\mathscr{L}_\omega^m)^{\mathrm{T}}$ are the same as the eigenvalues of $M(\omega, \mu)^m[M(\omega, \mu)^m]^{\mathrm{T}}$ where

$$M(\omega, \mu) = \begin{pmatrix} 1 & 0 \\ \omega\mu & 1 \end{pmatrix} \begin{pmatrix} 1 - \omega & \omega\mu \\ 0 & 1 - \omega \end{pmatrix}$$
$$= \begin{pmatrix} 1 - \omega & \omega\mu \\ (1 - \omega)\omega\mu & \omega^2\mu^2 + 1 - \omega \end{pmatrix} \tag{4.3}$$

We now develop an expression for $M(\omega, \mu)^m$. Let us define the polynomials $S_0(\mu)$, $S_1(\mu)$, $S_2(\mu)$, ... by the recursion formula

$$\begin{aligned} S_k(\mu) &= \omega\mu S_{k-1}(\mu) + (1 - \omega)S_{k-2}(\mu), \qquad k = 2, 3, \ldots \\ S_0(\mu) &= 1, \qquad S_1(\mu) = \omega\mu \end{aligned} \tag{4.4}$$

We show that

$$M(\omega, \mu)^m = \begin{pmatrix} (1 - \omega)S_{2m-2} & S_{2m-1} \\ (1 - \omega)S_{2m-1} & S_{2m} \end{pmatrix} \tag{4.5}$$

Evidently, if $m = 1$, then (4.5) holds by (4.3) and (4.4). Suppose now that (4.5) holds for all $k \leq m$. We seek to show that it is true for $k = m+1$.

But by (4.5) and (4.3) we have

$$
\begin{aligned}
&M(\omega, \mu)^m M(\omega, \mu)\\
&= \begin{pmatrix} (1-\omega)[(1-\omega)S_{2m-2}+\omega\mu S_{2m-1}] & \omega\mu(1-\omega)S_{2m-2}+(\omega^2\mu^2+1-\omega)S_{2m-1} \\ (1-\omega)[(1-\omega)S_{2m-1}+\omega\mu S_{2m}] & \omega\mu(1-\omega)S_{2m-1}+(\omega^2\mu^2+1-\omega)S_{2m} \end{pmatrix}\\
&= \begin{pmatrix} (1-\omega)S_{2m} & S_{2m+1} \\ (1-\omega)S_{2m+1} & S_{2m+2} \end{pmatrix}
\end{aligned}
\tag{4.6}
$$

by (4.4). Therefore (4.5) follows by induction.

Next, we show that for $k \geq 0$ we have

$$
S_k(\mu) = \sum_{i=0}^{k} \alpha_1^{k-i}\alpha_2^{\,i} = \begin{cases} \dfrac{\alpha_1^{k+1} - \alpha_2^{k+1}}{\alpha_1 - \alpha_2} & \text{if} \quad \alpha_1 \neq \alpha_2 \\ (k+1)\alpha^k & \text{if} \quad \alpha_1 = \alpha_2 = \alpha \end{cases}
\tag{4.7}
$$

where α_1 and α_2 are the solutions of the quadratic equation

$$
\alpha^2 - \omega\mu\alpha + \omega - 1 = 0
\tag{4.8}
$$

The result is obviously true for $k = 0$, and since

$$
\alpha_1 + \alpha_2 = \omega\mu, \qquad \alpha_1\alpha_2 = \omega - 1
\tag{4.9}
$$

it is also true for $k = 1$. For if $\alpha_1 \neq \alpha_2$, we have

$$
(\alpha_1^2 - \alpha_2^2)/(\alpha_1 - \alpha_2) = \alpha_1 + \alpha_2 = \omega\mu = S_1(\mu)
$$

and if $\alpha_1 = \alpha_2 = \alpha = \omega\mu/2$, then

$$
2\alpha = \omega\mu = S_1(\mu)
$$

Let us assume that (4.7) holds for $\alpha_1 \neq \alpha_2$ for $i = 0, 1, 2, \ldots, k$. Then by (4.4), (4.9), and (4.7) we have

$$
\begin{aligned}
S_{k+1}(\mu) &= \omega\mu S_k(\mu) + (1 - \omega)S_{k-1}(\mu)\\
&= \frac{(\alpha_1+\alpha_2)(\alpha_1^{k+1} - \alpha_2^{k+1}) - \alpha_1\alpha_2(\alpha_1^{\,k} - \alpha_2^{\,k})}{\alpha_1 - \alpha_2} = \frac{\alpha_1^{k+2} - \alpha_2^{k+2}}{\alpha_1 - \alpha_2}
\end{aligned}
$$

If $\alpha_1 = \alpha_2 = \alpha = \omega\mu/2$, then, since $\omega - 1 = \alpha^2$ we have

$$S_{k+1}(\mu) = 2\alpha(k+1)\alpha^k - \alpha^2 k\alpha^{k-1} = (k+2)\alpha^{k+1}$$

Thus (4.7) holds for $k+1$ and by induction it is true for all k.

We now show that if $\omega = \omega_b$, then

$$\max_{-\bar{\mu}\leq\mu\leq\bar{\mu}} |S_k(\mu)| = S_k(\bar{\mu}) = (k+1)(r^{1/2})^k \tag{4.10}$$

This is obvious for $k = 0, 1$. For $k \geq 0$, we have by (4.7)

$$S_k(\mu) = \sum_{i=0}^{k} \alpha_1^{k-i}\alpha_2^{\,i} \tag{4.11}$$

Since the discriminant of the quadratic (4.8) is nonpositive for any μ in the range $-\bar{\mu} \leq \mu \leq \bar{\mu}$ (since $\omega_b^2\mu^2 - 4(\omega_b - 1) \leq \omega_b^2\bar{\mu}^2 - 4(\omega_b - 1) = 0$), we have $|\alpha_1| = |\alpha_2| = (\omega_b - 1)^{1/2} = r^{1/2}$ and by (4.11)

$$|S_k(\mu)| \leq \sum_{i=0}^{k} (r^{1/2})^k = (k+1)(r^{1/2})^k$$

On the other hand, if $\mu = \bar{\mu}$, then the discriminant of (4.8) vanishes and we have $\alpha_1 = \alpha_2 = r^{1/2}$. By (4.11) we have $S_k(\bar{\mu}) = (k+1)(r^{1/2})^k$ and (4.10) follows.

Evidently, the characteristic equation for $M(\omega,\mu)^m(M(\omega,\mu)^m)^{\mathrm{T}}$ is

$$\Gamma^2 - T_m(\omega,\mu)\Gamma + \Delta = 0 \tag{4.12}$$

where

$$\begin{aligned} T_m(\omega,\mu) &= \operatorname{trace}[M(\omega,\mu)^m(M(\omega,\mu)^m)^{\mathrm{T}}] \\ &= (1-\omega)^2 S_{2m-2}^2 + S_{2m-1}^2 + (1-\omega)^2 S_{2m-1}^2 + S_{2m}^2 \end{aligned} \tag{4.13}$$

by (4.5), and

$$\Delta = \det[M(\omega,\mu)^m(M(\omega,\mu)^m)^{\mathrm{T}}] = (1-\omega)^{4m} \tag{4.14}$$

by (4.3). Since $T(\omega,\mu)^2 - 4\Delta \geq 0$, because the eigenvalues of the symmetric matrix $M(\omega,\mu)^m(M(\omega,\mu)^m)^{\mathrm{T}}$ are real, it follows that for fixed Δ the modulus of the root of (4.12) of largest modulus is maximized when $|T_m(\omega,\mu)|$ considered as a function of μ is maximized. But, by (4.10), $|T_m(\omega_b,\mu)|$ is maximized when $\mu = \bar{\mu}$. Hence, $\|\overline{\mathscr{L}_{\omega_b}^m}\|_{D^{1/2}} = \|\mathscr{L}_{\omega_b}^m\|_{D^{1/2}}$ and we have

$$T_m(\omega_b,\bar{\mu}) = 2r^{2m}(2m^2(r^{1/2} + r^{-1/2})^2 + 1) \tag{4.15}$$

and

$$\Delta = r^{4m} \tag{4.16}$$

Solving for the largest root of (4.12) we obtain

$$(\Gamma - r^{2m})^2 = 4m^2(r^{-1/2} + r^{1/2})^2 \Gamma r^{2m} \tag{4.17}$$

and

$$\Gamma^{1/2} = \{m(r^{-1/2} + r^{1/2}) + [m^2(r^{-1/2} + r^{1/2})^2 + 1]^{1/2}\} r^m \tag{4.18}$$

Since $\| \mathscr{L}_{\omega_b}^m \|_{D^{1/2}}$ is the square root of the largest eigenvalue of $M(\omega_b, \bar{\mu})^m \times (M(\omega_b, \bar{\mu})^m)^{\mathrm{T}}$ the theorem now follows.

The result (4.1) is consistent with what one might expect from Theorems 3-7.1 and 3-8.1. The eigenvalues of $M(\omega_b, \mu)^m[M(\omega_b, \mu)^m]^{\mathrm{T}}$ are maximized when $\mu = \bar{\mu}$. Therefore, by Theorem 3-7.1 we have

$$\| M(\omega_b, \bar{\mu})^m \| \sim m \mathscr{G}(M(\omega_b, \bar{\mu})) r^{m-1} \tag{4.19}$$

But by Theorem 3-8.1 we have

$$\mathscr{J}(M(\omega_b, \bar{\mu})) = \omega_b \bar{\mu} + (\omega_b - 1)\omega_b \bar{\mu} = \omega_b{}^2 \bar{\mu} = 2r^{1/2}(1+r) \sim 4r^{1/2} \tag{4.20}$$

and

$$\| M(\omega_b, \bar{\mu})^m \| \sim 4mr^{m-1/2} \tag{4.21}$$

On the other hand, by (4.1) we have

$$\| \mathscr{L}_{\omega_b}^m \|_{D^{1/2}} \sim 4mr^{m-1/2} \tag{4.22}$$

For values of r close to unity the function $F_1(m)$ defined by (4.1) increases initially before eventually decreasing. A graph of $F_1(m)$ for the case $r = 0.73$ is given in Figure 7.1. Also shown on the same graph is $\| \mathscr{L}_{\omega_b}^m \|_{A^{1/2}}$ which is a monotone-decreasing function of m and is considerably less than $F_1(m)$. Formulas for the $A^{1/2}$-norm will be developed in Section 7.6.

We now seek to estimate the number of iterations needed to reduce the $D^{1/2}$-norm of the error vector to a specified fraction ϱ of the $D^{1/2}$-norm of the initial error vector. To do this we determine m such that $\| \mathscr{L}_{\omega_b}^m \|_{D^{1/2}} \le \varrho$. We prove

Theorem 4.2. If $0 < \varrho < 1$ and $0 < r < 1$, then there exists a unique solution $m = m(r)$ of the equation

$$F_1(m) = \varrho \tag{4.23}$$

where $F_1(m) = \| \mathscr{L}_{\omega_b}^m \|_{D^{1/2}}$ is given by (4.1), such that $m > 0$. Moreover, if $r \geq 3 - 2(2)^{1/2} \doteq 0.172$ or if

$$\varrho < r\{(r^{-1/2} + r^{1/2}) + [(r^{-1/2} + r^{1/2})^2 + 1]^{1/2}\} \tag{4.24}$$

then $m > 1$. Also,

$$\lim_{r \to 1-} [m(r) \log (1/r)] = \infty \tag{4.25}$$

and

$$\lim_{\varrho \to 0} \left[m(r) \log \frac{1}{r} - \log\left(\frac{2\gamma}{\varrho} \log \frac{2\gamma}{\varrho} \right) \right] = 0 \tag{4.26}$$

where

$$\gamma = (r^{-1/2} + r^{1/2})/(\log 1/r) \tag{4.27}$$

Proof. By (4.1) we have

$$\log F_1(m) = m \log r + \sinh^{-1}[m(r^{-1/2} + r^{1/2})] \tag{4.28}$$

since $\sinh^{-1} x = \log[x + (x^2 + 1)^{1/2}]$. Therefore,

$$\frac{F_1'(m)}{F_1(m)} = \log r + \frac{r^{-1/2} + r^{1/2}}{[m^2(r^{-1/2} + r^{1/2})^2 + 1]^{1/2}} \tag{4.29}$$

which is positive for $0 < m < \tilde{m}$ where

$$\tilde{m} = \frac{1}{\log(1/r)} \left[1 - \left(\frac{\log(1/r)}{r^{-1/2} + r^{1/2}} \right)^2 \right]^{1/2} \tag{4.30}$$

We remark that $\log(1/r) < (1/r^{1/2}) + r^{1/2}$ since if $\theta = \log(1/r^{1/2})$ we have $\log(1/r) - (1/r^{1/2}) - r^{1/2} = 2\theta - 2 \cosh \theta \leq 0$. Thus the function $F_1(m)$ has the value unity at $m = 0$, increases until $m = \tilde{m}$, and then decreases, approaching zero as $m \to \infty$. Consequently, for any ϱ in the range $0 < \varrho < 1$ there is a unique value of $m > 0$ such that (4.23) holds.

We now show that if $r \geq 3 - 2(2)^{1/2}$ or if (4.24) holds, then $m > 1$. Obviously, if (4.24) holds, then $\varrho < F_1(1)$ and the unique solution of (4.23) is greater than unity. Next, we note that $F_1(1)$ is an increasing function of r. When $r = 3 - 2(2)^{1/2}$ we have $r^{1/2} = 2^{1/2} - 1$, $1/r^{1/2} = 2^{1/2} + 1$ and $r^{1/2} + (1/r^{1/2}) = 2(2)^{1/2}$. Therefore,

$$F_1(1) = (3 - 2(2)^{1/2})(2(2)^{1/2} + 3) = 1 \tag{4.31}$$

and again we have $\varrho < F_1(1)$.

To prove (4.25) we have from (4.1)

$$m(r) = \frac{\log[m(r^{-1/2} + r^{1/2}) + (m^2(r^{-1/2} + r^{1/2})^2 + 1)^{1/2}] + \log(1/\varrho)}{\log(1/r)}$$
$$\geq \frac{\log(1/\varrho)}{\log(1/r)} \tag{4.32}$$

and for fixed ϱ

$$\lim_{r \to 1-} m(r) = \infty \tag{4.33}$$

Moreover,

$$\lim_{r \to 1-} [m(r)\log(1/r)] = \lim_{r \to 1-} \{\log[m(r^{-1/2}+r^{1/2})+(m^2(r^{-1/2}+r^{1/2})^2+1)^{1/2}]\}$$
$$+\log(1/\varrho) = \infty \tag{4.34}$$

Thus (4.25) follows.

To prove (4.26) we introduce the function

$$y = m(r)\log(1/r) \tag{4.35}$$

and by (4.1) and (4.23) we have

$$K(y) = e^{-y}[y\gamma + (y^2\gamma^2 + 1)^{1/2}] = \varrho \tag{4.36}$$

This can be written in the form

$$y = \log \frac{y\gamma + (y^2\gamma^2 + 1)^{1/2}}{\varrho} \tag{4.37}$$

Consider now the related equation

$$w = \log(2w\gamma/\varrho) \tag{4.38}$$

which is equivalent to

$$\hat{K}(w) = 2w\gamma e^{-w} = \varrho \tag{4.39}$$

We let

$$w_0 = \log\frac{2\gamma}{\varrho}, \qquad w_1 = \log\left(\frac{2\gamma}{\varrho}\log\frac{2\gamma}{\varrho}\right) \tag{4.40}$$

Since $\hat{K}(w)$ is a decreasing function of w for $w \geq 1$ and since $\hat{K}(1) = 2\gamma/e$ it follows that if $0 < \varrho < 2/e$, then $\varrho < 2\gamma/e$ and hence (4.39) has a unique solution w such that $w \geq 1$. (We have already shown that $\gamma > 1$.)

Since

$$\hat{K}(w_0) = \varrho \log \frac{2\gamma}{\varrho} > \varrho \tag{4.41}$$

it follows that $w > w_0$. Evidently, we have

$$\begin{aligned} w - w_0 &= \log w = \log(w_0 + w - w_0) \\ &= \log w_0 + \log\left(1 + \frac{w - w_0}{w_0}\right) \\ &\leq \log w_0 + \frac{w - w_0}{w_0} \end{aligned} \tag{4.42}$$

so that

$$w - w_0 \leq \log w_0/(1 - w_0^{-1}) \tag{4.43}$$

Similarly, since $w_1 = \log(2\gamma w_0/\varrho)$ we have

$$w - w_1 = \log \frac{w}{w_0} = \log\left(1 + \frac{w - w_0}{w_0}\right) \leq \frac{w - w_0}{w_0} \tag{4.44}$$

and

$$0 \leq w - w_1 \leq \log w_0/(w_0 - 1) \tag{4.45}$$

From (4.37) and (4.38) we have

$$\begin{aligned} y - w &= \log\left(\frac{y\gamma + (y^2\gamma^2 + 1)^{1/2}}{2\gamma w}\right) \\ &= \log\left(1 + \frac{y\gamma + (y^2\gamma^2 + 1)^{1/2} - 2\gamma w}{2\gamma w}\right) \\ &\leq \frac{y\gamma + (y^2\gamma^2 + 1)^{1/2} - 2\gamma w}{2\gamma w} \\ &= \frac{(y - w)}{w} + \frac{(y^2\gamma^2 + 1)^{1/2} - y\gamma}{2\gamma w} \\ &\leq \frac{y - w}{w} + \frac{1}{4\gamma^2 y w} \end{aligned} \tag{4.46}$$

Since $K(y)$ is a decreasing function of y and since

$$K(w) \geq \hat{K}(w) = \varrho \tag{4.47}$$

it follows that $y \geq w \geq w_0$. Hence, we have by (4.46)

$$0 \leq y - w \leq 1/[4\gamma^2 w_0(w_0 - 1)] \tag{4.48}$$

Therefore,

$$0 \le y - w_1 \le \frac{\log w_0}{w_0 - 1} + \frac{1}{4\gamma^2 w_0(w_0 - 1)} \tag{4.49}$$

Since the right-hand side of the above inequality converges to zero as $\varrho \to 0$, the result (4.26) follows. This completes the proof of Theorem 4.2.

From Theorem 4.2 we use the following approximate value for the solution of (4.23).

$$m_0 = \frac{\log((2\gamma/\varrho) \log(2\gamma/\varrho))}{\log(1/r)}, \qquad \gamma = \frac{r^{1/2} + r^{-1/2}}{\log(1/r)} \tag{4.50}$$

We shall see in Section 7.6 that this approximation gives reasonably good values for the exact solution of (4.23).

7.5. DETERMINATION OF $\| \mathscr{L}_\omega \|_{A^{1/2}}$

Let us consider the problem of determining $\| \mathscr{L}_\omega \|_{A^{1/2}}$ under the assumption that A is positive definite. We first prove

Theorem 5.1. If A is a positive definite matrix, then

$$(\mathscr{L}_\omega')^{\mathrm{T}} = \mathscr{U}_\omega' \tag{5.1}$$

and

$$\| \overline{\mathscr{L}_\omega} \|_{A^{1/2}}^2 = \| \mathscr{L}_\omega \|_{A^{1/2}}^2 = S(\mathscr{U}_\omega \mathscr{L}_\omega) = S(\mathscr{S}_\omega) = \| \mathscr{S}_\omega \|_{A^{1/2}} \tag{5.2}$$

where

$$\begin{aligned} \mathscr{L}_\omega &= (I - \omega L)^{-1}(\omega U + (1 - \omega)I), \qquad & \mathscr{L}_\omega' &= A^{1/2} \mathscr{L}_\omega A^{-1/2} \\ \mathscr{U}_\omega &= (I - \omega U)^{-1}(\omega L + (1 - \omega)I), \qquad & \mathscr{U}_\omega' &= A^{1/2} \mathscr{U}_\omega A^{-1/2} \end{aligned} \tag{5.3}$$

$$\mathscr{S}_\omega = \mathscr{U}_\omega \mathscr{L}_\omega, \qquad \mathscr{S}_\omega' = A^{1/2} \mathscr{S}_\omega A^{-1/2} \tag{5.4}$$

Here, as usual, L and U are strictly lower and strictly upper triangular matrices, respectively, such that $L + U = I - (\operatorname{diag} A)^{-1} A$.

Proof. By (5.3) we have

$$\begin{aligned} \mathscr{L}_\omega &= I - \omega(I - \omega L)^{-1} D^{-1} A = I - \omega(D - \omega C_L)^{-1} A \\ \mathscr{U}_\omega &= I - \omega(I - \omega U)^{-1} D^{-1} A = I - \omega(D - \omega C_U)^{-1} A \end{aligned} \tag{5.5}$$

where $C_L = DL$, $C_U = DU$ and $C_L{}^{\mathrm{T}} = C_U$. Hence

$$(\mathscr{L}_\omega')^{\mathrm{T}} = (I - \omega A^{1/2}(D - \omega C_L)^{-1}A^{1/2})^{\mathrm{T}} = I - \omega A^{1/2}(D - \omega C_U)^{-1}A^{1/2} = \mathscr{U}_\omega' \tag{5.6}$$

and (5.1) follows.

By (5.4) we have

$$\mathscr{S}_\omega' = \mathscr{U}_\omega' \mathscr{L}_\omega' \tag{5.7}$$

Therefore, by (5.1), it follows that

$$\begin{aligned} \| \mathscr{L}_\omega \|_{A^{1/2}}^2 &= \| \mathscr{L}_\omega' \|^2 = S(\mathscr{L}_\omega'(\mathscr{L}_\omega')^{\mathrm{T}}) = S(\mathscr{L}_\omega' \mathscr{U}_\omega') \\ &= S(\mathscr{U}_\omega' \mathscr{L}_\omega') = S(\mathscr{S}_\omega') = S(\mathscr{S}_\omega) \end{aligned} \tag{5.8}$$

by Theorem 2-1.11. Finally, since $(\mathscr{S}_\omega')^{\mathrm{T}} = \mathscr{S}_\omega'$ it follows that $\| \mathscr{S}_\omega \|_{A^{1/2}} = \| \mathscr{S}_\omega' \| = S(\mathscr{S}_\omega') = S(\mathscr{S}_\omega)$. Hence (5.2) follows and the proof of Theorem 5.1 is complete.

We remark that the matrix $\mathscr{S}_\omega$ is associated with the symmetric SOR method, which will be studied further in Chapter 15.

Let us now assume that A has the form (1). We prove

Theorem 5.2. If A is a positive definite matrix of the form (1), then

$$\| \mathscr{L}_\omega \|_{A^{1/2}}^2 = S(\mathscr{L}_{\omega(2-\omega)}) \leq 1 - \tfrac{1}{2}\omega^2(2 - \omega)^2(1 - \bar{\mu}^2) \tag{5.9}$$

and unless $\omega = 1$ we have

$$\| \mathscr{L}_\omega \|_{A^{1/2}} > \| \mathscr{L}_1 \|_{A^{1/2}} = \bar{\mu} = S(B) \tag{5.10}$$

where $B = I - (\operatorname{diag} A)^{-1}A$.

Proof. We first prove

Lemma 5.3. If A is a matrix of the form (1) with nonvanishing diagonal elements, then the eigenvalues of $\mathscr{S}_\omega$ are the same as those of $\mathscr{L}_{\omega(2-\omega)}$ and

$$S(\mathscr{S}_\omega) = S(\mathscr{L}_{\omega(2-\omega)}) \tag{5.11}$$

Proof. Let us define the matrices

$$\mathscr{L}_{\omega,0} = \begin{pmatrix} (1-\omega)I_1 & \omega F \\ 0 & I_2 \end{pmatrix}, \qquad \mathscr{L}_{0,\omega} = \begin{pmatrix} I_1 & 0 \\ \omega G & (1-\omega)I_2 \end{pmatrix} \tag{5.12}$$

Evidently by (1.12) we have

$$\mathscr{L}_\omega = \mathscr{L}_{0,\omega} \mathscr{L}_{\omega,0} \tag{5.13}$$

Therefore, by (5.4) we have

$$\mathscr{S}_\omega = \mathscr{U}_\omega \mathscr{L}_\omega = \mathscr{L}_{\omega,0} \mathscr{L}_{0,\omega} \mathscr{L}_{0,\omega} \mathscr{L}_{\omega,0} \tag{5.14}$$

which, by Theorem 2-1.11 has the same eigenvalues as

$$\mathscr{L}_{0,\omega}^2 \mathscr{L}_{\omega,0}^2 = \mathscr{L}_{0,\hat{\omega}} \mathscr{L}_{\hat{\omega},0} = \mathscr{L}_{\hat{\omega}} \tag{5.15}$$

where

$$\hat{\omega} = \omega(2 - \omega) \tag{5.16}$$

This follows since

$$\mathscr{L}_{\omega,0}^2 = \mathscr{L}_{\hat{\omega},0}, \qquad \mathscr{L}_{0,\omega}^2 = \mathscr{L}_{0,\hat{\omega}} \tag{5.17}$$

The lemma now follows.

Since $\| \mathscr{L}_\omega \|_{A^{1/2}}^2 = S(\mathscr{S}_\omega)$ by Theorem 5.1, it follows that

$$\| \mathscr{L}_\omega \|_{A^{1/2}}^2 = S(\mathscr{L}_{\hat{\omega}}) = S(\mathscr{L}_{\omega(2-\omega)}) \tag{5.18}$$

Moreover, since $0 < \hat{\omega} \le 1 \le \omega_b$, we have by (6-2.10),

$$S(\mathscr{L}_{\hat{\omega}}) = \left(\frac{\hat{\omega}\bar{\mu} + (\hat{\omega}^2 \mu^2 - 4(\hat{\omega} - 1))^{1/2}}{2} \right)^2 \tag{5.19}$$

and

$$\begin{aligned} 1 - S(\mathscr{L}_{\hat{\omega}}) &= \frac{2\hat{\omega} - \hat{\omega}^2 \bar{\mu}^2 - \hat{\omega}\bar{\mu}(\hat{\omega}^2 \bar{\mu}^2 - 4(\hat{\omega} - 1))^{1/2}}{2} \\ &\ge \frac{2\hat{\omega} - \hat{\omega}^2 \bar{\mu}^2 - \hat{\omega}(\hat{\omega}^2 - 4(\hat{\omega} - 1))^{1/2}}{2} \\ &= \frac{2\hat{\omega} - \hat{\omega}^2 \bar{\mu}^2 - \hat{\omega}(2 - \hat{\omega})}{2} = \frac{\hat{\omega}^2(1 - \bar{\mu}^2)}{2} \end{aligned} \tag{5.20}$$

Therefore, (5.9) follows.

Since $\hat{\omega}$ varies between 0 and 1 as ω varies from 0 to 2 and since, by Theorem 6-2.3, $S(\mathscr{L}_{\hat{\omega}})$ is a decreasing function of $\hat{\omega}$ for $0 < \hat{\omega} < \omega_b$, it follows that the optimum value of $\hat{\omega}$ is unity. Therefore, if $\omega \ne 1$ we have $\hat{\omega} < 1$ and

$$\| \mathscr{L}_\omega \|_{A^{1/2}}^2 = S(\mathscr{L}_{\hat{\omega}}) > S(\mathscr{L}_1) = \bar{\mu}^2 \tag{5.21}$$

and since $\| \mathscr{L}_1 \|_{A^{1/2}}^2 = S(\mathscr{L}_1) = \bar{\mu}^2$ by (5.18) and Theorem 5-3.4, the result (5.10) follows.

From Theorem 5.1 it follows that $\| \mathscr{L}_\omega \|_{A^{1/2}}^2 = S(\mathscr{S}_\omega)$ for any positive definite matrix, not necessarily one of the form (1). From Theorem 5.2 it follows that $\min_\omega S(\mathscr{S}_\omega) = \bar{\mu}^2$; hence the symmetric SOR method with the optimum ω converges no more rapidly than the Gauss–Seidel method. Moreover, we have $\| \mathscr{L}_{\omega_b} \|_{A^{1/2}}^2 = S(\mathscr{S}_{\omega_b}) \geq \bar{\mu}^2$. On the other hand, as we shall see in Chapter 15, with other orderings of the equations which do not correspond to the form (1), one can often achieve a more rapid convergence for the symmetric SOR method. Hence, for these orderings one can expect to be able to obtain a smaller value of $\| \mathscr{L}_\omega \|_{A^{1/2}}^2$. Thus, there seems to be some reason to expect that orderings other than the red–black ordering, which corresponds to (1), might have advantages in some cases.

7.6. DETERMINATION OF $\| \mathscr{L}_{\omega_b}^m \|_{A^{1/2}}$

For r sufficiently close to unity, the $D^{1/2}$ norm of $\mathscr{L}_{\omega_b}^m$ increases initially with m before it eventually decreases. On the other hand, we know by Theorem 4-3.1 that if A is positive definite, then $\| \mathscr{L}_{\omega_b} \|_{A^{1/2}} < 1$. Hence since

$$\| \mathscr{L}_{\omega_b}^{m+1} \|_{A^{1/2}} \leq \| \mathscr{L}_{\omega_b}^m \|_{A^{1/2}} \| \mathscr{L}_{\omega_b} \|_{A^{1/2}} < \| \mathscr{L}_{\omega_b}^m \|_{A^{1/2}} \tag{6.1}$$

it follows that $\| \mathscr{L}_{\omega_b}^m \|_{A^{1/2}}$ is a steadily decreasing† function of m. We propose to develop an explicit expression for $\| \mathscr{L}_{\omega_b}^m \|_{A^{1/2}}$ under the assumption that A is positive definite and has the form (1). We first prove

Theorem 6.1. If A is a positive definite matrix of the form (1), then

$$\begin{aligned} \| \overline{\mathscr{L}_{\omega_b}^m} \|_{A^{1/2}} = \| \mathscr{L}_{\omega_b}^m \|_{A^{1/2}} &= r^m \{ m(r^{-1/2} - r^{1/2}) + [m^2(r^{-1/2} - r^{1/2})^2 + 1]^{1/2} \} \\ &= H_1(m) \end{aligned} \tag{6.2}$$

Proof. By Lemma 4-4.2 it is sufficient to consider the case $A = I - L - U$ where $L = U^{\mathrm{T}}$. In this case we have

$$\begin{aligned} \| \mathscr{L}_\omega^m \|_{A^{1/2}} = \| A^{1/2} \mathscr{L}_\omega^m A^{-1/2} \| &= [S(A^{1/2} \mathscr{L}_\omega^m A^{-1} (\mathscr{L}_\omega^m)^{\mathrm{T}} A^{1/2})]^{1/2} \\ &= [S(\mathscr{L}_\omega^m A^{-1} (\mathscr{L}_\omega^m)^{\mathrm{T}} A)]^{1/2} \end{aligned} \tag{6.3}$$

† We are assuming that $\bar{\mu} > 0$ so that $\| \mathscr{L}_{\omega_b} \|_{A^{1/2}} > 0$.

By Theorem 2.1, the eigenvalues of $\mathscr{L}_\omega^m A^{-1}(\mathscr{L}_\omega^m)^{\mathrm{T}} A$ are the eigenvalues of

$$Q = M(\omega, \mu)^m N(\mu)^{-1}(M(\omega, \mu)^m)^{\mathrm{T}} N(\mu) \tag{6.4}$$

where $M(\omega, \mu)^m$ is given by (4.5) and where

$$N(\mu) = \begin{pmatrix} 1 & -\mu \\ -\mu & 1 \end{pmatrix} \tag{6.5}$$

But we also have

$$S(Q) = \| N(\mu)^{1/2} M(\omega, \mu)^m N(\mu)^{-1/2} \|^2 \tag{6.6}$$

where

$$\begin{aligned} N(\mu)^{1/2} &= (\tfrac{1}{2}\mu/p)^{1/2} \begin{pmatrix} 1 & -p \\ -p & 1 \end{pmatrix} \\ N(\mu)^{-1/2} &= (2p/\mu)^{1/2}(1 - p^2)^{-1} \begin{pmatrix} 1 & p \\ p & 1 \end{pmatrix} \end{aligned} \tag{6.7}$$

and where

$$p = [1 - (1 - \mu^2)^{1/2}]/\mu = \mu/[1 + (1 - \mu^2)^{1/2}] \tag{6.8}$$

From (4.4) and (4.5) we have

$$M(\omega, \mu)^m = S_{2m} I + S_{2m-1} \begin{pmatrix} -\omega\mu & 1 \\ 1 - \omega & 0 \end{pmatrix} \tag{6.9}$$

Therefore by (6.7) we have

$$\begin{aligned} V_m(\mu) &= N(\mu)^{1/2} M(\omega, \mu)^m N(\mu)^{-1/2} \\ &= S_{2m} I + \frac{S_{2m-1}}{1 - p^2} \begin{pmatrix} 1 & -p \\ -p & 1 \end{pmatrix} \begin{pmatrix} -\omega\mu & 1 \\ 1 - \omega & 0 \end{pmatrix} \begin{pmatrix} 1 & p \\ p & 1 \end{pmatrix} \\ &= S_{2m} I + S_{2m-1} \begin{pmatrix} \dfrac{-\omega\mu}{2} & 1 - \dfrac{\omega p \mu}{2} \\ 1 - \dfrac{\omega\mu}{2p} & -\dfrac{\omega\mu}{2} \end{pmatrix} \\ &= \begin{pmatrix} S_{2m} - \dfrac{\omega\mu}{2} S_{2m-1} & \left(1 - \dfrac{\omega p \mu}{2}\right) S_{2m-1} \\ \left(1 - \dfrac{\omega\mu}{2p}\right) S_{2m-1} & S_{2m} - \dfrac{\omega\mu}{2} S_{2m-1} \end{pmatrix} \end{aligned} \tag{6.10}$$

since by (6.8) we have

$$\begin{aligned} p\mu - 1 &= -(1-\mu^2)^{1/2} = -(\tfrac{1}{2}\mu/p)(1-p^2) \\ p - \mu &= -(\mu/2)(1-p^2) \end{aligned} \tag{6.11}$$

Evidently, $\varDelta = \det(V_m(\mu)V_m(\mu)^{\mathrm{T}}) = \det V_m(\mu)^2 = \det(M(\omega, \mu)^m)^2 = (1-\omega)^{4m}$, by (4.3). Moreover,

$$\begin{aligned} T_m(\omega, \mu) &= \operatorname{trace}(V_m(\mu)V_m(\mu)^{\mathrm{T}}) \\ &= 2\left(S_{2m} - \frac{\omega\mu}{2} S_{2m-1}\right)^2 + \left[\left(1 - \frac{\omega p\mu}{2}\right)^2 \right. \\ &\quad \left. + \left(1 - \frac{\omega\mu}{2p}\right)^2\right] S_{2m-1}^2 \\ &= 2S_{2m}^2 - 2\omega\mu S_{2m}S_{2m-1} + \frac{\omega^2\mu^2}{2} S_{2m-1}^2 \\ &\quad + S_{2m-1}^2\left[2 - \omega\mu\left(p + \frac{1}{p}\right) + \frac{\omega^2\mu^2}{4}\left(p^2 + \frac{1}{p^2}\right)\right] \end{aligned} \tag{6.12}$$

But from (6.7) and (6.11) we have

$$p + (1/p) = 2/\mu, \qquad p^2 + (1/p^2) = 4\mu^{-2} - 2 \tag{6.13}$$

and by (6.12) we have

$$\begin{aligned} T_m(\omega, \mu) &= 2S_{2m}^2 - 2\omega\mu S_{2m}S_{2m-1} + (2 - 2\omega + \omega^2)S_{2m-1}^2 \\ &= 2S_{2m}(S_{2m} - \omega\mu S_{2m-1}) + (2 - 2\omega + \omega^2)S_{2m-1}^2 \\ &= 2S_{2m}S_{2m-2}(1-\omega) + S_{2m-1}^2(2 - 2\omega + \omega^2) \end{aligned} \tag{6.14}$$

Moreover, since

$$\det M(\omega, \mu)^m = (1-\omega)^{2m} = (1-\omega)(S_{2m}S_{2m-2} - S_{2m-1}^2) \tag{6.15}$$

by (4.3) and (4.5), we have

$$\begin{aligned} T_m(\omega, \mu) &= S_{2m-1}^2(4 - 4\omega + \omega^2) + 2(1-\omega)^{2m} \\ &= (2-\omega)^2 S_{2m-1}^2 + 2(1-\omega)^{2m} \end{aligned} \tag{6.16}$$

Since the characteristic equation of $V_m(\mu)V_m(\mu)^{\mathrm{T}}$ is

$$\varGamma^2 - T_m(\omega, \mu)\varGamma + \varDelta = 0 \tag{6.17}$$

and since $T_m(\omega, \mu) \geq 4\Delta$ (since the eigenvalues of $V_m(\mu)V_m(\mu)^T$ are real), it follows that the largest root of (6.15) is an increasing function of $T_m(\omega, \mu)$. Since $T(\omega_b, \mu)$ is maximized for $\mu = \bar{\mu}$ by (4.10), it follows that $\overline{\| \mathscr{L}_{\omega_b}^m \|_{A^{1/2}}} = \| \mathscr{L}_{\omega_b}^m \|_{A^{1/2}}$. Moreover, $\| \mathscr{L}_{\omega_b} \|_{A^{1/2}}^2$ is the largest root of

$$\Gamma^2 - [(2 - \omega_b)^2 4m^2 r^{2m-1} + 2r^{2m}]\Gamma + r^{4m} = 0 \tag{6.18}$$

or the largest root of

$$\Gamma - r^{2m} = 2(1 - r)mr^{m-1/2}\Gamma^{1/2} \tag{6.19}$$

which is

$$\Gamma^{1/2} = r^m \{m(r^{-1/2} - r^{1/2}) + [m^2(r^{-1/2} - r^{1/2})^2 + 1]^{1/2}\} \tag{6.20}$$

hence (6.2) follows and the proof of Theorem 6.1 is complete.

We remark that the expression for $\| \mathscr{L}_{\omega_b}^m \|_{A^{1/2}}$ is the same as that for $\| \mathscr{L}_{\omega_b}^m \|_{D^{1/2}}$ given by (4.1) except that $r^{-1/2} + r^{1/2}$ is replaced by $r^{-1/2} - r^{1/2}$. Therefore,

$$\| \mathscr{L}_{\omega_b}^m \|_{A^{1/2}} \leq \| \mathscr{L}_{\omega_b}^m \|_{D^{1/2}} \tag{6.21}$$

We now prove

Theorem 6.2. If $0 < \varrho < 1$ and $0 < r < 1$, then there exists a unique solution $m = m(r)$ of the equation

$$H_1(m) = \varrho \tag{6.22}$$

where $H_1(m) = \| \mathscr{L}_{\omega_b}^m \|_{A^{1/2}}$ is given by (6.2) such that $m \geq \frac{1}{2}$. Moreover,

$$\lim_{r \to 1-} [m(r) \log(1/r)] = z \tag{6.23}$$

where z is the unique positive solution of

$$h_1(z) = e^{-z}(z + [1 + z^2]^{1/2}) = \varrho \tag{6.24}$$

Also,

$$\lim_{\varrho \to 0} \left[m(r) \log \frac{1}{r} - \log\left(\frac{2\gamma}{\varrho} \log \frac{2\gamma}{\varrho} \right) \right] = 0 \tag{6.25}$$

where

$$\gamma = (r^{-1/2} - r^{1/2})/\log(1/r) \tag{6.26}$$

Finally, we have

$$\lim_{\varrho\to 0}\left[z(\varrho) - \log\left(\frac{2}{\varrho}\log\frac{2}{\varrho}\right)\right] = 0 \tag{6.27}$$

where $z = z(\varrho)$ is the solution of (6.24).

Proof. Evidently, by (6.2) we have

$$\log H_1(m) = m\log r + \sinh^{-1}[m(r^{-1/2} - r^{1/2})] \tag{6.28}$$

and

$$\frac{H_1'(m)}{H_1(m)} = -\log\frac{1}{r} + \frac{r^{-1/2} - r^{1/2}}{[1 + m^2(r^{-1/2} - r^{1/2})^2]^{1/2}} \tag{6.29}$$

If $m \geq \frac{1}{2}$, then

$$\begin{aligned}\frac{H_1'(m)}{H_1(m)} &\leq -\log\frac{1}{r} + \frac{r^{-1/2} - r^{1/2}}{[1 + \frac{1}{4}(r^{-1/2} - r^{1/2})^2]^{1/2}}\\ &= -\log\frac{1}{r} + \frac{2(r^{-1/2} - r^{1/2})}{r^{-1/2} + r^{1/2}} = -2\theta + 2\tanh\theta < 0\end{aligned} \tag{6.30}$$

where $\theta = -\log r^{1/2}$. Therefore $H_1(m)$ is a decreasing function of m for $m \geq \frac{1}{2}$. Moreover,

$$H_1(\tfrac{1}{2}) = r^{1/2}\{\tfrac{1}{2}(r^{-1/2} - r^{1/2}) + [\tfrac{1}{4}(r^{-1/2} - r^{1/2})^2 + 1]^{1/2}\} = 1 \tag{6.31}$$

Therefore, since $H_1(m) \to 0$ as $m \to \infty$ it follows that there is a unique solution $m(r) \geq \frac{1}{2}$ such that (6.22) holds.

We note that since

$$h_1'(z)/h_1(z) = -1 + (1 + z^2)^{-1/2} < 0 \tag{6.32}$$

since $h_1(0) = 1$, and since $h(z) \to 0$ as $z \to \infty$ there is a unique positive solution of (6.24).

Let $y = m(r)\log(1/r)$. Then

$$H_1(m) = e^{-y}\left\{y\frac{r^{-1/2} - r^{1/2}}{\log(1/r)} + \left[y^2\left(\frac{r^{-1/2} - r^{1/2}}{\log(1/r)}\right)^2 + 1\right]^{1/2}\right\} = \varrho \tag{6.33}$$

and if (6.22) holds, then

$$\begin{aligned}y &= \log\frac{1}{\varrho} + \log\left\{y\frac{r^{-1/2} - r^{1/2}}{\log(1/r)} + \left[y^2\left(\frac{r^{-1/2} - r^{1/2}}{\log(1/r)}\right)^2 + 1\right]^{1/2}\right\}\\ &= \log\frac{1}{\varrho} + \sinh^{-1}\left(y\frac{r^{-1/2} - r^{1/2}}{\log(1/r)}\right) = \psi(y, r)\end{aligned} \tag{6.34}$$

Moreover,

$$\frac{\partial \psi}{\partial y}(y, r) = \frac{(r^{-1/2} - r^{1/2})/\log(1/r)}{\left[y^2\left(\frac{r^{-1/2} - r^{1/2}}{\log(1/r)}\right)^2 + 1\right]^{1/2}} \tag{6.35}$$

Evidently the equation

$$y = \psi(y, r) \tag{6.36}$$

and the condition $y \geq \frac{1}{2}\log(1/r)$ define the function $y = y(r)$. By (6.24) and the fact that

$$\lim_{r \to 1-} \frac{r^{-1/2} - r^{1/2}}{\log(1/r)} = 1 \tag{6.37}$$

we have

$$z = \psi(z, 1) \tag{6.38}$$

Therefore,

$$y(1) = z \tag{6.39}$$

The conclusion (6.23) will follow if we can show that $y(r)$ is continuous for $r = 1$. By the implicit function theorem,† to do this it is sufficient to show that $\psi(y, r)$ and $\partial\psi/\partial y$ are continuous in a neighborhood of the point $y = z$, $r = 1$, and that $\partial\psi/\partial y \neq 1$ at that point. The continuity of ψ and $\partial\psi/\partial y$ is obvious. Moreover,

$$\partial\psi(z, 1)/\partial y = (1 + z^2)^{-1/2} < 1 \tag{6.40}$$

since $z > 0$. For, if $z = 0$, then (6.24) would not be satisfied. Therefore, the function $y(r)$ is continuous for $r = 1$ and (6.23) follows.

The proof of (6.25) is identical with the proof of (4.26) given in Theorem 4.2 with γ defined by (6.26) instead of (4.27). We note that $\gamma > 1$ since

$$(r^{-1/2} - r^{1/2})/\log(1/r) = 2 \sinh \theta/2\theta > 1 \tag{6.41}$$

where $\theta = \log(r^{-1/2})$.

To complete the proof of Theorem 6.2, we need only prove (6.27). The details are left as an exercise.

† We use the following form of the implicit function theorem: Let $F(x, y)$ and $F_y(x, y)$ be continuous in a neighborhood of (x_0, y_0) and let $F_y(x_0, y_0) \neq 0$. For some sufficiently small $\delta > 0$ there exists a unique continuous function $\phi(x)$ defined in I, $x_0 - \delta \leq x \leq x_0 + \delta$, such that $y_0 = \phi(x_0)$ and such that $F(x, \phi(x)) = 0$ for $x \in I$.

From Theorems 6.1 and 6.2 we use the following approximate solution of (6.22)

$$m_A = \frac{\log\left(\dfrac{2\gamma}{\varrho}\log\dfrac{2\gamma}{\varrho}\right)}{\log(1/r)}, \qquad \gamma = \frac{r^{-1/2} - r^{1/2}}{\log(1/r)} \tag{6.42}$$

In Section 7.7 we shall show that this formula gives a good approximation to the exact solution of (6.22).

7.7. COMPARISON OF $\| \mathscr{L}_{\omega_b}^m \|_{D^{1/2}}$ AND $\| \mathscr{L}_{\omega_b}^m \|_{A^{1/2}}$

Figure 7.1 gives a graph of $F_1(m) = \| \mathscr{L}_{\omega_b}^m \|_{D^{1/2}}$ and $H_1(m) = \| \mathscr{L}_{\omega_b}^m \|_{A^{1/2}}$ versus m for the case $r = 0.729$. The function $F_1(m)$ increases from the value unity at $m = 0$ to about 4.8 at $m = 3$ before decreasing. The function $H_1(m)$ is a monotone decreasing function of m. However, $H_1(m)$ is considerably larger than the spectral radius $S(\mathscr{L}_{\omega_b}^m) = r^m$.

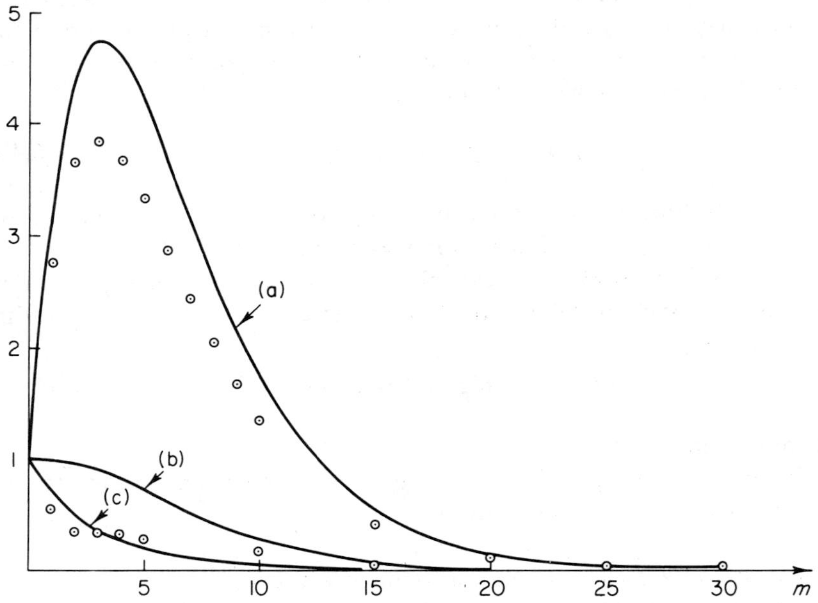

Figure 7.1. (a) $\| \mathscr{L}_{\omega_b}^m \|_{D^{1/2}}$; (b) $\| \mathscr{L}_{\omega_b}^m \|_{A^{1/2}}$; and (c) $S(\mathscr{L}_{\omega_b}^m)$ for $\omega_b = 1.729$, $r = 0.729$. (⊙: observed values.)

The values of $F_1(m)$ and $H_1(m)$ were compared with the ratios

$$\| \varepsilon^{(m)} \|_{D^{1/2}} / \| \varepsilon^{(0)} \|_{D^{1/2}} \qquad \text{and} \qquad \| \varepsilon^{(m)} \|_{A^{1/2}} / \| \varepsilon^{(0)} \|_{A^{1/2}}$$

as determined by solving by the SOR method with $\omega = \omega_b = 1.729$ for the discrete analogue of the Dirichlet problem in the unit square with $h = \frac{1}{20}$. Zero boundary values were assumed; hence the exact solution vanishes identically. The red–black ordering was used and the initial values of 0 at the "red points" (where $(x + y)/h$ is even) and of $\sin \pi x \sin \pi y$ at the "black point" (where $(x + y)/h$ is odd) were assumed. Closer agreement might be expected with different starting values.

SUPPLEMENTARY DISCUSSION

Section 7.1. Young [1954] gave a formula for the principal vector of grade 2 associated with $\lambda = \omega_b - 1$ for $\mathscr{L}_{\omega_b}$. This formula is valid for all consistently ordered matrices. If $\omega = 1$, there are in general many principal vectors associated with $\lambda = 0$, and the grades can be quite large. (See the discussion of Miles *et al.*, 1964.) However, if A is positive definite and has the form (1) then using the basis (1.22) one can show that the Jordan canonical form of $\mathscr{L}$ is diagonal. This result was proved by Young [1950].

Section 7.2. Theorem 2.1 was proved by Young and Kincaid [1969] and is a generalization of a result stated by Sheldon [1959] where the matrix $\mathscr{G}_m$ is related to the SOR method. Sheldon assumed B to be similar to a diagonal matrix and did not give a proof of (c). Wachspress [1966, Section 4.9] gave a proof of (c) for the SOR method under the assumption that B is similar to a diagonal matrix. Special forms of Theorem 2.1 were stated without proof by Golub and Varga [1961] and by Young, *et al.* [1965]. The relation (2.41) between $\mathscr{L}_\omega$ and B was also given by Young and Kincaid [1969].

Section 7.4. Young [1950, 1954] showed that $\| \mathscr{L}_{\omega_b}^m \|_{D^{1/2}}$ is usually considerably larger than $S(\mathscr{L}_{\omega_b}^m)$. A more precise result was obtained by Sheldon [1959]. The relation (4.10) was proved by Golub [1959].

Section 7.5. Theorem 5.1 was proved by Young [1969, 1970]. Lemma 5.3 was proved by Wachspress [1966, pp. 162–163]. The matrix $\mathscr{U}_\omega$ defined by (5.3) corresponds to the "backward" SOR method.

Section 7.6. The analysis of the $A^{1/2}$-norm of $\mathscr{L}_{\omega_b}^m$ was given by Young and Kincaid [1969].

EXERCISES

Section 7.1

1. Construct a basis of eigenvectors of B for the matrix

$$A = \begin{pmatrix} 2 & 0 & 0 & -1 & 0 \\ 0 & 2 & 0 & -1 & -1 \\ 0 & 0 & 2 & 0 & -1 \\ -1 & -1 & 0 & 2 & 0 \\ 0 & -1 & -1 & 0 & 2 \end{pmatrix}$$

Also, find the eigenvalues and construct a basis of eigenvectors for $\mathscr{L}_\omega$ in the following cases: $\omega = 1$, $\omega = \frac{4}{3}$.

2. Verify that the vector $\hat{z}$ given by (1.19) satisfies (1.20), and that $\hat{z}$ and w are linearly independent.

3. Show that the $w^{(i)}$ of (1.22) are linearly independent.

4. Show that if A is a positive definite matrix of the form (1), then the Jordan canonical form of $\mathscr{L}$ is diagonal.

5. Develop a basis of eigenvectors for B in the following way. Use the fact that μ is a k-fold eigenvalue of B if and only if $-\mu$ is; then there must be $2p$ nonzero eigenvectors of B for some integer p. Let

$$v^{(i)} \equiv \begin{pmatrix} x^{(i)} \\ y^{(i)} \end{pmatrix}, \qquad i = 1, 2, \ldots, p$$

be those eigenvectors associated with eigenvalues μ_i with positive real parts. Then show that the $x^{(i)}$ and the $y^{(i)}$ are linearly independent and that the vectors

$$v^{(i+p)} \equiv \begin{pmatrix} x^{(i)} \\ -y^{(i)} \end{pmatrix}, \qquad i = 1, 2, \ldots, p$$

are linearly independent and are associated with the eigenvalues $-\mu_i$. The vectors associated with $\mu = 0$ are $v^{(2p+1)}, \ldots, v^{(N)}$ and if $x^{(j)} \neq 0$ or $y^{(j)} \neq 0$, then

$$v^{(j)} = \begin{pmatrix} x^{(j)} \\ 0 \end{pmatrix} \qquad \text{or} \qquad v^{(j)} = \begin{pmatrix} 0 \\ y^{(j)} \end{pmatrix}$$

respectively, where $j = 2p + 1, \ldots, N$. Show that there are $r - p$ of the former and $s - p$ of the latter eigenvectors where r and s are the orders of D_1 and D_2 in (1), respectively.

6. Show that if A is a real matrix of the form (1) and if the order of D_1 or of D_2 is unity, then each eigenvalue of B is either real or purely imaginary.

Section 7.2

1. Show that (2.4) has the form (2.1) if all $\nu_k = 1$.

2. Show that if Q is nonsingular, then $T(B^2)$ is nonsingular.

3. Show by an example that if $\mu = 0$ is an eigenvalue of B, then both of the eigenvalues of $M_m(0)$ defined by (2.6) are not necessarily eigenvalues of $\mathcal{G}_m$, defined by (2.4).

4. Show that if Q is defined by (2.1) and if $T(FG)$ and $T(GF)$ are nonsingular, then Q is nonsingular and

$$Q^{-1} = \begin{pmatrix} T^{-1}(FG)\alpha_{2,2}(FG) & -T^{-1}(FG)\alpha_{1,2}(FG)F \\ -T^{-1}(GF)\alpha_{2,1}(GF)G & T^{-1}(GF)\alpha_{1,1}(GF) \end{pmatrix}$$

Apply this result to show that in the general case (2.4) has the form (2.1) where in (2.1) the $\alpha_{i,j}$ are rational functions of their arguments.

5. Suppose $p(x)$ and $g(x)$ are two polynomials. If B is an upper triangular matrix with diagonal elements $\mu_1, \mu_2, \ldots, \mu_N$ and if $g(\mu_i) \neq 0$, $i = 1, 2, \ldots, N$ show that

$$q(B)^{-1}p(B)$$

is an upper triangular matrix whose diagonal elements are

$$q(\mu_i)^{-1}p(\mu_i)$$

6. Construct a basis of principal vectors for $\mathcal{G}_m$ and thus give an alternate proof of (c) of Theorem 2.1.

7. Find $\mathcal{H}_m$ in (2.20).

Section 7.3

1. Let A be a positive definite matrix of the form (1). Find $\| \overline{\mathcal{L}} \|_{D^{1/2}}$ in terms of $\bar{\mu} = S(B)$. Also, find $\| \overline{\mathcal{L}_\omega} \|_{D^{1/2}}$ for $\omega = 1.0, 1.2$, and 1.4, for the case $\bar{\mu} = 0.85$. What is the largest value of ω for which $\| \overline{\mathcal{L}_\omega} \|_{D^{1/2}} < 1$? Find ω which minimizes $\| \overline{\mathcal{L}_\omega} \|_{D^{1/2}}$.

Section 7.4

1. Compute M^3 where $M = M(\omega, \mu)$ is given by (4.3). Verify that (4.5) holds.

2. Compute $S_5(\mu)$ for $\omega = 1.5$, $\mu = 0.6$ where $S_k(\mu)$ is defined by (4.4).

3. Compute $\overline{\| \mathscr{L}_{\omega_b}^m \|_{D^{1/2}}}$ and $\bar{S}(\mathscr{L}_{\omega_b}^m)$ if $m = 5$, $\bar{\mu} = 0.9$. Find m such that $\| \mathscr{L}_{\omega_b}^m \|_{D^{1/2}} = 0.001 = \varrho$ by solving (4.23) and also from (4.50). Find an approximate value of m such that $F_1(m) = \| \mathscr{L}_{\omega_b}^m \|_{D^{1/2}}$ is maximized and find the corresponding value of $F_1(m)$. Also, do the last part of $r = 0.729$.

Section 7.5

1. If A is a positive definite matrix of the form (1.1), find $\| \mathscr{L}_\omega \|_{A^{1/2}}$ for $\omega = 1.5$ given that $\bar{\mu} = 0.8$.

2. Consider the problem of solving

$$\begin{pmatrix} 4 & 0 & -1 & -1 \\ 0 & 4 & -1 & -1 \\ -1 & -1 & 4 & 0 \\ -1 & -1 & 0 & 4 \end{pmatrix} \begin{pmatrix} u_1 \\ u_2 \\ u_3 \\ u_4 \end{pmatrix} \quad \begin{pmatrix} 3 \\ 7 \\ 1 \\ -3 \end{pmatrix}$$

(a) Obtain the exact solution, $\bar{u}$.

(b) If $u^{(0)} = 0$, compute the $D^{1/2}$-norm and the $A^{1/2}$-norm of $\varepsilon^{(0)} = u^{(0)} - \bar{u}$.

(c) Perform one iteration with the GS method and compute $\varepsilon^{(1)}$ as well as the $D^{1/2}$-norm and the $A^{1/2}$-norm of $\varepsilon^{(1)}$. Compare the ratios of these norms with $\| \mathscr{L} \|_{D^{1/2}}$ and $\| \mathscr{L} \|_{A^{1/2}}$.

Section 7.6

1. Let

$$N = \begin{pmatrix} 1 & -\mu \\ -\mu & 1 \end{pmatrix}$$

where μ is real and $|\mu| < 1$. Show that

$$N^{1/2} = (\tfrac{1}{2}\mu/p)^{1/2} \begin{pmatrix} 1 & -p \\ -p & 1 \end{pmatrix}$$

where

$$p = [1 - (1 - \mu^2)^{1/2}]/\mu = \mu/[1 + (1 - \mu^2)^{1/2}]$$

Verify the result in the case $\mu = 0.8$.

2. Carry out the details of the proof of Theorem 6.2.

3. Compute $\| \mathscr{L}_{\omega_b}^m \|_{A^{1/2}}$ and $S(\mathscr{L}_{\omega_b}^m)$ if $\bar{\mu} = 0.9$, $m = 5$. Find m such that $\| \mathscr{L}_{\omega_b}^m \|_{A^{1/2}} \leq 0.001 = \varrho$ by solving (6.22) and also using (6.42). Also compute z from (6.24) and compare with $m(r) \cdot \log(1/r)$.

4. Consider the linear system

$$\begin{pmatrix} 1 & -0.8 \\ -0.8 & 1 \end{pmatrix} \begin{pmatrix} u_1 \\ u_2 \end{pmatrix} = \begin{pmatrix} 1.2 \\ -0.6 \end{pmatrix}$$

(a) Find the exact solution.

(b) Carry out three iterations using the SOR method with $\omega = 1.25$. Let the starting values be $u_1^{(0)} = u_2^{(0)} = 0$.

(c) Compute $\| u^{(n)} - u \|_{D^{1/2}}$ and $\| u^{(n)} - u \|_{A^{1/2}}$ for $n = 0, 1, 2, 3$ and compare

$$\| u^{(n)} - u \|_{D^{1/2}} / \| u^{(0)} - u \|_{D^{1/2}} \qquad \text{and} \qquad \| u^{(n)} - u \|_{A^{1/2}} / \| u^{(0)} - u \|_{A^{1/2}}$$

with the theoretical values of $\| \mathscr{L}_{\omega}^{n} \|_{D^{1/2}}$ and $\| \mathscr{L}_{\omega}^{n} \|_{A^{1/2}}$.

5. Consider the model problem for Laplace's equation using the usual five-point difference equation with $h = \frac{1}{20}$. If the red–black ordering and the "optimum" ω is used, compute the smallest values of n and n' required using the SOR method so that we are sure that

(a) $\| e^{(n)} \| \leq 10^{-6} \| e^{(0)} \|$

(b) $\| e^{(n')} \|_{A^{1/2}} \leq 10^{-6} \| e^{(0)} \|_{A^{1/2}}$

where $e^{(n)} = u^{(n)} - \bar{u}$ and $\bar{u}$ is the exact solution of the difference equation.

6. Compute the *exact* value of $z(\varrho)$ for $\varrho = 10^{-2}, 10^{-3}, 10^{-4}$ using (6.24). Also compute

$$z_0 = \log 2/\varrho, \qquad z_1 = \log(2/\varrho \log(2/\varrho))$$

Solve $2ze^{-z} = \varrho$ and show that the limit of the sequence

$$z_0 = \log(2/\varrho)$$
$$z_{n+1} = \log[(2/\varrho)z_n], \qquad n \geq 0$$

is the solution of $2ze^{-z} = \varrho$. Assume $\varrho < 2/e$.

7. Show directly that the function

$$r^m\{(r^{-1/2} - r^{1/2})m + [(r^{-1/2} - r^{1/2})^2 m^2 + 1]^{1/2}\}$$

is less than unity for all $m \geq 1$ and is a strictly decreasing function of m for $m \geq 1$. Here r is a fixed number in the interval $0 < r < 1$.

Chapter 8 / THE MODIFIED SOR METHOD: FIXED PARAMETERS

8.1. INTRODUCTION

In applying the SOR method to the system

$$Au = b \tag{1.1}$$

where A has the form

$$A = \begin{pmatrix} D_1 & H \\ K & D_2 \end{pmatrix} \tag{1.2}$$

where D_1 and D_2 are square nonsingular diagonal elements, it is natural to consider the possibility of using two different relaxation factors. One relaxation factor, say ω, could be used for the "red" equations, which correspond to D_1, and the other relaxation factor, say ω', could be used for the "black" equations, which correspond to D_2.

If we partition u and b in accordance with the partitioning of A, we can write the system (1.1) in the form

$$\begin{pmatrix} D_1 & H \\ K & D_2 \end{pmatrix} \begin{pmatrix} u_1 \\ u_2 \end{pmatrix} = \begin{pmatrix} b_1 \\ b_2 \end{pmatrix} \tag{1.3}$$

or

$$\begin{aligned} D_1 u_1 + H u_2 &= b_1 \\ K u_1 + D_2 u_2 &= b_2 \end{aligned} \tag{1.4}$$

Evidently (1.4) is equivalent to the system

$$\begin{aligned} u_1 &= Fu_2 + c_1 \\ u_2 &= Gu_1 + c_2 \end{aligned} \tag{1.5}$$

where

$$F = -D_1^{-1}H, \quad G = -D_2^{-1}K, \quad c_1 = D_1^{-1}b_1, \quad c_2 = D_2^{-1}b_2 \tag{1.6}$$

We can also write (1.1) in the form

$$u = Bu + c \tag{1.7}$$

where

$$B = \begin{pmatrix} O_1 & F \\ G & O_2 \end{pmatrix} \tag{1.8}$$

The *modified SOR method* (MSOR method) is the same as the SOR method with the red–black ordering except that we use the relaxation factor ω for the "red" equations and the relaxation factor ω' for the "black" equations. Thus we have

$$\begin{aligned} u_1^{(n+1)} &= \omega(Fu_2^{(n)} + c_1) + (1-\omega)u_1^{(n)} \\ u_2^{(n+1)} &= \omega'(Gu_1^{(n+1)} + c_2) + (1-\omega')u_2^{(n)} \end{aligned} \tag{1.9}$$

Evidently, we may write (1.9) in the form

$$u^{(n+1)} = \mathscr{L}_{\omega,\omega'}u^{(n)} + k_{\omega,\omega'} \tag{1.10}$$

where

$$\begin{aligned} \mathscr{L}_{\omega,\omega'} &= \begin{pmatrix} I_1 & 0 \\ -\omega'G & I_2 \end{pmatrix}^{-1} \begin{pmatrix} (1-\omega)I_1 & \omega F \\ 0 & (1-\omega')I_2 \end{pmatrix} \\ &= \begin{pmatrix} I_1 & 0 \\ \omega'G & I_2 \end{pmatrix} \begin{pmatrix} (1-\omega)I_1 & \omega F \\ 0 & (1-\omega')I_2 \end{pmatrix} \\ &= \begin{pmatrix} (1-\omega)I_1 & \omega F \\ \omega'(1-\omega)G & \omega\omega'GF + (1-\omega')I_2 \end{pmatrix} \end{aligned} \tag{1.11}$$

and

$$k_{\omega,\omega'} = \begin{pmatrix} \omega c_1 \\ \omega\omega'Gc_1 + \omega'c_2 \end{pmatrix} \tag{1.12}$$

We note that $\mathscr{L}_{\omega,0}$ and $\mathscr{L}_{0,\omega'}$ determined from (1.11) agree with (7-5.12) and, moreover, that

$$\mathscr{L}_{\omega,\omega'} = \mathscr{L}_{0,\omega'}\mathscr{L}_{\omega,0} \tag{1.13}$$

and by (7-5.13)

$$\mathscr{L}_{\omega,\omega} = \mathscr{L}_{0,\omega}\mathscr{L}_{\omega,0} = \mathscr{L}_{\omega} \tag{1.14}$$

Thus, for $\omega = \omega'$, the MSOR method reduces to the SOR method.

The MSOR method was first considered by DeVogelaere [1958] for the case where ω and ω' are fixed. Young *et al.* [1965] considered the case where ω and ω' vary with n, which will be discussed in Chapter 10. McDowell [1967] and Taylor [1969] considered the case where the relaxation factor varies from equation to equation in a more general way.

In Section 8.2 we shall develop a relation between the eigenvalues of $\mathscr{L}_{\omega,\omega'}$ and those of B as well as a relation between the matrices themselves. These relations will be used in Section 8.3 to find necessary and sufficient conditions for strong convergence, as defined in Section 6.1. The virtual $D^{1/2}$-norm and the virtual $A^{1/2}$-norm of $\mathscr{L}_{\omega,\omega'}$ will be studied in Sections 8.4 and 8.5, respectively.

Many of the results obtained in the study of the convergence properties of $\mathscr{L}_{\omega,\omega'}$ will be used later, both in Chapter 15, where we shall consider the symmetric SOR method, and in Chapter 10, where we shall consider the MSOR method with variable ω and ω'.

8.2. EIGENVALUES OF $\mathscr{L}_{\omega,\omega'}$

If we apply Theorem 7-2.1 to the matrix $\mathscr{L}_{\omega,\omega'}$ given by (1.11) we obtain

Theorem 2.1. Let A be a matrix with nonvanishing diagonal elements which has the form (1.2) and let $\mathscr{L}_{\omega,\omega'}$ be defined by (1.11). Then

(a) If μ is a nonzero eigenvalue of B and if λ satisfies

$$(\lambda + \omega - 1)(\lambda + \omega' - 1) = \omega\omega'\mu^2\lambda \tag{2.1}$$

then λ is an eigenvalue of $\mathscr{L}_{\omega,\omega'}$. If $\mu = 0$ is an eigenvalue of B, then $\lambda = 1 - \omega$ and/or $\lambda = 1 - \omega'$ is an eigenvalue of $\mathscr{L}_{\omega,\omega'}$.

(b) If λ is an eigenvalue of $\mathscr{L}_{\omega,\omega'}$, then there exists an eigenvalue μ of B such that (2.1) holds.

We remark that if $\omega = \omega'$, then (2.1) reduces to

$$(\lambda + \omega - 1)^2 = \omega^2\mu^2\lambda \tag{2.2}$$

which is (5-2.5). Thus in this case Theorem 2.1 and Theorem 5-3.4 are consistent.

One can obtain a relation between $\mathscr{L}_{\omega,\omega'}$ and B from Theorem 7-2.1. By (1.11) and (7-2.22) we have

$$\mathscr{J}_1 = \begin{pmatrix} (1-\omega)I & \omega B \\ \omega'(1-\omega)B & \omega\omega' B^2 + (1-\omega')I \end{pmatrix} \tag{2.3}$$

From (7-2.33) we obtain

$$((1-\omega)I - \mathscr{L}_{\omega,\omega'})((1-\omega')I - \mathscr{L}_{\omega,\omega'}) = \omega\omega' \mathscr{L}_{\omega,\omega'} B^2 \tag{2.4}$$

which reduces to (7-2.45) when $\omega = \omega'$.

Using the analysis of the eigenvectors of B given in Section 7.1, we can replace (a) by a more precise statement when B is similar to a diagonal matrix. Indeed, we have

Theorem 2.2. Under the hypotheses of Theorem 2.1 if B is similar to a diagonal matrix, then we can replace the second statement of (a) by the following.

(a′) If $\mu = 0$ is an eigenvalue of B, then $1 - \omega$ or $1 - \omega'$ is an eigenvalue of $\mathscr{L}_{\omega,\omega'}$ if FG or GF is singular, respectively. If $r > s$ or $r < s$, where r and s are the orders of FG and GF, respectively, then $\mu = 0$ is an eigenvalue of B and FG or GF is singular, respectively. If $r = s$ and if $\mu = 0$ is an eigenvalue of B, then both FG and GF are singular and both $1 - \omega$ and $1 - \omega'$ are eigenvalues of $\mathscr{L}_{\omega,\omega'}$.

Proof. If FG is singular, then $FGx = 0$ for some $x \neq 0$. Hence $B^2(x^T, 0)^T = 0$ and since B is similar to a diagonal matrix $B(x^T, 0)^T = 0$ and $Gx = 0$. (See Theorem 2-1.15.) But by (1.11) we have

$$\mathscr{L}_{\omega,\omega'}\begin{pmatrix} x \\ 0 \end{pmatrix} = (1-\omega)\begin{pmatrix} x \\ 0 \end{pmatrix}$$

so that $1 - \omega$ is an eigenvalue of $\mathscr{L}_{\omega,\omega'}$. Similarly, if GF is singular, $1 - \omega'$ is an eigenvalue of $\mathscr{L}_{\omega,\omega'}$. If $r > s$, then as shown in Section 7.1, FG is singular. Similarly GF is singular if $r < s$. The rest of the proof follows easily.

As a consequence of Theorem 2.1 we have

Theorem 2.3. Under the hypotheses of Theorem 2.1 we have

(a) λ is an eigenvalue of $\mathscr{L}_{\omega,\omega'}$ if and only if λ is an eigenvalue of $\mathscr{L}_{\omega',\omega}$ provided one of the following conditions holds:

(i) $\omega = 0$ or $\omega' = 0$

(ii) $\lambda = 0$

(iii) $\lambda \neq 1 - \omega$ and $\lambda \neq 1 - \omega'$

(b) If $\omega \neq 1$ and if $1 - \omega$ is an eigenvalue of $\mathscr{L}_{\omega,\omega'}$, then $1 - \omega'$ is an eigenvalue of $\mathscr{L}_{\omega',\omega}$. If $\omega' \neq 1$ and $1 - \omega'$ is an eigenvalue of $\mathscr{L}_{\omega,\omega'}$, then $1 - \omega$ is an eigenvalue of $\mathscr{L}_{\omega',\omega}$.

Proof. The proof of (a) for the case $\omega = 0$ or $\omega' = 0$ follows at once from (1.11). Thus, for $\omega = 0$ the eigenvalues of $\mathscr{L}_{\omega,\omega'}$ and those of $\mathscr{L}_{\omega',\omega}$ are $1 - \omega'$ and 1. If $\lambda = 0$ is an eigenvalue of $\mathscr{L}_{\omega,\omega'}$ then $\omega = 1$ or $\omega' = 1$ by (2.1). It is obvious that $\lambda = 0$ is an eigenvalue of $\mathscr{L}_{1,\omega'}$. Also, $\lambda = 0$ is an eigenvalue of $\mathscr{L}_{1,1}$. To show $\lambda = 0$ is an eigenvalue of $\mathscr{L}_{\omega',1}$ if $\omega' \neq 1$, we have by (1.11) for any $y \neq 0$

$$\mathscr{L}_{\omega',1}\begin{pmatrix} -\dfrac{\omega'}{1-\omega'} Fy \\ y \end{pmatrix} = 0$$

and hence $\lambda = 0$ is an eigenvalue of $\mathscr{L}_{\omega',1}$. Next, if $\lambda \neq 1 - \omega$, $\lambda \neq 1 - \omega'$, and λ is an eigenvalue of $\mathscr{L}_{\omega,\omega'}$, then (2.1) holds for some eigenvalue μ of B, which does not vanish. Hence λ is an eigenvalue of $\mathscr{L}_{\omega',\omega}$.

Suppose now that $\omega \neq 0$, $\omega' \neq 0$, $\omega \neq 1$, $\omega' \neq 1$ and $1 - \omega$ is an eigenvalue of $\mathscr{L}_{\omega,\omega'}$. Then for some x and y not both zero we have by (1.11)

$$\begin{aligned} \omega F y &= 0 \\ \omega'(1-\omega)Gx + \omega\omega' GFy + (\omega - \omega')y &= 0 \end{aligned} \tag{2.5}$$

And hence $Fy = 0$. Thus, we have

$$\begin{aligned} (\mathscr{L}_{\omega',\omega} - (1-\omega')I)&\begin{pmatrix} -\dfrac{\omega'(1-\omega)}{\omega(1-\omega')} x \\ y \end{pmatrix} \\ &= \begin{pmatrix} \omega' F y \\ -\omega'(1-\omega)Gx + (\omega' - \omega)y \end{pmatrix} = 0 \end{aligned} \tag{2.6}$$

by (2.5). Hence $1 - \omega'$ is an eigenvalue of $\mathscr{L}_{\omega',\omega}$.

Suppose now that $\omega \neq 0$, $\omega' \neq 0$, $\omega \neq 1$, $\omega' \neq 1$, and $1 - \omega'$ is an eigenvalue of $\mathscr{L}_{\omega,\omega'}$. For some x and y not both zero we have

$$(\mathscr{L}_{\omega,\omega'} - (1-\omega')I)\begin{pmatrix} x \\ y \end{pmatrix} = \begin{pmatrix} (\omega' - \omega)x + \omega F y \\ \omega'(1-\omega)Gx + \omega\omega' GFy \end{pmatrix} = 0 \tag{2.7}$$

and therefore

$$\begin{aligned}(\omega' - \omega)x + \omega F y &= 0 \\ (1 - \omega)Gx + \omega GFy &= 0\end{aligned} \tag{2.8}$$

But if $\omega' \neq 1$, then we must have

$$Gx = 0, \qquad GFy = 0 \tag{2.9}$$

Therefore

$$(\mathscr{L}_{\omega',\omega} - (1 - \omega)I)\begin{pmatrix} x \\ -(\omega/\omega')y \end{pmatrix} = 0 \tag{2.10}$$

and $1 - \omega$ is an eigenvalue of $\mathscr{L}_{\omega',\omega}$.

We remark that if $\omega \neq 0$, $\omega' \neq 0$, and $\omega \neq 1$ but $\omega' = 1$, then the result (b) is still true since $1 - \omega' = 0$ is an eigenvalue of $\mathscr{L}_{1,\omega}$ for any ω. This is also true for the case $\omega \neq 0$, $\omega' \neq 0$, $\omega' \neq 1$, $\omega = 1$. The proof for the case $\omega = 0$ or $\omega' = 0$ follows from the fact that $1 - \omega$ and $1 - \omega'$ are eigenvalues both of $\mathscr{L}_{\omega,\omega'}$ and $\mathscr{L}_{\omega',\omega}$. This completes the proof of Theorem 2.3.

The following example shows that the eigenvalues of $\mathscr{L}_{\omega,\omega'}$ are not necessarily the same as those of $\mathscr{L}_{\omega',\omega}$. If

$$A = \begin{pmatrix} 2 & 0 & -1 \\ 0 & 2 & -1 \\ -1 & -1 & 2 \end{pmatrix}, \qquad B = \begin{pmatrix} 0 & 0 & \frac{1}{2} \\ 0 & 0 & \frac{1}{2} \\ \frac{1}{2} & \frac{1}{2} & 0 \end{pmatrix} \tag{2.11}$$

then the eigenvalues of B are $\mu_1 = 2^{-1/2}$, $\mu_2 = -2^{-1/2}$, $\mu_3 = 0$. By Theorem 2.1, two of the eigenvalues of $\mathscr{L}_{\omega,\omega'}$ and of $\mathscr{L}_{\omega',\omega}$ are λ_1 and λ_2 the roots of

$$(1 - \omega - \lambda)(1 - \omega' - \lambda) = \tfrac{1}{2}\omega\omega'\lambda \tag{2.12}$$

Since B is similar to a diagonal matrix and since FG is singular it follows by Theorem 2.2 that $1 - \omega$ is an eigenvalue of $\mathscr{L}_{\omega,\omega'}$ and that $1 - \omega'$ is an eigenvalue of $\mathscr{L}_{\omega',\omega}$. Thus we have

eigenvalues of $\mathscr{L}_{\omega,\omega'}$: $\quad \lambda_1, \ \lambda_2, \ 1 - \omega$

eigenvalues of $\mathscr{L}_{\omega',\omega}$: $\quad \lambda_1, \ \lambda_2, \ 1 - \omega'$

If $\omega = 1$ and $\omega' = 7/4$, then

$$\lambda_1 = \tfrac{1}{8}, \qquad \lambda_2 = 0, \qquad 1 - \omega = 0, \qquad 1 - \omega' = -\tfrac{3}{4}$$

and

$$S(\mathscr{L}_{\omega,\omega'}) = \max(\tfrac{1}{8}, 0, 0) = \tfrac{1}{8}$$
$$S(\mathscr{L}_{\omega',\omega}) = \max(\tfrac{1}{8}, 0, |-\tfrac{3}{4}|) = \tfrac{3}{4}$$

8.3. CONVERGENCE AND SPECTRAL RADIUS

We now study the virtual spectral radius of $\mathscr{L}_{\omega,\omega'}$ as defined in Section 6.1. Thus, if for fixed ω and ω' we let

$$\psi(\mu) = \varrho(\omega, \omega', \mu) \tag{3.1}$$

where $\varrho(\omega, \omega', \mu)$ is the root radius of (2.1), then we have

$$\bar{S}(\mathscr{L}_{\omega,\omega'}) = \max_{\mu \in \bar{S}_B} \psi(\mu) \tag{3.2}$$

where $\bar{S}_B$ is the smallest convex set containing the set S_B of all eigenvalues of B.

As an immediate consequence of (3.2) we have

Theorem 3.1. Under the hypotheses of Theorem 2.1 we have

$$\bar{S}(\mathscr{L}_{\omega,\omega'}) = \bar{S}(\mathscr{L}_{\omega',\omega}) \tag{3.3}$$

We recall that $\mathscr{L}_{\omega,\omega'}$ is strongly convergent if and only if $\bar{S}(\mathscr{L}_{\omega,\omega'}) < 1$. Thus Theorem 3.1 shows that $\mathscr{L}_{\omega,\omega'}$ is strongly convergent if and only if $\mathscr{L}_{\omega',\omega}$ is strongly convergent.

We now prove

Theorem 3.2. Let A be a matrix with nonvanishing diagonal elements of the form (1.2). If A is positive definite or, more generally, if the eigenvalues of B are real and if $\bar{\mu} = S(B) < 1$, then the MSOR method is strongly convergent if

$$0 < \omega < 2, \qquad 0 < \omega' < 2. \tag{3.4}$$

Conversely, if the eigenvalues of B are real and if the MSOR method is strongly convergent, then

$$\bar{\mu} = S(B) < 1 \tag{3.5}$$

and (3.4) holds. Also, if A is symmetric and has positive diagonal elements and if the MSOR method is strongly convergent, then A is positive definite and (3.4) holds.

Proof. Evidently, by (2.1), $\varrho(\omega, \omega', \mu)$ is the root radius of the quadratic equation

$$\lambda^2 - b\lambda + c = 0 \tag{3.6}$$

where

$$c = (\omega - 1)(\omega' - 1) \tag{3.7}$$

and

$$b = b(\mu) = \omega\omega'\mu^2 - \omega - \omega' + 2 = 1 + c - \omega\omega'(1 - \mu^2) \tag{3.8}$$

By (3.4) we have

$$|c| < 1 \tag{3.9}$$

By Lemma 6-2.5, since A is positive definite and has the form (1.2), then the eigenvalues of B are real and $\bar{\mu} = S(B) < 1$. By (3.7) and (3.8) we have $1 + c - b(\mu) = \omega\omega'(1 - \mu^2) > 0$ and $1 + c + b(\mu) = 2(1+c) - \omega\omega'(1 - \mu^2) = (2 - \omega)(2 - \omega') + \omega\omega'\mu^2 > 0$. Therefore,

$$|b(\mu)| < 1 + c \tag{3.10}$$

By Lemma 6-2.1 it follows from (3.9) and (3.10) that $\varrho(\omega, \omega', \mu) < 1$ for all μ and hence the MSOR method is strongly convergent.

If the MSOR method is strongly convergent, then $\varrho(\omega, \omega', \mu) < 1$ for all μ. By Lemma 6-2.1 we must have (3.9), and for all μ we must also have (3.10). From (3.10), (3.7), and (3.8) we have

$$\begin{aligned} \omega\omega'(1 - \mu^2) &> 0 \\ (2 - \omega)(2 - \omega') + \omega\omega'\mu^2 &> 0 \end{aligned} \tag{3.11}$$

If the eigenvalues of B are real, then $\mu = 0$ is an element of $\bar{S}_B$ (since μ is an eigenvalue of B if and only if $-\mu$ is). Hence we must have $\omega\omega' > 0$. This implies that $\mu^2 < 1$ for all μ and $\bar{\mu} < 1$. Moreover, we must also have $(2 - \omega)(2 - \omega') > 0$. The conditions $\omega\omega' > 0$, (3.9), and $(2 - \omega)(2 - \omega') > 0$ imply (3.4).

If A is a symmetric matrix of the form (1.2) with positive diagonal elements, and if $\bar{\mu} < 1$, then by Theorem 5-3.5, A is positive definite. This completes the proof of Theorem 3.2.

We now prove

Theorem 3.3. If A is a matrix with nonvanishing diagonal elements of the form (1.2) such that the eigenvalues of B are real and $\bar{\mu} = S(B) < 1$, then

$$\bar{S}(\mathscr{L}_{\omega_b,\omega_b}) = S(\mathscr{L}_{\omega_b,\omega_b}) = S(\mathscr{L}_{\omega_b}) = \omega_b - 1 \tag{3.12}$$

and unless $\omega = \omega' = \omega_b$ we have

$$\bar{S}(\mathscr{L}_{\omega,\omega'}) > \bar{S}(\mathscr{L}_{\omega_b,\omega_b}) \tag{3.13}$$

Proof. We have already observed that $\mathscr{L}_{\omega,\omega} = \mathscr{L}_\omega$ and that if A is consistently ordered and the eigenvalues of B are real, then $\bar{S}(\mathscr{L}_\omega) = S(\mathscr{L}_\omega)$. Hence by (6-2.8) we have (3.12).

To prove (3.13) we can, by Theorem 3.2, restrict our consideration to values of ω and ω' satisfying (3.4). If we define b and c by (3.8) and (3.7), respectively, we note that $\bar{S}(\mathscr{L}_{\omega,\omega'})$ is the root radius of

$$\lambda^2 - \bar{b}\lambda + c = 0 \tag{3.14}$$

where

$$\begin{aligned}\bar{b} &= \max_{-\bar{\mu}\le\mu\le\bar{\mu}} |\, b(\mu)\,| = \max_{-\bar{\mu}\le\mu\le\bar{\mu}} |\, 1 + c - \omega\omega'(1-\mu^2)\,| \\ &= \max\{|\, 1 + c - \omega\omega'(1-\bar{\mu}^2)\,|,\ \ |\, 1 + c - \omega\omega'\,|\}\end{aligned} \tag{3.15}$$

The latter equality holds since $1 + c - \omega\omega'(1-\mu^2)$ is a linear function of μ^2 for fixed ω and ω' and thus assumes extreme values for $\mu^2 = 0$ and $\mu^2 = \bar{\mu}^2$.

If $\omega\omega' \le 2(1+c)/(2-\bar{\mu}^2)$, then since $|\,c\,| < 1$ by (3.4) we have $1 + c - \omega\omega'(1-\bar{\mu}^2) \ge (1+c)\bar{\mu}^2/(2-\bar{\mu}^2)$. On the other hand, if $\omega\omega' \ge 2(1+c)/(2-\bar{\mu}^2)$, then $1 + c - \omega\omega' \le -\bar{\mu}^2(1+c)/(2-\bar{\mu}^2)$. Therefore, by (3.15) for all ω and ω' such that $(1-\omega)(1-\omega') = c$, we have

$$\bar{b} \ge \frac{\bar{\mu}^2}{2-\bar{\mu}^2}(1+c) = \frac{2r}{1+r^2}(1+c) \tag{3.16}$$

Here

$$r = \omega_b - 1 \tag{3.17}$$

and ω_b is given by (6-1). We have used the fact that, by (6-2.11), $\omega_b{}^2\bar{\mu}^2 = 4(\omega_b - 1)$ and hence

$$\bar{\mu}^2 = 4r/(1+r)^2 \tag{3.18}$$

By Lemma 6-2.9 and (3.16) if $(1-\omega)(1-\omega')=c$, then the root radius of (3.14) is at least as large as the root radius $\varrho(c)$ of

$$P(\lambda, c)=\lambda^2-\frac{2r}{1+r^2}(1+c)\lambda+c=0 \tag{3.19}$$

If we let

$$c_0=r^2 \tag{3.20}$$

then we have

$$\varrho(c_0)=r \tag{3.21}$$

Moreover, if $c>c_0$, then $\varrho(c)\geq c^{1/2}>c_0^{1/2}=r$. If $c<c_0$, then

$$P(r, c)<0 \tag{3.22}$$

Hence, since $P(\lambda, c)>0$ for λ large enough, there is a root of (3.19) greater than r. Therefore we have

$$\varrho(c)>\varrho(c_0), \qquad c\neq c_0 \tag{3.23}$$

If $\bar{\mu}=0$, then $r=0$ and since $\bar{S}(\mathscr{L}_{\omega,\omega'})\geq\psi(0)=\max(|1-\omega|, |1-\omega'|)$ it follows that unless $\omega=\omega'=1$, then $\bar{S}(\mathscr{L}_{\omega,\omega'})>S(\mathscr{L}_{1,1})=0$. Assuming that $r>0$ we show that if $(1-\omega)(1-\omega')=c_0$, then unless $\omega\omega'=(1+r)^2$ we have

$$\bar{b}>2r \tag{3.24}$$

If $\omega\omega'<(1+r)^2$, then since $1-\bar{\mu}^2=((1-r)/(1+r))^2$, by (3.18), we have

$$1+c_0-\omega\omega'(1-\bar{\mu}^2)>1+r^2-(1-r)^2=2r \tag{3.25}$$

On the other hand, if $\omega\omega'>(1+r)^2$, then

$$\omega\omega'-(1+c_0)>(1+r)^2-(1+r^2)=2r \tag{3.26}$$

so that (3.24) follows. If $\bar{b}>2r$, then by Lemma 6-2.9 the root radius of

$$Q(\lambda)=\lambda^2-\bar{b}\lambda+c_0=0 \tag{3.27}$$

is greater than r, since by (3.24), we have

$$Q(r)=r^2-\bar{b}r+r^2<0 \tag{3.28}$$

and hence there is a root of (3.27) greater than r. In order that the root radius of (3.27) shall equal r, we must have $\bar{b} = 2r$ and hence

$$\omega\omega' = (1 + r)^2 \tag{3.29}$$

If $(1 - \omega)(1 - \omega') = c_0$, then by (3.20) and (3.29) we have

$$\omega + \omega' = 2(1 + r) \tag{3.30}$$

and

$$(\omega + \omega')^2 - 4\omega\omega' = (\omega - \omega')^2 = 0 \tag{3.31}$$

Thus we have

$$\omega = \omega' = 1 + r = \omega_b \tag{3.32}$$

This completes the proof of Theorem 3.3.

Later we shall need to minimize $\bar{S}(\mathscr{L}_{\omega,\omega'})$ subject to the condition that $0 \leq \omega \leq 1$, $0 \leq \omega' \leq 1$. We prove

Theorem 3.4. Under the hypotheses of Theorem 3.3, if $0 < \omega \leq 1$, $0 < \omega' \leq 1$, we have

$$\bar{S}(\mathscr{L}_{\omega,\omega'}) = S(\mathscr{L}_{\omega,\omega'}) \leq 1 - \tfrac{1}{2}\omega\omega'(1 - \bar{\mu}^2) \tag{3.33}$$

and unless $\omega = \omega' = 1$,

$$\bar{S}(\mathscr{L}_{\omega,\omega'}) = S(\mathscr{L}_{\omega,\omega'}) > S(\mathscr{L}_{1,1}) = \bar{S}(\mathscr{L}_{1,1}) = \bar{\mu}^2 \tag{3.34}$$

Proof. Since $0 < \omega \leq 1$, $0 < \omega' \leq 1$ it follows that for fixed ω and ω', the function $b = b(\mu)$ defined by (3.8) is maximized when $\mu = \bar{\mu}$. Thus we have

$$\max_{-\bar{\mu} \leq \mu \leq \bar{\mu}} b(\mu) = b(\bar{\mu}) = 1 + c - \omega\omega'(1 - \bar{\mu}^2) \tag{3.35}$$

where c is given by (3.7). Thus, by Lemma 6-2.9, $\bar{S}(\mathscr{L}_{\omega,\omega'})$ is the root radius $\varrho(\omega, \omega', \bar{\mu})$ of

$$\lambda^2 - b(\bar{\mu})\lambda + c = 0 \tag{3.36}$$

Since $\bar{\mu}$ is an eigenvalue of B, we also have

$$\bar{S}(\mathscr{L}_{\omega,\omega'}) = S(\mathscr{L}_{\omega,\omega'}) = \varrho(\omega, \omega', \bar{\mu}) \tag{3.37}$$

We now seek a bound on $\varrho(\omega, \omega', \bar{\mu})$. Letting $\theta = 1 - \lambda$ we have from (3.35) and (3.36),

$$\theta^2 - (1 + \omega\omega'(1 - \bar{\mu}^2) - c)\theta + \omega\omega'(1 - \bar{\mu}^2) = 0 \tag{3.38}$$

But the discriminant of (3.38) is, by (3.7),

$$(\omega - \omega')^2 + \omega\omega'\bar{\mu}^2[4 - 2(\omega + \omega') + \omega\omega'\bar{\mu}^2] \tag{3.39}$$

which is nonnegative since $0 \leq \omega \leq 1$, $0 \leq \omega' \leq 1$. Since $0 \leq c < 1$, the smallest root of (3.38) is not less than $\omega\omega'(1 - \bar{\mu}^2)/2$. The result (3.33) now follows.

We now prove

Lemma 3.5. Let c be a given number in the interval $0 \leq c \leq 1$. If $(1 - \omega)(1 - \omega') = c$ and if $0 \leq \omega \leq 1$, $0 \leq \omega' \leq 1$, then

$$\omega\omega' \leq (1 - c^{1/2})^2 \tag{3.40}$$

and unless

$$\omega = \omega' = 1 - c^{1/2} \tag{3.41}$$

$$\omega\omega' < (1 - c^{1/2})^2 \tag{3.42}$$

Proof. If $c = 0$, then either $\omega = 1$ or $\omega' = 1$. In either case we maximize $\omega\omega'$ by choosing $\omega = \omega' = 1$. If $c \neq 0$, then $\omega \neq 1$ and we have

$$\omega' = (c + \omega - 1)/(\omega - 1) \tag{3.43}$$

Evidently

$$\omega\omega' - (1 - c^{1/2})^2 = [\omega - (1 - c^{1/2})]^2/(\omega - 1) \leq 0 \tag{3.44}$$

since $\omega < 1$, with the strict inequality holding unless $\omega = \omega' = 1 - c^{1/2}$. The lemma now follows.

From Lemma 3.5 it follows that $1 + c - \omega\omega'(1 - \bar{\mu}^2)$ is minimized for fixed positive $c < 1$ and for $(1 - \omega)(1 - \omega') = c$ if $\omega = \omega' = 1 - c^{1/2}$. Therefore, by Lemma 6-2.9, it follows that $\varrho(\omega, \omega', \bar{\mu}) \geq \varrho(1 - c^{1/2}, 1 - c^{1/2}, \bar{\mu})$. Moreover, $\varrho(1 - c^{1/2}, 1 - c^{1/2}, \bar{\mu})$ is the root radius of

$$\lambda^2 - (1 + c - (1 - c^{1/2})^2(1 - \bar{\mu}^2))\lambda + c = 0 \tag{3.45}$$

i.e.,

$$(\lambda - c^{1/2})^2 = \bar{\mu}^2(1 - c^{1/2})^2\lambda \tag{3.46}$$

But if we let $\omega_1 = 1 - c^{1/2}$, we have

$$(\lambda + \omega_1 - 1)^2 = \omega_1{}^2\bar{\mu}^2\lambda \tag{3.47}$$

By Theorem 6-2.3, the root radius of (3.47) is a strictly decreasing function of ω_1 for $0 < \omega_1 \leq 1$. Thus it is minimized when $\omega_1 = 1$, i.e., when $c = 0$ and when $\omega = \omega' = 1$. Since $\varrho(1, 1, \bar{\mu}^2) = \bar{\mu}^2$ the result (3.34) follows.

8.4. DETERMINATION OF $\| \overline{\mathscr{L}_{\omega,\omega'}} \|_{D^{1/2}}$

We now study the virtual $D^{1/2}$-norm of $\mathscr{L}_{\omega,\omega'}$ under the assumption that A is positive definite and has the form (1.2). We first show that we can, without loss of generality, assume that diag $A = I$. If we let $\hat{A} = D^{-1/2}AD^{-1/2}$, where $D =$ diag A, then by (1.2) we have

$$\hat{A} = \begin{pmatrix} I_1 & D_1^{-1/2}HD_2^{-1/2} \\ D_1^{-1/2}KD_2^{-1/2} & I_2 \end{pmatrix} \tag{4.1}$$

and $\hat{B} = B[\hat{A}] = I - (\text{diag } \hat{A})^{-1}\hat{A}$ is given by

$$B[\hat{A}] = \begin{pmatrix} O_1 & \tilde{F} \\ \tilde{G} & O_2 \end{pmatrix} = \tilde{B}[A] \tag{4.2}$$

where $\tilde{B}[A] = D^{1/2}B[A]D^{-1/2}$, $\tilde{F} = D_1^{1/2}FD_2^{-1/2}$, $\tilde{G} = D_2^{1/2}GD_1^{-1/2} = \tilde{F}^{\mathrm{T}}$, $F = -D_1^{-1}H$, $G = -D_2^{-1}K$. Moreover, by (1.11) we have

$$\begin{aligned} \mathscr{L}_{\omega,\omega'}[\hat{A}] &= \begin{pmatrix} I_1 & 0 \\ -\omega'\tilde{G} & I_2 \end{pmatrix}^{-1} \begin{pmatrix} (1-\omega)I_1 & \omega\tilde{F} \\ 0 & (1-\omega')I_2 \end{pmatrix} \\ &= D^{1/2}\mathscr{L}_{\omega,\omega'}[A]D^{-1/2} = \tilde{\mathscr{L}}_{\omega,\omega'}[A] \end{aligned} \tag{4.3}$$

Moreover, we also have

$$\| \mathscr{L}_{\omega,\omega'}[A] \|_{D^{1/2}} = \| \tilde{\mathscr{L}}_{\omega,\omega'}[A] \| = \| \mathscr{L}_{\omega,\omega'}[\hat{A}] \| = \| \mathscr{L}_{\omega,\omega'}[\hat{A}] \|_{D[\hat{A}]^{1/2}} \tag{4.4}$$

Thus the problem of finding $\| \mathscr{L}_{\omega,\omega'}[A] \|_{D^{1/2}}$ is reduced to that of finding $\| \mathscr{L}_{\omega,\omega'}[\hat{A}] \|_{D[\hat{A}]^{1/2}} = \| \mathscr{L}_{\omega,\omega'}[\hat{A}] \|$ where diag $\hat{A} = I$. This is also true for the virtual norm of $\mathscr{L}_{\omega,\omega'}[A]$. Hence we may, without loss of generality, assume that diag $A = I$.

Let us now consider the problem of determining $\| \overline{\mathscr{L}_{\omega,\omega'}} \|$ in the case where

$$A = \begin{pmatrix} I_1 & -F \\ -G & I_2 \end{pmatrix}, \qquad B = \begin{pmatrix} O_1 & F \\ G & O_2 \end{pmatrix} \tag{4.5}$$

and where

$$F = G^{\mathrm{T}} \tag{4.6}$$

By (1.11) and by Theorem 7-2.1, the eigenvalues of $\mathscr{L}_{\omega,\omega'}\mathscr{L}^{\mathrm{T}}_{\omega,\omega'}$ are the eigenvalues of

$$M = \begin{pmatrix} 1 & 0 \\ \omega'\mu & 1 \end{pmatrix}\begin{pmatrix} 1-\omega & \omega\mu \\ 0 & 1-\omega' \end{pmatrix}\begin{pmatrix} 1-\omega & 0 \\ \omega\mu & 1-\omega' \end{pmatrix}\begin{pmatrix} 1 & \omega'\mu \\ 0 & 1 \end{pmatrix}$$
$$= \begin{pmatrix} 1-\omega & \omega\mu \\ \omega'(1-\omega)\mu & \omega\omega'\mu^2+1-\omega' \end{pmatrix}\begin{pmatrix} 1-\omega & \omega\mu \\ \omega'(1-\omega)\mu & \omega\omega'\mu^2+1-\omega' \end{pmatrix}^{\mathrm{T}} \tag{4.7}$$

The characteristic equation of M is

$$\lambda^2 - T\lambda + C = 0 \tag{4.8}$$

where

$$\begin{aligned} T &= T(\mu^2) = (1-\omega)^2+\omega^2\mu^2+(\omega'(1-\omega)\mu)^2+(\omega\omega'\mu^2+1-\omega')^2 \\ C &= (\omega-1)^2(\omega'-1)^2 \end{aligned} \tag{4.9}$$

We now prove

Theorem 4.1. If A is a positive definite matrix of the form (1.2) and if $\bar{\mu} = S(B)$, where $B = I - (\operatorname{diag} A)^{-1}A$ satisfies

$$\bar{\mu} \geq (\tfrac{1}{3})^{1/2} \doteq 0.577 \tag{4.10}$$

then

$$\| \overline{\mathscr{L}_{\omega_0,\omega_0'}} \|_{D^{1/2}} = (1+\bar{\mu}^2)/(3-\bar{\mu}^2) \tag{4.11}$$

where

$$\omega_0 = 4/(5+\bar{\mu}^2), \qquad \omega_0' = 4/(3-\bar{\mu}^2) \tag{4.12}$$

and unless $\omega = \omega_0$, $\omega' = \omega_0'$

$$\| \overline{\mathscr{L}_{\omega,\omega'}} \|_{D^{1/2}} > \| \overline{\mathscr{L}_{\omega_0,\omega_0'}} \|_{D^{1/2}} \tag{4.13}$$

Proof. We first note that $T(\mu^2) \geq 0$ for all μ. Moreover,

$$d^2T(\mu^2)/d(\mu^2)^2 = 2\omega^2(\omega')^2 > 0 \tag{4.14}$$

Hence, if

$$dT(\mu^2)/d(\mu^2) = 0 \tag{4.15}$$

for some μ_c^2 such that $0 \le \mu_c^2 \le \bar{\mu}^2$, then $T(\mu_c^2)$ must be a *minimum* value. Hence we have

$$\max_{-\bar{\mu} \le \mu \le \bar{\mu}} T(\mu^2) = \max(T(0), T(\bar{\mu}^2)) \tag{4.16}$$

Evidently $T(0) = T(\bar{\mu}^2)$ if

$$\omega^2[(\omega')^2\bar{\mu}^2 + (\omega')^2 + 1] + \omega[2\omega' - 4(\omega')^2] + (\omega')^2 = 0 \tag{4.17}$$

This equation has two, one, or no real roots depending on whether the discriminant

$$\Delta = 4(\omega')^3(3 - \bar{\mu}^2)[\omega' - 4/(3 - \bar{\mu}^2)] \tag{4.18}$$

is positive, zero, or negative. Thus, as indicated in Figure 4.1, if $\omega' > \omega_0'$, then there are two real roots of (4.17) and if $\omega' = \omega_0'$, then there is only one, namely ω_0. We let Region I and Region II be such that $T(0) \ge T(\bar{\mu}^2)$ and $T(0) \le T(\bar{\mu}^2)$, respectively.

We remark that if $\omega = \omega'$, then (ω, ω') lies in Region II. This is consistent with our analysis of Section 7.3.

In Region I, the characteristic equation (4.8) becomes

$$\lambda^2 - [(1 - \omega)^2 + (1 - \omega')^2]\lambda + (\omega - 1)^2(\omega' - 1)^2 = 0 \tag{4.19}$$

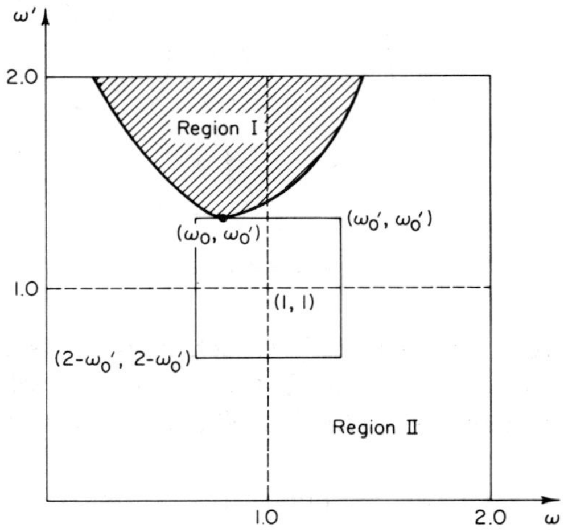

Figure 4.1.

whose roots are

$$\lambda = (\omega - 1)^2, \qquad (\omega' - 1)^2 \tag{4.20}$$

Hence we have

$$\| \overline{\mathscr{L}_{\omega,\omega'}} \|_{D^{1/2}}^2 = \max\{(\omega - 1)^2, (\omega' - 1)^2\} \tag{4.21}$$

Since the line $\omega' = \omega_0'$ is tangent to the boundary of Region I and since $\omega' > \omega$ in Region I it follows that the optimum values of ω and ω' in Region I are ω_0 and ω_0', respectively.

Let us now consider contours in the (ω, ω')-plane such that $\bar{\lambda}(\bar{\mu}^2) = \lambda$ where λ is a nonnegative number and where $\bar{\lambda}(\bar{\mu}^2)$ is the root radius of (4.8) corresponding to $\mu = \bar{\mu}$. If $\bar{\mu} = 0$, these contours are concentric squares with center at (1,1). Based on the results of numerical computations we state without proof the following. For $\bar{\mu} > 0$ there exists $\alpha(\bar{\mu}) > 0$ such that for each $\lambda \geq \alpha(\bar{\mu})$ there is a closed contour with $\lambda(\bar{\mu}^2) = \lambda$. Moreover, the contours are nested in the sense that if $\lambda_1 > \lambda_2$, then the contour corresponding to λ_1 surrounds the curve corresponding to λ_2. Moreover, the contour corresponding to $\alpha(\bar{\mu})$ is a single point.

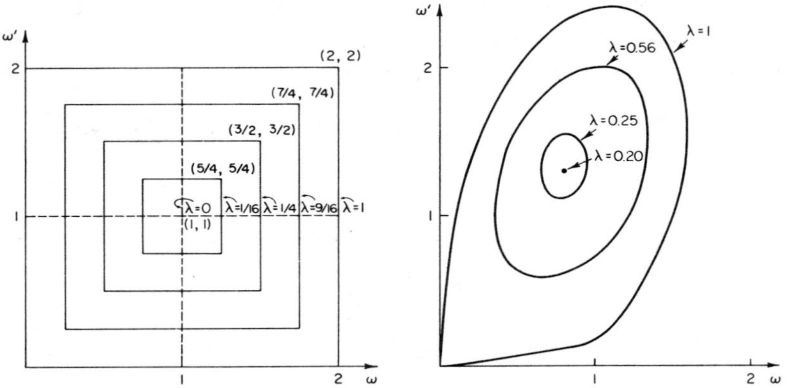

Figure 4.2.

Thus in Figure 4.2 we have shown schematically some contours for the cases $\bar{\mu} = 0$ and $\bar{\mu} = \frac{1}{2}$. In the first case $\alpha(\bar{\mu}) = 0$, which corresponds to (1,1). In the second case $\alpha(\bar{\mu}) = 0.447$, which corresponds to (0.800, 1.333).

Let us now consider the contour corresponding to

$$\lambda = [(1 + \bar{\mu}^2)/(3 - \bar{\mu}^2)]^2 \tag{4.22}$$

We determine the intersection of this contour and the line $\omega = \omega_0$. From (4.8) and (4.9) we have

$$(\omega')^2\{(\omega-1)^2 - \lambda[(1-\omega)^2\bar{\mu}^2 + (\omega\bar{\mu}^2 - 1)^2]\} + \omega'\{-2(\omega-1)^2 - \lambda[2(\omega\bar{\mu}^2-1)]\} + \{(\omega-1)^2 + \lambda^2 - \lambda[(1-\omega)^2 + \omega^2\bar{\mu}^2 + 1]\} = 0 \tag{4.23}$$

If $\lambda = (1+\bar{\mu}^2)^2/(3-\bar{\mu}^2)^2$ and $\omega_0 = 4(5+\bar{\mu}^2)^{-1}$, then one root is $\omega_0' = 4(3-\bar{\mu}^2)^{-1}$ since $\bar{\lambda}(0) = \bar{\lambda}(\bar{\mu}^2)$ when $\omega = \omega_0$, $\omega' = \omega_0'$. The other root is given by

$$\begin{aligned}\hat{\omega}_0' &= \frac{2[(\omega-1)^2 + \lambda(\omega\bar{\mu}^2 - 1)]}{(\omega-1)^2 - \lambda[(1-\omega)^2\bar{\mu}^2 + (\omega\bar{\mu}^2-1)^2]} - \frac{4}{3-\bar{\mu}^2} \\ &= \frac{8(4-\bar{\mu}^2-\bar{\mu}^4)}{16 - 23\bar{\mu}^2 + 10\bar{\mu}^4 + \bar{\mu}^6} - \frac{4}{3-\bar{\mu}^2}\end{aligned} \tag{4.24}$$

Hence

$$\hat{\omega}_0' - \omega_0' = -4(1-3\bar{\mu}^2)(1-\bar{\mu}^2)/[(16-23\bar{\mu}^2+10\bar{\mu}^4+\bar{\mu}^6)(3-\bar{\mu}^2)] \tag{4.25}$$

which is nonnegative if $\bar{\mu}^2 \geq \frac{1}{3}$. Let us first consider the case $\bar{\mu}^2 = \frac{1}{3}$. From (4.8) we have

$$\frac{\partial\lambda}{\partial\omega} = \frac{\partial C/\partial\omega - \lambda\,\partial T/\partial\omega}{2\lambda - T}, \qquad \frac{\partial\lambda}{\partial\omega'} = \frac{\partial C/\partial\omega' - \lambda\,\partial T/\partial\omega'}{2\lambda - T} \tag{4.26}$$

Moreover, if $\omega = \frac{3}{4}$, $\omega' = \frac{3}{2}$, $\lambda = \frac{1}{4}$, we have

$$\begin{aligned}\partial T/\partial\omega &= -2(1-\omega) + 2\omega\bar{\mu}^2 - 2(\omega')^2(1-\omega)\bar{\mu}^2 + 2(\omega\omega'\bar{\mu}^2 + 1 - \omega')\omega'\bar{\mu}^2 \\ &= -\tfrac{1}{2} \\ \partial T/\partial\omega' &= 2\omega'(1-\omega)^2\bar{\mu}^2 + 2(\omega\omega'\bar{\mu}^2 + 1 - \omega')(\omega\bar{\mu}^2 - 1) = \tfrac{1}{4} \\ \partial C/\partial\omega &= 2(\omega-1)(\omega'-1)^2 = -\tfrac{1}{8} \\ \partial C/\partial\omega' &= 2(\omega-1)^2(\omega'-1) = \tfrac{1}{16}\end{aligned} \tag{4.27}$$

Consequently,

$$\partial\lambda/\partial\omega = \partial\lambda/\partial\omega' = 0 \tag{4.28}$$

and we have a relative minimum at $(\omega_0, \omega_0') = (\frac{3}{4}, \frac{3}{2})$.

Suppose now that $\bar{\mu}^2 > \frac{1}{3}$ and the contour $\bar{\lambda}(\bar{\mu}^2) = (1+\bar{\mu}^2)^2(3-\bar{\mu}^2)^{-2}$ contains a point of Region II other than (ω_0, ω_0'). Then since it contains

(ω_0, ω_0') and $(\omega_0, \hat{\omega}_0')$ where $\hat{\omega}_0' > \omega_0'$ it must cut the boundary of Region I at some point, say $(\tilde{\omega}, \tilde{\omega}')$. But at such a point $\bar{\lambda}(0) = (\tilde{\omega}' - 1)^2 > (1 + \bar{\mu}^2)^2(3 - \bar{\mu}^2)^{-2}$. But since $\bar{\lambda}(0) = \bar{\lambda}(\bar{\mu}^2)$ at $(\tilde{\omega}, \tilde{\omega}')$ we have a contradiction. Since the contour corresponding to $(1 + \bar{\mu}^2)^2(3 - \bar{\mu}^2)^{-2}$ lies wholly in Region I it follows that all contours corresponding to smaller values of $\bar{\lambda}(\bar{\mu}^2)$ must also lie in Region I. Hence (ω_0, ω_0') gives the smallest value of $\| \overline{\mathscr{L}_{\omega,\omega'}} \|_{D^{1/2}}$, and the theorem follows.

8.5. DETERMINATION OF $\| \overline{\mathscr{L}_{\omega,\omega'}} \|_{A^{1/2}}$

In this section we show that if A is a positive definite matrix of the form (1.2) then the virtual $A^{1/2}$-norm of $\mathscr{L}_{\omega,\omega'}$ is minimized for $\omega = \omega' = 1$. We also obtain a bound on $\| \overline{\mathscr{L}_{\omega,\omega'}} \|_{A^{1/2}}$ for any given ω and ω' in the region $0 \leq \omega \leq 2$, $0 \leq \omega' \leq 2$. This bound will be useful in Chapter 10 in studies of the MSOR method with varying ω and ω'.

We now prove

Theorem 5.1. Let A be a positive definite matrix of the form (1.2) and let $B = I - (\text{diag } A)^{-1}A$. Then

$$\| \overline{\mathscr{L}_{1,1}} \|_{A^{1/2}} = \| \mathscr{L}_{1,1} \|_{A^{1/2}} = \bar{\mu} = S(B) \tag{5.1}$$

and unless $\omega = \omega' = 1$ we have

$$\| \overline{\mathscr{L}_{\omega,\omega'}} \|_{A^{1/2}} > \| \mathscr{L}_{1,1} \|_{A^{1/2}} \tag{5.2}$$

Moreover, for any ω and ω' in the range $0 \leq \omega \leq 2$, $0 \leq \omega' \leq 2$ we have

$$\begin{aligned} \| \overline{\mathscr{L}_{\omega,\omega'}} \|_{A^{1/2}}^2 &= \| \mathscr{L}_{\omega,\omega'} \|_{A^{1/2}}^2 \\ &= S(\mathscr{L}_{\hat{\omega},\hat{\omega}'}) \leq 1 - \tfrac{1}{2}\hat{\omega}\hat{\omega}'(1 - \bar{\mu}^2) < 1 \end{aligned} \tag{5.3}$$

where

$$\hat{\omega} = \omega(2 - \omega), \qquad \hat{\omega}' = \omega'(2 - \omega') \tag{5.4}$$

Proof. By Lemma 4-4.2 we have

$$\| \mathscr{L}_{\omega,\omega'}[A] \|_{A^{1/2}} = \| \tilde{\mathscr{L}}_{\omega,\omega'}[A] \|_{(\hat{A})^{1/2}}$$

where $\hat{A} = D^{-1/2}AD^{-1/2}$. By (4.3) we have $\tilde{\mathscr{L}}_{\omega,\omega'}[A] = \mathscr{L}_{\omega,\omega'}[\hat{A}]$ and hence

$$\|\mathscr{L}_{\omega,\omega'}[A]\|_{A^{1/2}} = \|\mathscr{L}_{\omega,\omega'}[\hat{A}]\|_{(\hat{A})^{1/2}}. \tag{5.5}$$

This result also holds for the virtual norms. Thus it is sufficient to consider the case where A has the form (4.5) where (4.6) holds.

From (1.11) and (4.5)

$$\mathscr{L}_{\omega'\omega,} = I - \begin{pmatrix} \omega I_1 & 0 \\ \omega\omega' G & \omega' I_2 \end{pmatrix} A \tag{5.6}$$

and hence

$$\mathscr{L}'_{\omega,\omega'} = A^{1/2}\mathscr{L}_{\omega,\omega'}A^{-1/2} = I - A^{1/2}\begin{pmatrix} \omega I_1 & 0 \\ \omega\omega' G & \omega' I_2 \end{pmatrix} A^{1/2} \tag{5.7}$$

and by (4.6)

$$(\mathscr{L}'_{\omega,\omega'})^{\mathrm{T}} = I - A^{1/2}\begin{pmatrix} \omega I_1 & \omega\omega' F \\ 0 & \omega' I_2 \end{pmatrix} A^{1/2} \tag{5.8}$$

Let us now define for any ω and ω' the matrix

$$\mathscr{U}_{\omega,\omega'} = \mathscr{L}_{\omega,0}\mathscr{L}_{0,\omega'} = \begin{pmatrix} (1-\omega)I_1 + \omega\omega' FG & \omega(1-\omega')F \\ \omega' G & (1-\omega')I_2 \end{pmatrix} \tag{5.9}$$

where $\mathscr{L}_{\omega,0}$ and $\mathscr{L}_{0,\omega'}$ are given by (7-5.12). Evidently, $\mathscr{U}_{\omega,\omega'}$ corresponds to the "backward" MSOR method where one first applies the SOR method with relaxation factor ω' to the black equations and then uses ω on the red equations. (For $\mathscr{L}_{\omega,\omega'}$, we use ω for the red equations and then ω' for the black equations.)

Evidently, we have, by (5.9) and (1.2),

$$\mathscr{U}_{\omega,\omega'} = I - \begin{pmatrix} \omega I_1 & \omega\omega' F \\ 0 & \omega' I_2 \end{pmatrix} A \tag{5.10}$$

Hence, by (5.8) it follows that

$$(\mathscr{L}'_{\omega,\omega'})^{\mathrm{T}} = A^{1/2}\,\mathscr{U}_{\omega,\omega'}A^{-1/2} = \mathscr{U}'_{\omega,\omega'} \tag{5.11}$$

Therefore,

$$\begin{aligned} \|\overline{\mathscr{L}_{\omega,\omega'}}\|^2_{A^{1/2}} &= \|\overline{\mathscr{L}'_{\omega,\omega'}}\|^2 \\ &= \bar{S}(\mathscr{L}'_{\omega,\omega'}(\mathscr{L}'_{\omega,\omega'})^{\mathrm{T}}) = \bar{S}(\mathscr{L}'_{\omega,\omega'}\mathscr{U}'_{\omega,\omega'}) \\ &= \bar{S}(\mathscr{L}_{\omega,\omega'}\mathscr{U}_{\omega,\omega'}) = \bar{S}(\mathscr{U}_{\omega,\omega'}, \mathscr{L}_{\omega,\omega'}) \end{aligned} \tag{5.12}$$

We now prove the following generalization of Lemma 7-5.3.

Lemma 5.2. If A is a matrix of the form (1.2) with nonvanishing diagonal elements, then the eigenvalues of

$$\mathscr{W}_{\omega,\omega',\bar{\omega},\bar{\omega}'} = \mathscr{U}_{\bar{\omega},\bar{\omega}'}\mathscr{L}_{\omega,\omega'} \tag{5.13}$$

are the same as those of

$$\mathscr{L}_{\phi(\omega,\bar{\omega}),\phi(\omega',\bar{\omega}')} \tag{5.14}$$

where for any ω_1 and ω_2 we let

$$\phi(\omega_1, \omega_2) = \omega_1 + \omega_2 - \omega_1\omega_2 \tag{5.15}$$

Moreover,

$$\bar{S}(\mathscr{W}_{\omega,\omega',\bar{\omega},\bar{\omega}'}) = \bar{S}(\mathscr{L}_{\phi(\omega,\bar{\omega}),\phi(\omega',\bar{\omega}')}) \tag{5.16}$$

Proof. By (1.13) and (5.9) we have

$$\mathscr{U}_{\bar{\omega},\bar{\omega}'}\mathscr{L}_{\omega,\omega'} = \mathscr{L}_{\bar{\omega},0}\mathscr{L}_{0,\bar{\omega}'}\mathscr{L}_{0,\omega'}\mathscr{L}_{\omega,0} \tag{5.17}$$

which, by Theorem 2-1.11, has the same eigenvalues as

$$\mathscr{L}_{0,\bar{\omega}'}\mathscr{L}_{0,\omega'}\mathscr{L}_{\omega,0}\mathscr{L}_{\bar{\omega},0} = \mathscr{L}_{0,\phi(\omega',\bar{\omega}')}\mathscr{L}_{\phi(\omega,\bar{\omega})0} = \mathscr{L}_{\phi(\omega,\bar{\omega}),\phi(\omega',\bar{\omega}')} \tag{5.18}$$

since by (1.11) we have

$$\begin{aligned}\mathscr{L}_{\omega,0}\mathscr{L}_{\bar{\omega},0} &= \begin{pmatrix}(1-\omega)I_1 & \omega F\\ 0 & I_2\end{pmatrix}\begin{pmatrix}(1-\bar{\omega})I_1 & \bar{\omega}F\\ 0 & I_2\end{pmatrix}\\ &= \begin{pmatrix}(1-\phi(\omega,\bar{\omega}))I_1 & \phi(\omega,\bar{\omega})F\\ 0 & I_2\end{pmatrix} = \mathscr{L}_{\phi(\omega,\bar{\omega}),0}\end{aligned} \tag{5.19}$$

and similarly

$$\mathscr{L}_{0,\bar{\omega}'}\mathscr{L}_{0,\omega'} = \mathscr{L}_{0,\phi(\omega',\bar{\omega}')} \tag{5.20}$$

This completes the proof of Lemma 5.2.

Applying Lemma 5.2 and using (5.12) we obtain

$$\| \overline{\mathscr{L}_{\omega,\omega'}} \|^2_{A^{1/2}} = \bar{S}(\mathscr{L}_{\hat{\omega},\hat{\omega}'}) \tag{5.21}$$

where $\hat{\omega}$ and $\hat{\omega}'$ are given by (5.4). By Theorem 3.4 we have $\bar{S}(\mathscr{L}_{1,1}) = S(\mathscr{L}_{1,1}) = \bar{\mu}^2$ so that $\| \overline{\mathscr{L}_{1,1}} \|^2_{A^{1/2}} = \| \mathscr{L}_{1,1} \|^2_{A^{1/2}} = S(\mathscr{L}_{1,1}) = \bar{\mu}^2$. Therefore, (5.1) holds. Moreover, in order that $\| \overline{\mathscr{L}_{\omega,\omega'}} \|^2_{A^{1/2}} < 1$ we must have $\bar{S}(\mathscr{L}_{\hat{\omega},\hat{\omega}'}) < 1$ by (5.3) and (5.4); hence, by Theorem 3.2 we must have $0 < \hat{\omega} < 2$, $0 < \hat{\omega}' < 2$. Since ω and ω' are real, we must have $0 < \hat{\omega} \le 1$ and $0 < \hat{\omega}' < 1$ by (5.4). By Theorem 3.4 the results (5.2) and (5.3) follow, and the proof of Theorem 5.1 is complete.

SUPPLEMENTARY DISCUSSION

Section 8.2. Much of the material in this section is included in Young [1969].

Section 8.3. McDowell [1967] and Taylor [1969] analyzed the convergence of the MSOR method. They obtained slightly better convergence by considering $S(\mathscr{L}_{\omega,\omega'})$ instead of $\bar{S}(\mathscr{L}_{\omega,\omega'})$. Russell [1963] considered the use of the SOR method with ω varying from equation to equation for a linear system associated with the Navier–Stokes equation.

Theorem 3.3 is a special case of a result of Young *et al.* [1965].

Section 8.4. The material in this section represents an extension of earlier unpublished work by Mary F. Wheeler and James Downing in collaboration with the author.

Section 8.5. Most of the results in this section were obtained by the author in collaboration with David R. Kincaid. (See Young [1969].)

EXERCISES

Section 8.1

1. Carry out two iterations of the MSOR method with $\omega = \frac{3}{2}$, $\omega' = \frac{1}{2}$ for the linear system

$$\begin{pmatrix} 4 & 0 & -1 & -1 \\ 0 & 4 & -1 & -1 \\ -1 & -1 & 4 & 0 \\ -1 & -1 & 0 & 4 \end{pmatrix} \begin{pmatrix} u_1 \\ u_2 \\ u_3 \\ u_4 \end{pmatrix} = \begin{pmatrix} 1 \\ 0 \\ 0 \\ 0 \end{pmatrix}$$

with the starting vector $u^{(0)} = (0, 0, 0, 0)^T$.

2. Verify that

$$\mathscr{L}_{\omega,\omega'} = \begin{pmatrix} I_1 & 0 \\ \omega' G & (1-\omega')I_2 \end{pmatrix} \begin{pmatrix} (1-\omega)I_1 & \omega F \\ 0 & I_2 \end{pmatrix} = \mathscr{L}_{0,\omega'} \mathscr{L}_{\omega,0}$$

Show that this representation corresponds to the following pair of iterations

$$\begin{aligned} u_1^{(n+1/2)} &= \omega(Fu_2^{(n)} + c_1) + (1-\omega)u_1^{(n)} \\ u_2^{(n+1/2)} &= u_2^{(n)} \\ u_1^{(n+1)} &= u_1^{(n+1/2)} \\ u_2^{(n+1)} &= \omega'(Gu_1^{(n+1/2)} + c_2) + (1-\omega')u_2^{(n+1/2)} \end{aligned}$$

Section 8.2

1. Work out the details of the proof of Theorem 2.1.

2. Find the eigenvalues of $\mathscr{L}_{\omega,\omega'}$ and those of $\mathscr{L}_{\omega',\omega}$ for the matrix

$$A = \begin{pmatrix} 2 & 0 & 0 & -1 & 0 \\ 0 & 2 & 0 & -1 & -1 \\ 0 & 0 & 2 & 0 & -1 \\ -1 & -1 & 0 & 2 & 2 \\ 0 & -1 & -1 & 0 & 2 \end{pmatrix}$$

where

$$\omega = \tfrac{3}{2}, \qquad \omega' = \tfrac{3}{4}$$

Also find all solutions λ of (2.1) corresponding to the eigenvalues μ of B.

3. Complete the proof of Theorem 2.2.

4. Show that the first statement of Theorem 2.3(b) is not necessarily true if $\omega = 1$.

5. Let A be a matrix of the form (1.2) and let the order of D_1 and D_2 be r and s, respectively. Show that if $\mu = 0$ is an eigenvalue of B of multiplicity greater than $|r - s|$, then both $1 - \omega$ and $1 - \omega'$ are eigenvalues of $\mathscr{L}_{\omega,\omega'}$.

Section 8.3

1. Give a proof of Theorem 3.3 using the methods employed in the alternative proof of Theorem 6-2.3.
(*Hint*: First find, for given c, the value of $\omega\omega'$ which maximizes $\bar{b}$ as defined by (3.15).)

2. Under the hypothesis of Theorem 2.1, what value of ω minimizes $\bar{S}(\mathscr{L}_{\omega,1})$? What is the corresponding value of $\bar{S}(\mathscr{L}_{\omega,1})$?

3. Under the hypothesis of Theorem 2.1 if $\bar{\mu} = 0.8$, find the values of ω and ω' which minimize $S(\mathscr{L}_{\omega,\omega'})$ subject to the condition $(\omega - 1)(\omega' - 1) = 0.05$.

Section 8.4.

1. If A is a positive definite matrix of the form (1.2) and $\bar{\mu} = 0.8$, find $\| \overline{\mathscr{L}_{\omega,\omega'}} \|_{D^{1/2}}$ if $\omega = \frac{3}{2}$, $\omega' = \frac{1}{2}$. Also, find the values of ω and ω' which minimize $\| \overline{\mathscr{L}_{\omega,\omega'}} \|_{D^{1/2}}$, and the corresponding value of $\| \overline{\mathscr{L}_{\omega,\omega'}} \|_{D^{1/2}}$.

2. Prove the following: If A is a positive definite matrix of the form (1.2), then

$$\min_{\omega'} \| \overline{\mathscr{L}_{1,\omega'}} \|_{D^{1/2}}$$

$$= \begin{cases} \| \overline{\mathscr{L}_{1,\omega_1'}} \|_{D^{1/2}} = \bar{\mu} & \text{if } \bar{\mu} < \dfrac{3^{1/2} - 1}{2} \\ \| \overline{\mathscr{L}_{1,\omega_2'}} \|_{D^{1/2}} = \dfrac{(3 - \bar{\mu}^2)^{1/2} + \bar{\mu}^2 - 1}{2 - \bar{\mu}^2} & \text{if } \bar{\mu} > \dfrac{3^{1/2} - 1}{2} \end{cases}$$

where

$$\omega_1' = \frac{1}{1 - \bar{\mu}^2}, \qquad \omega_2' = \frac{(3 - \bar{\mu}^2)^{1/2} + 1}{2 - \bar{\mu}^2}$$

Also, prove that

$$\min_{\omega} \| \overline{\mathscr{L}_{\omega,1}} \|_{D^{1/2}} = \| \mathscr{L}_{\omega_1,1} \|_{D^{1/2}} = \bar{\mu}$$

where

$$\omega_1 = \frac{1}{1 - \bar{\mu}^2}$$

Evaluate the above quantities for the case $\bar{\mu} = 0.6$.

3. Show that if $\omega = \omega'$, then $T(0) \leq T(\bar{\mu}^2)$ where $T(\mu^2)$ is defined by (4.9).

4. Work out the details for (4.24), (4.25), and (4.27).

5. Show that $\omega_0' = 4(3 - \bar{\mu}^2)^{-1}$ is a root of (4.23) if λ is given by (4.22) and if $\omega_0 = 4(5 + \bar{\mu}^2)^{-1}$.

Section 8.5

1. Verify that $\mathscr{U}_{\omega,\omega'}$ as defined by (5.9) corresponds to the backward MSOR method. (Assume A is a matrix of the form (1.2) with nonvanishing diagonal elements.)

2. If A is a positive definite matrix of the form (1.2) with $\bar{\mu} = S(B) = 0.8$, find a bound on $\| \overline{\mathscr{L}_{\omega,\omega'}} \|_{A^{1/2}}$ if $\omega = \frac{3}{2}$, $\omega' = \frac{1}{2}$.

3. Find $S(\mathscr{W}_{\omega,\omega',\tilde{\omega},\tilde{\omega}'})$ where $\omega = \frac{3}{2}$, $\omega' = \frac{1}{2}$, $\tilde{\omega} = 1$, $\tilde{\omega}' = \frac{1}{4}$ if A is a positive definite matrix of the form (1.2) and if $\bar{\mu} = 0.8$.

Chapter 9 / NONSTATIONARY LINEAR ITERATIVE METHODS

We now consider the use of nonstationary linear iterative methods. We define the concepts of consistency, reciprocal consistency, complete consistency, and convergence for such methods. For linear stationary methods of first degree these definitions reduce to those given in Chapter 3. We also define various measures for the rapidity of convergence on nonstationary methods.

In Section 9.2, we shall consider "periodic" methods which are constructed by the periodic use of m linear stationary methods of first degree. The study of such methods can be reduced to the study of stationary methods. In Section 9.3, we shall consider various properties of Chebyshev polynomials, which are very useful in the study of nonstationary methods.

9.1. CONSISTENCY, CONVERGENCE, AND RATES OF CONVERGENCE

Let us first consider a nonstationary linear iterative method where at each step one uses the formula

$$u^{(n+1)} = G_{n+1}u^{(n)} + k_{n+1}, \qquad n \geq 0 \tag{1.1}$$

Evidently we can express $u^{(n)}$ in the form

$$u^{(n)} = \mathscr{G}_n u^{(0)} + \hat{k}_n, \qquad n \geq 1 \tag{1.2}$$

where

$$\mathscr{G}_n = G_n G_{n-1} \cdots G_1 \tag{1.3}$$

and

$$\hat{k}_n = k_n + G_n k_{n-1} + G_n G_{n-1} k_{n-2} + \cdots + G_n G_{n-1} \cdots G_2 k_1 \tag{1.4}$$

We also obtain a result of the form (1.2) by the use of more general formulas than (1.1) at each step. Thus the more general method defined by

$$u^{(n+1)} = \sum_{j=0}^{n} G_{n+1,j} u^{(j)} + k_{n+1}, \qquad n \geq 0 \tag{1.5}$$

also leads to (1.2) where $\mathscr{G}_1 = G_{1,0}$

$$\mathscr{G}_n = G_{n,0} + G_{n,1}\mathscr{G}_1 + G_{n,2}\mathscr{G}_2 + \cdots + G_{n,n-1}\mathscr{G}_{n-1}, \qquad n \geq 2 \tag{1.6}$$

and where $\hat{k}_n$ is suitably defined.

We now define various kinds of consistency. As in Section 3.1, we let $\mathscr{S}(A, b)$ denote the set of solutions of (3-1.1).

Definition 1.1. The method (1.2) is *consistent* with (3-1.1) if the following condition holds: If $u^{(n)} \in \mathscr{S}(A, b)$, then $u^{(n')} \in \mathscr{S}(A, b)$ for all $n' \geq n$.

Definition 1.2. The method (1.2) is *reciprocally consistent* with (3-1.1) if the following condition holds: If the sequence $u^{(0)}, u^{(1)}, \ldots$ defined by (1.2) converges, it converges to an element of $\mathscr{S}(A, b)$.

Definition 1.3. The method (1.2) is *completely consistent* if it is both consistent and reciprocally consistent.

By Theorems 3-2.1 and 3-2.3 the above definitions reduce to those given in Section 3.1 for linear stationary methods of first degree.

Theorem 1.1. If the nonstationary method (1.2) is obtained by the use of (1.1) and if for each $j = 1, 2, \ldots$ the linear stationary iterative method defined by

$$u^{(n+1)} = G_j u^{(n)} + k_j, \qquad n \geq 0 \tag{1.7}$$

is consistent, then the method (1.2) is consistent. Conversely, if the method (1.2) is consistent, then for each j the method (1.7) is consistent.

Proof. Suppose $u^{(n)}$ is an element of $\mathcal{S}(A, b)$, say $\bar{u}$. Then by the definition of consistency given in Section 3.1 we have

$$\bar{u} = G_j \bar{u} + k_j \tag{1.8}$$

for each j. Therefore, by (1.1), $u^{(n+1)} = u^{(n+2)} = \cdots = \bar{u}$ and the method (1.2) is consistent.

If the method (1.2) is consistent, then if $u^{(0)} = \bar{u}$, where $\bar{u} \in \mathcal{S}(A, b)$, every iterant $u^{(n)}$ must equal $\bar{u}$. But this implies (1.8) holds for each j. Since this is true for each element of $\mathcal{S}(A, b)$, the method (1.7) is consistent for each j. This completes the proof of Theorem 1.1.

By analogy with Theorem 3-2.2 we have

Theorem 1.2. If A is nonsingular and the method (1.2) is consistent, then for each n we have

$$\hat{k}_n = (I - \mathcal{G}_n)A^{-1}b \tag{1.9}$$

Evidently the definitions of convergence and weak convergence given in Section 3.5 can be applied to nonstationary iterative methods. Necessary and sufficient conditions for convergence are that $\{\hat{k}_n\}$ converges and

$$\lim_{n\to\infty} \mathcal{G}_n = 0 \tag{1.10}$$

A necessary and sufficient condition for weak convergence is that $\{\hat{k}_n\}$ converges and for some matrix, say $\mathcal{G}$, we have

$$\lim_{n\to\infty} \mathcal{G}_n = \mathcal{G} \tag{1.11}$$

The proof of (1.10) is left as an exercise. To prove (1.11) let u be the limit of the sequence defined by (1.2) corresponding to $u^{(0)}$. For a given $\hat{u}^{(0)}$ let the limit be $\hat{u}$. Let $v^{(0)} = \hat{u}^{(0)} - u^{(0)}$ and $v = \hat{u} - u$. Evidently we have

$$\lim_{n\to\infty} \mathcal{G}_n v^{(0)} = v \tag{1.12}$$

But we can show that for some matrix $\mathcal{G}$ we have

$$v = \mathcal{G} v^{(0)} \tag{1.13}$$

(The matrix $\mathcal{G}$ can be constructed by successively letting $v^{(0)}$ be the unit vectors $(1, 0, 0, \ldots, 0)^T$, $(0, 1, 0, \ldots, 0)^T$, etc. and letting $\mathcal{G}$ be the matrix whose columns are the corresponding vectors v.) One can also look upon (1.12) as defining a linear transformation from E^N onto a subset of E^N, and $\mathscr{G}$ is the matrix corresponding to the linear transformation. In any case, we have by (1.13) and (1.12)

$$\lim_{n\to\infty} (\mathscr{G}_n - \mathscr{G})v^{(0)} = 0 \tag{1.14}$$

for all $v^{(0)}$ which implies (1.11).

We now show that in order for (1.2) to be convergent it is neither necessary nor sufficient that each individual method be convergent. For example, consider the system (3-1.1) where

$$A = \begin{pmatrix} 2 & -1 \\ -1 & 2 \end{pmatrix}, \qquad b = \begin{pmatrix} 1 \\ 0 \end{pmatrix} \tag{1.15}$$

whose solution is

$$\bar{u} = \begin{pmatrix} \frac{2}{3} \\ \frac{1}{3} \end{pmatrix} \tag{1.16}$$

Let us consider the nonstationary iterative method

$$\begin{aligned} u^{(n+1)} &= G_1 u^{(n)} + k_1, \qquad n \quad \text{even} \\ u^{(n+1)} &= G_2 u^{(n)} + k_2, \qquad n \quad \text{odd} \end{aligned} \tag{1.17}$$

where

$$\begin{aligned} G_1 &= \begin{pmatrix} 0 & 0 \\ 2 & 0 \end{pmatrix}, \qquad k_1 = \begin{pmatrix} \frac{2}{3} \\ -1 \end{pmatrix} \\ G_2 &= \begin{pmatrix} 0 & 2 \\ 0 & 0 \end{pmatrix}, \qquad k_2 = \begin{pmatrix} 0 \\ \frac{1}{3} \end{pmatrix} \end{aligned} \tag{1.18}$$

Evidently each of the methods of (1.12) is convergent; in fact, $S(G_1) = S(G_2) = 0$. However, we have

$$\mathscr{G}_n = \begin{cases} G_1(G_2G_1)^{(n-1)/2} = \begin{pmatrix} 0 & 0 \\ 2^n & 0 \end{pmatrix}, & n \quad \text{odd} \\ (G_2G_1)^{n/2} \qquad\quad = \begin{pmatrix} 2^n & 0 \\ 0 & 0 \end{pmatrix}, & n \quad \text{even} \end{cases} \tag{1.19}$$

Clearly (1.10) does not hold and the method is not convergent. Since (1.11) does not hold for any $\mathscr{G}$ the method is not weakly convergent.

Suppose, on the other hand, that

$$G_1 = \begin{pmatrix} 2 & 0 \\ 0 & 2 \end{pmatrix}, \qquad k_1 = \begin{pmatrix} -\frac{2}{3} \\ -\frac{1}{3} \end{pmatrix}$$
$$G_2 = \begin{pmatrix} \frac{1}{4} & 0 \\ 0 & \frac{1}{4} \end{pmatrix}, \qquad k_2 = \begin{pmatrix} \frac{1}{2} \\ \frac{1}{4} \end{pmatrix} \tag{1.20}$$

Evidently the first method of (1.20) is not convergent. However, we have

$$G_2 G_1 = \begin{pmatrix} \frac{1}{2} & 0 \\ 0 & \frac{1}{2} \end{pmatrix} \tag{1.21}$$

and

$$\mathscr{G}_n = \begin{cases} G_1(G_2G_1)^{(n-1)/2} = \begin{pmatrix} \frac{1}{2}^{(n+1)/2} & 0 \\ 0 & \frac{1}{2}^{(n+1)/2} \end{pmatrix}, & n \ \text{odd} \\ (G_2G_1)^{n/2} = \begin{pmatrix} \frac{1}{2}^{n/2} & 0 \\ 0 & \frac{1}{2}^{n/2} \end{pmatrix}, & n \ \text{even} \end{cases} \tag{1.22}$$

Clearly $\mathscr{G}_n \to 0$ as $n \to \infty$ and the method is convergent.

We now introduce, in the accompanying tabulation, certain measures of convergence for nonstationary methods. We apply each convergence measure to the stationary case, where $\mathscr{G}_n = G^n$. It can be seen that these definitions are consistent with those given in Section 3.7. In the tabulation

Measure of convergence	General case	Stationary case
1. Average spectral radius	$\mathscr{S}_n(\mathscr{G}_n) = [S(\mathscr{G}_n)]^{1/n}$	$S(G)$
2. Asymptotic average spectral radius	$\mathscr{S}(\mathscr{G}_n) = \lim_{n\to\infty} \mathscr{S}_n(\mathscr{G}_n)$	$S(G)$
3. Average rate of convergence	$\mathscr{R}_n(\mathscr{G}_n) = -\frac{1}{n} \log \lVert \mathscr{G}_n \rVert$	$R_n(G) = -\frac{1}{n} \log \lVert G^n \rVert$
4. Asymptotic average rate of convergence	$\mathscr{R}(\mathscr{G}_n) = \lim_{n\to\infty} \mathscr{R}_n(\mathscr{G}_n)$	$R(G) = -\log S(G)$
5. Quasi-average rate of convergence	$\mathscr{R}_n'(\mathscr{G}_n) = -\log \mathscr{S}_n(\mathscr{G}_n)$	$R(G) = -\log S(G)$
6. Asymptotic quasi-average rate of convergence	$\mathscr{R}'(\mathscr{G}_n) = \lim_{n\to\infty} \mathscr{R}_n'(\mathscr{G}_n) = -\log \mathscr{S}(\mathscr{G}_n)$	$R(G) = -\log S(G)$

it is understood that the spectral norm $\| \cdot \|_2$ is used, although other norms could also be used.

If the matrix $\mathscr{G}_n$ is a function of n and if the eigenvalues of $\mathscr{G}_n$, or those of $\mathscr{G}_n \mathscr{G}_n^{\mathrm{T}}$ if we are considering $\| \mathscr{G}_n \|$, are related to those of B as in Section 6.1, then we can define the *virtual average spectral radius* by

$$\bar{\mathscr{S}}_n(\mathscr{G}_n) = [\bar{S}(\mathscr{G}_n)]^{1/n}$$

where $\bar{S}(\mathscr{G}_n)$ is the virtual spectral radius of $\mathscr{G}_n$. Similarly, we can define the *virtual asymptotic average spectral radius*, the *virtual average rate of convergence*, etc.

9.2. PERIODIC NONSTATIONARY METHODS

For some $m \geq 1$ we define a *periodic* nonstationary method by letting

$$G_n = G_{\phi(n)}, \qquad k_n = k_{\phi(n)} \tag{2.1}$$

where

$$\begin{aligned} \phi(n) &\equiv n \pmod m \\ 1 \leq \phi(n) &\leq m \end{aligned} \tag{2.2}$$

Let us consider the linear stationary method

$$u^{(n+1)} = \mathscr{G}_m u^{(n)} + \hat{k}_m, \qquad n \geq 0 \tag{2.3}$$

where

$$\begin{aligned} \mathscr{G}_m &= G_m G_{m-1} \cdots G_1 \\ \hat{k}_m &= k_m + G_m k_{m-1} + \cdots + G_m G_{m-1} \cdots G_2 k_1 \end{aligned} \tag{2.4}$$

We shall refer to this method as the *composite method.*

It follows immediately from Theorem 1.1 that the composite method is consistent if and only if the individual methods (1.7) are consistent for $j = 1, 2, \ldots, m$. However, even if the individual methods are reciprocally consistent, the composite method need not be. Thus consider the system $Au = b$ with A and b given by (1.15). Let $m = 2$ and

$$\begin{aligned} G_1 &= \begin{pmatrix} 0 & 0 \\ 2 & 0 \end{pmatrix}, \qquad k_1 = \begin{pmatrix} \frac{2}{3} \\ -1 \end{pmatrix} \\ G_2 &= \begin{pmatrix} 0 & \frac{1}{2} \\ 0 & 0 \end{pmatrix}, \qquad k_2 = \begin{pmatrix} \frac{1}{2} \\ \frac{1}{3} \end{pmatrix} \end{aligned} \tag{2.5}$$

Evidently each method is completely consistent but the composite method is not since $I - G_2 G_1$ is singular. It should be noted, however, that the nonstationary method is indeed reciprocally consistent since if $u^{(n)} \to \bar{u}$ for some $\bar{u}$, then $\bar{u} \in \mathscr{S}(I - G_1, k_1)$ and $\bar{u} \in \mathscr{S}(I - G_2, k_2)$ and hence $\bar{u} \in \mathscr{S}(A, b)$.

The periodic nonstationary method is convergent if and only if the composite method is convergent. For given n we have

$$\mathscr{G}_n = G_s G_{s-1} \cdots G_1 (\mathscr{G}_m)^p$$

for some s and p such that $1 \leq s < m$. Thus

$$\| \mathscr{G}_n \| \leq \left(\prod_{i=1}^{s} \| G_i \| \right) \| \mathscr{G}_m^p \|$$

Hence $\mathscr{G}_n \to 0$ as $n \to \infty$ if $\mathscr{G}_m^p \to 0$ as $p \to 0$, i.e., if $S(\mathscr{G}_m) < 1$. On the other hand, we have

$$\| \mathscr{G}_m^{p+1} \| \leq \| \mathscr{G}_n \| \left(\prod_{i=1}^{m} \| G_i \| \right)$$

so that the convergence of $\mathscr{G}_n$ implies the convergence of $\mathscr{G}_m^p$.

Let us now consider the measures of convergence defined in Section 13.1 as applied to periodic nonstationary methods. If we let $n = mt$, we have

$$\begin{aligned}
\mathscr{S}_n(\mathscr{G}_n) &= [S(\mathscr{G}_m^t)]^{1/mt} = [S(\mathscr{G}_m)]^{1/m} = \mathscr{S}(\mathscr{G}_n) \\
\mathscr{R}_n(\mathscr{G}_n) &= -\frac{1}{mt} \log \| \mathscr{G}_m^t \| \\
\mathscr{R}(\mathscr{G}_n) &= \lim_{t \to \infty} \left(-\frac{1}{mt} \log \| \mathscr{G}_m^t \| \right) = -\frac{1}{m} \log S(\mathscr{G}_m) \\
\mathscr{R}_n'(\mathscr{G}_n) &= \mathscr{R}'(\mathscr{G}_n) = \mathscr{R}(\mathscr{G}_n)
\end{aligned} \tag{2.6}$$

9.3. CHEBYSHEV POLYNOMIALS

A basic tool in the study of nonstationary linear iterative methods is the set of Chebyshev polynomials. For any nonnegative integer n we define the *Chebyshev polynomials of degree* n by the recurrence relation

$$\begin{aligned}
T_0(x) &= 1, \qquad T_1(x) = x \\
T_{n+1}(x) &= 2xT_n(x) - T_{n-1}(x), \qquad n \geq 1
\end{aligned} \tag{3.1}$$

Thus we have

$$T_2(x) = 2x^2 - 1, \quad T_3(x) = 4x^3 - 3x, \quad T_4(x) = 8x^4 - 8x^2 + 1, \quad \cdots \tag{3.2}$$

It can be verified by mathematical induction that

$$\begin{aligned} T_n(x) = \cos(n \cos^{-1} x) &= \tfrac{1}{2}[(x + (x^2 - 1)^{1/2})^n + (x + (x^2 - 1)^{1/2})^{-n}] \\ &= \cosh(n \cosh^{-1} x) \end{aligned} \tag{3.3}$$

We remark that the above formulas can be used for all x even though in some cases, e.g., $\cosh^{-1} x$ and $(x^2 - 1)^{1/2}$ for $|x| < 1$, the intermediate quantities may be complex.

It is evident that $T_n(1) = 1$ for all n and that $T_n(x)$ assumes the values $(-1)^k$ at the $n + 1$ points

$$x_k = \cos(k\pi/n), \qquad k = 0, 1, \ldots, n \tag{3.4}$$

Moreover, since $\cos^{-1} x$ is real for $|x| \leq 1$, it follows from (3.3) that $|T_n(x)| \leq 1$ for $|x| \leq 1$; hence we have

$$\max_{-1 \leq x \leq 1} |T_n(x)| = 1 \tag{3.5}$$

On the other hand, $|T_n(x)| > 1$ if $|x| > 1$.

The zeros of $T_n(x)$ occur at the points

$$x_k = \cos\left(\frac{2k - 1}{2n}\right)\pi, \qquad k = 1, 2, \ldots, n \tag{3.4'}$$

We also note that $T_n(-x) = (-1)^n T_n(x)$ and hence $T_n(x)$ is an *even* polynomial if n is even and an *odd* polynomial if n is odd.

We now prove a fundamental property of Chebyshev polynomials.

Theorem 3.1. Let n be a fixed nonnegative integer and let z be any fixed real number such that $z > 1$. If we let

$$P_n(x) = T_n(x)/T_n(z) \tag{3.6}$$

where $T_n(x)$ is the Chebyshev polynomial of degree n given by (3.3), then

$$P_n(z) = 1 \tag{3.7}$$

and

$$\max_{-1 \leq x \leq 1} |P_n(x)| = 1/T_n(z) \tag{3.8}$$

Moreover, if $Q(x)$ is any polynomial of degree n or less such that $Q(z) = 1$ and

$$\max_{-1 \le x \le 1} |Q(x)| \le \max_{-1 \le x \le 1} |P_n(x)| \tag{3.9}$$

then

$$Q(x) \equiv P(x) \tag{3.10}$$

Proof. Evidently (3.8) follows from (3.5). To prove (3.10), for each $k = 1, 2, \ldots, n$ let I_k denote the interval $x_k \le x \le x_{k-1}$ where the x_k are given by (3.4). Suppose that $Q(x)$ is any polynomial with the properties described and let

$$R(x) = Q(x) - P_n(x) \tag{3.11}$$

Evidently we have

$$R(x_k) \le 0, \qquad k \text{ even}$$
$$R(x_k) \ge 0, \qquad k \text{ odd}$$

Hence for $k = 1, 2, \ldots, n$ we have

$$R(x_k)R(x_{k-1}) \le 0$$

If $R(x_k)R(x_{k-1}) < 0$, then there is a zero of $R(x)$ in the open interval I_k. If $R(x_k)R(x_{k-1}) = 0$, then either $R(x_k) = 0$ or $R(x_{k-1}) = 0$. If $R(x_k) = 0$ for $k = 1, 2, \ldots, n - 1$, then $R'(x_k) = 0$ and we have a double zero at x_k. This follows since x_k is an extreme value for both $P_n(x)$ and for $Q(x)$. Hence $P_n'(x_k) = Q'(x_k) = 0$.

We proceed to construct n zeros of $R(x)$ in the interval $-1 \le x \le 1$ as follows. If $R(x_0) = 0$, we assign x_0 to I_1. If $R(x_0) < 0$ and $R(x_1) > 0$, there is a zero in the interior of I_1 which we assign to I_1. If $R(x_1) = 0$, then there is a double zero of $R(x)$, one of which we assign to I_1 and the other to I_2. For $k > 1$, if $R(x_{k-1}) = 0$, x_{k-1} is a double zero of $R(x)$ and we assign one zero to I_k. If $R(x_{k-1}) \ne 0$ and $R(x_k) \ne 0$, then we have a zero in the interior of I_k. Otherwise $R(x_k) = 0$ and we can assign one of the two zeros at x_k to I_k. Continuing this process we obtain n zeros of $R(x)$ in the interval $-1 \le x \le 1$. Since $R(z) = 0$ it follows that $R(x)$, which is a polynomial of degree n at most and which has $n + 1$ distinct zeros, must vanish identically. Therefore (3.10) holds and the proof of Theorem 3.1 is complete.

SUPPLEMENTARY DISCUSSION

Section 9.3. Theorem 3.1 is due to A. Markoff [1892], but the proof given is based on that of Flanders and Shortley [1950].

EXERCISES

Section 9.1

1. If each method (1.1) is reciprocally consistent (completely consistent), is the method (1.2) necessarily reciprocally consistent (completely consistent)?

2. Consider the nonstationary method

$$u^{(n+1)} = B_{\omega_{n+1}} u^{(n)} + c^{(n)}$$

where $B_\omega = \omega B + (1 - \omega)I$ and $c^{(n)} = \omega_{n+1}c$. Here $\omega_n = n/(n+1)$. If

$$A = \begin{pmatrix} 2 & -1 \\ -1 & 2 \end{pmatrix}$$

find the average spectral radius for five iterations, the asymptotic average spectral radius, etc.

3. Prove that (1.2) is convergent if and only if $\{\hat{k}_n\}$ converges and (1.10) holds.

Section 9.2

1. Suppose that for the example of Exercise 2, Section 9.1, we have

$$\omega_n = 1 + (-1)^n/2$$

Evaluate the various measures of convergence for the associated periodic nonstationary method. Also evaluate the corresponding virtual measures of convergence.

2. Consider the periodic nonstationary method

$$u^{(n+1)} = -(H + \varrho I)^{-1}(V - \varrho I)u^{(n)} + (H + \varrho I)^{-1}b, \qquad n \text{ even}$$
$$u^{(n+1)} = -(V + \varrho I)^{-1}(H - \varrho I)u^{(n)} + (V + \varrho I)^{-1}b, \qquad n \text{ odd}$$

for solving the system

$$(H + V)\begin{pmatrix} u_1 \\ u_2 \\ u_3 \\ u_4 \end{pmatrix} = b$$

where

$$H = \begin{pmatrix} 2 & -1 & 0 & 0 \\ -1 & 2 & 0 & 0 \\ 0 & 0 & 2 & -1 \\ 0 & 0 & -1 & 2 \end{pmatrix}, \qquad V = \begin{pmatrix} 2 & 0 & -1 & 0 \\ 0 & 2 & 0 & -1 \\ -1 & 0 & 2 & 0 \\ 0 & -1 & 0 & 2 \end{pmatrix}$$

Test whether the individual methods converge and whether the composite method converges for the case $\varrho = 1$.

Section 9.3

1. Find the polynomial $P_n(x)$ of degree 4 or less such that $P_n(z) = 1$ for some $z > 1$ and such that

$$\max_{-1 \le x \le 1} | P_n(x) |$$

is minimized. Evaluate the polynomial for $x = \frac{1}{4}$ and $z = \frac{3}{2}$. Find its zeros and extreme values in the interval $[-1, 1]$.

2. Verify Theorem 3.1 for the cases $n = 1$ and $n = 2$ by assuming $P(x) = \alpha + \beta x$ and $P(x) = \alpha + \beta x + \gamma x^2$, respectively, and choosing the coefficients to minimize

$$\max_{-1 \le x \le 1} | P(x) |$$

3. Verify (3.3).

4. Compute the zeros and extreme points of $T_{10}(x)$ in the interval $[-1, 1]$.

5. Prove that $| T_n(x) | > 1$ for $| x | > 1$.

Chapter 10 / THE MODIFIED SOR METHOD: VARIABLE PARAMETERS

In Chapter 8 we considered the MSOR method with fixed iteration parameters ω and ω' for solving (3-1.1) when A has the form (8-1.2). We now study the convergence properties of the MSOR method where ω and ω' are allowed to vary from iteration to iteration. In this case the formulas (8-1.9) are replaced by

$$\begin{aligned} u_1^{(n+1)} &= \omega_{n+1}(Fu_2^{(n)} + c_1) + (1 - \omega_{n+1})u_1^{(n)} \\ u_2^{(n+2)} &= \omega'_{n+1}(Gu_1^{(n+1)} + c_2) + (1 - \omega'_{n+1})u_2^{(n)} \end{aligned} \tag{1}$$

In Section 10.1 we shall give sufficient conditions on the iteration parameters so that the method will converge. In Section 10.2 it is shown that, in a sense, there is no advantage to be gained in varying ω and ω' if the matrix A is positive definite and has the form (8-1.2). However, as shown in Section 10.3, there are parameter choices which are as good, though not better than, the choice $\omega_1 = \omega_1' = \omega_2 = \omega_2' = \cdots = \omega_b$, where ω_b is the best single parameter.

Certain other choices of iteration parameters are more effective if one measures the effectiveness in terms of certain norms. In Sections 10.4–10.7 we study the $D^{1/2}$-norms and the $A^{1/2}$-norms of various parameter choices including Sheldon's method, where $\omega_1 = \omega_1' = 1$, $\omega_2 = \omega_2' = \omega_3 = \omega_3' = \cdots = \omega_b$, the modified Sheldon's method where $\omega_1 = 1$, $\omega_1' = \omega_2 = \omega_2' = \cdots = \omega_b$, and the cyclic Chebyshev semi-iterative method.

10.1. CONVERGENCE OF THE MSOR METHOD

The following theorem follows from Theorem 8-5.1. The proof is similar to that of Theorem 4-4.1.

Theorem 1.1. If A be a positive definite matrix of the form (8-1.2), the MSOR method based on the use of the relaxation factors ω_1, ω_1', $\omega_2, \omega_2', \ldots$ converges provided at least one of the following conditions holds:

(a) For some $\varepsilon > 0$, we have

$$\varepsilon \le \omega_i \le 2 - \varepsilon, \qquad \varepsilon \le \omega_i' \le 2 - \varepsilon, \qquad i = 1, 2, \ldots, \tag{1.1}$$

for all i sufficiently large.

(b) For all i sufficiently large $0 \le \omega_i \le 2$, $0 \le \omega_i' \le 2$, and the series

$$\sum_{i=1}^{\infty} \omega_i \omega_i'(2 - \omega_i)(2 - \omega_i') \tag{1.2}$$

diverges.

10.2. OPTIMUM CHOICE OF RELAXATION FACTORS

Let us define the matrix $\mathscr{G}_m$ by

$$\mathscr{G}_m = \prod_{k=m}^{1} \mathscr{L}_{\omega_k, \omega_k'} \tag{2.1}$$

By Theorem 7-2.1 and (8-1.11) it follows that the eigenvalues of $\mathscr{G}_m$ are the same as the eigenvalues of $M_m(\mu)$ where

$$M_m(\mu) = \prod_{k=m}^{1} M(\omega_k, \omega_k', \mu) \tag{2.2}$$

for all eigenvalues μ of B. Here

$$\begin{aligned} M(\omega, \omega', \mu) &= \begin{pmatrix} 1 & 0 \\ -\omega'\mu & 1 \end{pmatrix}^{-1} \begin{pmatrix} 1 - \omega & \omega\mu \\ 0 & 1 - \omega' \end{pmatrix} \\ &= \begin{pmatrix} 1 - \omega & \omega\mu \\ (1 - \omega)\omega'\mu & \omega\omega'\mu^2 + 1 - \omega' \end{pmatrix} \end{aligned} \tag{2.3}$$

More precisely, if $\mu \ne 0$ is an eigenvalue of B, then each eigenvalue of $M_m(\mu)$ is an eigenvalue of $\mathscr{G}_m$, while if $\mu = 0$ is an eigenvalue of B, then at least one of the two eigenvalues of $M_m(0)$ is an eigenvalue of $\mathscr{G}_m$. Conversely, if λ is an eigenvalue of $\mathscr{G}_m$, then there exists an eigenvalue of B such that λ is an eigenvalue of $M_m(\mu)$.

Evidently for the virtual spectral radius of $\mathscr{G}_m$ as defined in Section 6.1 we have

$$\bar{S}(\mathscr{G}_m) = \max_{\mu \in \bar{S}_B} S(M_m(\mu)) \tag{2.4}$$

where $\bar{S}_B$ is the smallest convex set containing the set S_B of eigenvalues of B. Normally, we shall assume that all eigenvalues of B are real and that

$$\bar{\mu} = S(B) < 1 \tag{2.5}$$

For this it is sufficient, though not necessary, that A be positive definite, by Lemma 6-2.5. Hence, since the eigenvalues μ of B occur in pairs $(\mu, -\mu)$ it follows that $\bar{S}_B$ is the interval

$$-\bar{\mu} \le \mu \le \bar{\mu} \tag{2.6}$$

We now prove

Theorem 2.1. Let A be a matrix with nonvanishing diagonal elements which has the form (8-1.2) and let $B = I - (\operatorname{diag} A)^{-1}A$. If the eigenvalues of B are real and if $S(B) = \bar{\mu} < 1$, then for any set of parameters $\omega_1, \omega_1', \omega_2, \omega_2', \ldots, \omega_m, \omega_m'$ we have

$$\bar{S}\left(\prod_{k=m}^{1} \mathscr{L}_{\omega_k, \omega_k'}\right) \ge S(\mathscr{L}^m_{\omega_b, \omega_b}) = S(\mathscr{L}^m_{\omega_b}) = (\omega_b - 1)^m = r^m \tag{2.7}$$

where

$$\omega_b = \frac{2}{1 + (1 - \bar{\mu}^2)^{1/2}}, \qquad r = \omega_b - 1 \tag{2.8}$$

Moreover, if equality holds in (2.7) then

$$\prod_{k=1}^{m} (1 - \omega_k)(1 - \omega_k') = r^{2m} \tag{2.8$'$}$$

Proof. Using (2.3) and mathematical induction we have

Lemma 2.2. If $M_m(\mu)$ is given by (2.2), then for $m = 1, 2, \ldots,$ we have

$$M_m(\mu) = \begin{pmatrix} (1 - \omega_1)P_{2m-2}(\mu) & Q_{2m-1}(\mu) \\ (1 - \omega_1)P_{2m-1}(\mu) & Q_{2m}(\mu) \end{pmatrix} \tag{2.9}$$

where $P_k(\mu)$, $Q_k(\mu)$, $k = 0, 1, 2, \ldots$ are polynomials of degree k in μ such that

$$\begin{aligned} P_0(\mu) &= 1, \\ P_1(\mu) &= \omega_1'\mu \\ P_{2k}(\mu) &= \omega_{k+1}\mu P_{2k-1}(\mu) + (1 - \omega_{k+1})P_{2k-2}(\mu), \qquad k = 1, 2, \ldots \\ P_{2k+1}(\mu) &= \omega'_{k+1}\mu P_{2k}(\mu) + (1 - \omega'_{k+1})P_{2k-1}(\mu), \qquad k = 1, 2, \ldots \end{aligned} \tag{2.10}$$

and

$$\begin{aligned} Q_0(\mu) &= 1, \\ Q_1(\mu) &= \omega_1\mu \\ Q_{2k}(\mu) &= \omega_k'\mu Q_{2k-1}(\mu) + (1 - \omega_k')Q_{2k-2}(\mu), \qquad k = 1, 2, \ldots \\ Q_{2k+1}(\mu) &= \omega_{k+1}\mu Q_{2k}(\mu) + (1 - \omega_{k+1})Q_{2k-1}(\mu), \qquad k = 1, 2, \ldots \end{aligned} \tag{2.11}$$

The representation of $M_m(\mu)$ given by (2.9) will not only be used to prove Theorem 2.1 but will also be useful later on. An alternative proof of Theorem 2.1, not involving the use of (2.9), is given by Young *et al.* [1965].

Evidently the characteristic equation of $M_m(\mu)$ is

$$\lambda^2 - b\lambda + c = 0 \tag{2.12}$$

where

$$b = b(\mu) = \text{trace } M_m(\mu) \tag{2.13}$$

$$c = \det M_m(\mu) = \prod_{k=1}^{m} \det M(\omega_k, \omega_k', \mu) = \prod_{k=1}^{m} (1 - \omega_k)(1 - \omega_k') \tag{2.14}$$

We first show that

$$b(1) = 1 + c \tag{2.15}$$

But if $\mu = 1$, we have from (2.3) that

$$M(\omega_k, \omega_k', 1)\begin{pmatrix}1\\1\end{pmatrix} = \begin{pmatrix}1\\1\end{pmatrix} \tag{2.16}$$

so that by (2.2), $\lambda = 1$ is an eigenvalue of $M_m(\mu)$. Hence, by (2.12) we have (2.15).

From Lemma 2.2 and (2.10)–(2.11) for some polynomial $H_m(\mu^2)$ of degree m in μ^2 we have

$$b(\mu) = H_m(\mu^2) \tag{2.17}$$

If we let

$$\theta = (2\mu^2/\bar{\mu}^2) - 1 \tag{2.18}$$

then the interval $0 \leq \mu^2 \leq \bar{\mu}^2$ corresponds to the interval $-1 \leq \theta \leq 1$. Moreover, if $\mu^2 = 1$, then

$$\theta = 2/\bar{\mu}^2 - 1 = z > 1 \tag{2.19}$$

Therefore, we have

$$b(\mu) = H_m(\tfrac{1}{2}\bar{\mu}^2(\theta + 1)) = K_m(\theta) \tag{2.20}$$

where $K_m(\theta)$ is a polynomial of degree m. By (2.19) we have

$$\bar{b} = \max_{-\bar{\mu}\leq\mu\leq\bar{\mu}} | b(\mu) | = \max_{-1\leq\theta\leq 1} | K_m(\theta) | \tag{2.21}$$

Since $K_m(z) = 1 + c$ we can apply Theorem 9-3.1 to the function $K_m(\theta)/(1 + c)$ obtaining

$$\bar{b} = \max_{-1\leq\theta\leq 1} | K_m(\theta) | \geq | 1 + c |/T_m(z) \tag{2.22}$$

where $T_m(z)$ is the Chebyshev polynomial of degree m.

If $| c | \geq 1$, then the root radius of (2.12) is at least equal to unity. If $| c | < 1$, then $1 + c > 0$ and by Lemma 6-2.9, for given c, the maximum root radius of (2.12) for $-\bar{\mu} \leq \mu \leq \bar{\mu}$ is at least as large as that of

$$\lambda^2 - \frac{2(1 + c)\bar{\lambda}}{1 + \bar{\lambda}^2}\lambda + c = 0 \tag{2.23}$$

where

$$\bar{\lambda} = r^m \tag{2.24}$$

This follows since by (8-3.18), $\bar{\mu}^2 = 4r(1 + r)^{-2}$ and hence

$$z = (2/\bar{\mu}^2) - 1 = (1 + r^2)/2r \tag{2.25}$$

and

$$\begin{aligned} T_m(z) &= \tfrac{1}{2}[(z + [z^2 - 1]^{1/2})^m + (z + [z^2 - 1]^{1/2})^{-m}] \\ &= \tfrac{1}{2}(1 + r^{2m})/r^m = \tfrac{1}{2}(1 + \bar{\lambda}^2)/\bar{\lambda} \end{aligned} \tag{2.26}$$

As in the proof of Theorem 8-3.3, if $\varrho(c)$ is the root radius of (2.23), then

$$\varrho(c) > \varrho(c_0) = \bar{\lambda}, \qquad c \neq c_0 \tag{2.27}$$

where $c_0 = \bar{\lambda}^2 = r^{2m}$. Hence (2.7) and (2.8′) follow and the proof of Theorem 2.1 is complete.

10.3. ALTERNATIVE OPTIMUM PARAMETER SETS

We now seek to determine, for given m, the parameter sets ω_1, ω_1', $\omega_2, \omega_2', \ldots, \omega_m, \omega_m'$ such that equality holds in (2.7). From Theorem 2.1 it follows that in order for (2.7) to hold we must have

$$c = \prod_{k=1}^{m} (1 - \omega_k)(1 - \omega_k') = \bar{\lambda}^2 \tag{3.1}$$

By Lemma 6-2.9, if $c = \bar{\lambda}^2$, the maximum of the root radii of (2.12) for $-\bar{\mu} \leq \mu \leq \bar{\mu}$ is not less than that of

$$\lambda^2 - \bar{b}\lambda + \bar{\lambda}^2 = 0 \tag{3.2}$$

By (3.1), (2.22), and (2.26) we have

$$\bar{b} \geq 2\bar{\lambda} \tag{3.3}$$

On the other hand, if $\bar{b} > 2\bar{\lambda}$ and if $\bar{\lambda} > 0$, then the left member of (3.2) is negative for $\lambda = \bar{\lambda}$; hence, there is a root greater than $\bar{\lambda}$. If $\bar{\lambda} = 0$ and if

$$\bar{b} > 2\bar{\lambda} = 0 \tag{3.4}$$

then the root radius of (3.2) is positive. In any case it follows that equality holds in (2.7) if and only if we have (3.1) and

$$\bar{b} = 2\bar{\lambda} \tag{3.5}$$

By Theorem 9-3.1, if (3.1) holds, then we have $\bar{b} > 2\bar{\lambda}$ unless $K_m(\theta)$ is the polynomial

$$K_m(\theta) = T_m(\theta)(1 + \bar{\lambda}^2)/T_m(z) = 2\bar{\lambda}T_m(\theta) \tag{3.6}$$

i.e., unless

$$b(\mu) = 2\bar{\lambda}T_m(2\mu^2/\bar{\mu}^2 - 1) \tag{3.7}$$

We seek to choose $\omega_1, \omega_1', \ldots, \omega_m, \omega_m'$ such that (3.7) holds. If we succeed in doing so, then we have by (2.15), $c = b(1) - 1 = \bar{\lambda}^2$. Thus by (2.9) and (3.7) we have an optimum parameter set if and only if

$$b(\mu) = \text{trace}\, M_m(\mu) = (1 - \omega_1)P_{2m-2} + Q_{2m} = 2\bar{\lambda}T_m(2\mu^2/\bar{\mu}^2 - 1) \tag{3.8}$$

If $m = 1$, the condition (3.8) becomes

$$(1 - \omega_1) + (\omega\omega'\mu^2 + 1 - \omega') = 2\bar{\lambda}(2\mu^2/\bar{\mu}^2 - 1) \tag{3.9}$$

which is satisfied for all μ if and only if

$$\omega_1 + \omega_1' = 2(1 + \bar{\lambda}) = 2\omega_b, \qquad \omega_1\omega_1' = 4\bar{\lambda}/\bar{\mu}^2 = \omega_b^2 \tag{3.10}$$

Since $\bar{\lambda} = \omega_b - 1$, these conditions imply

$$\omega_1 = \omega_1' = \omega_b \tag{3.11}$$

In the case $m = 2$, we have

$$\begin{aligned} b(\mu) &= (1 - \omega_1)P_2 + Q_4 \\ &= \mu^4(\omega_1\omega_1'\omega_2\omega_2') + \mu^2[\omega_1\omega_2'(1 - \omega_2) + \omega_2\omega_2'(1 - \omega_1') \\ &\quad + \omega_1\omega_1'(1 - \omega_2') + \omega_1'\omega_2(1 - \omega_1)] \\ &\quad + [(1 - \omega_1)(1 - \omega_2) + (1 - \omega_1')(1 - \omega_2')] \end{aligned} \tag{3.12}$$

Equating the power of μ^2 in the above expression to those of

$$2\bar{\lambda}T_2(2\mu^2/\bar{\mu}^2 - 1) = 2\bar{\lambda}[8(\mu^4/\bar{\mu}^4) - 8(\mu^2/\bar{\mu}^2) + 1] \tag{3.13}$$

we have

$$\begin{aligned} pp' &= \omega_b^4 \\ -ps' - sp' + ss' &= -16\bar{\lambda}/\bar{\mu}^2 \\ p - s + p' - s' &= 2\bar{\lambda} - 2 \end{aligned} \tag{3.14}$$

where

$$s = \omega_1 + \omega_2, \qquad s' = \omega_1' + \omega_2', \qquad p = \omega_1\omega_2, \qquad p' = \omega_1'\omega_2' \tag{3.15}$$

It is easy to verify that for any $p \neq 0$ the values

$$p' = \omega_b^4/p, \qquad s = p - \omega_b(\omega_b - 2), \qquad s' = p' - \omega_b(\omega_b - 2) \tag{3.16}$$

satisfy (3.14). The optimum parameters are

$$\begin{aligned} \omega_1 &= \tfrac{1}{2}[s + (s^2 - 4p)^{1/2}], & \omega_2 &= \tfrac{1}{2}[s - (s^2 - 4p)^{1/2}] \\ \omega_1' &= \tfrac{1}{2}[s' + ((s')^2 - 4p')^{1/2}], & \omega_2' &= \tfrac{1}{2}[s' - ((s')^2 - 4p')^{1/2}] \end{aligned} \tag{3.17}$$

Thus to each real or complex $p \neq 0$ there corresponds an optimum parameter set.

In the case $\omega_1 = \omega_1'$, $\omega_2 = \omega_2'$, there are two different parameter sets. Since $p = p'$ and $s = s'$ we have from (3.16)

$$p = \omega_b^2 \qquad \text{or} \qquad p = -\omega_b^2 \tag{3.18}$$

In the first case, we have $s = 2\omega_b$ and

$$\omega_1 = \omega_2 = \omega_b \tag{3.19}$$

In the second case, $s = 2\omega_b(1 - \omega_b)$ and

$$\begin{aligned} \omega_1 &= \omega_b[1 - \omega_b + [1 + (\omega_b - 1)^2]^{1/2}] \\ \omega_2 &= \omega_b[1 - \omega_b - [1 + (\omega_b - 1)^2]^{1/2}] \end{aligned} \tag{3.20}$$

or

$$\begin{aligned} \omega_1 &= \omega_b[1 - \omega_b - [1 + (\omega_b - 1)^2]^{1/2}] \\ \omega_2 &= \omega_b[1 - \omega_b + [1 + (\omega_b - 1)^2]^{1/2}] \end{aligned} \tag{3.20$'$}$$

As an example, consider the case where $r = \omega_b - 1 = 0.7$. In addition to the optimal parameter set

$$\omega_1 = \omega_2 = 1.7$$

we have the alternative optimal parameter set

$$\omega_1 \doteq 0.885, \qquad \omega_2 \doteq -3.27$$

This is perhaps somewhat surprising since the size of the negative parameter alone would result in a divergent process. Moreover, the use of the other parameter by itself would not yield as rapid convergence as would the use of $\omega_1 = \omega_2 = \omega_b = 1.7$. However, when used together, the two parameters give a convergence rate which is as good as the choice $\omega_1 = \omega_2 = 1.7$. This has been confirmed by actual numerical cases.

Let us now consider the case $m = 3$ where $\omega_k = \omega_k'$, $k = 1, 2, 3$. We obtain the following conditions from (3.8)

$$\begin{aligned} 2(1 - s + t - p) &= -2\bar{\lambda} = -2(\omega_b - 1)^3 \\ s(s - 2t + 3p) &= 36\bar{\lambda}/\bar{\mu}^2 \\ t^2 - sp - 2tp &= -96\bar{\lambda}/\bar{\mu}^4 \\ p^2 &= \omega_b^6 \end{aligned} \tag{3.21}$$

Here

$$s = \omega_1 + \omega_2 + \omega_3, \quad t = \omega_1\omega_2 + \omega_1\omega_3 + \omega_2\omega_3, \quad p = \omega_1\omega_2\omega_3 \tag{3.22}$$

In the first case we choose $p = \omega_b^3$. This leads to $s = 3\omega_b$ or $s = 3\omega_b \times (\omega_b - 1)^2$. If we let $s = 3\omega_b$, then $t = 3\omega_b^2$ and we obtain the solution $\omega_1 = \omega_2 = \omega_3 = \omega_b$. If $s = 3\omega_b(\omega_b - 1)^2$, then $t = 3\omega_b^2(\omega_b - 1)$ and the optimal parameters satisfy the cubic equation

$$\omega^3 - 3\omega_b(\omega_b - 1)^2\omega^2 + 3\omega_b^2(\omega_b - 1)\omega - \omega_b^3 = 0 \tag{3.23}$$

Solving, we obtain the roots

$$\omega_k = \omega_b/[(\omega_b - 1) - [(\omega_b - 1)^3 - 1]^{1/3}e^{2\pi ik/3}], \qquad k = 1, 2, 3 \tag{3.24}$$

Thus, there are, in general, two complex roots and one real root.

If we choose $p = -\omega_b^3$ we obtain the quadratic

$$s^2 - \omega_b(\omega_b^2 - 6\omega_b + 6)s + 9(\omega_b - 1)^2\omega_b^2 = 0 \tag{3.25}$$

Solving for s, we can obtain t from

$$t = s - 2\omega_b^3 + 3\omega_b^2 - 3\omega_b \tag{3.26}$$

We then determine for each t the optimum parameters from the cubic

$$\omega^3 - s\omega^2 + t\omega - p = 0 \tag{3.27}$$

Thus, in general, one obtains a total of four optimal parameter sets.

In the case $m = 4$ and $\omega_k = \omega_k'$, $k = 1, 2, 3, 4$, one can determine equations analogous to (3.21). Unfortunately, however, the equations do not involve only symmetric functions in the ω_k. However, numerical solutions have been obtained in specific cases. In each case eight sets were found. This leads to the conjecture that there are in general 2^{m-1} such sets.

Even in cases involving complex parameters, the alternative optimum parameter sets were found to be as effective, in terms of the number of iterations required, as the set $\omega_1 = \omega_1' = \cdots = \omega_b$. However, it appears that the existence of the alternative sets is primarily of academic rather than practical interest.

10.4. NORMS OF THE MSOR METHOD: SHELDON'S METHOD

In Chapter 7 we derived the following formulas for the $D^{1/2}$-norm and the $A^{1/2}$-norm of $\mathscr{L}_{\omega_b}^m$.

$$\| \mathscr{L}_{\omega_b}^m \|_{D^{1/2}} = r^m \{m(r^{-1/2} + r^{1/2}) + [m^2(r^{-1/2} + r^{1/2})^2 + 1]^{1/2}\} = F_1(m) \tag{4.1}$$

$$\| \mathscr{L}_{\omega_b}^m \|_{A^{1/2}} = r^m \{m(r^{-1/2} - r^{1/2}) + [m^2(r^{-1/2} - r^{1/2})^2 + 1]^{1/2}\} = H_1(m) \tag{4.2}$$

While the parameter choice $\omega_1 = \omega_1' = \cdots = \omega_m = \omega_m' = \omega_b$ minimizes the spectral radius of $\prod_{k=m}^{1} \mathscr{L}_{\omega_k,\omega_k'}$, we now consider other variants of the MSOR method which have smaller norms. The first such variant was considered by Sheldon [1959] and will be referred to as Sheldon's method. The parameters are

$$\omega_1 = \omega_1' = 1; \qquad \omega_2 = \omega_2' = \omega_3 = \omega_3' = \cdots = \omega_b \tag{4.3}$$

We denote the iteration matrix corresponding to Sheldon's method by $\mathscr{H}_m$. Evidently we have

$$\mathscr{H}_m = \mathscr{L}_{\omega_b}^{m-1}\mathscr{L} \tag{4.4}$$

By Theorem 2.1 we know that $\bar{S}(\mathscr{L}_{\omega_b}^m) \leq \bar{S}(\mathscr{H}_m)$; nevertheless, we do hope to obtain smaller $D^{1/2}$ or $A^{1/2}$ norms for $\mathscr{H}_m$ than for $\mathscr{L}_{\omega_b}^m$.

Theorem 4.1. If A is a positive definite matrix of the form (8-1.2), then

$$S(\mathscr{H}_m) = \frac{4r^m}{(1+r)^2}[1 + (1-r)(m-1)] = E_2(m) \tag{4.5}$$

$$\begin{aligned}\| \mathscr{H}_m \|_{D^{1/2}} &= \frac{4r^m}{(1+r)^2} \\ &\quad \times \left\{[1+(1-r)(m-1)]^2 + \frac{1}{r}\left[1 + (1-r)\left(m - \frac{3}{2}\right)\right]^2\right\}^{1/2} \\ &= F_2(m)\end{aligned} \tag{4.6}$$

$$\begin{aligned}\| \mathscr{H}_m \|_{A^{1/2}} &= \frac{2r^{m-1/2}}{1+r}\left\{1+4(m-1)\left(\frac{1-r}{1+r}\right)+4(m-1)^2\frac{(1-r)^2}{1+r}\right\}^{1/2} \\ &= H_2(m)\end{aligned} \tag{4.7}$$

where $r = \omega_b - 1$.

Proof. The matrix $M_m(\mu)$ corresponding to $\mathscr{H}_m$ is by Lemma 2.2

$$M_m(\mu) = \begin{pmatrix} 0 & R_{2m-1} \\ 0 & R_{2m} \end{pmatrix} \tag{4.8}$$

where the polynomials $R_0(\mu)$, $R_1(\mu)$, ... are given by

$$\begin{aligned} &R_0(\mu) = 1, \qquad R_1(\mu) = \mu, \qquad R_2(\mu) = \mu^2 \\ &R_k(\mu) = \omega_b \mu R_{k-1}(\mu) + (1 - \omega_b) R_{k-2}(\mu), \qquad k = 3, 4, \ldots \end{aligned} \tag{4.9}$$

We seek to show that

$$\max_{-\bar{\mu} \le \mu \le \bar{\mu}} | R_k(\mu) | = R_k(\bar{\mu}) \tag{4.10}$$

and that for $k \ge 1$

$$R_k(\bar{\mu}) = \frac{2[2 + (1 - r)(k - 2)]}{(1 + r)^2} r^{k/2} \tag{4.11}$$

Clearly, (4.10) holds for $k = 0, 1, 2$. By induction we can show that for $k = 3, 4, \ldots$ we have

$$R_k(\mu) = \mu^2 S_{k-2}(\mu) + \mu(1 - \omega_b) S_{k-3}(\mu) \tag{4.12}$$

where the $S_k(\mu)$ are given by (7-4.4). To prove (4.10) for $k \ge 3$, it is clearly sufficient to show that for $k \ge 1$ and $\omega = \omega_b$ we have

$$\max_{-\bar{\mu} \le \mu \le \bar{\mu}} | \mu S_k(\mu) + (1 - \omega) S_{k-1}(\mu) | = \bar{\mu} S_k(\bar{\mu}) + (1 - \omega) S_{k-1}(\bar{\mu}) \tag{4.13}$$

But by (7-4.7), we have, since $\alpha_1 \alpha_2 = \omega - 1$, $\alpha_1 + \alpha_2 = \omega\mu$, by (7-4.9),

$$\begin{aligned} \mu S_k(\mu) + (1 - \omega) S_{k-1}(\mu) &= \mu \sum_{i=0}^{k} \alpha_1^{k-i} \alpha_2^{\,i} - \alpha_1 \alpha_2 \sum_{i=0}^{k-1} \alpha_1^{k-1-i} \alpha_2^{\,i} \\ &= \frac{1}{\omega} (\alpha_1^{k+1} + \alpha_2^{k+1}) \\ &\quad + \left(\frac{2}{\omega} - 1\right) \alpha_1 \alpha_2 \sum_{i=0}^{k-1} \alpha_1^{k-1-i} \alpha_2^{\,i} \end{aligned} \tag{4.14}$$

If $\omega = \omega_b = 2(1 + [1 - \bar{\mu}^2]^{1/2})^{-1}$ and if $0 \le \mu^2 \le \bar{\mu}^2$, then the discriminant of the quadratic (7-4.8) is nonpositive since $\omega_b^{\,2}\mu^2 - 4(\omega_b - 1) \le \omega_b^{\,2}\bar{\mu}^2 - 4(\omega_b - 1) = 0$ and we have $| \alpha_1 | = | \alpha_2 | = r^{1/2}$. Moreover,

$$\begin{aligned} | \mu S_k(\mu) + (1 - \omega_b) S_{k-1}(\mu) | &\le \left[\left(\frac{2}{\omega_b} - 1\right) k + \frac{2}{\omega_b}\right] r^{(k+1)/2} \\ &= \frac{2 + (1 - r)k}{1 + r} r^{(k+1)/2} \end{aligned} \tag{4.15}$$

On the other hand, if $\mu = \bar{\mu}$, then $\alpha_1 = \alpha_2 = r^{1/2}$ and $S_k(\bar{\mu}) = (k+1)r^{k/2}$ by (7-4.7). Therefore, since $\bar{\mu} = 2r^{1/2}(1 + r)^{-1}$ we have

$$\bar{\mu}S_k(\bar{\mu}) + (1 - \omega_b)S_{k-1}(\bar{\mu}) = \frac{2 + (1 - r)k}{1 + r} r^{(k+1)/2} \tag{4.16}$$

and (4.13) follows. Hence (4.10) follows. It is easy to show that (4.11) holds for $k = 1, 2$. For $k \geq 3$, (4.11) follows by (4.16) and (4.13). From (4.11) and (4.8) the result (4.5) follows. Moreover, (4.6) follows since by (4.8) we have

$$\| M_m(\bar{\mu}) \| = [R_{2m}^2(\mu) + R_{2m-1}^2(\mu)]^{1/2}$$

To prove (4.7) we seek, as in the proof of Theorem 7-6.1, the eigenvalues of RR^{T} where

$$R = N(\mu)^{1/2} M_m(\mu) N(\mu)^{-1/2} \tag{4.17}$$

and where $N(\mu)$ and $M_m(\mu)$ are given by (7-6.5) and by (4.8), respectively. For the case $m = 1$, we have by (4.8) and (7-6.7)

$$R = \frac{1}{1 - p^2} \begin{pmatrix} 1 & -p \\ -p & 1 \end{pmatrix} \begin{pmatrix} 0 & \mu \\ 0 & \mu^2 \end{pmatrix} \begin{pmatrix} 1 & p \\ p & 1 \end{pmatrix} = \frac{\mu^2}{2} \begin{pmatrix} 1 & 1/p \\ p & 1 \end{pmatrix} \tag{4.18}$$

by (7-6.11). Since $\det R = 0$, the characteristic equation of RR^{T} is

$$\begin{aligned} \Gamma^2 - \operatorname{trace}(RR^{\mathrm{T}})\Gamma &= \Gamma^2 - \frac{\mu^4}{4}\left(2 + p^2 + \frac{1}{p^2}\right)\Gamma \\ &= \Gamma^2 - \frac{\mu^4}{4}(4\mu^{-2})\Gamma = \Gamma^2 - \mu^2\Gamma = 0 \end{aligned} \tag{4.19}$$

by (7-6.13) or

$$\Gamma = 0, \quad \mu^2$$

Hence,

$$\| \mathscr{H}_1 \|_{A^{1/2}} = \bar{\mu} = 2r^{1/2}(1 + r)^{-1} \tag{4.20}$$

and (4.7) holds for $m = 1$.

From $m \geq 2$ we have from (4.8), (4.12), and (7-4.4)

$$\begin{aligned} M_m(\mu) &= \begin{pmatrix} 0 & \mu^2 S_{2m-3} + \mu(1 - \omega)S_{2m-4} \\ 0 & \mu^2 S_{2m-2} + \mu(1 - \omega)S_{2m-3} \end{pmatrix} \\ &= \mu S_{2m-2}\begin{pmatrix} 0 & 1 \\ 0 & \mu \end{pmatrix} + \mu(1 - \omega)S_{2m-3}\begin{pmatrix} 0 & \mu \\ 0 & 1 \end{pmatrix} \end{aligned} \tag{4.21}$$

Therefore,

$$R = N(\mu)^{1/2}M_m(\mu)M(\mu)^{-1/2}$$
$$= \frac{\mu^2}{2}\begin{pmatrix} S_{2m-2} + (1-\omega)pS_{2m-3} & \frac{1}{p}S_{2m-2} + (1-\omega)S_{2m-3} \\ pS_{2m-2} + (1-\omega)S_{2m-3} & S_{2m-2} + (1-\omega)\frac{1}{p}S_{2m-3} \end{pmatrix} \tag{4.22}$$

and

$$\begin{aligned} T_m(\mu) &= \text{trace } RR^{\mathrm{T}} \\ &= \frac{\mu^4}{4}\left(2 + p^2 + \frac{1}{p^2}\right)(S_{2m-2}^2 + (1-\omega)^2 S_{2m-3}^2) \\ &\quad + \mu^4 S_{2m-2}S_{2m-3}\left(p + \frac{1}{p}\right)(1-\omega) \\ &= \mu^2(S_{2m-2}^2 + (1-\omega)^2 S_{2m-3}^2) + 2\mu^3(1-\omega)S_{2m-2}S_{2m-3} \end{aligned} \tag{4.23}$$

by (7-6.13). Moreover, by (7-6.15), we have

$$\begin{aligned} (1-\omega)^{2m-2} &= (1-\omega)S_{2m-2}S_{2m-4} - (1-\omega)S_{2m-3}^2 \\ &= S_{2m-2}[S_{2m-2} - \omega\mu S_{2m-3}] - (1-\omega)S_{2m-3}^2 \end{aligned} \tag{4.24}$$

by (7-6.12). Therefore,

$$\omega\mu S_{2m-2}S_{2m-3} = S_{2m-2}^2 + (\omega - 1)S_{2m-3}^2 - (1-\omega)^{2m-2}$$

and

$$\frac{1}{\mu^2}T_m(\mu) = \frac{2-\omega}{\omega}[S_{2m-2}^2 - (1-\omega)^2 S_{2m-3}^2] + \frac{2}{\omega}(\omega - 1)^{2m-1} \tag{4.25}$$

Let $\omega = \omega_b$ and let $\mu = \bar{\mu}\cos\theta$ where $0 \le \theta \le \pi$. By (7-4.8) we have

$$\alpha_1 = r^{1/2}e^{i\theta}, \qquad \alpha_2 = r^{1/2}e^{-i\theta} \tag{4.26}$$

so that by (7-4.7)

$$S_k = \begin{cases} (r^{1/2})^k \dfrac{\sin(k+1)\theta}{\sin\theta} & \text{if } \mu \ne \bar{\mu}, -\bar{\mu}; \quad \theta \ne 0, \pi \\ (k+1)(r^{1/2})^k & \text{if } \mu = \bar{\mu}; \quad \theta = 0 \\ (-1)^k(k+1)(r^{1/2})^k & \text{if } \mu = -\bar{\mu}; \quad \theta = \pi \end{cases} \tag{4.27}$$

If $\theta \neq 0, \pi$ we have

$$S_{2m-2}^2 - (1-\omega_b)^2 S_{2m-3}^2 = r^{2m-2}\left\{\frac{\sin^2(2m-1)\theta - r\sin^2(2m-2)\theta}{\sin^2\theta}\right\}$$

$$= \frac{r^{2m-2}}{\sin^2\theta}\{(1-r)\sin^2(2m-2)\theta$$

$$+ [\sin^2(2m-1)\theta - \sin^2(2m-2)\theta]\} \quad (4.28)$$

But $\sin^2(2m-1)\theta - \sin^2(2m-2)\theta = \frac{1}{2}[\cos(4m-4)\theta - \cos(4m-2)\theta] = \sin(4m-3)\theta \sin\theta$; hence

$$S_{2m-2}^2 - (1-\omega_b)^2 S_{2m-3}^2 = r^{2m-2}\left\{(1-r)\frac{\sin^2(2m-2)\theta}{\sin^2\theta} + \frac{\sin(4m-3)\theta}{\sin\theta}\right\}$$

$$= (1-r)rS_{2m-3}^2 + S_{4m-4} \quad (4.29)$$

Clearly, by (7-4.10) the maximum value of the above expression is assumed when $\mu = \bar{\mu}$, and we have

$$S_{2m-2}^2(\bar{\mu}) - r^2 S_{2m-3}^2(\bar{\mu}) = r^{2m-2}(2m-1)^2 - r^2 r^{2m-3}(2m-2)^2$$

$$= r^{2m-2}\{(2m-1)^2 - r(2m-2)^2\} \quad (4.30)$$

Therefore, by (4.25) we have, since $r = \omega_b - 1$,

$$\max_{0 \le \mu^2 \le \bar{\mu}^2} T_m(\mu) = \bar{\mu}^2\left\{\frac{1-r}{1+r}r^{2m-2}[(2m-1)^2 - r(2m-2)^2] + \frac{2}{1+r}r^{2m-1}\right\}$$

$$= \bar{\mu}^2\frac{r^{2m-2}}{1+r}\{1+r+4(1-r)(m-1)+4(1-r)^2(m-1)^2\}$$

$$(4.31)$$

Since $\det RR^{\mathrm{T}} = 0$, it follows that $\| \mathscr{H}_m \|_{A^{1/2}} = (T_m(\bar{\mu}))^{1/2}$, and the result (4.7) follows since $\bar{\mu}^2 = 4r(1+r)^{-2}$. This completes the proof of Theorem 4.1.

10.5. THE MODIFIED SHELDON METHOD

Golub and Varga [1961] considered a modification of Sheldon's method wherein $\omega_1 = 1$ and $\omega_1' = \omega_2 = \omega_2' = \cdots = \omega_b$. We define the matrix $\mathscr{K}_m$ by

$$\mathscr{K}_m = \mathscr{L}_{\omega_b}^{m-1}\mathscr{L}_{1,\omega_b} \quad (5.1)$$

We now prove

Theorem 5.1. If A is a positive definite matrix of the form (8-1.2), then

$$S(\mathscr{K}_m) = \frac{2r^m}{1+r}\left\{1 + (1-r)\left(m - \frac{1}{2}\right)\right\} = E_4(m) \tag{5.2}$$

$$\begin{aligned}\| \mathscr{K}_m \|_{D^{1/2}} &= \frac{2r^m}{1+r}\left\{\left[1+(1-r)\left(m-\frac{1}{2}\right)\right]^2 + \frac{1}{r}\,[1+(1-r)(m-1)]^2\right\}^{1/2} \\ &= F_4(m)\end{aligned} \tag{5.3}$$

$$\begin{aligned}\| \mathscr{K}_m \|_{A^{1/2}} &= r^{m-1/2}\left\{1 + 4\left(m - \frac{1}{2}\right)\left(\frac{1-r}{1+r}\right) + 4\left(m - \frac{1}{2}\right)^2 \frac{(1-r)^2}{1+r}\right\}^{1/2} \\ &= H_4(m)\end{aligned} \tag{5.4}$$

where $r = \omega_b - 1$.

Proof. By Lemma 2.2, the matrix $M_m(\mu)$ corresponding to $\mathscr{K}_m$ is

$$M_m(\mu) = \begin{pmatrix} 0 & U_{2m-1} \\ 0 & U_{2m} \end{pmatrix} \tag{5.5}$$

where the polynomials $U_0(\mu)$, $U_1(\mu)$, ... satisfy

$$\begin{aligned} U_0(\mu) &= 1 \\ U_1(\mu) &= \mu \\ U_k(\mu) &= \omega_b \mu U_{k-1}(\mu) + (1-\omega_b)U_{k-2}(\mu), \qquad k = 2, 3, \ldots \end{aligned} \tag{5.6}$$

It is easy to show that for $k = 2, 3, \ldots$, we have

$$U_k(\mu) = \mu S_{k-1}(\mu) + (1-\omega_b)S_{k-2}(\mu) \tag{5.7}$$

Hence by (4.13) and (4.27) we have

$$\begin{aligned}\max_{|\mu| \le \bar{\mu}} | U_k(\mu) | &= r^{k/2}\left[1 + k\left(\frac{1-r}{1+r}\right)\right] \\ &= \frac{2r^{k/2}}{1+r}\left[1 + \left(\frac{k-1}{2}\right)(1-r)\right]\end{aligned} \tag{5.8}$$

The results (5.2) and (5.3) follow.

From (7-4.4), (5.5), and (5.7) we have

$$M_m(\mu) = S_{2m-1}\begin{pmatrix} 0 & 1 \\ 0 & \mu \end{pmatrix} + (1-\omega_b)S_{2m-2}\begin{pmatrix} 0 & \mu \\ 0 & 1 \end{pmatrix} \tag{5.9}$$

Letting $R = N(\mu)^{1/2}M_m(\mu)N(\mu)^{-1/2}$ we have by the methods used to prove (4.31)

$$T_{\max} = \max_{|\mu| \le \bar{\mu}} \operatorname{trace} RR^{\mathrm{T}}$$
$$= \frac{r^{2m-1}}{1+r} \{1 + r + (4m-2)(1-r) + (2m-1)^2(1-r)^2\} \tag{5.10}$$

The result now follows since $\|\mathscr{K}_m\|_{A^{1/2}} = (T_{\max})^{1/2}$.

We remark that since $\bar{\mu} = 2r^{1/2}(1+r)^{-1}$ we have

$$E_4(m) = \frac{1}{\bar{\mu}} E_2(m+\tfrac{1}{2}) = \frac{1+r}{2r^{1/2}} E_2(m+\tfrac{1}{2})$$
$$F_4(m) = \frac{1}{\bar{\mu}} F_2(m+\tfrac{1}{2}) = \frac{1+r}{2r^{1/2}} F_2(m+\tfrac{1}{2}) \tag{5.11}$$
$$H_4(m) = \frac{1}{\bar{\mu}} H_2(m+\tfrac{1}{2}) = \frac{1+r}{2r^{1/2}} H_2(m+\tfrac{1}{2})$$

10.6. CYCLIC CHEBYSHEV SEMI-ITERATIVE METHOD

We now consider the MSOR method with the following parameter choice

$$\begin{aligned} \omega_1 &= 1, & \omega_1' &= 2/(2-\bar{\mu}^2) \\ \omega_k &= (1 - \tfrac{1}{4}\omega_{k-1}'\bar{\mu}^2)^{-1}, & \omega_k' &= (1 - \tfrac{1}{4}\omega_k\bar{\mu}^2)^{-1}, \quad k = 2, 3, \ldots \end{aligned} \tag{6.1}$$

These parameters correspond to the "cyclic Chebyshev semi-iterative method" which we shall also study in Chapter 11. Let us define $\mathscr{C}_m$ by

$$\mathscr{C}_m = \prod_{k=m}^{1} \mathscr{L}_{\omega_k, \omega_k'} \tag{6.2}$$

where the ω_k and ω_k' are given by (6.1).

We now prove

Theorem 6.1. If A is a positive definite matrix of the form (8-1.2), then

$$S(\mathscr{C}_m) = \frac{2r^m}{1+r^{2m}} = E_3(m) \tag{6.3}$$

$$\| \mathscr{C}_m \|_{D^{1/2}} = \frac{2r^m}{1 + r^{2m}} \left[1 + \frac{1}{r} \left(\frac{1 + r^{2m}}{1 + r^{2m-1}} \right)^2 \right]^{1/2} = F_3(m) \tag{6.4}$$

$$\| \mathscr{C}_m \|_{A^{1/2}} = 2r^{m-1/2} \frac{[(1 + r)(1 + r^{4m-1})]^{1/2}}{(1 + r^{2m})(1 + r^{2m-1})} = H_3(m) \tag{6.5}$$

where $r = \omega_b - 1$.

Proof. The matrix $M_m(\mu)$ corresponding to $\mathscr{C}_m$ is by Lemma 2.2

$$M_m(\mu) = \begin{pmatrix} 0 & C_{2m-1} \\ 0 & C_{2m} \end{pmatrix} \tag{6.6}$$

where the polynomials $C_0(\mu)$, $C_1(\mu)$, ... are the $Q_0(\mu)$, $Q_1(\mu)$, ... given by (2.11) for the particular choise of ω_k and ω_k' given by (6.1). Let us introduce the sequence $\hat{\omega}_1$, $\hat{\omega}_2$, ... where

$$\omega_k = \hat{\omega}_{2k-1}, \qquad \omega_k' = \hat{\omega}_{2k}, \qquad k = 1, 2, \ldots \tag{6.7}$$

From (6.1) we have

$$\begin{aligned} \hat{\omega}_1 &= 1, \qquad \hat{\omega}_2 = \left(1 - \frac{\bar{\mu}^2}{2}\right)^{-1} \\ \hat{\omega}_k &= \left(1 - \frac{\hat{\omega}_{k-1}\bar{\mu}^2}{4}\right)^{-1}, \qquad k = 3, 4, \ldots \end{aligned} \tag{6.8}$$

and from (2.11) we obtain

$$\begin{aligned} C_0 &= 1, \qquad C_1 = \mu \\ C_{2k} &= \hat{\omega}_{2k}\mu C_{2k-1} + (1 - \hat{\omega}_{2k})C_{2k-2}, \qquad k = 1, 2, \ldots \\ C_{2k+1} &= \hat{\omega}_{2k+1}\mu C_{2k} + (1 - \hat{\omega}_{2k+1})C_{2k-1}, \qquad k = 1, 2, \ldots \end{aligned} \tag{6.9}$$

Evidently the above relations can be replaced by

$$\begin{aligned} C_0 &= 1, \qquad C_1 = \mu \\ C_k &= \hat{\omega}_k\mu C_{k-1} + (1 - \hat{\omega}_k)C_{k-2}, \qquad k = 2, 3, \ldots \end{aligned} \tag{6.10}$$

We now seek to show that $C_k(\mu) \equiv W_k(\mu)$ for $k = 0, 1, 2, \ldots$ where

$$W_k(\mu) = \frac{T_k(\mu/\bar{\mu})}{T_k(1/\bar{\mu})} \tag{6.11}$$

where $T_k(x)$ is the Chebyshev polynomial of degree k defined in Section 9.3 by (9-3.1). We seek to show that the $W_k(\mu)$ satisfy the recurrence relation (6.10). By (9-3.1) and (6.11) we have

$$\begin{aligned} W_k(\mu) &= \frac{2(\mu/\bar{\mu})T_{k-1}(\mu/\bar{\mu}) - T_{k-2}(\mu/\bar{\mu})}{T_k(1/\bar{\mu})} \\ &= \varrho_k \mu W_{k-1}(\mu) + (1 - \varrho_k) W_{k-2}(\mu) \end{aligned} \tag{6.12}$$

where we define the ϱ_k by

$$\begin{aligned} \varrho_1 &= 1 \\ \varrho_k &= \frac{2}{\bar{\mu}} \frac{T_{k-1}(1/\bar{\mu})}{T_k(1/\bar{\mu})}, \qquad k = 2, 3, \ldots \end{aligned} \tag{6.13}$$

For $k \geq 2$ we have

$$\begin{aligned} \varrho_k^{-1} &= \frac{\bar{\mu}}{2} \frac{(2/\bar{\mu})T_{k-1}(1/\bar{\mu}) - T_{k-2}(1/\bar{\mu})}{T_{k-1}(1/\bar{\mu})} \\ &= 1 - \frac{\bar{\mu}^2}{4} \left(\frac{2}{\bar{\mu}} \frac{T_{k-2}(1/\bar{\mu})}{T_{k-1}(1/\bar{\mu})} \right) \end{aligned} \tag{6.14}$$

Thus if $k \geq 3$, we have by (6.13)

$$\varrho_k^{-1} = 1 - \tfrac{1}{4}\bar{\mu}^2 \varrho_{k-1} \tag{6.15}$$

and

$$\varrho_2 = \frac{2}{\bar{\mu}} \frac{T_1(1/\bar{\mu})}{T_2(1/\bar{\mu})} = \frac{2}{\bar{\mu}} \frac{1/\bar{\mu}}{(2/\bar{\mu}^2) - 1} = \frac{2}{2 - \bar{\mu}^2} = \omega_2 \tag{6.16}$$

Hence $\varrho_1 = \hat{\omega}_1$, $\varrho_2 = \hat{\omega}_2$ and for $k \geq 3$ both ϱ_k and $\hat{\omega}_k$ satisfy the same recurrence relation (6.15). Therefore $\varrho_k = \hat{\omega}_k$ for all k. Moreover, since $W_0(\mu) = 1$, $W_1(\mu) = \mu$ and since both $C_k(\mu)$ and $W_k(\mu)$ satisfy the same recurrence relation (6.12) for $k \geq 2$, it follows that $C_k(\mu) \equiv W_k(\mu)$ for all k. Thus we have

$$C_k(\mu) = \frac{T_k(\mu/\bar{\mu})}{T_k(1/\bar{\mu})}, \qquad k = 0, 1, 2, \ldots \tag{6.17}$$

Evidently

$$\max_{-\bar{\mu} \leq \mu \leq \bar{\mu}} | C_k(\mu) | = C_k(\bar{\mu}) = \frac{1}{T_k(1/\bar{\mu})} \tag{6.18}$$

since $\max_{|x| \leq 1} | T_k(x) | = \max_{|x| \leq 1} | \cos(k \cos^{-1} x) | = 1$. Moreover, by (9-3.3) and (6-1) we have

$$[T_k(1/\bar{\mu})]^{-1} = 2r^{k/2}/(1 + r^k) \tag{6.19}$$

where $r = \omega_b - 1$. Therefore

$$\max_{-\bar\mu \le \mu \le \bar\mu} | C_k(\mu) | = C_k(\bar\mu) = 2r^{k/2}/(1 + r^k) \tag{6.20}$$

From (6.6) it follows at once that

$$S(T_m) = \max_{-\bar\mu \le \mu \le \bar\mu} S(M_m(\mu)) = \max_{-\bar\mu \le \mu \le \bar\mu} | C_{2m}(\mu) | = 2r^m/(1+r^{2m}) \tag{6.21}$$

and hence (6.3) follows. Similarly

$$\begin{aligned} \| \mathscr{C}_m \|_{D^{1/2}}^{2} &= \max_{-\bar\mu \le \mu \le \bar\mu} [S(M_m(\mu)M_m(\mu)^{\mathrm{T}})]^{1/2} \\ &= \max_{-\bar\mu \le \mu \le \bar\mu} [C_{2m}(\mu)^2 + C_{2m-1}^2(\mu)]^{1/2} \end{aligned} \tag{6.22}$$

and (6.4) follows from (6.20).

To prove (6.5) we seek, as in the proof of Theorem 4.1, the eigenvalues of RR^{T} where

$$R = N(\mu)^{1/2} M_m(\mu) N(\mu)^{-1/2} \tag{6.23}$$

and where $N(\mu)$ and $M_m(\mu)$ are given by (7-6.5) and (6.6), respectively. Evidently by (7-6.7) we have

$$R = \frac{1}{1-p^2}\begin{pmatrix} pC_{2m-1} - p^2 C_{2m} & C_{2m-1} - pC_{2m} \\ -p^2 C_{2m-1} + pC_{2m} & -pC_{2m-1} + C_{2m} \end{pmatrix} \tag{6.24}$$

Moreover,

$$\begin{aligned} T_m(\mu) &= \operatorname{trace} RR^{\mathrm{T}} \\ &= \frac{1}{(1-p^2)^2}[(1+p^2)^2(C_{2m-1}^2 + C_{2m}^2) - 4p(1+p^2)C_{2m}C_{2m-1}] \\ &= \frac{p^2}{(1-p^2)^2}\left[\left(p + \frac{1}{p}\right)^2 (C_{2m-1}^2 + C_{2m}^2) - 4\left(p + \frac{1}{p}\right) C_{2m}C_{2m-1}\right] \\ &= \frac{p^2}{(1-p^2)^2}\left[\frac{4}{\mu^2}(C_{2m-1}^2 + C_{2m}^2) - \frac{8}{\mu} C_{2m}C_{2m-1}\right] \\ &= \frac{4p^2}{\mu^2(1-p^2)^2}[C_{2m-1}^2 + C_{2m}^2 - 2\mu C_{2m}C_{2m-1}] \end{aligned} \tag{6.25}$$

by (7-6.13). But

$$\frac{4p^2}{\mu^2(1-p^2)^2} = \frac{4}{\mu^2\left(\dfrac{1}{p} - p\right)^2} = \frac{1}{1-\mu^2} \tag{6.26}$$

since by (7-6.8) we have

$$\begin{aligned} p &= \frac{\mu}{1+(1-\mu^2)^{1/2}} = \frac{1-(1-\mu^2)^{1/2}}{\mu} \\ \frac{1}{p} &= \frac{1+(1-\mu^2)^{1/2}}{\mu} \\ \frac{1}{p} - p &= \frac{2(1-\mu^2)^{1/2}}{\mu} \end{aligned} \tag{6.27}$$

Therefore,

$$T_m(\mu) = \frac{1}{1-\mu^2}(C_{2m-1}^2 + C_{2m}^2 - 2\mu C_{2m}C_{2m-1}) = \phi_{2m}(\mu) \tag{6.28}$$

where

$$\phi_k(\mu) = \frac{1}{1-\mu^2}(C_k^2 + C_{k-1}^2 - 2\mu C_k C_{k-1}) \tag{6.29}$$

Evidently

$$\phi_k(\mu) = \frac{1}{1-\mu^2}(C_k - \mu C_{k-1})^2 + C_{k-1}^2 \tag{6.30}$$

But by (6.10) we have, for $k \geq 2$

$$\begin{aligned} \phi_k(\mu) &= \frac{(\hat{\omega}_k - 1)^2}{1-\mu^2}(\mu C_{k-1} - C_{k-2})^2 + C_{k-1}^2 \\ &= \frac{(\hat{\omega}_k - 1)^2}{1-\mu^2}(C_{k-2}^2 + C_{k-1}^2 - 2\mu C_{k-1}C_{k-2}) + (1-(\hat{\omega}_k - 1)^2)C_{k-1}^2 \\ &= (\hat{\omega}_k - 1)^2\phi_{k-1} + (1-(\hat{\omega}_k - 1)^2)C_{k-1}^2 \end{aligned} \tag{6.31}$$

We now seek to show that $\phi_k(\mu)$ is maximized in the range $-\bar{\mu} \leq \mu \leq \bar{\mu}$ when $\mu = \bar{\mu}$. Since $C_0 = 1$ and $C_1 = \mu$, by (6.10), we have by (6.29)

$$\phi_1(\mu) = \frac{1}{1-\mu^2}(1 + \mu^2 - 2\mu^2) = 1 \tag{6.32}$$

Moreover, by (6.31)

$$\begin{aligned} \phi_2(\mu) &= (\hat{\omega}_2 - 1)^2\phi_1 + (1-(\hat{\omega}_2 - 1)^2)C_1^2 \\ &= (\hat{\omega}_2 - 1)^2 + (1-(\hat{\omega}_2 - 1)^2)\mu^2 \end{aligned} \tag{6.33}$$

which is clearly maximized when $\mu = \bar{\mu}$ since $1 \leq \hat{\omega}_2 < 2$. Since $|C_k(\mu)|$

is clearly maximized when $\mu = \bar{\mu}$, it follows from (6.31) and the use of mathematical induction that $\phi_k(\mu)$ is also maximized. Therefore

$$\max_{-\bar{\mu}\le\mu\le\bar{\mu}} |\phi_k(\mu)| = \phi_k(\bar{\mu}) \tag{6.34}$$

But by (6.29) and (6.20) we have

$$\begin{aligned}
\phi_k(\bar{\mu}) &= \frac{1}{1-\bar{\mu}^2}\Big[\frac{1}{T_k(1/\bar{\mu})^2} + \frac{1}{T_{k-1}(1/\bar{\mu})^2} \\
&\quad -2\bar{\mu}\Big[\frac{1}{T_k(1/\bar{\mu})}\Big]\Big[\frac{1}{T_{k-1}(1/\bar{\mu})}\Big]\Big] \\
&= \frac{1}{(1-\bar{\mu}^2)}\,\frac{1}{T_k(1/\bar{\mu})^2 T_{k-1}(1/\bar{\mu})^2} \\
&\quad \times\Big[T_k\Big(\frac{1}{\bar{\mu}}\Big)^2 + T_{k-1}\Big(\frac{1}{\bar{\mu}}\Big)^2 - 2\bar{\mu}T_k\Big(\frac{1}{\bar{\mu}}\Big)T_{k-1}\Big(\frac{1}{\bar{\mu}}\Big)\Big]
\end{aligned} \tag{6.35}$$

But for $\bar{\mu} < 1$ we have

$$T_k(1/\bar{\mu}) = \cosh k\theta$$

where $\theta = \cosh^{-1}(1/\bar{\mu})$. Therefore

$$\begin{aligned}
&T_k(1/\bar{\mu})^2 + T_{k-1}(1/\bar{\mu})^2 - 2\bar{\mu}T_k(1/\bar{\mu})T_{k-1}(1/\bar{\mu}) \\
&\quad = \cosh^2 k\theta + \cosh^2(k-1)\theta - 2\bar{\mu}\cosh k\theta\cosh(k-1)\theta \\
&\quad = 1 + \tfrac{1}{2}\cosh 2k\theta + \tfrac{1}{2}\cosh[2(k-1)\theta] - 2\bar{\mu}\cosh k\theta\cosh[(k-1)\theta] \\
&\quad = 1 + \cosh[(2k-1)\theta]\cosh\theta - \bar{\mu}[\cosh[(2k-1)\theta] + \cosh\theta] \\
&\quad = (1/\bar{\mu} - \bar{\mu})\cosh[(2k-1)\theta] \\
&\quad = (1/\bar{\mu} - \bar{\mu})T_{2k-1}(1/\bar{\mu})
\end{aligned} \tag{6.36}$$

Since $\det R = \det M_m(\mu) = 0$ by (6.23) and (6.6) it follows that

$$S(RR^{\mathrm{T}}) = \max_{-\bar{\mu}\le\mu\le\bar{\mu}} (\text{trace } RR^{\mathrm{T}}) = \max_{-\bar{\mu}\le\mu\le\bar{\mu}} (T_m(\mu)) = T_m(\bar{\mu}) \tag{6.37}$$

by (6.25), (6.28), and (6.34). Hence, by (6.35) and (6.36) we have

$$\mathscr{E}_m(\bar{\mu}) = \phi_{2m}(\bar{\mu}) = \frac{1}{\bar{\mu}}\,\frac{T_{4m-1}(1/\bar{\mu})}{T_{2m}(1/\bar{\mu})^2 T_{2m-1}(1/\bar{\mu})^2} \tag{6.38}$$

and the result (6.5) follows from (6.19). This completes the proof of Theorem 6.1.

10.7. COMPARISON OF NORMS

In this section we compare the various methods on the basis of the formulas for the $D^{1/2}$-norms and the $A^{1/2}$-norms. For a given method we define the *theoretical number of iterations* as the solution m of the equation

$$J(m) = \varrho \tag{7.1}$$

where $J(m)$ is the $D^{1/2}$-norm or the $A^{1/2}$-norm of the method and where ϱ is a number in the interval $0 < \varrho < 1$. We shall show that for each method, m exists if ϱ is small enough. We shall also show that

$$m \geq \log \varrho^{-1} / \log r^{-1} \tag{7.2}$$

Hence $m \to \infty$ as $r \to 1-$ for fixed ϱ or as $\varrho \to 0$ for fixed r.

To summarize the results obtained in Sections 7.4 and 7.6, and in Sections 10.3–10.6, we have

(1) SOR $(\mathscr{L}_{\omega_b}^m)$

$$\begin{aligned} E_1(m) &= r^m \\ F_1(m) &= r^m\{m(r^{-1/2} + r^{1/2}) + [m^2(r^{-1/2} + r^{1/2})^2 + 1]^{1/2}\} \\ H_1(m) &= r^m\{m(r^{-1/2} - r^{1/2}) + [m^2(r^{-1/2} - r^{1/2})^2 + 1]^{1/2}\} \end{aligned} \tag{7.3}$$

(2) Sheldon $(\mathscr{H}_m)$

$$\begin{aligned} E_2(m) &= \frac{4r^m}{(1+r)^2}[1 + (1-r)(m-1)] \\ F_2(m) &= \frac{4r^m}{(1+r)^2}\left\{[1 + (1-r)(m-1)]^2 + \frac{1}{r}\left[1 + (1-r)\left(m - \frac{3}{2}\right)\right]^2\right\}^{1/2} \\ H_2(m) &= \frac{2r^{m-1/2}}{1+r}\left\{1 + 4(m-1)\left(\frac{1-r}{1+r}\right) + 4(m-1)^2\frac{(1-r)^2}{1+r}\right\}^{1/2} \end{aligned} \tag{7.4}$$

(3) CCSI $(\mathscr{C}_m)$

$$\begin{aligned} E_3(m) &= \frac{2r^m}{1+r^{2m}} \\ F_3(m) &= \frac{2r^m}{1+r^{2m}}\left[1 + \frac{1}{r}\left(\frac{1+r^{2m}}{1+r^{2m-1}}\right)^2\right]^{1/2} \\ H_3(m) &= 2r^{m-1/2}\frac{[(1+r)(1+r^{4m-1})]^{1/2}}{(1+r^{2m})(1+r^{2m-1})} \end{aligned} \tag{7.5}$$

(4) Modified Sheldon ($\mathscr{K}_m$)

$$
\begin{aligned}
E_4(m) &= \frac{2r^m}{1+r}\left[1 + (1-r)\left(m - \frac{1}{2}\right)\right] \\
F_4(m) &= \frac{2r^m}{1+r}\left\{\left[1 + (1-r)\left(m - \frac{1}{2}\right)\right]^2 \right. \\
&\qquad \left. + \frac{1}{r}\,[1 + (1-r)(m-1)]^2\right\}^{1/2} \qquad (7.6) \\
H_4(m) &= r^{m-1/2}\left[1 + 4\left(m - \frac{1}{2}\right)\left(\frac{1-r}{1+r}\right) + 4\left(m - \frac{1}{2}\right)^2 \frac{(1-r)^2}{1+r}\right]^{1/2}
\end{aligned}
$$

Let us first show the existence of the theoretical number of iterations. We prove

Theorem 7.1. If $0 < \varrho < \varrho_0$ and if $0 < r < 1$, then there exists a unique solution m of the equation (7.1) such that $m \geq m_0$. Moreover, (7.2) holds. Here ϱ_0 and m_0 are given by the following table.

Method	Norm	ϱ_0	m_0
SOR ($\mathscr{L}_{\omega_b}^m$)	$D^{1/2}$	1	0
	$A^{1/2}$	1	$\frac{1}{2}$
Sheldon ($\mathscr{H}_m$)	$D^{1/2}$	$(\bar{\mu}^2 + \bar{\mu}^4)^{1/2}$	1
	$A^{1/2}$	$\bar{\mu}$	1
CCSI ($\mathscr{C}_m$)	$D^{1/2}$	1	$\frac{1}{2}$
	$A^{1/2}$	1	$\frac{1}{2}$
Modified Sheldon ($\mathscr{K}_m$)	$D^{1/2}$	1	$\frac{1}{2}$
	$A^{1/2}$	1	$\frac{1}{2}$

Proof. We have already shown the existence and uniqueness of m for $\mathscr{L}_{\omega_b}^m$ in Sections 7.4 and 7.6. For the $D^{1/2}$-norms of the other methods we have

$$F_i(m) = [E_i(m)^2 + E_i(m - \tfrac{1}{2})^2]^{1/2}, \qquad i = 2, 3, 4 \tag{7.7}$$

We can show that $F_i(m)$ is a decreasing function of m if we can show that $E_i(m)$ and $E_i(m - \frac{1}{2})$ are decreasing. But for the case $i = 2$ we have by (7.4)

$$E_2'(m) = \frac{4r^m}{(1+r)^2}\,\{\log r + 1 - r + (m-1)(1-r)\log r\} \tag{7.8}$$

Since $\log r = \log(1-(1-r)) = -[(1-r)+\frac{1}{2}(1-r)^2+\cdots] \leq -(1-r)$, we have $\log r + 1 - r < 0$. Hence $E_2'(m) < 0$ for $m \geq 1$. Also, if $m \geq \frac{1}{2}$ we have

$$E_2'(m) \leq \frac{4r^m}{(1+r)^2}\,[\log r + 1 - r - \tfrac{1}{2}(1-r)\log r] \qquad (7.9)$$

But

$$\log r + 1 - r - \tfrac{1}{2}(1-r)\log r = -[(\tfrac{1}{3}-\tfrac{1}{4})(1-r)^3+(\tfrac{1}{4}-\tfrac{1}{5})(1-r)^4 + (\tfrac{1}{5}-\tfrac{1}{6})(1-r)^5+\cdots] < 0 \qquad (7.10)$$

Thus $E_2'(m) \leq 0$ for $m \geq \frac{1}{2}$ and $F_2(m)$ decreases for $m \geq 1$. Since $F_4(m) = \bar{\mu}^{-1}F_2(m+\frac{1}{2})$ it follows that $F_4(m)$ decreases for $m \geq \frac{1}{2}$.

Evidently

$$\begin{aligned} &E_2(1) = 4r/(1+r)^2 = \bar{\mu}^2, && E_4(\tfrac{1}{2}) = \bar{\mu} \\ &E_2(\tfrac{1}{2}) = (4r^{1/2}/(1+r)^2)(\tfrac{1}{2}(1+r)) = \bar{\mu}, && E_4(0) = 1 \end{aligned} \qquad (7.11)$$

and hence

$$F_2(1) = (\bar{\mu}^2 + \bar{\mu}^4)^{1/2}, \qquad F_4(\tfrac{1}{2}) = (1+\bar{\mu}^2)^{1/2} > 1 \qquad (7.12)$$

Since $F_2(m) \to 0$ as $m \to \infty$ it follows that if $\varrho < (\bar{\mu}^2 + \bar{\mu}^4)^{1/2}$, there is a unique solution m of (7.1) such that $m \geq 1$. Similarly, since $F_4(\frac{1}{2}) \geq 1$, there is a unique solution of (7.1) with $J(m) = F_4(m)$ such that $m \geq \frac{1}{2}$.

It is easy to show that $E_3(m)$ is decreasing for $m \geq 0$ since

$$\frac{d}{dx}\left(\frac{2x}{1+x^2}\right) = \frac{2(1-x^2)}{(1+x)^2} > 0$$

if $|x| < 1$, and since r^m is a decreasing function of m. Moreover,

$$F_3(\tfrac{1}{2}) = [E_3(\tfrac{1}{2})^2 + E_3(0)^2]^{1/2} = [(2r^{1/2}/[1+r])^2 + 1]^{1/2} = (1+\bar{\mu}^2)^{1/2} > 1 \qquad (7.13)$$

Hence (7.1) with $J(m) = F_3(m)$ has a unique solution such that $m \geq \frac{1}{2}$.

For the $A^{1/2}$-norm we can show that $H_2'(m) < 0$ for $m \geq 1$ and since $H_4(m) = \bar{\mu}^{-1}H_2(m+\frac{1}{2})$ it follows that $H_4'(m) < 0$ for $m \geq \frac{1}{2}$. Details can be found in the report by Young and Kincaid [1969]. Since $H_2(1) = \bar{\mu}$ and $H_4(\frac{1}{2}) = 1$ the unique solvability of (7.1), with $J(m) = H_2(m)$ or $H_4(m)$, follows. Similarly, we can show that $H_3'(m) < 0$ and $H_3(\frac{1}{2}) = 1$. Hence (7.1) can be solved uniquely if $J(m) = H_3(m)$.

To show that (7.2) holds we need only show that $J(m) \geq r^m$. But this follows since $\mathscr{L}_{\omega_b}^m$ has the smallest spectral radius of any MSOR method, namely r^m, and since any norm is at least as large as the spectral radius, the proof of Theorem 7.1 is complete.

Some information in the comparison of the norms can be obtained from the formulas themselves. For example, for the $D^{1/2}$-norms we have, as $m \to \infty$,

$$\begin{aligned} &F_1/F_2 \to \tfrac{1}{2}(1+r)^{5/2}/(1-r) \\ &\text{which is greater than one if } r > 0.204 \\ &F_1/F_4 \to (1+r)^{3/2}/(1-r) > 1 \\ &F_1/F_3 \to \infty, \qquad F_2/F_4 \to 2/(1+r) > 1, \qquad F_2/F_3 \to \infty, \qquad F_4/F_2 \to \infty \end{aligned} \tag{7.14}$$

Moreover, as shown by Golub and Varga [1961] we have

$$\begin{aligned} F_1(m) &\geq F_3(m) && \text{for} \quad m \geq 2 \\ F_3(m) &\leq F_4(m) \leq F_2(m) && \text{for} \quad m \geq 1 \end{aligned} \tag{7.15}$$

Thus, the CCSI method appears to be by far the best method with respect to the $D^{1/2}$-norm, followed by the modified Sheldon method, Sheldon's method, and the SOR method.

For the $A^{1/2}$-norms we have, as $m \to \infty$,

$$\begin{aligned} &H_1/H_2 \to (1+r)^{3/2}/2 \\ &\text{which is greater than one if } r > 4^{1/3} - 1 \doteq 0.587 \\ &H_1/H_4 \to (1+r)^{1/2} > 1, \qquad H_1/H_3 \to \infty \\ &H_2/H_4 \to 2/(1+r) > 1, \qquad H_2/H_3 \to \infty, \qquad H_4/H_3 \to \infty \end{aligned} \tag{7.16}$$

One can also prove (see Young and Kincaid, 1969) that

$$H_2(m) \geq H_4(m) \tag{7.17}$$

for $m \geq \frac{3}{2}$ and $r \leq 0.840$. Again, the CCSI method appears to be the best method with respect to the $A^{1/2}$-norm. While the SOR method is not quite as good as the two Sheldon methods, the difference is much less, for r near to unity, than was the case with the $D^{1/2}$-norms.

Even though the ratios $F_1(m)/F_3(m)$ and $H_1(m)/H_3(m)$ approach ∞ as $m \to \infty$, nevertheless, one cannot conclude that the CCSI method is better by an order of magnitude than the SOR method or the Sheldon methods. It would seem that a more significant comparison should be based on the theoretical number of iterations, m. We shall now obtain asymptotic formulas for m and show that for any two methods the ratio of the values of m approach unity as $\varrho \to 0$, for fixed r. Moreover, the ratios approach a constant for fixed ϱ as $r \to 1-$ except for the SOR method and the $D^{1/2}$-norm. The ratio of the theoretical number of iterations corresponding to $F_1(m)$ to the number corresponding to any other method approaches ∞ as $r \to 1-$, though the approach to ∞ is very slow—on the order of $|\log \log r^{-1}|$.

We now seek to develop asymptotic formulas for the theoretical number of iterations m. We prove

Theorem 7.2. If m is the unique solution of (7.1), then

$$\lim_{\varrho \to 0}(m \log r^{-1} - \theta(\varrho, r)) = 0 \tag{7.18}$$

where $\theta(\varrho, r)$ is given in Table 7.1. Moreover, for each function $F_i(m)$ and $H_i(m)$, $i = 1, 2, 3, 4$, except for $F_1(m)$, we have

$$\lim_{r \to 1-} m \log r^{-1} = z \tag{7.19}$$

where $z = z(\varrho)$ satisfies the equation

$$\Gamma(z) = \varrho \tag{7.20}$$

Here $\Gamma(z)$ is given in Table 7.1. Finally, we have

$$\lim_{\varrho \to 0}(z(\varrho) - \theta(\varrho, 1)) = 0 \tag{7.21}$$

where

$$\theta(\varrho, 1) = \lim_{r \to 1-} \theta(\varrho, r) \tag{7.22}$$

Proof. We have already proved (7.18) for $J(m) = F_1(m)$ and $J(m) = H_1(m)$ in Chapter 7. We generalize the argument used there. Letting $y = m \log r^{-1}$ we write (7.1) in the form

$$y = \log[\phi(y)/\varrho] \tag{7.23}$$

where

$$\phi(y) = J(y(\log r^{-1})^{-1})r^{-m}$$

For the SOR, Sheldon, and modified Sheldon methods we determine a constant $\alpha \geq 1$ such that

$$\phi(y) \geq \alpha y \qquad \text{for} \quad y \geq 0 \tag{7.24}$$

TABLE 7.1.

Method	Norm	$\theta(\varrho, r)$	γ	$\theta(\varrho, 1)$	$\Gamma(z)$
SOR	$D^{1/2}$	$\log\left(\frac{2\gamma}{\varrho}\log\frac{2\gamma}{\varrho}\right)$	$\frac{r^{-1/2}+r^{1/2}}{\log r^{-1}}$	∞	—
	$A^{1/2}$		$\frac{r^{-1/2}-r^{1/2}}{\log r^{-1}}$	$\log\left(\frac{2}{\varrho}\log\frac{2}{\varrho}\right)$	$h_1(z) = e^{-z}(z + [1 + z^2]^{1/2})$
Sheldon	$D^{1/2}$	$\log\left(\frac{2^{1/2}\gamma}{\varrho}\log\frac{2^{1/2}\gamma}{\varrho}\right)$	$\frac{2(2)^{1/2}r^{-1/2}(1-r)}{(1+r)^{3/2}\log r^{-1}}$	$\log\left(\frac{2^{1/2}}{\varrho}\log\frac{2^{1/2}}{\varrho}\right)$	$f_2(z) = 2^{1/2}e^{-z}(1 + z)$
	$A^{1/2}$				$h_2(z) = e^{-z}(1 + 2z + 2z^2)^{1/2}$
Modified Sheldon	$D^{1/2}$	$\log\left(\frac{2^{1/2}\gamma}{\varrho}\log\frac{2^{1/2}\gamma}{\varrho}\right)$	$\frac{2^{1/2}r^{-1/2}(1-r)}{(1+r)^{1/2}\log r^{-1}}$	$\log\left(\frac{2^{1/2}}{\varrho}\log\frac{2^{1/2}}{\varrho}\right)$	$f_4(z) = 2^{1/2}e^{-z}(1 + z)$
	$A^{1/2}$				$h_4(z) = e^{-z}(1 + 2z + 2z^2)^{1/2}$
CCSI	$D^{1/2}$	$\log\frac{2(2)^{1/2}\gamma}{\varrho}$	$\left(\frac{1+r}{2r}\right)^{1/2}$	$\log\left(\frac{2(2)^{1/2}}{\varrho}\right)$	$f_3(z) = 2(2)^{1/2}e^{-z}(1 + e^{-2z})^{-1}$
	$A^{1/2}$				$h_3(z) = 2(2)^{1/2}e^{-z}(1 + e^{-4z})^{1/2}(1 + e^{-2z})^{-2}$

and such that

$$\lim_{y\to\infty} [(\phi(y) - \alpha y)/y] = 0 \tag{7.25}$$

We let

$$\theta(\varrho, r) = w_1 \tag{7.26}$$

where

$$w_0 = \log \frac{\alpha}{\varrho}, \qquad w_1 = \log \frac{\alpha w_0}{\varrho} = \log\left(\frac{\alpha}{\varrho} \log \frac{\alpha}{\varrho}\right) \tag{7.27}$$

Evidently if $y \to \infty$, then $w_0 \to \infty$ and $w_1 \to \infty$.

For the CCSI method we determine a constant α such that

$$\lim_{y\to\infty} \phi(y) = \alpha \tag{7.28}$$

and we let

$$\theta(\varrho, r) = \log(\alpha/\varrho) \tag{7.29}$$

In this case we have

$$y - \theta(\varrho, r) = \log \frac{\phi(y)}{\alpha} = \log\left(1 + \frac{\phi(y) - \alpha}{\alpha}\right) \tag{7.30}$$

and (7.18) follows by (7.28).

The values of the constant α are as follows:

Norm	SOR	Sheldon	Modified Sheldon	CCSI
$D^{1/2}$	2γ	$2^{1/2}\gamma$	$2^{1/2}\gamma$	$2(2)^{1/2}\gamma$
$A^{1/2}$	2γ	$2^{1/2}\gamma$	$2^{1/2}\gamma$	$2(2)^{1/2}\gamma$

where for each method γ is defined in the table given with Theorem 7.2.

We now consider the equation

$$w = \log(\alpha w/\varrho) \tag{7.31}$$

Evidently the function

$$\hat{F}(w) = \alpha w e^{-w} \tag{7.32}$$

is a decreasing function of w for $w \geq 1$ and $\hat{F}(1) = \alpha/e$. Therefore, if $\varrho < \alpha/\varepsilon$, then (7.31) has a unique solution such that $w \geq 1$.

By the methods used in the proof of Theorem 7-4.2, we can show that

$$w - w_0 \leq \log w_0/(1 - w_0^{-1}) \tag{7.33}$$

where w_0 is defined by (7.27).

Assuming that $\varrho < \varrho_0$, there is a unique solution of (7.1) such that $y \geq y_0 = m_0 \log r^{-1}$. One can show that $\phi(y)$ is a decreasing function of y for $y \geq y_0$. (This follows since the $F_i(m)$ and $H_i(m)$ are decreasing.) Also, since $\phi(y) \geq \alpha y$, it follows that $y \geq w$ where w is the solution of (7.31).

We now seek to show that

$$\lim_{y\to\infty} (y - w) = 0 \tag{7.34}$$

But by (7.23) and (7.31) we have

$$\begin{aligned} y - w &= \log \frac{\phi(y)}{\alpha w} = \log\left[\left(\frac{\phi(y)}{\alpha y}\right)\left(\frac{\alpha y}{\alpha w}\right)\right] \\ &= \log \frac{\phi(y)}{\alpha y} + \log \frac{y}{w} \\ &= \log\left(1 + \frac{\phi(y) - \alpha y}{\alpha y}\right) + \log\left(1 + \frac{y - w}{w}\right) \\ &\leq \frac{\phi(y) - \alpha y}{\alpha y} + \frac{y - w}{w} \end{aligned} \tag{7.35}$$

and

$$y - w \leq \frac{1}{1 - w^{-1}} \, \frac{\phi(y) - \alpha y}{\alpha y} \tag{7.36}$$

By (7.25), (7.34) follows. Moreover, by (7.33) and (7.27) it follows that

$$\lim_{y\to\infty} (y - w_1) = 0 \tag{7.37}$$

For $F_1(m)$ and $H_1(m)$ we have

$$\begin{aligned} \phi(y) - \alpha y &= \gamma y + (\gamma^2 y^2 + 1)^{1/2} - 2\gamma y \\ &= (\gamma^2 y^2 + 1)^{1/2} - \gamma y = \frac{1}{(\gamma^2 y^2 + 1)^{1/2} + \gamma y} \end{aligned} \tag{7.38}$$

Hence $\phi(y) - \alpha y \to 0$ as $y \to \infty$ and (7.25) holds. We leave it to the reader to verify (7.25) for the other methods. The proof of (7.18) is complete.

We have already proved the remainder of Theorem 7.2 for $H_1(m)$, (see Theorem 7-6.2). The reader should carry out the details for the other cases. This completes the proof of Theorem 7.2. (See Young and Kincaid, 1969.)

On the basis of Theorem 7.2, we conclude that, based on the $D^{1/2}$-norm and the $A^{1/2}$-norm, the CCSI method is better than the other methods. Moreover, the SOR method is not as good as the other methods. However, for the $A^{1/2}$-norm the SOR method is practically as good as the two Sheldon methods. Let m_i and m_i^* denote, respectively, the theoretical number of iterations for the $D^{1/2}$-norm and the $A^{1/2}$-norm corresponding to $F_i(m)$ and $H_i(m)$. Evidently we have for fixed r

$$\lim_{\varrho\to 0} [m_i \log r^{-1}/\log \varrho] = \lim_{\varrho\to 0} [m_i^* \log r^{-1}/\log \varrho] = 1 \qquad (7.39)$$

Thus $m_i/m_j \to 1$ and $m_i^*/m_j^* \to 1$ as $\varrho \to 0$. On the other hand, we have for fixed ϱ

$$\lim_{r\to 1-} (m_1/m_i) = \infty, \qquad \lim_{r\to 1-} (m_j/m_i) = z_i, \qquad i, j \neq 1 \qquad (7.40)$$

and

$$\lim_{r\to 1-} (m_i^*/m_j^*) = z_i^*(\varrho)/z_j^*(\varrho) \qquad (7.41)$$

where $z_i(\varrho)$ and $z_i^*(\varrho)$ are defined by (7.20). Thus for the $A^{1/2}$-norm the theoretical numbers of iterations are of the same order-of-magnitude as $r \to 1-$. The same is true for the $D^{1/2}$-norm except for the SOR method.

Table 7.2 gives the exact theoretical and the approximate theoretical values of $m \log r^{-1}$ for each method and for each norm. The last two columns refer to a method to be discussed in Chapter 11. The CCSI method is best both for the $D^{1/2}$-norm and for the $A^{1/2}$-norm. For the $D^{1/2}$-norm the SOR method is considerably slower than the other methods. In fact, $m \log r^{-1}$ becomes arbitrarily large as $r \to 1-$ whereas for the other methods $m \log r^{-1}$ converges to a finite limit, namely $z(\varrho)$, as $r \to 1-$. (The values of $z(\varrho)$ are given as the exact values of $m \log r^{-1}$ for $r = 1$.) For $r = 0.9$ and $\varrho = 10^{-6}$ the SOR method requires about 36.5% more iterations than the CCSI method.

It should be noted that for the SOR method with the $D^{1/2}$-norm the values of $m \log r^{-1}$ are not excessively large until r is very close to unity,

TABLE 7.2. Theoretical Values of $m \log r^{-1}$

		SOR		Sheldon		M-Sheldon		CCSI		GS-SSI	
		Exact	Approx.	Exact	Approx.	Exact	Approx.	Exact	Approx.	Exact	Approx.
$\varrho = 10^{-3}$											
$r = 0.5$	$D^{1/2}$	11.129	10.885	10.059	9.748	9.422	9.778	8.150	8.150	8.553	8.553
	$A^{1/2}$	9.918	9.652	10.011	9.748	9.729	9.422	8.150	8.150	8.235	8.235
$r = 0.6$	$D^{1/2}$	11.435	11.195	9.957	9.628	9.738	9.375	8.091	8.091	8.410	8.410
	$A^{1/2}$	9.908	9.641	9.907	9.628	9.688	9.375	8.091	8.091	8.079	8.079
$r = 0.7$	$D^{1/2}$	11.810	11.575	9.863	9.519	9.703	9.334	8.045	8.045	8.280	8.280
	$A^{1/2}$	9.901	9.635	9.812	9.519	9.651	9.334	8.045	8.045	7.942	7.942
$r = 0.8$	$D^{1/2}$	12.311	12.083	9.776	9.418	9.671	9.298	8.006	8.006	8.161	8.161
	$A^{1/2}$	9.898	9.632	9.724	9.418	9.619	9.298	8.006	8.006	7.818	7.818
$r = 0.9$	$D^{1/2}$	13.120	12.902	9.694	9.323	9.643	9.266	7.975	7.975	8.051	8.051
	$A^{1/2}$	9.898	9.632	9.724	9.418	9.619	9.298	8.006	8.006	7.818	7.818
$r = 1.0$	$D^{1/2}$	∞	∞	9.617	9.236	9.617	9.236	7.947	7.947	7.947	7.947
	$A^{1/2}$	9.896	9.629	9.565	9.236	9.565	9.236	7.947	7.947	7.601	7.601
$\varrho = 10^{-6}$											
$r = 0.5$	$D^{1/2}$	18.548	18.376	17.502	17.296	17.218	16.988	15.058	15.058	15.461	15.461
	$A^{1/2}$	17.385	17.205	17.475	17.296	17.191	16.988	15.058	15.058	15.143	15.143
$r = 0.6$	$D^{1/2}$	18.842	18.672	17.398	17.182	17.177	16.944	14.999	14.999	15.318	15.318
	$A^{1/2}$	17.375	17.195	17.370	17.182	17.149	16.944	14.999	14.999	14.987	14.987
$r = 0.7$	$D^{1/2}$	19.204	19.037	17.303	17.079	17.142	16.905	14.952	14.952	15.188	15.188
	$A^{1/2}$	17.369	17.189	17.274	17.079	17.113	16.905	14.952	14.952	14.850	14.850

TABLE 7.2. *(continued)*

		SOR		Sheldon		M-Sheldon		CCSI		GS-SSI	
		Exact	Approx.	Exact	Approx.	Exact	Approx.	Exact	Approx.	Exact	Approx.
$r = 0.8$	$D^{1/2}$	19.688	19.524	17.215	16.984	17.110	16.871	14.914	14.914	15.069	15.069
	$A^{1/2}$	17.366	17.186	17.186	16.984	17.081	16.871	14.914	14.914	14.726	14.726
$r = 0.9$	$D^{1/2}$	20.473	20.313	17.133	16.896	17.082	16.841	14.882	14.882	14.959	14.959
	$A^{1/2}$	17.364	17.184	17.104	16.896	17.052	16.841	14.882	14.882	14.613	14.613
$r = 1.0$	$D^{1/2}$	∞	∞	17.056	16.813	17.056	16.813	14.855	14.855	14.855	14.855
	$A^{1/2}$	17.364	17.364	17.026	16.813	17.026	16.813	14.855	14.855	14.509	14.509
$\varrho = 10^{-9}$											
$r = 0.5$	$D^{1/2}$	25.785	25.650	24.749	24.590	24.463	24.289	21.966	21.966	22.369	22.369
	$A^{1/2}$	24.641	24.501	24.729	24.590	24.444	24.289	21.966	21.966	22.051	22.051
$r = 0.6$	$D^{1/2}$	26.075	25.941	24.644	24.480	24.422	24.246	21.907	21.907	22.226	22.226
	$A^{1/2}$	24.632	24.492	24.625	24.480	24.403	24.246	21.907	21.907	21.895	21.895
$r = 0.7$	$D^{1/2}$	26.431	26.299	24.549	24.379	24.387	24.208	21.860	21.860	22.096	22.096
	$A^{1/2}$	24.626	24.486	24.529	24.379	24.367	24.208	21.860	21.860	21.757	21.757
$r = 0.8$	$D^{1/2}$	26.908	26.778	24.460	24.285	24.355	24.175	21.822	21.822	21.977	21.977
	$A^{1/2}$	24.623	24.483	24.440	24.285	24.335	24.175	21.822	21.822	21.633	21.633
$r = 0.9$	$D^{1/2}$	27.682	27.554	24.378	24.199	24.327	24.145	21.790	21.790	21.866	21.866
	$A^{1/2}$	24.621	24.481	24.357	24.199	24.306	24.145	21.790	21.790	21.520	21.520
$r = 1.0$	$D^{1/2}$	∞	∞	24.301	24.118	24.301	24.118	21.763	21.763	21.763	21.763
	$A^{1/2}$	24.620	24.481	24.280	24.118	24.280	24.118	21.763	21.763	21.416	21.416

probably closer than would normally occur in practice. Thus for the case $\varrho = 10^{-6}$ we have

r	0.90	0.95	0.96	0.97	0.98	0.99	1.00
$m_1 \log r^{-1}$	20.473	21.228	21.467	21.774	22.204	22.934	∞
$m_3 \log r^{-1}$	14.882	14.868	14.865	14.863	14.860	14.858	14.855
m_1/m_3	1.376	1.428	1.444	1.465	1.494	1.535	∞

The very slow approach to infinity of the ratio m_1/m_3 as $r \to 1-$ is like that of

$$\frac{\log\left(\dfrac{2(r^{-1/2} + r^{1/2})}{\varrho \log r^{-1}} \log \dfrac{2(r^{-1/2} + r^{1/2})}{\varrho \log r^{-1}}\right)}{\log \dfrac{2^{3/2}}{\varrho} \left(\dfrac{1 + r}{2r}\right)^{1/2}} \sim \frac{\log\left(\dfrac{4}{\varrho \log r^{-1}} \log \dfrac{4}{\varrho \log r^{-1}}\right)}{\log \dfrac{2^{3/2}}{\varrho}}$$

which approaches infinity like $|\log \log r^{-1}|$.

For the $A^{1/2}$-norm the SOR method is still not as good as the other methods but the difference is much less than in the case of the $D^{1/2}$-norms. As a matter of fact, the SOR method is practically as good as the two Sheldon methods. For $\varrho = 10^{-6}$, the SOR methods requires about 18.7% more iterations than the CCSI method.

The agreement between the exact theoretical values and the approximate values is very good. In the worst case involving $\varrho = 10^{-3}$ and $r = 0.5$, the differences were not more than 3%.

We now describe the results obtained by solving the problem studied in Section 7.7 by the various methods. Both the case $h = \frac{1}{20}$ and the case $h = \frac{1}{40}$ were solved. As shown in Table 7.3, the comparison between the observed and theoretical numbers of iterations is very good for the $D^{1/2}$-norm and for the $A^{1/2}$-norm for the SOR method. In other cases the observed number of iterations m_O is considerably less than m_T, the exact theoretical number. This difference is due to the choice of starting vector.

TABLE 7.3. Comparison of Observed and Theoretical Number of Iterations[a]

	Norm	SOR	Sheldon	M-Sheldon	CCSI	GS-SSI
$\varrho = 10^{-3}$	($h = \frac{1}{20}$; $r = 0.729$)					
m_O	$D^{1/2}$	37	32	31	26	27
	$A^{1/2}$	31	25	24	20	20
m_T	$D^{1/2}$	37.8	31.1	30.7	25.5	26.1
	$A^{1/2}$	31.4	31.0	30.6	25.5	25.1
m_A	$D^{1/2}$	37.1	30.1	29.6	25.5	26.1
	$A^{1/2}$	30.5	30.1	29.6	25.5	25.1
$\varrho = 10^{-6}$	($h = \frac{1}{20}$; $r = 0.729$)					
m_O	$D^{1/2}$	61	55	55	48	49
	$A^{1/2}$	55	49	48	42	42
m_T	$D^{1/2}$	61.3	54.7	54.3	47.4	48.0
	$A^{1/2}$	55.0	54.7	54.3	47.4	47.0
m_A	$D^{1/2}$	60.7	54.0	53.6	47.4	48.0
	$A^{1/2}$	54.5	54.0	53.6	47.4	47.0
$\varrho = 10^{-9}$	($h = \frac{1}{20}$; $r = 0.729$)					
m_O	$D^{1/2}$	84	78	78	70	70
	$A^{1/2}$	78	72	72	64	63
m_T	$D^{1/2}$	84.2	77.7	77.3	69.3	69.9
	$A^{1/2}$	78.1	77.7	77.3	69.3	68.9
m_A	$D^{1/2}$	83.8	77.2	76.7	69.3	69.9
	$A^{1/2}$	77.6	77.2	76.7	69.3	68.9
$\varrho = 10^{-3}$	($h = \frac{1}{40}$; $r = 0.854$)					
m_O	$D^{1/2}$	79	62	62	51	52
	$A^{1/2}$	61	44	43	35	34
m_T	$D^{1/2}$	80.7	61.7	61.4	50.8	51.5
	$A^{1/2}$	62.9	61.6	61.3	51.0	49.3
m_A	$D^{1/2}$	79.3	59.6	59.0	50.8	51.5
	$A^{1/2}$	61.2	59.6	59.0	50.8	49.3

TABLE 7.3. *(continued)*

	Norm	SOR	Sheldon	M-Sheldon	CCSI	GS-SSI
$\varrho = 10^{-6}$	$(h = \frac{1}{40};\ r = 0.854)$					
m_O	$D^{1/2}$	126	110	109	95	96
	$A^{1/2}$	109	92	92	79	78
m_T	$D^{1/2}$	127.5	109.0	108.7	94.7	95.4
	$A^{1/2}$	110.4	109.0	108.8	94.9	93.3
m_A	$D^{1/2}$	126.5	107.7	107.2	94.7	95.4
	$A^{1/2}$	109.3	107.7	107.2	94.7	93.3
$\varrho = 10^{-9}$	$(h = \frac{1}{40};\ r = 0.854)$					
m_O	$D^{1/2}$	172	156	155	139	140
	$A^{1/2}$	155	139	138	123	121
m_T	$D^{1/2}$	173.4	155.1	154.8	138.7	139.4
	$A^{1/2}$	156.6	155.1	154.9	138.9	137.2
m_A	$D^{1/2}$	172.6	154.1	153.6	138.7	139.4
	$A^{1/2}$	155.7	154.1	153.6	138.7	137.2

[a] The symbols used are: m_O, observed number of iterations; m_T, exact theoretical number of iterations; m_A, approximate theoretical number of iterations.

SUPPLEMENTARY DISCUSSION

Theorem 1.1 is given in Young [1969]. Much of the material of Sections 10.2 and 10.3 was given by Young *et al.* [1965]. The formulas for the $D^{1/2}$-norms were given by Sheldon [1959] for Sheldon's method, by Golub [1959] and by Golub and Varga [1961] for the modified Sheldon method, and by Golub and Varga [1961] for the CCSI method. Formulas for the $A^{1/2}$-norms of these methods and the comparison of norms were given by Young and Kincaid [1969].

EXERCISES

Section 10.1

1. For which of the following choices of $\omega_1, \omega_1', \omega_2, \omega_2', \ldots$ does the MSOR method converge?

(a) $(1, 0), (0, \frac{1}{2}), (\frac{1}{3}, 0), (0, \frac{1}{4}), \ldots$
(b) $(1, 0), (0, \frac{1}{2}), (\frac{1}{4}, 0), (0, \frac{1}{8}), \ldots$
(c) $(1, 1), (1, \frac{1}{2}), (1, \frac{1}{3}), \ldots$
(d) $(1, 1), (1, \frac{1}{2}), (1, \frac{1}{4}), \ldots$

Section 10.2

1. Work out the details of the proof of Lemma 2.2.

2. Compute the first five of the polynomials $P_k(x)$ and $Q_k(x)$ from (2.10) and (2.11) and verify that $Q_{2m}(x)$ is a polynomial of degree m in μ^2.

3. Under the hypothesis of Theorem 2.1 if $\bar{\mu} = 0.8$ and $m = 2$, find $S(\mathscr{L}_{\omega_1}\mathscr{L}_{\omega_2})$ in the following cases

(a) $\omega_1 = \omega_2 = 1.25$
(b) $\omega_1 = 1.0,\ \omega_2 = 1.25$
(c) $\omega_1 = 1.5,\ \omega_2 = 1.3$
(d) $\omega_1 = 0.8,\ \omega_2 = 0.5$

4. Show that the left member of (2.23) is negative if $c < \bar{\lambda}^2$ and $\lambda = \bar{\lambda}$.

Section 10.3

1. Find alternative optimum parameter sets for the cases $m = 2$ and $m = 3$ with $\omega_b = 2(1 + [1 - \bar{\mu}^2]^{1/2})^{-1}$ and $\bar{\mu} = 0.95$. For $m = 3$, assume $\omega_k = \omega_k'$, $k = 1, 2, 3$. For $m = 2$, give solutions both for the case $\omega_1 = \omega_1'$, $\omega_2 = \omega_2'$ and for the more general case. (In the latter case assume $\omega_1 = 1.2$.)

2. Consider the linear system $u_1 - 0.95u_2 = 1$, $-0.95u_1 + u_2 = 0$. Using the alternative optimum parameter set (3.20) for $m = 2$ with $\omega_1 = \omega_1'$, $\omega_2 = \omega_2'$ as determined in the previous exercise, carry out five double iterations with the MSOR method. Use starting values $u_1^{(0)} = u_2^{(0)} = 0$. Do the same if $\omega_1 = \omega_1' = \omega_2 = \omega_2' = \omega_b$.

Section 10.4

1. For a matrix satisfying the conditions of Theorem 4.1, compute $S(\mathscr{H}_m)$, $\| \mathscr{H}_m \|_{D^{1/2}}$ and $\| \mathscr{H}_m \|_{A^{1/2}}$ if $\bar{\mu} = 0.95$ and $m = 5$.

2. Carry out five iterations with Sheldon's method for the linear system

$$\begin{pmatrix} 1 & -0.95 \\ -0.95 & 1 \end{pmatrix} \begin{pmatrix} u_1 \\ u_2 \end{pmatrix} = \begin{pmatrix} 1 \\ 0 \end{pmatrix}$$

with starting values $u_1^{(0)} = u_2^{(0)} = 0$. Compute $\| u^{(0)} - \bar{u} \|_{D^{1/2}}$ and $\| u^{(5)} - \bar{u} \|_{D^{1/2}}$ as well as $\| u^{(0)} - \bar{u} \|_{A^{1/2}}$ and $\| u^{(5)} - \bar{u} \|_{A^{1/2}}$. In each case compare the ratios of these norms with the corresponding theoretical values of $\| \mathscr{H}_5 \|_{D^{1/2}}$ and $\| \mathscr{H}_5 \|_{A^{1/2}}$.

Section 10.5

1. For a matrix satisfying the conditions of Theorem 5.1, compute $S(\mathscr{K}_m)$, $\| \mathscr{K}_m \|_{D^{1/2}}$ and $\| \mathscr{K}_m \|_{A^{1/2}}$ if $\bar{\mu} = 0.95$ and $m = 5$.

2. Work out Exercise 2 of Section 10.4 for the modified Sheldon method.

3. Write out the complete details for the proof of Theorem 5.1.

4. Verify (5.7).

Section 10.6

1. For a matrix satisfying the conditions of Theorem 6.1, compute $S(\mathscr{C}_m)$, $\| \mathscr{C}_m \|_{D^{1/2}}$ and $\| \mathscr{C}_m \|_{A^{1/2}}$ if $\bar{\mu} = 0.95$ and $m = 5$.

2. Work out Exercise 2 of Section 10.4 for the CCSI method.

Section 10.7

1. Consider the linear system $Au = b$ where A is a positive definite matrix of the form (8-1.2). For each of the methods corresponding to $\mathscr{L}_{\omega_b}^m$, $\mathscr{H}_m$, $\mathscr{C}_m$, and $\mathscr{K}_m$ find the smallest value of m such that the $D^{1/2}$-norm is less than $\varrho = 10^{-3}$. Do the same for the $A^{1/2}$-norm. Assume $\bar{\mu} = 0.95$. Also compute an approximate value of m by

(a) $m \sim \theta(\varrho, r)/\log r^{-1}$,
(b) $m \sim z/\log r^{-1}$ (except for $\| \mathscr{L}_{\omega_b}^m \|_{D^{1/2}}$) where z satisfies (7.20),
(c) $m \sim \theta(\varrho, 1)/\log r^{-1}$.

2. Using the results of the exercises in Sections 10.4–10.6, verify (7.15) and (7.17) for the case $m = 5$, $\bar{\mu} = 0.95$.

3. Give the details of the derivation of (7.33).

4. Verify (7.25) for those methods not included in the analysis given in the text.

5. Verify that if y satisfies (7.23) and if $\phi(y)$ satisfies (7.24) and (7.25) then $\varrho \to 0$ as $y \to \infty$. Thus show that as $y \to \infty$, we have $w_0 \to \infty$, $w_1 \to \infty$, and $w \to \infty$.

6. Complete the proof of Theorem 7.2.

Chapter 11 / SEMI-ITERATIVE METHODS

Given a linear stationary iterative method of first degree, one can often find an associated nonstationary method which will converge faster than the given method. The associated method is known as a "semi-iterative method" (SI method) with respect to, or based on, the given method.

In Section 11.1, we shall describe the construction of SI methods and show that they are analogous to certain summability methods for accelerating the convergence of sequences of real numbers. In Section 11.2, we shall consider the case where the matrix of the given iterative method has real eigenvalues and show that an order-of-magnitude improvement in the convergence can often be achieved by using the proper SI method. SI methods based on the J, JOR, and RF methods will be studied in Section 11.3. Two related methods, namely, Richardson's method and the cyclic Chebyshev semi-iterative method will be treated in Sections 11.4 and 11.5. SI methods based on the GS, SOR, and MSOR methods will be considered in Sections 11.6–11.8. A comparison of some of the methods based on the $D^{1/2}$- and $A^{1/2}$-norms will be given in Section 11.9.

If A is positive definite but does not have Property A and is not a L-matrix, then, as we will show in Chapter 12, the SOR theory does not apply and the method may not be effective. However, the use of a semi-iterative method based on the J, RF, or GS methods may still be effective. The fact that these methods, along with Richardson's method which is essentially an SI method based on the RF method, are often effective in cases where the SOR method is not, does not seem to be universally recognized.

For systems where the matrix A has the form (8.12) the cyclic Chebyshev semi-iterative method is less effective than the SOR method as far as spectral radius is concerned. However, as we have seen in Section 10.7, it is somewhat more effective as far as the $D^{1/2}$-norm and, to a lesser extent, as far as the $A^{1/2}$-norm is concerned.

There seems to be little gained in applying semi-iterative methods to the SOR or MSOR methods. However, as will be shown in Chapter 15, one can often achieve a substantial gain with the use of the SSOR method provided A does *not* have the form (8-1.2).

11.1. GENERAL CONSIDERATIONS

Given a sequence of real numbers $x_1, x_2, \ldots$ one can often develop another sequence $y_1, y_2, \ldots$ by a *method of summability* so that either the new sequence converges when the old one does not or else the new one converges faster than the old one if the old one converges. As an example, consider the sequence

$$\begin{aligned} y_1 &= x_1 \\ y_2 &= \tfrac{1}{2}(x_1 + x_2) \\ y_3 &= \tfrac{1}{3}(x_1 + x_2 + x_3) \\ &\vdots \end{aligned}$$

It is easy to show that if $\{x_i\}$ converges, then $\{y_i\}$ converges; hence the method of summability is *regular*. Even if $\{x_i\}$ does not converge, $\{y_i\}$ may converge. If $\{y_i\}$ converges, the original sequence is said to be Cesàro summable,[†] or Cesàro (C.1) summable.

A more general method of summability is defined by the triangular array of coefficients

$$\begin{array}{lll} \alpha_{0,0} & & \\ \alpha_{1,0}, & \alpha_{1\,1} & \\ \alpha_{2,0}, & \alpha_{2,1}, & \alpha_{2,2} \\ \vdots & & \end{array}$$

[†] See, for instance, Widder [1947, p. 264].

where

$$\sum_{k=0}^{n} \alpha_{n,k} = 1, \qquad n = 0, 1, 2, \ldots \tag{1.1}$$

Given a sequence $\{x_i\}$ one can construct a new sequence $\{y_i\}$ by

$$y_n = \sum_{k=0}^{n} \alpha_{n,k} x_k, \qquad n = 0, 1, 2, \ldots$$

If $\alpha_{n,k} = 0$ for $k < n$, and $\alpha_{n,n} = 1$ for all n, then we obtain the original sequence. If $\alpha_{n,k} = (n+1)^{-1}$, $k = 0, 1, \ldots, n$ for all n, then we have the Cesàro (C.1) method of summability.

Let us now apply the method of summability to linear stationary iterative methods of first degree for solving the linear system

$$Au = b \tag{1.2}$$

where the matrix A is nonsingular. Given the sequence $u^{(0)}, u^{(1)}, \ldots$ defined by the completely consistent iterative method

$$u^{(n+1)} = Gu^{(n)} + k \tag{1.3}$$

and given the coefficients $\alpha_{n,k}$ satisfying (1.1), we define the sequence

$$v^{(n)} = \sum_{k=0}^{n} \alpha_{n,k} u^{(k)} \tag{1.4}$$

We say that the process defined by (1.4) is a *semi-iterative method* (SI method) with respect to the linear stationary iterative method (1.3).

To analyze the convergence of an SI method we let

$$\varepsilon^{(n)} = u^{(n)} - u, \qquad \eta^{(n)} = v^{(n)} - u \tag{1.5}$$

where u is the solution of (1.2). From (1.4) we have

$$\begin{aligned} \eta^{(n)} = v^{(n)} - u &= \sum_{k=0}^{n} \alpha_{n,k} u^{(k)} - u \\ &= \sum_{k=0}^{n} \alpha_{n,k} u^{(k)} - u \sum_{k=0}^{n} \alpha_{n,k} \\ &= \sum_{k=0}^{n} \alpha_{n,k} (u^{(k)} - u) = \sum_{k=0}^{n} \alpha_{n,k} \varepsilon^{(k)} \end{aligned} \tag{1.6}$$

But since the method (1.3) is consistent, we have by Theorem 3-5.2,

$$\varepsilon^{(k)} = G^k \varepsilon^{(0)} \tag{1.7}$$

and

$$\eta^{(n)} = P_n(G)\varepsilon^{(0)} = P_n(G)\eta^{(0)} \tag{1.8}$$

where

$$P_n(G) = \sum_{k=0}^{n} \alpha_{n,k} G^k \tag{1.9}$$

As in Chapter 9, we consider as a measure of the convergence of (1.4) the virtual average spectral radius

$$\bar{\mathscr{S}}_n(P_n(G)) = [\bar{S}(P_n(G))]^{1/n} = \left[\max_{\mu \in \bar{S}_G} | P_n(\mu) |\right]^{1/n} \tag{1.10}$$

where $\bar{S}_G$ is the smallest convex set containing S_G, the set of all eigenvalues of G. We shall attempt to choose $\{\alpha_{n,k}\}$ to minimize $\bar{\mathscr{S}}_n(P_n(G))$.

11.2. THE CASE WHERE G HAS REAL EIGENVALUES

If the matrix G corresponding to a linear stationary iterative method has real eigenvalues, then we can often achieve a large improvement in the convergence by using a suitable SI method based on the given method. Let us first suppose that the eigenvalues μ of G lie in the range

$$\alpha \leq \mu \leq \beta < 1 \tag{2.1}$$

where $\beta > \alpha$. We remark that we do not require that $S(G)$ be less than unity; hence α may be less than -1. Later we shall consider the case where (2.1) is replaced by a weaker condition. The requirement that $\beta > \alpha$ is not essential. If $\beta = \alpha$, then the method which is derived assuming $\beta > \alpha$ reduces to the method

$$u^{(n+1)} = \frac{1}{1-\alpha}(G - \alpha I)u^{(n)} + \frac{1}{1-\alpha}k \tag{2.2}$$

which is a completely consistent method which converges rapidly since

$$S(G - \alpha I) = 0 \tag{2.3}$$

Let us introduce the new variable $\gamma = \gamma(\mu)$ defined by

$$\gamma = \frac{2\mu - (\alpha + \beta)}{\beta - \alpha} \tag{2.4}$$

Evidently $\gamma(\alpha) = -1$, $\gamma(\beta) = 1$. If we let

$$z = \gamma(1) = \frac{2 - (\alpha + \beta)}{\beta - \alpha} \tag{2.5}$$

then

$$z > 1 \tag{2.6}$$

since $2 - (\alpha + \beta) > 0$, $\beta - \alpha > 0$ and $[2 - (\alpha + \beta)] - (\beta - \alpha) = 2 - 2\beta > 0$.

Given the polynomial $P_n(\mu)$ such that $P_n(1) = 1$, let us define the polynomial $Q_n(\gamma)$ by

$$Q_n(\gamma) = P_n\left(\frac{(\beta - \alpha)\gamma + \beta + \alpha}{2}\right) \tag{2.7}$$

Evidently

$$P_n(\mu) = Q_n\left(\frac{2\mu - (\alpha + \beta)}{\beta - \alpha}\right) = Q_n(\gamma) \tag{2.8}$$

Therefore

$$\max_{\alpha \le \mu \le \beta} | P_n(\mu) | = \max_{-1 \le \gamma \le 1} | Q_n(\gamma) | \tag{2.9}$$

By Theorem 9-3.1, since $Q_n(z) = P_n(1) = 1$ and $z > 1$, the polynomial $Q_n(\gamma)$ which minimizes the right member of (2.9) is given by

$$Q_n(\gamma) = T_n(\gamma)/T_n(z) \tag{2.10}$$

Moreover,

$$\max_{-1 \le \gamma \le 1} | Q_n(\gamma) | = 1/T_n(z) = 1 \Big/ T_n\left(\frac{2 - (\beta + \alpha)}{\beta - \alpha}\right) \tag{2.11}$$

Therefore, we have

$$P_n(\mu) = Q_n\left(\frac{2\mu - (\beta+\alpha)}{\beta - \alpha}\right) = T_n\left(\frac{2\mu - (\beta+\alpha)}{\beta - \alpha}\right) \Big/ T_n\left(\frac{2 - (\beta + \alpha)}{\beta - \alpha}\right) \tag{2.12}$$

Since $z > 1$ and $T_n(z) > 1$ it follows that the method is convergent—even though the basic method (1.3) may not converge.

Since $T_0(x) = 1,\ T_1(x) = x,\ T_2(x) = 2x^2 - 1$ we have

$$P_0(\mu) = 1 \tag{2.13}$$

and $\alpha_{0,0} = 1$;

$$P_1(\mu) = \frac{2\mu - (\beta + \alpha)}{2 - (\beta + \alpha)} \tag{2.14}$$

and $\alpha_{1,0} = -[\beta + \alpha]/[2 - (\beta + \alpha)],\ \alpha_{1,1} = 2/[2 - (\beta + \alpha)]$;

$$\begin{aligned} P_2(\mu) &= \left[2\left(\frac{2\mu - (\beta + \alpha)}{\beta - \alpha}\right)^2 - 1\right] \Big/ \left[2\left(\frac{2 - (\beta + \alpha)}{\beta - \alpha}\right)^2 - 1\right] \\ &= \alpha_{2,2}\mu^2 + \alpha_{2,1}\mu + \alpha_{2,0} \end{aligned} \tag{2.15}$$

and

$$\begin{aligned} \alpha_{2,0} &= \frac{(\alpha + \beta)^2 + 4\alpha\beta}{(\alpha + \beta)^2 + 8(1 - \alpha - \beta) + 4\alpha\beta} \\ \alpha_{2,1} &= \frac{-8(\alpha + \beta)}{(\alpha + \beta)^2 + 8(1 - \alpha - \beta) + 4\alpha\beta} \\ \alpha_{2,2} &= \frac{8}{(\alpha + \beta)^2 + 8(1 - \alpha - \beta) + 4\alpha\beta} \end{aligned} \tag{2.16}$$

One could determine the $\{\alpha_{n,k}\}$ in this way for any n. However, it is more convenient to develop a relation between $v^{(n+1)}$, $v^{(n)}$, and $v^{(n-1)}$ and to compute the $v^{(n)}$ directly without first computing the $u^{(n)}$. We use the three-term recurrence relation for the Chebyshev polynomials given by (9-3.1).

From (2.12) we may write†

$$P_n(G) = T_n\left(\frac{2G - (\beta + \alpha)I}{\beta - \alpha}\right) \Big/ T_n(z) \tag{2.17}$$

and hence, by (1.8),

$$\eta^{(n+1)} = P_{n+1}(G)\varepsilon^{(0)} = \left[T_{n+1}\left(\frac{2G - (\beta + \alpha)I}{\beta - \alpha}\right) \Big/ T_{n+1}(z)\right]\varepsilon^{(0)} \tag{2.18}$$

† Strictly speaking, we should write

$$P_n(G) = \frac{1}{T_n(z)}\, T_n\left[\left(\frac{2}{\beta - \alpha}\right)G - \left(\frac{\beta + \alpha}{\beta - \alpha}\right)I\right]$$

but the form used is more convenient, and no confusion should arise.

Moreover, by (9-3.1) we have for $n \geq 1$

$$\eta^{(n+1)} = \left\{\left[2\left(\frac{2G - (\beta+\alpha)I}{\beta-\alpha}\right)T_n\left(\frac{2G - (\beta+\alpha)I}{\beta-\alpha}\right)\right.\right.$$
$$\left.\left. -T_{n-1}\left(\frac{2G - (\beta+\alpha)I}{\beta-\alpha}\right)\right] \Big/ T_{n+1}(z)\right\}\varepsilon^{(0)} \tag{2.19}$$

But we also have

$$\eta^{(n)} = \left[T_n\left(\frac{2G - (\beta+\alpha)I}{\beta-\alpha}\right) \Big/ T_n(z)\right]\varepsilon^{(0)}$$
$$\eta^{(n-1)} = \left[T_{n-1}\left(\frac{2G - (\beta+\alpha)I}{\beta-\alpha}\right) \Big/ T_{n-1}(z)\right]\varepsilon^{(0)} \tag{2.20}$$

Hence

$$\eta^{(n+1)} = 2\left[\frac{2G - (\beta+\alpha)I}{\beta-\alpha}\right]\frac{T_n(z)}{T_{n+1}(z)}\eta^{(n)} - \frac{T_{n-1}(z)}{T_{n+1}(z)}\eta^{(n-1)} \tag{2.21}$$

Therefore, by (1.5) we have

$$v^{(n+1)} = 2\left(\frac{2G - (\beta+\alpha)I}{\beta-\alpha}\right)\frac{T_n(z)}{T_{n+1}(z)}v^{(n)} - \frac{T_{n-1}(z)}{T_{n+1}(z)}v^{(n-1)}$$
$$+\left[I - 2\left(\frac{2G - (\beta+\alpha)I}{\beta-\alpha}\right)\frac{T_n(z)}{T_{n+1}(z)} + \frac{T_{n-1}(z)}{T_{n+1}(z)}\right]u \tag{2.22}$$

But the last term of the above expression is

$$2\left[zI - \left(\frac{2G - (\beta+\alpha)I}{\beta-\alpha}\right)\right]\frac{T_n(z)}{T_{n+1}(z)}u = \frac{4(I-G)}{\beta-\alpha}\,\frac{T_n(z)}{T_{n+1}(z)}u$$
$$= \frac{4}{\beta-\alpha}\,\frac{T_n(z)}{T_{n+1}(z)}k \tag{2.23}$$

since $u = Gu + k$ because of the complete consistency of the original iterative method. Therefore, for $n \geq 1$ we have

$$v^{(n+1)} = 2\left[\frac{2}{\beta-\alpha}G - \left(\frac{\beta+\alpha}{\beta-\alpha}\right)I\right]\frac{T_n(z)}{T_{n+1}(z)}v^{(n)} - \frac{T_{n-1}(z)}{T_{n+1}(z)}v^{(n-1)}$$
$$+ \frac{4}{\beta-\alpha}\,\frac{T_n(z)}{T_{n+1}(z)}k \tag{2.24}$$

By (2.17) we have, since $T_1(x) = x$,

$$P_1(G) = \frac{2G - (\beta + \alpha)I}{2 - (\beta + \alpha)} \tag{2.25}$$

and, by (1.8) and (1.5),

$$v^{(1)} - u = \frac{2G - (\beta + \alpha)I}{2 - (\beta + \alpha)} (u^{(0)} - u) \tag{2.26}$$

Therefore,

$$\begin{aligned} v^{(1)} &= \frac{2G - (\beta + \alpha)I}{2 - (\beta + \alpha)} u^{(0)} + \left[I - \frac{2G - (\beta + \alpha)I}{2 - (\beta + \alpha)}\right]u \\ &= \frac{2G - (\beta + \alpha)I}{2 - (\beta + \alpha)} u^{(0)} + \frac{2(I - G)}{2 - (\beta + \alpha)} u \\ &= \frac{2}{2 - (\beta + \alpha)} (Gu^{(0)} + k) - \frac{\beta + \alpha}{2 - (\beta + \alpha)} u^{(0)} \end{aligned} \tag{2.27}$$

We now show that (2.24) and (2.27) can be written as a single formula

$$v^{(n+1)} = \frac{\varrho_{n+1}}{2 - (\alpha + \beta)} \{[2G - (\beta + \alpha)I]v^{(n)} + 2k\} + (1 - \varrho_{n+1})v^{(n-1)} \tag{2.28}$$

where

$$\begin{aligned} &\varrho_1 = 1, \qquad \varrho_2 = \frac{2z^2}{2z^2 - 1} \\ &\varrho_{n+1} = \left(1 - \frac{1}{4z^2} \varrho_n\right)^{-1}, \qquad n = 2, 3, \ldots \end{aligned} \tag{2.29}$$

Let us define the numbers $\hat{\varrho}_1, \hat{\varrho}_2, \ldots$ by

$$\begin{aligned} &\hat{\varrho}_1 = 1 \\ &\hat{\varrho}_n = \frac{2zT_{n-1}(z)}{T_n(z)}, \qquad n = 2, 3, \ldots \end{aligned} \tag{2.30}$$

Evidently, by (9-3.1) we have

$$T_{n+1}(z) = 2zT_n(z) - T_{n-1}(z), \qquad n = 1, 2, \ldots \tag{2.31}$$

and hence

$$\hat{\varrho}_{n+1}^{-1} = 1 - \frac{1}{4z^2} \hat{\varrho}_n, \qquad n = 2, 3, \ldots \tag{2.32}$$

Evidently by (2.30) we have $\hat{\varrho}_2 = 2zT_1(z)/T_2(z) = 2z^2(2z^2 - 1)^{-1}$. Thus we have $\hat{\varrho}_1 = \varrho_1$, $\hat{\varrho}_2 = \varrho_2$, $\ldots$.

Upon substituting (2.30) in (2.24) we have, for $n \geq 1$,

$$v^{(n+1)} = \frac{1}{z}\,\varrho_{n+1}\left[\frac{2}{\beta-\alpha}\,G - \frac{\beta+\alpha}{\beta-\alpha}\,I\right]v^{(n)} + (1-\varrho_{n+1})v^{(n-1)} + \frac{2\varrho_{n+1}}{z(\beta-\alpha)}\,k \tag{2.33}$$

by (9-3.1). By (2.5) we have (2.28). Evidently from (2.27) and (2.30) it follows that (2.28) holds when $n = 0$, since $u^{(0)} = v^{(0)}$.

We now seek to determine $\bar{S}(P_n(G))$ which by (1.10), (2.9), and (2.11) is given by

$$\bar{S}(P_n(G)) = \max_{\alpha \leq \mu \leq \beta} |\, P_n(\mu) \,| = (T_n(z))^{-1} \tag{2.34}$$

But by (9-3.3), we have

$$(T_n(z))^{-1} = 2\tau^n/(1 + \tau^{2n}) \tag{2.35}$$

where

$$\tau = \frac{\sigma}{1 + (1-\sigma^2)^{1/2}}, \qquad \sigma = \frac{1}{z} \tag{2.36}$$

Therefore, by (2.34) and (2.35) we have

$$\bar{S}(P_n(G)) = 2\tau^n/(1 + \tau^{2n}) \tag{2.37}$$

By Section 9.1, the virtual quasi-average rate of convergence $\bar{\mathscr{R}}_n'(P_n(G))$ is given by

$$\bar{\mathscr{R}}_n'(P_n(G)) = -\frac{1}{n}\log \bar{S}(P_n(G)) = -\frac{1}{n}\log\frac{2\tau^n}{1+\tau^{2n}} \tag{2.38}$$

In particular, by (2.36) we have

$$\bar{\mathscr{R}}_1'(P_1(G)) = -\log\frac{2\tau}{1+\tau^2} = -\log\sigma \tag{2.39}$$

The virtual asymptotic quasi-average rate of convergence $\bar{\mathscr{R}}(P_n(G))$ is given by

$$\bar{\mathscr{R}}'(P_n(G)) = \lim_{n\to\infty} \bar{\mathscr{R}}'(P_n(G)) = -\log\tau \tag{2.40}$$

Moreover, as in Theorem 6-3.1, we have

$$\lim_{\sigma\to 1-} [-\log \tau/(-\log \sigma)^{1/2}] = 2^{1/2} \tag{2.41}$$

Thus, for n sufficiently large and σ sufficiently close to unity, we have

$$\bar{\mathscr{R}}_n'(P_n(G))/[\bar{\mathscr{R}}_1'(P_1(G))]^{1/2} \sim 2^{1/2} \tag{2.42}$$

Hence the use of the optimum SI method results in an order-of-magnitude improvement in the convergence rate as compared to the repeated use of the linear stationary iterative method

$$\begin{aligned} u^{(n+1)} &= \frac{1}{2-(\alpha+\beta)}([2G-(\beta+\alpha)I]u^{(n)} + 2k) \\ &= P_1(G)u^{(n)} + \frac{2k}{2-(\alpha+\beta)} \end{aligned} \tag{2.43}$$

This latter procedure is clearly at least as good as the repeated use of the basic method itself. Hence we obtain an even greater improvement with the optimum SI method as compared with the basic method.

We remark that the above analysis also holds if all eigenvalues of G are real and greater than unity. In this case the basic iterative method will not converge. However, the linear stationary method defined by $P_1(G)$ will converge but can be substantially improved upon by using the SI method. We use the same formulas as above but assume that the eigenvalues μ of G lie in the range $1 < \beta \leq \mu \leq \alpha$.

Even if some of the eigenvalues of G are less than unity and some are greater than unity, there still exists a convergent SI method, provided G is completely consistent and has only real eigenvalues. We construct one such method, though in general not the best method, as follows. Since G is completely consistent, $I - G$ is nonsingular and hence $\mu = 1$ is not an eigenvalue of G. Therefore there exists α and β such that $\alpha < \beta < 1$ and such that each eigenvalue of G lies in one of the intervals

$$\alpha \leq \mu \leq \beta \quad \text{or} \quad 2-\beta \leq \mu \leq 2-\alpha \tag{2.44}$$

Therefore, the eigenvalues μ' of $G_1 = G(2I - G)$ lie in the interval

$$\alpha(2-\alpha) \leq \mu' \leq \beta(2-\beta) < 1 \tag{2.45}$$

The linear stationary iterative method

$$u^{(n+1)} = G_1 u^{(n)} + k_1 \tag{2.46}$$

where

$$k_1 = (I - G_1)A^{-1}b = (I - G)^2A^{-1}b \tag{2.47}$$

is completely consistent since $(I - G_1) = (I - G)^2$ is nonsingular. We have already seen that since the eigenvalues of G_1 are real and lie in the range (2.45) a convergent SI method can be constructed based on G_1. Suppose the $\{\beta_{n,k}\}$ are the coefficients corresponding to this semi-iterative method. Let us define the coefficients $\{\gamma_{2n,k}\}$ as follows:

$$\sum_{k=0}^{n} \beta_{n,k}G_1^k = \sum_{k=0}^{n} \beta_{n,k}[G(2I - G)]^k = \sum_{k=0}^{2n} \gamma_{2n,k}G^k \tag{2.48}$$

We note that $\sum_{k=0}^{2n} \gamma_{2n,k} = 1$ since $\sum_{k=0}^{n} \beta_{n,k} = 1$. If for each n the sequence $\{\alpha_{n,k}\}$ is defined by

$$\begin{aligned} &\alpha_{n,k} = \gamma_{n,k}, && k = 0, 1, 2, \ldots, n, && n \ \text{even} \\ &\alpha_{n,k} = \alpha_{n-1,k}, && k = 0, 1, 2, \ldots, n-1 \quad \text{and} \quad \alpha_{n,n} = 0, && n \ \text{odd} \end{aligned} \tag{2.49}$$

then $\sum_{k=0}^{n}\alpha_{n,k} = 1$ for all n. Moreover, the SI method based on $\{\alpha_{n,k}\}$ and G yields the same iterants as the SI method based on $\{\beta_{n,k}\}$ and G_1 on even iterations and with $v^{(n)} = v^{(n-1)}$ on odd iterations. Since the latter method converges, the former does also.

Let us now consider the sensitivity of the choice of α and β on the convergence of the method. Actually, we are concerned with the choice of σ where, by (2.36) and (2.5),

$$\sigma = \frac{\beta - \alpha}{2 - (\beta + \alpha)} \tag{2.50}$$

If we choose σ' instead of the true value σ and if $\sigma' \geq \sigma$, then for large n we have

$$[\bar{S}(P_n(G))]^{1/n} = [1/T_n(1/\sigma')]^{1/n} \sim (\omega - 1)^{1/2} \tag{2.51}$$

where

$$\omega = \frac{2}{1 + [1 - (\sigma')^2]^{1/2}} \tag{2.52}$$

On the other hand, if $\sigma' \leq \sigma$, then

$$\bar{S}(P_n(G)) = T_n(\sigma/\sigma')/T_n(1/\sigma') \tag{2.53}$$

and, for large n,

$$[\bar{S}(P_n(G))]^{1/n} \sim (\omega - 1)^{1/2}\{(\sigma/\sigma')^2 + [(\sigma/\sigma')^2 - 1]^{1/2}\} \tag{2.54}$$

Let us compare the effect on the convergence with that of the SOR method where A is consistently ordered and where $\sigma = S(B)$. We suppose that one uses σ' instead of σ to determine ω. By Theorem 6-2.3, if $\sigma' \geq \sigma$, then

$$S(\mathscr{L}_\omega) = \omega - 1 \tag{2.55}$$

while, if $\sigma' \leq \sigma$, then

$$\begin{aligned} S(\mathscr{L}_\omega) &= \left\{\frac{\omega\sigma + [\omega^2\sigma^2 - 4(\omega - 1)]^{1/2}}{2}\right\}^2 \\ &= \left(\frac{\omega}{2}\right)^2\left\{\sigma + \left[\sigma^2 - \frac{4(\omega-1)}{\omega^2}\right]^{1/2}\right\}^2 = \frac{\omega}{2}\{\sigma+[\sigma^2 - (\sigma')^2]^{1/2}\}^2 \\ &= \left(\frac{\omega\sigma'}{2}\right)^2\left\{\left(\frac{\sigma}{\sigma'}\right)^2 + \left[\left(\frac{\sigma}{\sigma'}\right)^2 - 1\right]^{1/2}\right\}^2 \\ &= (\omega - 1)\left\{\left(\frac{\sigma}{\sigma'}\right)^2 + \left[\left(\frac{\sigma}{\sigma'}\right)^2 - 1\right]^{1/2}\right\}^2 \end{aligned} \tag{2.56}$$

since $\omega^2(\sigma')^2 = 4(\omega - 1)$. Thus the behavior of $[\bar{S}(P_n(G))]^{1/n}$ is very similar to the behavior of $[S(\mathscr{L}_\omega)]^{1/n}$ when ω is chosen based on σ' instead of on the true value σ. Therefore, as observed in Section 6.5, a small overestimation of σ results in a small decrease in the rate of convergence, but a comparable underestimation of σ results in a much larger decrease.

11.3. J, JOR, AND RF SEMI-ITERATIVE METHODS

If A is a positive definite matrix, then the eigenvalues of the J, JOR, and RF methods are real and the analysis of Section 11.3 is applicable. We refer to the optimum SI methods for the J, JOR, and RF methods as the J–SI, JOR–SI, and RF–SI methods, respectively.

J-SI Method

For the J–SI and JOR–SI methods we need only assume that the matrix $B = I - (\text{diag } A)^{-1}A$, corresponding to the J method has real eigenvalues all of which are less than unity. These assumptions hold if A is positive definite since B is similar to $\tilde{B} = D^{1/2}BD^{-1/2} = I - D^{-1/2}AD^{-1/2}$ where $D = \text{diag } A$. Since $D^{-1/2}AD^{-1/2}$ is positive definite, it follows that the eigenvalues of $\tilde{B}$, and hence those of B, are real and less than unity.

It follows from the considerations of Section 11.2 that even though the eigenvalues of B may be less than -1 and hence $S(B)$ may exceed unity, nevertheless, a convergent SI method based on the J method can be constructed. If the eigenvalues of B lie in the range $\alpha \le \mu \le \beta < 1$, then by (2.28) the optimum such method is defined by

$$v^{(n+1)} = \frac{\varrho_{n+1}}{2-(\alpha+\beta)}\{[2B-(\beta+\alpha)I]v^{(n)}+2c\}+(1-\varrho_{n+1})v^{(n-1)} \tag{3.1}$$

where the $\{\varrho_n\}$ are given by (2.29) and where $c = D^{-1}b$.

We remark that (3.1) is valid even if $\alpha = \beta$. In that case, the ϱ_n are all unity and the method reduces to

$$v^{(n+1)} = \frac{1}{1-\alpha}[(B-\alpha I)v^{(n)} + c] \tag{3.2}$$

The convergence will be extremely rapid since $S(B - \alpha I) = 0$.

In the case where $\beta \ne \alpha$, we denote the iteration matrix corresponding to the J–SI matrix by $\mathscr{B}_n$ where

$$\mathscr{B}_n = p_n(B) = T_n\left(\frac{2B-(\beta+\alpha)I}{\beta-\alpha}\right)\Big/ T_n\left(\frac{2-(\beta-\alpha)}{\beta-\alpha}\right) \tag{3.3}$$

If A has Property A, then $\alpha = -\beta$ by Theorem 5-3.4 and $\bar{\mu} = S(B) = \beta < 1$, $\alpha = -\bar{\mu}$. By (2.5) we have

$$z = 1/\bar{\mu} \tag{3.4}$$

and (3.1) becomes

$$v^{(n+1)} = \varrho_{n+1}(Bv^{(n)} + c) + (1 - \varrho_{n+1})v^{(n-1)}, \qquad n = 0, 1, 2, \ldots \tag{3.5}$$

where

$$\begin{aligned} \varrho_1 &= 1, \qquad \varrho_2 = \frac{2}{2-\bar{\mu}^2} \\ \varrho_{n+1} &= (1 - \tfrac{1}{4}\bar{\mu}^2\varrho_n)^{-1}, \qquad n = 2, 3, \ldots \end{aligned} \tag{3.6}$$

By (2.34), (2.35), and (2.36) we have

$$\bar{S}(\mathscr{B}_n) = \bar{S}(p_n(B)) = 2r^{n/2}/(1 + r^n) \tag{3.7}$$

since $\sigma = \bar{\mu}$ and, by (6-1), $\tau = \sigma(1 + [1 - \sigma^2]^{1/2})^{-1} = (\omega_b - 1)^{1/2} = r^{1/2}$.

If A is positive definite, then

$$\| \overline{\mathscr{B}_n} \|_{D^{1/2}} = \| \overline{p_n(B)} \|_{D^{1/2}} = \| \overline{D^{1/2} p_n(B) D^{-1/2}} \| = \| \overline{p_n(\tilde{B})} \| = \bar{S}(p_n(\tilde{B})) \tag{3.8}$$

since $\tilde{B} = D^{1/2} B D^{-1/2}$ is symmetric. Therefore

$$\bar{S}(p_n(\tilde{B})) = \bar{S}(D^{1/2} p_n(B) D^{-1/2}) = \bar{S}(p_n(B)) = \bar{S}(\mathscr{B}_n) \tag{3.9}$$

and hence

$$\| \mathscr{B}_n \|_{D^{1/2}} = \bar{S}(\mathscr{B}_n) \tag{3.10}$$

Similarly,

$$\| \mathscr{B}_n \|_{A^{1/2}} = \bar{S}(\mathscr{B}_n) \tag{3.11}$$

Thus we have

Theorem 3.1. If A is positive definite and has Property A, then

$$\bar{S}(\mathscr{B}_m) = \| \overline{\mathscr{B}_m} \|_{D^{1/2}} = \| \overline{\mathscr{B}_m} \|_{A^{1/2}} = 2r^{m/2}/(1 + r^m) \tag{3.12}$$

Since $\bar{S}(\mathscr{B}_m) \geq r^{m/2}$ it follows that the virtual quasi-average rate of convergence of the J–SI method is less than one-half that of $\mathscr{L}_{\omega_b}^m$ if A is consistently ordered. The advantage of $\mathscr{L}_{\omega_b}^m$ is somewhat less if one considers the $D^{1/2}$-norm or the $A^{1/2}$-norm instead of the spectral radius. Later we will show how if A has the form (8-1.2) one can modify the J–SI method to obtain the CCSI method and thereby improve the convergence rate by a factor of two. The CCSI method, as we have seen, is very competitive with $\mathscr{L}_{\omega_b}^m$.

JOR–SI Method

The matrix B_ω corresponding to the JOR method is given by $B_\omega = \omega B + (1 - \omega)I$. If A is positive definite, then, as before, the eigenvalues of B are real and lie in the range $\alpha \leq \mu \leq \beta < 1$. Therefore, if $\omega > 0$, the eigenvalues μ' of B_ω lie in the range

$$\alpha' = \omega\alpha + 1 - \omega \leq \mu' \leq \omega\beta + 1 - \omega = \beta' < 1 \tag{3.13}$$

Therefore, by (2.5) we have

$$z = \frac{2 - (\alpha' + \beta')}{\beta' - \alpha'} = \frac{2 - (\alpha + \beta)}{\beta - \alpha} \tag{3.14}$$

which is independent of ω. Moreover, by (2.28) the method is defined by (3.1) since $k = \omega c$ by (3-3.12). Thus the iteration formula is independent of ω; hence the optimum SI method based on the JOR method is identical to the optimum SI method based on the J method.

RF–SI Method

For the RF method we have $R_p = I + pA$, where $p \neq 0$. Since A is positive definite, all of the eigenvalues of A are positive and there exist positive numbers $\bar{a}$ and $\bar{b}$ such that all eigenvalues ν of A lie in the range

$$0 < \bar{a} \leq \nu \leq \bar{b} \tag{3.15}$$

Therefore, all eigenvalues μ of R_p are real and lie in the range

$$\alpha = 1 + p\bar{b} \leq \mu \leq 1 + p\bar{a} = \beta < 1 \tag{3.16}$$

if $p < 0$ and

$$\alpha = 1 + p\bar{b} \geq \mu \geq 1 + p\bar{a} = \beta > 1 \tag{3.17}$$

if $p > 0$. In either case, the methods of Section 11.2 are applicable and by (2.5) we have

$$z = (\bar{b} + \bar{a})/(\bar{b} - \bar{a}) \tag{3.18}$$

By (2.28) and (3-3.27) the formula for the RF–SI method is

$$v^{(n+1)} = \varrho_{n+1}/(\bar{b} + \bar{a})\{-[2A - (\bar{b} + \bar{a})I]v^{(n)} + 2b\} + (1 - \varrho_{n+1})v^{(n-1)} \tag{3.19}$$

where the ϱ_n are given by (2.29). Evidently (3.19) is independent of p. Moreover, by (2.17) we have

$$\begin{aligned} P_n(I + pA) &= T_n\left(\frac{2(I + pA) - [2 + p(\bar{b} + \bar{a})]I}{p(\bar{a} - \bar{b})}\right) \Big/ T_n\left(\frac{\bar{b} + \bar{a}}{\bar{b} - \bar{a}}\right) \\ &= T_n\left(\frac{2A - (\bar{b} + \bar{a})I}{\bar{a} - \bar{b}}\right) \Big/ T_n\left(\frac{\bar{b} + \bar{a}}{\bar{b} - \bar{a}}\right) \end{aligned} \tag{3.20}$$

We denote the iteration matrix corresponding to the RF–SI method by $\mathscr{Q}_m$ where

$$\mathscr{Q}_m = f_m(A) = T_m\left(\frac{2A - (\bar{b} + \bar{a})I}{\bar{a} - \bar{b}}\right) \Big/ T_m\left(\frac{\bar{b} + \bar{a}}{\bar{b} - \bar{a}}\right) \tag{3.21}$$

We remark that (3.19) may be written in the alternative form

$$v^{(n+1)} = v^{(n-1)} + \varrho_{n+1}(v^{(n)} - v^{(n-1)}) - [2\varrho_{n+1}/(\bar{b}+\bar{a})](Av^{(n)} - b) \tag{3.22}$$

which involves the residual vector $Av^{(n)} - b$.

Theorem 3.2. If A is positive definite, then

$$S(\mathscr{Q}_m) = \| \mathscr{Q}_m \| = \| \mathscr{Q}_m \|_{A^{1/2}} = 1 \Big/ T_m\left(\frac{\bar{b}+\bar{a}}{\bar{b}-\bar{a}}\right) \tag{3.23}$$

where the eigenvalues ν of A lie in the range $\bar{a} \leq \nu \leq \bar{b}$.

We note that in this case we use the ordinary spectral norm instead of the $D^{1/2}$ norm.

We now show that the RF–SI method and the J–SI method are identical if the diagonal elements of A are constant. But in this case $A = dI - C$ for some constant $d > 0$ and $B = d^{-1}C$. Hence, the eigenvalues μ of B and the eigenvalues ν of A are related by

$$\nu = d(1 - \mu) \tag{3.24}$$

Thus we have

$$\begin{aligned} \alpha &= 1 - (\bar{b}/d), & \beta &= 1 - (\bar{a}/d) \\ \bar{b} &= d(1-\alpha), & \bar{a} &= d(1-\beta) \end{aligned} \tag{3.25}$$

so that

$$z = \frac{\bar{b}+\bar{a}}{\bar{b}-\bar{a}} = \frac{2-(\alpha+\beta)}{\beta-\alpha} \tag{3.26}$$

and

$$\frac{2A - (\bar{b}+\bar{a})I}{\bar{a}-\bar{b}} = \frac{2B - (\alpha+\beta)I}{\beta-\alpha} \tag{3.27}$$

Hence, by (3.3) and (3.21) we have

$$\mathscr{Q}_m = f_m(A) = p_m(B) = \mathscr{P}_m \tag{3.28}$$

and the two methods are the same.

Consider now the case where A is positive definite but the diagonal elements of A are not necessarily constant. Evidently the convergence of the J–SI method applied to the linear system

$$\hat{A}\hat{u} = \hat{b} \tag{3.29}$$

where $\hat{A} = D^{-1/2}AD^{-1/2}$, $\hat{u} = D^{1/2}u$, $\hat{b} = D^{-1/2}b$, $D = \operatorname{diag} A$, is the same as for the original system, and for any scaled system

$$(EAE)v = Eb \tag{3.30}$$

where E is a diagonal matrix with positive diagonal elements. On the other hand, the convergence of the RF–SI method will in general be different. As a matter of fact, by (3.23) the convergence rate of the RF–SI method for (3.30) depends on the *condition* of EAE, $K(EAE)$, i.e., the ratio of the largest to the smallest eigenvalue. But

$$EAE = ED^{1/2}\hat{A}D^{1/2}E \tag{3.31}$$

Moreover, if A has Property A, it follows from Theorem 6-7.1 that

$$K(EAE) \geq K(\hat{A}) \tag{3.32}$$

Consequently, the RF–SI method converges less rapidly than the J–SI method. Since the convergence of the GRF–SI method applied to the system $Au = b$ is the same as that of the RF–SI method applied to an appropriate scaled system, it follows that the J–SI method is also at least as good as the GRF–SI method if A has Property A.

Neither the RF–SI method nor the J–SI method is really competitive with the SOR method, $\mathscr{L}_{\omega_b}^m$, for consistently ordered positive definite matrices. However, as we shall show, in A has the form (8-1.2), the J–SI method can be made competitive. Also, if A is positive definite but does not have Property A (and is not an L-matrix), then the RF–SI and the J–SI methods may be superior. For example, let us consider the linear system $Au = b$ where

$$A = \begin{pmatrix} 1 & a & a \\ a & 1 & a \\ a & a & 1 \end{pmatrix} \tag{3.33}$$

where $0 \leq a < 1$. Evidently A is a positive matrix but does not have Property A and is not an L-matrix. As can be seen in Table 3.1, the SOR method is only slightly more effective than the GS method. However, if we apply the J–SI method, which in this case is equivalent to the RF–SI method, we achieve an order-of-magnitude improvement over the GS method in the sense that the ratio $-\log \tau/[R(\mathscr{L})]^{1/2}$ converges to a limit of approximately 0.940. (For the SOR method in the consistently ordered case we have $-\log S(\mathscr{L}_{\omega_b})/[R(\mathscr{L})]^{1/2} \rightarrow 2$.) Thus the RF–SI

TABLE 3.1. Comparison of the J–SI Method and the SOR Method[a]

a	0.90	0.92	0.94	0.96	0.98	0.99
$\bar{\omega}$	1.22344	1.23203	1.24062	1.24961	1.25859	$\sim$1.25
$R(\mathscr{L})$	0.15804	0.12507	0.09281	0.06123	0.03030	0.01508
$R(\mathscr{L}_{\bar{\omega}})$	0.17950	0.14252	0.10611	0.07024	0.03488	0.01737
$R(\mathscr{L}_{\bar{\omega}})/R(\mathscr{L})$	1.13577	1.13951	1.14327	1.14705	1.15086	1.15230
σ	0.93103	0.94520	0.95918	0.97297	0.98658	0.99331
τ	0.68211	0.71257	0.74774	0.79044	0.84809	0.89049
$-\log \tau$	0.38256	0.33888	0.29070	0.23516	0.16477	0.11599
$\bar{\mathscr{R}}_1$	0.07146	0.05635	0.04167	0.02740	0.01351	0.00671
$\bar{\mathscr{R}}_5$	0.24824	0.20688	0.16271	0.11472	0.06135	0.03190
$\bar{\mathscr{R}}_{10}$	0.31329	0.26968	0.22169	0.16675	0.09909	0.05605
$\bar{\mathscr{R}}_{20}$	0.34790	0.30422	0.25605	0.20051	0.13018	0.08181
$-\log \tau/[R(\mathscr{L})]^{1/2}$	0.960	0.959	0.954	0.949	0.945	0.944
$-\log \tau/[R(\mathscr{L})]^{1/2}$	2.42	2.71	3.14	3.84	5.44	7.70

[a] $\bar{\mathscr{R}}_n$ is the virtual quasi-average rate of convergence with n iterations and $\bar{\omega}$ is the omega which minimizes $S(\mathscr{L}_\omega)$.

method is approximately one-half as effective as the SOR method would be in the consistently ordered case.

11.4. RICHARDSON'S METHOD

In 1910 Richardson presented a method for solving the linear system (3.1.1). The method, known as Richardson's method, is defined by

$$u^{(n+1)} = u^{(n)} + \beta_{n+1}(Au^{(n)} - b) \tag{4.1}$$

where $\beta_1, \beta_2, \ldots$ are iteration parameters. Evidently (4.1) defines a nonstationary linear iterative method. Since the method

$$u^{(n+1)} = u^{(n)} + \beta(Au^{(n)} - b) \tag{4.2}$$

is completely consistent for each $\beta \neq 0$, it follows from Theorem 9-1.1 that Richardson's method is consistent though not necessarily completely consistent. Evidently, if A is nonsingular then Richardson's method is convergent in the sense of Section 9.1 if and only if

$$\lim_{n\to\infty} \prod_{k=1}^{n} (I + \beta_k A) = 0 \tag{4.3}$$

By (3-3.27) the RF method is defined by

$$u^{(n+1)} = u^{(n)} + p(Au^{(n)} - b) \tag{4.4}$$

where $p \neq 0$. Thus, the RF method is equivalent to Richardson's method with fixed β_{n+1}. We now show that there is a close correspondence between Richardson's method and SI methods based on the RF method.

Theorem 4.1. Let A be a nonsingular matrix. Given $\beta_1, \beta_2, \ldots$, there exist, for any $p \neq 0$, $\alpha_{n,k}$, $n = 0, 1, 2, \ldots$, $k = 0, 1, 2, \ldots, n$ such that $\sum_{k=0}^{n} \alpha_{n,k} = 1$ and such that the SI method based on (4.4) and the $\{\alpha_{n,k}\}$ yields the same iterants for the same starting vector $u^{(0)}$ as does Richardson's method based on the $\{\beta_k\}$.

Conversely, given $\alpha_{n,k}$, $n = 0, 1, 2, \ldots$, $k = 0, 1, 2, \ldots, n$ such that $\sum_{k=0}^{n} \alpha_{n,k} = 1$ and any $p \neq 0$, there exists, for any n, $\beta_1^{(n)}, \beta_2^{(n)}, \ldots$ such that $u^{(n)}$, as determined by Richardson's method starting with $u^{(0)}$ and based on the $\{\beta_k^{(n)}\}$, is the same as $v^{(n)}$, as determined by the SI method based on (4.4) and the $\{\alpha_{n,k}\}$, where $v^{(0)} = u^{(0)}$.

Proof. Let $\bar{u}$ denote the exact solution of (3-1.1). For given $\beta_1, \beta_2, \ldots$ let $\varepsilon^{(n)} = u^{(n)} - \bar{u}$ where the sequence $u^{(0)}, u^{(1)}, u^{(2)}, \ldots$ is determined by Richardson's method starting with $u^{(0)}$ and based on the $\{\beta_k\}$. Evidently we have

$$\varepsilon^{(n+1)} = (I + \beta_{n+1}A)\varepsilon^{(n)} \tag{4.5}$$

and

$$\varepsilon^{(n)} = \prod_{k=1}^{n} (I + \beta_k A)\varepsilon^{(0)} \tag{4.6}$$

Moreover, for any $p \neq 0$ we have

$$I + \beta A = I + \frac{\beta}{p}[(I + pA) - I] = \left(1 - \frac{\beta}{p}\right)I + \frac{\beta}{p}(I + pA) \tag{4.7}$$

so that $\prod_{k=1}^{n} (I + \beta_k A)$ is a polynomial in $I + pA$ of degree $n' \leq n$. We define the polynomial $P_n(I + pA)$ of degree n or less and the numbers $\gamma_0^{(n)}, \gamma_1^{(n)}, \ldots, \gamma_n^{(n)}$ by

$$\prod_{k=1}^{n} (I + \beta_k A) = \sum_{k=0}^{n} \gamma_k^{(n)}(I + pA)^k = P_n(I + pA) \tag{4.8}$$

We let

$$\alpha_{n,k} = \gamma_k^{(n)}, \qquad k = 0, 1, 2, \ldots, n \tag{4.9}$$

If $v^{(0)}, v^{(1)}, v^{(2)}, \ldots$ is the sequence determined by the SI method based on (4.4) and the $\{\alpha_{n,k}\}$, we have

$$\eta^{(n)} = P_n(I + pA)\eta^{(0)} \tag{4.10}$$

where $\eta^{(n)} = v^{(n)} - \bar{u}$. Since $\eta^{(0)} = \varepsilon^{(0)}$ we have

$$\eta^{(n)} = P_n(I + pA)\eta^{(0)} = \prod_{k=1}^{n} (I + \beta_k A)\varepsilon^{(0)} = \varepsilon^{(n)} \tag{4.11}$$

and therefore $v^{(n)} = u^{(n)}$, $n = 0, 1, 2, \ldots$. Thus Richardson's method yields the same iterants as the semi-iterative method based on the RF method.

Suppose now that $p \neq 0$ and the $\{\alpha_{n,k}\}$ are given. For each n we seek to determine $\beta_1^{(n)}, \beta_2^{(n)}, \ldots, \beta_n^{(n)}$ such that Richardson's method based on the $\{\beta_k^{(n)}\}$ will yield the same nth iterant as the semi-iterative method based on (4.4) and the $\{\alpha_{n,k}\}$. Let the roots of $P_n(x) = \sum_{k=0}^{n} \alpha_{n,k} x^k = 0$ be $a_1, a_2, \ldots, a_{n'}$ where $n' \leq n$ and let

$$\begin{aligned} \beta_k^{(n)} &= p(1 - a_k)^{-1}, & k &= 1, 2, \ldots, n' \\ \beta_k^{(n)} &= 0, & k &= n' + 1, n' + 2, \ldots \end{aligned} \tag{4.12}$$

Since $P_n(1) = 1$ it follows that none of the a_k equals unity. Evidently for some constant K we have

$$P_n(x) = K \prod_{k=1}^{n'} (x - a_k) \tag{4.13}$$

and hence

$$\begin{aligned} P_n(I + pA) &= K \prod_{k=1}^{n'} (I + pA - a_k I) \\ &= K \prod_{k=1}^{n'} (1 - a_k) \prod_{k=1}^{n'} \left(I + \left(\frac{p}{1 - a_k}\right)A\right) \end{aligned} \tag{4.14}$$

But since $P_n(1) = 1$ we have

$$K^{-1} = \prod_{k=1}^{n'} (1 - a_k) \tag{4.15}$$

so that

$$P_n(I + pA) = \prod_{k=1}^{n'} \left(I + \frac{p}{1 - a_k} A\right) = \prod_{k=1}^{n} (I + \beta_k^{(n)} A) \tag{4.16}$$

Evidently for given $u^{(0)}$ and n the sequence $v^{(1)}, v^{(2)}, \ldots$ defined by the semi-iterative method based on (4.4) and the $\{\alpha_{n,k}\}$ and the sequence $u^{(0)}, u^{(1)}, \ldots$ defined by Richardson's method with the $\{\beta_k^{(n)}\}$ are such that $v^{(n)} = u^{(n)}$. This follows since

$$\begin{aligned} v^{(n)} &= \bar{u} + P_n(I + pA)(v^{(0)} - \bar{u}) \\ &= \bar{u} + \prod_{k=1}^{n} (I + \beta_k^{(n)} A)(u^{(0)} - \bar{u}) = u^{(n)} \end{aligned} \tag{4.17}$$

The proof of Theorem 4.1 is now complete.

From the above considerations it is clear that for any $p \neq 0$ the optimum SI method with respect to the RF method (4.4) converges at least as fast as Richardson's method, since no matter what choice of $\beta_1, \beta_2, \ldots$ is made there is an SI method which produces identical results. On the other hand, given the optimum semi-iterative method based on the RF method, then for any n one can choose $\beta_1^{(n)}, \beta_2^{(n)}, \ldots, \beta_n^{(n)}$ so that the semi-iterative method and Richardson's method will agree on the nth, and in general *only* on the nth iteration.

In the actual application of Richardson's method one would normally choose a value of m and then use the parameters

$$\beta_1^{(m)}, \ \beta_2^{(m)}, \ \ldots, \ \beta_m^{(m)}$$

in a cyclic order. This periodic nonstationary method could be studied as a linear stationary method with the matrix

$$G_m = \prod_{k=1}^{m} (I + \beta_k^{(m)} A) \tag{4.18}$$

Using the optimum $\beta_k^{(m)}$ we obtain by (3.18) and (2.37)

$$\bar{S}(G_m) = 2\tau^m/(1 + \tau^{2m}) \tag{4.19}$$

where

$$\tau = \frac{\sigma}{1 + [1 - \sigma^2]^{1/2}}, \qquad \sigma = \frac{\bar{b} - \bar{a}}{\bar{b} + \bar{a}} \tag{4.20}$$

and where the eigenvalues ν of A lie in the range $0 < \bar{a} \leq \nu \leq \bar{b}$. From (3.20) the roots of $P_m(x) = 0$ are the roots a_k of

$$T_m\left(\frac{2x - [2 + p(\bar{b} + \bar{a})]}{p(\bar{a} - \bar{b})}\right) = 0 \tag{4.21}$$

Since the roots of $T_m(y) = 0$ are $y_k = \cos((2k - 1)\pi/2m)$, $k = 1$,

2, . . ., *m*, we have

$$a_k = [2 + p(\bar{a} - \bar{b})y_k + p(\bar{b} + \bar{a})]/2 \tag{4.22}$$

By (4.12) we have

$$\begin{aligned}\beta_k^{(m)} &= p/(1 - a_k)\\ &= -2/[(\bar{a}-\bar{b})\cos[(2k-1)\pi/2m]+\bar{b}+\bar{a}], \quad k = 1, 2, \ldots, m\end{aligned} \tag{4.23}$$

Thus we have an alternative derivation for the formula for the optimum $\{\beta_k^{(m)}\}$ given by Young [1954].

Numerical experiments, as described by Young [1954] and [1956] and by Young and Warlick [1953] show that unless great care is taken Richardson's method is very sensitive to rounding errors especially for large m. For this reason the use of the RF–SI method is recommended. Not only does this method result in less rounding errors but for each n one obtains at the nth iteration the same result as one would have obtained by Richardson's method with the parameters $\beta_1^{(n)}, \beta_2^{(n)}, \ldots, \beta_n^{(n)}$ designed for that n.

11.5. CYCLIC CHEBYSHEV SEMI-ITERATIVE METHOD

The cyclic Chebyshev semi-iterative method (CCSI method), introduced by Golub and Varga [1961], is applicable if A is a matrix of the form (8-1.2), with nonvanishing diagonal elements such that the eigenvalues of $B = I - (\text{diag } A)^{-1}A$ are real and $S(B) < 1$. We have already considered the CCSI method, in Section 10.6, as a special case of the MSOR method. We now show that the CCSI method can also be derived from the J–SI method. Indeed, from (3.5) we have, since B is given by (8-1.8)

$$\begin{pmatrix} v_1^{(n+1)} \\ v_2^{(n+1)} \end{pmatrix} = \varrho_{n+1}\left\{\begin{pmatrix} 0 & F \\ G & 0 \end{pmatrix}\begin{pmatrix} v_1^{(n)} \\ v_2^{(n)} \end{pmatrix} + \begin{pmatrix} c_1 \\ c_2 \end{pmatrix}\right\} + (1 - \varrho_{n+1})\begin{pmatrix} v_1^{(n-1)} \\ v_2^{(n-1)} \end{pmatrix} \tag{5.1}$$

where the $\{\varrho_{n+1}\}$ are given by (3.6) and

$$\begin{aligned} v_1^{(n+1)} &= \varrho_{n+1}(Fv_2^{(n)} + c_1) + (1 - \varrho_{n+1})v_1^{(n-1)} \\ v_2^{(n+1)} &= \varrho_{n+1}(Gv_1^{(n)} + c_2) + (1 - \varrho_{n+1})v_2^{(n-1)} \end{aligned} \tag{5.2}$$

Here $(c_1{}^{\mathrm{T}}, c_2{}^{\mathrm{T}})^{\mathrm{T}} = c$, $((v_1^{(n)})^{\mathrm{T}}, (v_2^{(n)})^{\mathrm{T}})^{\mathrm{T}} = v^{(n)}$, etc. Suppose now that $v_1^{(0)}$ and $v_2^{(0)}$ are arbitrary. Evidently since $\varrho_1 = 1$ we have

$$v_1^{(1)} = Fv_2^{(0)} + c_1 \tag{5.3}$$

Moreover,

$$v_2^{(2)} = \varrho_2(Gv_1^{(1)} + c_2) + (1 - \varrho_2)v_2^{(0)} \tag{5.4}$$

$$v_1^{(3)} = \varrho_3(Fv_2^{(2)} + c_1) + (1 - \varrho_3)v_1^{(1)} \tag{5.5}$$

etc. Thus, we can determine the circled vectors $v_1^{(1)}, v_2^{(2)}, v_1^{(3)}, v_2^{(4)}, \ldots$

$$\begin{pmatrix} \enclose{circle}{v_1^{(0)}} \\ \enclose{circle}{v_2^{(0)}} \end{pmatrix} \begin{pmatrix} \enclose{circle}{v_1^{(1)}} \\ v_2^{(1)} \end{pmatrix} \begin{pmatrix} v_1^{(2)} \\ \enclose{circle}{v_2^{(2)}} \end{pmatrix} \begin{pmatrix} \enclose{circle}{v_1^{(3)}} \\ v_2^{(3)} \end{pmatrix} \begin{pmatrix} v_1^{(4)} \\ \enclose{circle}{v_2^{(4)}} \end{pmatrix} \tag{5.6}$$

without determining the others such as $v_2^{(1)}, v_1^{(2)}, v_2^{(3)}, v_1^{(4)}, \ldots$. Let us consider the vectors

$$w^{(0)} = \begin{pmatrix} v_1^{(0)} \\ v_2^{(0)} \end{pmatrix}, \qquad w^{(1)} = \begin{pmatrix} v_1^{(1)} \\ v_2^{(2)} \end{pmatrix}, \qquad w^{(2)} = \begin{pmatrix} v_1^{(3)} \\ v_2^{(4)} \end{pmatrix}, \cdots \tag{5.7}$$

and in general for $n \geq 1$

$$w^{(n)} = \begin{pmatrix} v_1^{(2n-1)} \\ v_2^{(2n)} \end{pmatrix} \tag{5.8}$$

From (5.2) we have

$$\begin{aligned} w_1^{(n+1)} &= \varrho_{2n+1}(Fw_2^{(n)} + c_1) + (1 - \varrho_{2n+1})w_1^{(n)} \\ w_2^{(n+1)} &= \varrho_{2n+2}(Gw_1^{(n+1)} + c_2) + (1 - \varrho_{2n+2})w_2^{(n)} \end{aligned} \tag{5.9}$$

If we let

$$\omega_k = \varrho_{2k-1}, \qquad \omega_k' = \varrho_{2k}, \qquad k = 1, 2, \ldots \tag{5.10}$$

then we have

$$\begin{aligned} w_1^{(n+1)} &= \omega_{n+1}(Fw_2^{(n)} + c_1) + (1 - \omega_{n+1})w_1^{(n)} \\ w_2^{(n+1)} &= \omega_{n+1}'(Gw_1^{(n+1)} + c_2) + (1 - \omega_{n+1}')w_2^{(n)} \end{aligned} \tag{5.11}$$

which are precisely equivalent to the formula (10-1) for the MSOR method. From (5.10) and (2.29) we have, since $z = 1/\bar{\mu}$,

$$\begin{aligned} &\omega_1 = 1, && \omega_1' = \frac{2}{2 - \bar{\mu}^2} \\ &\omega_{k+1} = \left(1 - \frac{\bar{\mu}^2}{4}\,\omega_k'\right)^{-1}, && \omega_{k+1}' = \left(1 - \frac{\bar{\mu}^2}{4}\,\omega_{k+1}\right)^{-1}, \qquad k = 1, 2, \ldots \end{aligned} \tag{5.12}$$

But this is just the choice given by (10-6.1). Thus the CCSI method is indeed the method discussed in Section 10.6.

11.6. GS SEMI-ITERATIVE METHODS

We now consider the GS–SI method and a variant proposed by Sheldon [1959]. If A is a positive definite matrix, then by Theorem 4-3.1, the GS method converges. However, the eigenvalues of $\mathscr{L}$ need not be real and the results of Section 11.2 may not be applicable. On the other hand, even though the J method need not converge, nevertheless, it has real eigenvalues and we can effectively use either the J–SI method or the RF–SI method.

If A is a Stieltjes matrix, then by Theorems 2-7.3 and 4-5.1 both the J method and the GS method converge and $S(\mathscr{L}) \leq S(B)$. However, since the eigenvalues of the GS method may be complex we are not sure that the optimum GS–SI method will be as good as the J–SI method. If we also assume the eigenvalues of $\mathscr{L}$ are real, we are still not certain that the GS–SI method is better than the J–SI method since while the upper bound β_{GS} is smaller for the GS method than the upper bound β_{J} for the J method, the lower bound α_{GS} may be less than α_{J}. On the other hand, we do know that $\beta_{\mathrm{J}} - \alpha_{\mathrm{J}} \geq S(B)$ since the sum of the eigenvalues of B is zero.

If A is a consistently ordered positive definite matrix, however, then in terms of the spectral radius criterion the GS–SI method converges almost as fast as the SOR method and almost twice as fast as the J–SI method. In this case the eigenvalues λ of $\mathscr{L}$ lie in the range

$$0 \leq \lambda \leq \bar{\mu}^2 \tag{6.1}$$

by Theorem 5-3.4. Therefore, by (2.5) we have

$$z = \frac{2}{\bar{\mu}^2} - 1 \tag{6.2}$$

By (2.28) the GS–SI method is given by

$$\begin{aligned} v^{(n+1)} = {} & \frac{\varrho_{n+1}}{1 - (\bar{\mu}^2/2)} \left[\left(\mathscr{L} - \frac{\bar{\mu}^2}{2} I \right) v^{(n)} + k \right] \\ & + (1 - \varrho_{n+1}) v^{(n-1)}, \qquad n = 0, 1, 2, \ldots \end{aligned} \tag{6.3}$$

where the ϱ_i are given by (2.29) with z given by (6.2). By (2.17) the iteration matrix corresponding to the GS–SI method is given by

$$\mathscr{X}_m = q_m(\mathscr{L}) = T_m\left(\frac{2}{\bar{\mu}^2}\mathscr{L} - I\right) \Big/ T_m\left(\frac{2}{\bar{\mu}^2} - 1\right) \tag{6.4}$$

By (2.34) we have

$$\begin{aligned} S(\mathscr{X}_m) = S(q_m(\mathscr{L})) &= \max_{0 \le \lambda \le \bar{\mu}^2} |\, q_m(\lambda) \,| \\ &= \max_{0 \le \lambda \le \bar{\mu}^2} \left| T_m\left(\frac{2\lambda}{\bar{\mu}^2} - 1\right)\right| \Big/ \left| T_m\left(\frac{2}{\bar{\mu}^2} - 1\right)\right| \\ &= 1 \Big/ \left| T_m\left(\frac{2}{\bar{\mu}^2} - 1\right)\right| = q_m(\bar{\mu}^2) \\ &= \frac{2r^m}{1 + r^{2m}} = \frac{2(\omega_b - 1)^m}{1 + (\omega_b - 1)^{2m}} \end{aligned} \tag{6.5}$$

This follows by (2.35), (2.36), (6.2), and (6-1). Thus based on the spectral radius criterion, the GS–SI method is about twice as fast as the J–SI method. However, as we shall see, the comparison is less favorable with respect to the $D^{1/2}$- and $A^{1/2}$-norms.

At first glance, it might appear from (6.3) that twice as much storage would be required than, for example, with the SOR or MSOR methods. However, if A has the form (8-1.2), then we can reduce the number of calculations and the storage requirements as follows. By (8-1.8) and (3-3.18) we have

$$\mathscr{L} = \begin{pmatrix} 0 & F \\ 0 & GF \end{pmatrix}, \qquad k = \begin{pmatrix} c_1 \\ Gc_1 + c_2 \end{pmatrix} \tag{6.6}$$

Therefore,

$$\begin{pmatrix} v_1^{(n+1)} \\ v_2^{(n+1)} \end{pmatrix} = \frac{\varrho_{n+1}}{1 - \frac{1}{2}\bar{\mu}^2} \begin{pmatrix} (F - \frac{1}{2}\bar{\mu}^2 I)v_2^{(n)} + c_1 \\ (GF - \frac{1}{2}\bar{\mu}^2 I)v_2^{(n)} + Gc_1 + c_2 \end{pmatrix} + (1 - \varrho_{n+1}) \begin{pmatrix} v_1^{(n)} \\ v_2^{(n)} \end{pmatrix} \tag{6.7}$$

and

$$v_2^{(n+1)} = \frac{\varrho_{n+1}}{1 - \frac{1}{2}\bar{\mu}^2} \left(G(Fv_2^{(n)} + c_1) - \tfrac{1}{2}\bar{\mu}^2 v_2^{(n)} + c_2\right) + (1 - \varrho_{n+1})v_2^{(n-1)} \tag{6.8}$$

One can iterate using (6.8) until convergence has been achieved. With

$\bar{u}_2$ thus determined, one can compute $\bar{u}_1$ by

$$\bar{u}_1 = F\bar{u}_2 + c_1 \tag{6.9}$$

In actual practice, it seems best to assign two memory cells to each "black" point, i.e., to each point associated with $v_2^{(n)}$, plus a number of temporary locations. Thus, for a five-point difference equation we compute $Fv_2^{(n)} + c_2$ at the red points (indicated by ○ in Figure 6.1) using the ordinary Gauss–Seidel method. This is done for two consecutive diagonals. Then one can compute $v_2^{(n+1)}$ at the black points (indicated by □ in Figure 6.1) using the Gauss–Seidel values at the appropriate red points. Thus only two diagonals of red points are needed at any one time.

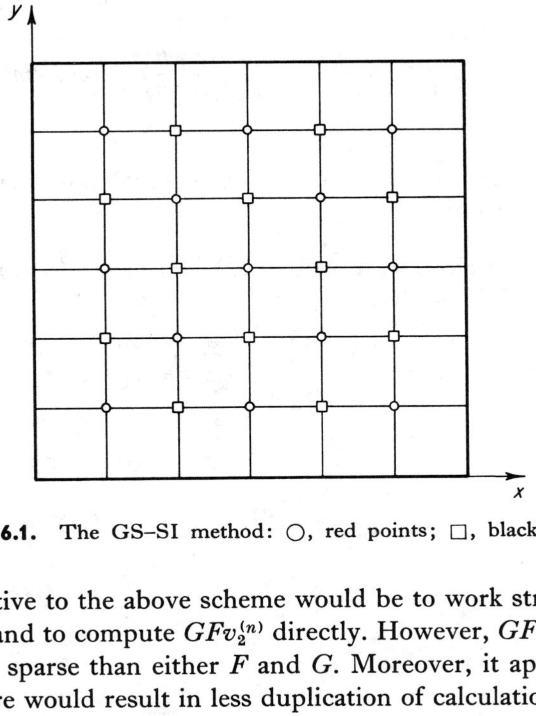

Figure 6.1. The GS–SI method: ○, red points; □, black points.

An alternative to the above scheme would be to work strictly with the black points and to compute $GFv_2^{(n)}$ directly. However, GF will normally be much less sparse than either F and G. Moreover, it appears that the first procedure would result in less duplication of calculations.

We now prove

Theorem 6.1. If A is a positive definite consistently ordered matrix, then

$$\bar{S}(\chi_m) = 2r^m/(1 + r^{2m}) = E_5(m) \tag{6.10}$$

Moreover, if A has the form (8-1.2), then

$$\overline{\| \chi_m \|_{D^{1/2}}} = F_5(m)$$
$$= \left(\frac{2r^m}{1+r^{2m}}\right) \max_{|\mu| \le \bar{\mu}} \begin{cases} (\sin^2 m\theta)\phi^{1/2} + [(\sin^4 m\theta)\phi + \cos 2m\theta]^{1/2}, & m \text{ even} \\ (\cos^2 m\theta)\phi^{1/2} + [(\cos^4 m\theta)\phi - \cos 2m\theta]^{1/2}, & m \text{ odd} \end{cases} \tag{6.11}$$

where

$$\phi = 1 + (\sec^2 \theta)/\bar{\mu}^2, \qquad \cos\theta = \mu/\bar{\mu}, \qquad 0 \le \theta \le \pi \tag{6.12}$$

Also,

$$\overline{\| \chi_m \|_{A^{1/2}}} = H_5(m)$$
$$= \left(\frac{2r^m}{1+r^{2m}}\right) \max_{|\mu| \le \bar{\mu}} \begin{cases} \left(\dfrac{\sin^2 m\theta}{\bar{\mu}\cos\theta}\right) + \left[\left(\dfrac{\sin^2 m\theta}{\bar{\mu}\cos\theta}\right)^2 + \cos 2m\theta\right]^{1/2}, & m \text{ even} \\ \left(\dfrac{\cos^2 m\theta}{\bar{\mu}\cos\theta}\right) + \left[\left(\dfrac{\cos^2 m\theta}{\bar{\mu}\cos\theta}\right)^2 - \cos 2m\theta\right]^{1/2}, & m \text{ odd} \end{cases} \tag{6.13}$$

Here $\overline{\| \chi_m \|_{D^{1/2}}}$ and $\overline{\| \chi_m \|_{A^{1/2}}}$ are the virtual $D^{1/2}$-norm and virtual $A^{1/2}$-norm of χ_m, respectively.

Proof. We have already proved (6.10). To prove (6.11) we note that by (6.4) and (6.5) we have

$$\begin{aligned} q_m(\mathscr{L}) = \sum_{k=0}^{m} \beta_{m,k}\mathscr{L}^k &= \beta I + \sum_{k=1}^{m} \beta_{m,k} \begin{pmatrix} 0 & F \\ 0 & GF \end{pmatrix}^k \\ &= \beta I + \sum_{k=1}^{m} \beta_{m,k} \begin{pmatrix} 0 & (FG)^{k-1}F \\ 0 & (GF)^k \end{pmatrix} \\ &= \begin{pmatrix} \beta I_1 & \sum_{k=1}^{m} \beta_{m,k}(FG)^{k-1}F \\ 0 & \beta I_2 + \sum_{k=1}^{m} \beta_{m,k}(GF)^k \end{pmatrix} \end{aligned} \tag{6.14}$$

where $\beta = \beta_{m,0}$. Let

$$\begin{aligned} M = M_m(\mu) &= \begin{pmatrix} \beta & \sum_{k=1}^{m} \beta_{m,k}\mu^{2k-1} \\ 0 & \beta + \sum_{k=1}^{m} \beta_{m,k}\mu^{2k} \end{pmatrix} \\ &= \begin{pmatrix} \beta & (q_m(\mu^2) - \beta)\mu^{-1} \\ 0 & q_m(\mu^2) \end{pmatrix} \end{aligned} \tag{6.15}$$

The characteristic equation for MM^{T} is

$$\Gamma^2 - T\Gamma + \Delta = 0 \tag{6.16}$$

where

$$\Delta = \det MM^{\mathrm{T}} = \beta^2 q_m(\mu^2)^2 \tag{6.17}$$

$$T = \operatorname{trace} MM^{\mathrm{T}} = \beta^2 + q_m(\mu^2)^2 + (q_m(\mu^2) - \beta)^2\mu^{-2} \tag{6.18}$$

Evidently, we can write (6.16) in the form

$$\Gamma - (|\,\beta - q_m(\mu^2)\,|)\phi^{1/2}\Gamma^{1/2} - \beta q_m(\mu^2) = 0 \tag{6.19}$$

and hence

$$\overline{\|\, M \,\|} = \max_{|\mu| \le \bar{\mu}} \{\tfrac{1}{2}\,|\, q_m(\mu^2) - \beta \,|\, \phi^{1/2} + [\tfrac{1}{4}\,|\, q_m(\mu^2) - \beta \,|^2\, \phi + \beta q_m(\mu^2)]^{1/2}\} \tag{6.20}$$

We now evaluate $q_m(\mu^2)$ and β in terms of μ and $\bar{\mu}$. From (6.4) we have

$$\beta = \beta_{m,0} = q_m(0) = (-1)^m / T_m(2/\mu^2 - 1) = (-1)^m 2r^m/(1 + r^{2m}) \tag{6.21}$$

Moreover, by (6.12) and (9-3.3), we have

$$T_m(2\mu^2/\bar{\mu}^2 - 1) = T_m(\cos 2\theta) = \cos 2m\theta \tag{6.22}$$

Therefore,

$$\begin{aligned} q_m(\mu^2) - \beta &= \left[T_m\left(\frac{2\mu^2}{\bar{\mu}^2} - 1\right) - (-1)^m\right] \Big/ T_m\left(\frac{2}{\bar{\mu}^2} - 1\right) \\ &= \left(\frac{2r^m}{1 + r^{2m}}\right)(\cos 2m\theta - (-1)^m) \\ &= \frac{4r^m}{1 + r^{2m}} \begin{cases} -\sin^2 m\theta, & m \text{ even} \\ \cos^2 m\theta, & m \text{ odd} \end{cases} \end{aligned} \tag{6.23}$$

Moreover, by (6.21) we have

$$\beta q_m(\mu^2) = (-1)^m \left(\frac{2r^m}{1 + r^{2m}}\right)^2 \cos 2m\theta \tag{6.24}$$

The result (6.11) now follows since by Theorem 7-2.1, we have

$$\overline{\|\, \chi_m \,\|_{D^{1/2}}} = \max_{|\mu| \le \bar{\mu}} \|\, M_m(\mu) \,\| \tag{6.25}$$

To prove (6.13) we note that as in the proof of Theorem 7-6.1 the eigenvalues of $(A^{1/2}\chi_m A^{-1/2})(A^{1/2}\chi_m A^{-1/2})^{\mathrm{T}}$ are the eigenvalues of VV^{T} where

$$V = N(\mu)^{1/2} M_m(\mu) N(\mu)^{-1/2} \tag{6.26}$$

and $N(\mu)^{1/2}$ and $N(\mu)^{-1/2}$ are given by (7-6.7). By (6.15) and (7-6.7) we have

$$V = N^{1/2} M N^{-1/2} = \frac{1}{2}\begin{pmatrix} q_m + \beta & \dfrac{1}{p}(q_m - \beta) \\ p(q_m - \beta) & q_m + \beta \end{pmatrix} \tag{6.27}$$

where $q_m = q_m(\mu^2)$ is given by (6.14). By (7-6.13) the eigenvalues of VV^{T} are the roots of the equation

$$\Gamma^2 - [2\beta q_m + (1/\mu^2)(q_m - \beta)^2]\Gamma + \beta^2 q_m{}^2 = 0 \tag{6.28}$$

or

$$\Gamma - |\,(1/\mu)(q_m - \beta)\,|\,\Gamma^{1/2} - \beta q_m = 0 \tag{6.29}$$

Therefore, we have

$$\begin{aligned} \overline{\|\, \chi_m \,\|_{A^{1/2}}} &= \max_{|\mu| \le \bar{\mu}} \|\, VV^{\mathrm{T}} \,\| \\ &= \max_{|\mu| \le \bar{\mu}} \left\{ \left| \frac{q_m - \beta}{2\mu} \right| + \left[\left(\frac{q_m - \beta}{2\mu} \right)^2 + \beta q_m \right]^{1/2} \right\} \end{aligned} \tag{6.30}$$

The result (6.13) now follows from (6.23) and (6.24).

Given $\bar{\mu}$ and m one could compute (6.11) and (6.13) numerically. It can be shown, see Sheldon [1959] for the $D^{1/2}$-norm and Young and Kincaid [1969] for the $A^{1/2}$-norm that each norm is approximately equal to $2cmr^m(1 + r^{2m})^{-1}$ for some constant c. We now show that by a simple modification of the GS–SI method proposed by Sheldon [1959], we are able to eliminate the factor of m in the norm formulas with virtually no increase in the spectral radius.

GS–SSI Method

Sheldon [1959] considered a variant of the GS–SI method whose iteration matrix is

$$\mathscr{D}_m = q_{m-1}(\mathscr{L})\mathscr{L} \tag{6.31}$$

Thus one determines $u^{(1)}$ using the GS method and then with $u^{(1)}$ as a starting vector proceeds with the GS–SI method. We shall refer to this modified GS–SI method as the "GS–SSI method."

Evidently, the eigenvalues of $\mathscr{D}_m$ are simply

$$q_{m-1}(\lambda) \cdot \lambda \tag{6.32}$$

where λ is any eigenvalue of $\mathscr{L}$. Therefore,

$$\begin{aligned}\bar{S}(\mathscr{D}_m) &= \max_{0\le\lambda\le\bar{\mu}^2} |\, \lambda q_{m-1}(\lambda)\,| \\ &= \bar{\mu}^2 \frac{2r^{m-1}}{1+r^{2m-2}} = \frac{8r^m}{(1+r)^2(1+r^{2m-2})}\end{aligned} \tag{6.33}$$

This follows since by (6.2) we have

$$\max_{0\le\lambda\le\bar{\mu}^2} |\, q_{m-1}(\lambda)\,| = q_{m-1}(\bar{\mu}^2) = \frac{2r^{m-1}}{1+r^{2m-2}} \tag{6.34}$$

and since $\bar{\mu}^2 = 4r(1+r)^{-2}$.

While the spectral radius of $\mathscr{D}_m$ is slightly larger than that of the GS–SI method, the $D^{1/2}$-norm and the $A^{1/2}$-norm are considerably smaller. Thus we have

Theorem 6.2. If A is a positive definite consistently ordered matrix, then

$$\bar{S}(\mathscr{D}_m) = \left(\frac{2r^{m-1}}{1+r^{2m-2}}\right)\bar{\mu}^2 = \frac{8r^m}{(1+r)^2(1+r^{2m-2})} = E_6(m) \tag{6.35}$$

Moreover, if A has the form (8-1.2), then

$$\overline{\|\, \mathscr{D}_m \,\|}_{D^{1/2}} = \left(\frac{2r^{m-1}}{1+r^{2m-2}}\right)\bar{\mu}(1+\bar{\mu}^2)^{1/2} = F_6(m) \tag{6.36}$$

and

$$\overline{\|\, \mathscr{D}_m \,\|}_{A^{1/2}} = \left(\frac{2r^{m-1}}{1+r^{2m-2}}\right)\bar{\mu} = \frac{4r^{m-1/2}}{(1+r)(1+r^{2m-2})} = H_6(m) \tag{6.37}$$

Proof. We have already proved (6.35). To prove (6.36) we note that by (6.31), (6.6), and (6.14) we have

$$\mathscr{D}_m = q_{m-1}(\mathscr{L})\mathscr{L} = \begin{pmatrix} \gamma I_1 & \sum_{k=1}^{m-1} \beta_{m-1,k}(FG)^{k-1}F \\ 0 & \gamma I_2 + \sum_{k=1}^{m-1} \beta_{m-1,k}(GF)^k \end{pmatrix} \begin{pmatrix} 0 & F \\ 0 & GF \end{pmatrix} \tag{6.38}$$

where $\gamma = \beta_{m-1,0}$. Therefore

$$M = M_m(\mu) = \begin{pmatrix} \gamma & (q_{m-1}(\mu^2) - \gamma)\mu^{-1} \\ 0 & q_{m-1}(\mu^2) \end{pmatrix} \begin{pmatrix} 0 & \mu \\ 0 & \mu^2 \end{pmatrix}$$

$$= \begin{pmatrix} 0 & \mu q_{m-1}(\mu^2) \\ 0 & \mu^2 q_{m-1}(\mu^2) \end{pmatrix} \tag{6.39}$$

The result (6.36) follows since by (6.4) we have

$$\max_{|\mu| \le \bar{\mu}} \| M_m(\mu) \| = \max_{|\mu| \le \bar{\mu}} (\mu^2 + \mu^4)^{1/2} q_{m-1}(\mu^2)$$

$$= (\bar{\mu}^2 + \bar{\mu}^4)^{1/2} q_{m-1}(\bar{\mu}^2) \tag{6.40}$$

To prove (6.37) we have from (7-6.7), (7-6.11), and (6.39)

$$V = N^{1/2} M N^{-1/2} = \tfrac{1}{2}\mu^2 q_{m-1}(\mu^2) \begin{pmatrix} 1 & 1/p \\ p & 1 \end{pmatrix} \tag{6.41}$$

The characteristic equation of VV^{T} is given by

$$\Gamma^2 - \mu^2 q_{m-1}(\mu^2)^2 \Gamma = 0 \tag{6.42}$$

Since the roots of (6.43) are 0 and $\mu^2 q_{m-1}(\mu^2)^2$ and since $\| \mathscr{D}_m \|_{A^{1/2}} = [S(VV^{\mathrm{T}})]^{1/2}$ the result (6.37) now follows, and the proof of Theorem 6.2 is complete.

Evidently the norms for the GS–SSI method are considerably less than for the GS–SI method. Actually, as will be shown in Section 11.9, the GS–SSI method is competitive with the other methods as well.

We remark that as observed by Miles *et al.* [1964], if A is consistently ordered and if we use the natural ordering rather than the red–black ordering, then the Jordan canonical form of $\mathscr{L}$ will not be diagonal. There are many principal vectors associated with the eigenvalue zero. While these are of little importance for the GS method and do not affect the spectral radius of the GS–SI method, nevertheless they do seriously impede the convergence in practice. (See Tee, 1963.) Thus it is strongly recommended that the red–black ordering be used with the GS–SI and with the GS–SSI methods.

11.7. SOR SEMI-ITERATIVE METHODS

If $\omega > 1$, then the eigenvalues of the SOR method are, in general, not all real and hence the methods of Section 11.2 are not applicable.

One could, of course, hope to accelerate the convergence using semi-iterative methods. However, for the case $\omega = \omega_b$ it seems intuitively clear that this is not possible since all eigenvalues of $\mathscr{L}_{\omega_b}$ are complex and have modulus $\omega_b - 1$. Hence it would seem unlikely that any semi-iterative method could result in improved convergence. This suspicion is confirmed by the following theorem, due to E. Zarantenello, which we state without proof. Details can be found in a paper by Varga [1957].

Theorem 7.1. Let $P(z)$ be any polynomial of degree m or less such that $P(1) = 1$. Then if $0 < r < 1$, we have

$$\max_{|z| \leq r} | P(z) | \geq r^m \tag{7.1}$$

To illustrate the theorem we give a proof for the case $m = 1$. Since $P(1) = 1$ we have

$$P(z) = (z - \alpha)/(1 - \alpha) \tag{7.2}$$

for some $\alpha \neq 1$. Letting $z = re^{i\theta}$, $\alpha = \sigma e^{i\phi}$, we obtain

$$| P(z) |^2 - r^2 = [\sigma^2(1 - r^2) + 2r\sigma(r \cos \phi - \cos(\theta - \phi))] | 1 - \alpha |^{-2} \tag{7.3}$$

We maximize the last expression for given α by letting $\theta = \pi + \phi$ and we obtain

$$| P(z) |^2 - r^2 = [\sigma^2(1 - r^2) + 2r\sigma(1 + r \cos \phi)] | 1 - \alpha |^{-2} \tag{7.4}$$

which is positive unless $\sigma = 0$.

From Theorem 7.1 we have (Varga, 1957)

Theorem 7.2. Let A be a positive definite consistently ordered matrix and let $P(x)$ be any polynomial of degree m or less such that $P(1) = 1$. Then

$$\bar{S}(P(\mathscr{L}_{\omega_b})) \geq (\omega_b - 1)^m = S(\mathscr{L}_{\omega_b}^m) \tag{7.5}$$

Hence the optimum semi-iterative method based on $\mathscr{L}_{\omega_b}$ is simply the basic method itself.

It is, of course, conceivable that for some ω between 1 and ω_b one might be able to accelerate the convergence by a semi-iterative method. In this case the eigenvalues of $\mathscr{L}_\omega$ lie in the union of a circle of radius $\omega - 1$ and the line segment $(\omega - 1)^2/S(\mathscr{L}_\omega) \leq \lambda \leq S(\mathscr{L}_\omega)$. It is conjectured, however, that such is not the case, and that for any ω in the range $0 < \omega < 2$ we have

$$\bar{S}(P(\mathscr{L}_\omega)) \geq S(\mathscr{L}_{\omega_b}^m) \tag{7.6}$$

11.8. MSOR SEMI-ITERATIVE METHODS

In this section we consider semi-iterative methods based on the MSOR method under the assumption that A is positive definite and has the form (8.12). First we consider the case where the relaxation factors ω and ω' are fixed. We consider those values of ω and ω' such that all eigenvalues of $\mathscr{L}_{\omega,\omega'}$ are real. We prove (Young, 1969)

Theorem 8.1. Let A be a positive definite matrix of the form (7-1). Let $\bar{\mu} = S(B)$ where $B = I - (\text{diag } A)^{-1}A$. If ω and ω' are such that the roots of

$$(\lambda + \omega - 1)(\lambda + \omega' - 1) = \omega\omega'\mu^2\lambda \tag{8.1}$$

are real for all μ in the interval $-\bar{\mu} \leq \mu \leq \bar{\mu} < 1$ and if $P_m(x)$ is any polynomial of degree m or less such that $P_m(1) = 1$, then

$$\bar{S}(P_m(\mathscr{L}_{\omega,\omega'})) \geq S(\mathscr{L}_{\omega_b}^m) = \omega_b - 1 \tag{8.2}$$

and the strict inequality holds unless either

$$\omega' = 1, \qquad 1 \leq \omega \leq 1/(1 - \bar{\mu}^2) \tag{8.3}$$

or else

$$\omega = 1, \qquad 1 \leq \omega' \leq 1/(1 - \bar{\mu}^2) \tag{8.4}$$

Evidently, if all roots of (8.1) are real, then all eigenvalues of $\mathscr{L}_{\omega,\omega'}$ are real by Theorem 8-2.1. If the roots of (8.1) are complex for some μ in the interval $-\bar{\mu} \leq \mu \leq \bar{\mu}$, then some eigenvalues of $\mathscr{L}_{\omega,\omega'}$ are complex provided μ actually is an eigenvalue of B.

Proof. We first prove

Lemma 8.2. Under the hypotheses of Theorem 8.1, Eq. (8.1) has complex roots for some μ in the interval $-\bar{\mu} \leq \mu \leq \bar{\mu}$ if and only if we have

$$\omega > 1, \qquad \omega' > 1 \tag{8.5}$$

and

$$([\omega - 1]^{1/2} - [\omega' - 1]^{1/2})^2/(\omega\omega') < \bar{\mu}^2 \tag{8.6}$$

Moreover, (8.6) holds if $\omega > 1$ and $\omega' > 1$ unless either

$$\omega \geq \frac{1}{1 - \bar{\mu}^2} \quad \text{and} \quad \omega' \leq \omega\left(\frac{1}{[\omega - 1]^{1/2}\bar{\mu} + [1 - \bar{\mu}^2]^{1/2}}\right)^2 \leq \omega \tag{8.7}$$

or else

$$\omega' \geq \frac{1}{1 - \bar{\mu}^2} \quad \text{and} \quad \omega \leq \omega'\left(\frac{1}{[\omega' - 1]^{1/2}\bar{\mu} + [1 - \bar{\mu}^2]^{1/2}}\right)^2 \leq \omega' \tag{8.8}$$

Proof. We write (8.1) in the form

$$\lambda^2 - b\lambda + c = 0 \tag{8.9}$$

where

$$\begin{aligned} c &= (\omega - 1)(\omega' - 1) \\ b &= b(\mu) = \omega\omega'\mu^2 - (\omega + \omega' - 2) = 1 + c - \omega\omega'(1 - \mu^2) \end{aligned} \tag{8.10}$$

Let μ_1^2 and μ_2^2 be the roots of

$$\Delta = b^2 - 4c = \omega^2(\omega')^2\mu^4 - 2\omega\omega'\mu^2(\omega + \omega' - 2) + (\omega - \omega')^2 = 0 \tag{8.11}$$

and let $\mu_1^2 \leq \mu_2^2$. Evidently μ_1^2 and μ_2^2 are real if and only if the discriminant

$$16\omega^2(\omega')^2(\omega - 1)(\omega' - 1) \tag{8.12}$$

is nonnegative. If μ_1^2 and μ_2^2 are real, then, in order that there exist complex roots of (8.9) for some μ, we must have $\mu_2^2 > \mu_1^2$ and $\bar{\mu}^2 > \mu_1^2$. The first condition is satisfied if $\omega > 1$ and $\omega' > 1$ or if $0 < \omega < 1$ and $0 < \omega' < 1$. The second condition is satisfied for $\omega > 1$, $\omega' > 1$ if

$$\bar{\mu}^2 > \mu_1^2 = ([\omega - 1]^{1/2} - [\omega' - 1]^{1/2})^2/\omega\omega' \tag{8.13}$$

On the other hand, if $0 < \omega < 1$ and $0 < \omega' < 1$, then

$$\mu_1^2 = -([1 - \omega]^{1/2} - [1 - \omega']^{1/2})^2/\omega\omega' \tag{8.14}$$

and Δ does not vanish for any real μ. Similarly, if $(\omega - 1)(\omega' - 1) \leq 0$, then $c \leq 0$ and $\Delta \geq 0$. Thus, the first statement of Lemma 8.2 follows.

If $\omega = \omega' > 1$, then the left member of (8.6) vanishes and both roots of (8.9) are complex for μ sufficiently small, but positive. The left member of (8.6) increases as ω' decreases. If $1 \le \omega \le (1 - \bar{\mu}^2)^{-1}$, then the left member of (8.6) is less than $\bar{\mu}^2$ even when $\omega' = 1$. Hence we cannot have equality in (8.6) for any ω' such that $1 < \omega' < \omega$. If $\omega \ge (1 - \bar{\mu}^2)^{-1}$, then for $\omega' = 1$ we have

$$([\omega - 1]^{1/2} - [\omega' - 1]^{1/2})^2/\omega\omega' \le \bar{\mu}^2 \tag{8.15}$$

Thus for $\omega \ge (1 - \bar{\mu}^2)^{-1}$ there is a unique value of ω' in the range $1 \le \omega' \le \omega$ for which equality holds in (8.15). One can verify directly that this value is $\omega([\omega - 1]^{1/2}\bar{\mu} + [1 - \bar{\mu}^2]^{1/2})^{-2}$. To complete the proof one can show that if $1 \le \omega \le (1 - \bar{\mu}^2)^{-1}$, then $[\omega - 1]^{1/2}\bar{\mu} + [1 - \bar{\mu}^2]^{1/2} \ge \omega^{1/2}$.

Let $\alpha = \alpha(\omega, \omega')$ and $\beta = \beta(\omega, \omega')$ denote the minimum and maximum, respectively, of the roots of (8.1) as μ varies over the interval $-\bar{\mu} \le \mu \le \bar{\mu}$, and let

$$\sigma(\omega, \omega') = [\beta(\omega, \omega') - \alpha(\omega, \omega')]/\{2 - [\alpha(\omega, \omega') + \beta(\omega, \omega')]\} \tag{8.16}$$

By (2.34) the theorem will follow if we can show that

$$\sigma(\omega, \omega') > \sigma(1, 1) = \bar{\mu}^2/(2 - \bar{\mu}^2) \tag{8.17}$$

unless either (8.3) or (8.4) hold.

We prove (8.17) by considering various subsets in the (ω, ω') plane where the roots of (8.1) are real for all μ in the interval $-\bar{\mu} \le \mu \le \bar{\mu}$ (see Figure 8.1). By symmetry it is sufficient to consider the following cases:

Case (a): $\omega' = 1$,
Case (b): $\omega = \omega' \le 1$,
Case (b′): $\omega' \le \omega \le 1$,
Case (c): $\omega \ge \omega' > 1$ and (8.6) holds,
Case (d): $\omega' \le 1$, $\omega \ge 1$.

Case (*a*): If $\omega' = 1$, then by (8.10) we have $c = 0$ and the roots of (8.1) are 0, $\omega\mu^2 + 1 - \omega$. Hence we have

$$\beta = \begin{cases} 0 & \text{if } \omega \ge 1/(1 - \bar{\mu}^2) \\ \omega\bar{\mu}^2 + 1 - \omega & \text{if } \omega \le 1/(1 - \bar{\mu}^2) \end{cases} \tag{8.18}$$

and

$$\alpha = \begin{cases} 1 - \omega & \text{if } \omega > 1 \\ 0 & \text{if } \omega < 1 \end{cases} \tag{8.19}$$

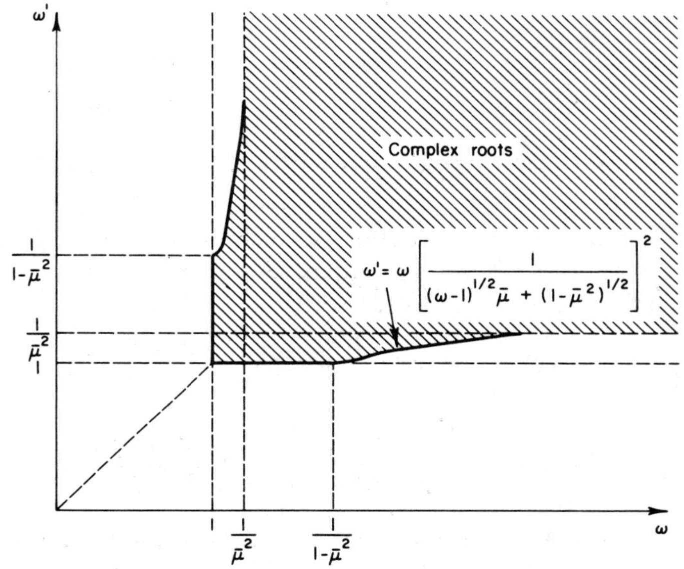

Figure 8.1. Regions of real and complex eigenvalues of $\mathscr{L}_{\omega,\omega'}$.

Thus we have

$$\sigma(\omega, \omega') = \frac{\beta - \alpha}{2-(\beta+\alpha)} = \begin{cases} \dfrac{1-\omega(1-\bar{\mu}^2)}{1+\omega(1-\bar{\mu}^2)} & \text{if } \omega < 1 \\ \dfrac{\bar{\mu}^2}{2-\bar{\mu}^2} & \text{if } 1 \leq \omega \leq \dfrac{1}{1-\bar{\mu}^2} \\ \dfrac{\omega - 1}{\omega + 1} & \text{if } \omega \geq \dfrac{1}{1-\bar{\mu}^2} \end{cases} \tag{8.20}$$

Evidently $\sigma > \bar{\mu}^2/(2 - \bar{\mu}^2)$ unless $1 \leq \omega \leq (1 - \bar{\mu}^2)^{-2}$, and hence (8.17) holds for Case (a).

Case (*b*): If $\omega = \omega' \leq 1$, then (8.1) becomes

$$\lambda^2 - [\omega^2\mu^2 - 2(\omega - 1)]\lambda + (\omega - 1)^2 = 0 \tag{8.21}$$

Evidently β is maximized and α is minimized, for given ω, if $\mu = \bar{\mu}$. Thus if we let

$$\bar{b} = b(\bar{\mu}) = \omega^2\bar{\mu}^2 - 2(\omega - 1) \tag{8.22}$$

we have

$$\beta = \frac{\bar{b} + (\bar{b}^2 - 4c)^{1/2}}{2}, \qquad \alpha = \frac{\bar{b} - (\bar{b}^2 - 4c)^{1/2}}{2} \tag{8.23}$$

and

$$\sigma(\omega, \omega) = \frac{\beta - \alpha}{2-(\beta+\alpha)} = \frac{(\bar{b}^2 - 4c)^{1/2}}{2 - \bar{b}} = \left[1 - \frac{4(1 - \bar{\mu}^2)}{(2 - \omega\bar{\mu}^2)^2}\right]^{1/2}. \tag{8.24}$$

Since $\sigma(\omega, \omega)$ is a decreasing function of ω for $0 < \omega \leq 1$, (8.17) holds for Case (b).

Case (b'): If $\omega' \leq \omega \leq 1$, then the largest root of (8.9) is maximized and the smallest root is minimized with respect to μ for $\mu = \bar{\mu}$. Thus β and α are given by (8.23) where

$$\bar{b} = b(\bar{\mu}) = \omega\omega'\bar{\mu}^2 - (\omega + \omega' - 2) = 1 + c - \omega\omega'(1 - \bar{\mu}^2) \tag{8.25}$$

and hence

$$\sigma(\omega, \omega') = \frac{\beta - \alpha}{2 - (\beta + \alpha)} = \frac{(\bar{b}^2 - 4c)^{1/2}}{2 - \bar{b}} \tag{8.26}$$

For a given value of c, $0 \leq c \leq 1$, $\sigma(\omega, \omega')$ is minimized if $\bar{b}$ is minimized. By (8.25) $\bar{b}$ is maximized if $\omega\omega'$ is maximized. By Lemma 8-3.5, $\omega\omega'$ is maximized for given $c = (\omega - 1)(\omega' - 1)$ if

$$\omega = \omega' = 1 - c^{1/2} \tag{8.27}$$

Therefore, by (8.26) and the result of Case (a) we have, unless $\omega = \omega' = 1$,

$$\sigma(\omega, \omega') \geq \sigma(\hat{\omega}, \hat{\omega}) > \bar{\mu}^2/(2 - \bar{\mu}^2) \tag{8.28}$$

where

$$\hat{\omega} = 1 - [(1 - \omega)(1 - \omega')]^{1/2} \tag{8.29}$$

Thus (8.17) holds for Case (b′).

Case (c): If $\omega \geq \omega' > 1$, then the function

$$g(\lambda) = \lambda^2 - b\lambda + c \tag{8.30}$$

has the behavior indicated in Figure 8.2. For $\mu = 0$, the roots are $1 - \omega$ and $1 - \omega'$. As μ increases, the roots, if they are real, lie in the interval $1 - \omega \leq \lambda \leq 1 - \omega'$. Thus we have $\alpha = 1 - \omega$, $\beta = 1 - \omega'$ and

$$\sigma = \frac{\beta - \alpha}{2 - (\beta + \alpha)} = \frac{\omega - \omega'}{\omega + \omega'} \tag{8.31}$$

For given ω, we minimize σ by letting ω' be as large as possible. By

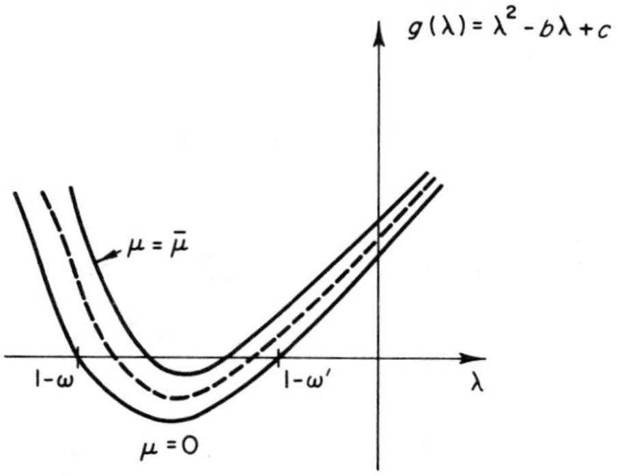

Figure 8.2. Case (c): $\omega \geq \omega' > 1$.

Lemma 8.2 the largest allowable value of ω' such that $\omega' \leq \omega$ is $\omega' = 1$ if $1 \leq \omega \leq (1 - \bar{\mu}^2)^{-1}$, and we have Case (a). If $\omega \geq (1 - \bar{\mu}^2)^{-1}$ we let ω' be given by

$$\omega' = \omega\left(\frac{1}{[\omega - 1]^{1/2}\bar{\mu} + [1 - \bar{\mu}^2]^{1/2}}\right)^2 \tag{8.32}$$

and we have

$$\sigma(\omega, \omega') = \frac{1 - ([\omega - 1]^{1/2}\bar{\mu} + [1 - \bar{\mu}^2]^{1/2})^{-2}}{1 + ([\omega - 1]^{1/2}\bar{\mu} + [1 - \bar{\mu}^2]^{1/2})^{-2}} \tag{8.33}$$

Since $[\omega - 1]^{1/2}\bar{\mu} + [1 - \bar{\mu}^2]^{1/2} \geq 1$ by (8.7), the right member of (8.31) is an increasing function of ω for $\omega \geq (1 - \bar{\mu}^2)^{-1}$. Thus we minimize it by letting $\omega = (1 - \bar{\mu}^2)^{-1}$. This gives $\sigma(\omega, \omega') = \bar{\mu}^2/(2-\bar{\mu}^2)$ and hence (8.17) holds for Case (c).

Case (d): If $\omega' < 1$, $\omega > 1$, then $g(\lambda)$ has the behavior indicated in Figure 8.3. Thus β is the largest root of (8.9) for $b = \bar{b} = b(\bar{\mu})$, while

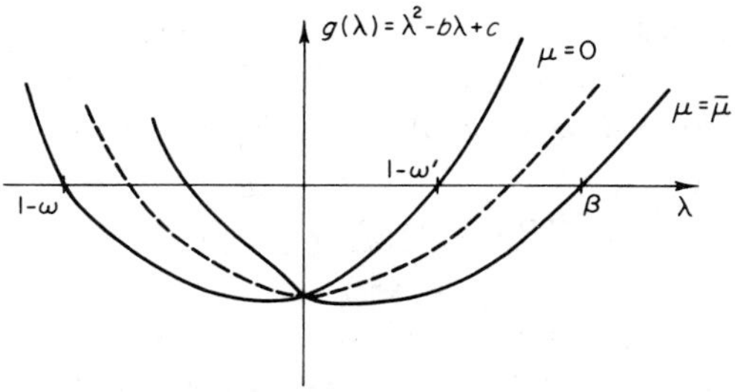

Figure 8.3. Case (d): $\omega' < 1$, $\omega > 1$.

α is the smallest root of (8.9) for $b = b(0)$, i.e., $\alpha = 1 - \omega$. Thus we have

$$(\beta + \omega - 1)(\beta + \omega' - 1) = \omega\omega'\bar{\mu}^2\beta \tag{8.34}$$

and

$$\sigma = \frac{\beta - (1 - \omega)}{2 - (\beta + 1 - \omega)} = \left(1 - \frac{1 - \beta}{\omega}\right) \Big/ \left(1 + \frac{1 - \beta}{\omega}\right) \tag{8.35}$$

Evidently, if we let

$$y = (1 - \beta)/\omega \tag{8.36}$$

then y is the smallest root of

$$z^2 - \left(1 + \frac{\omega'}{\omega} - \omega'\bar{\mu}^2\right)z + \frac{\omega'}{\omega}(1 - \bar{\mu}^2) = 0 \tag{8.37}$$

For fixed ω'/ω the smallest root of (8.37) increases as ω' increases. Thus the largest value of y and hence the smallest value of σ is attained for given ω'/ω when ω' is as large as possible, namely when $\omega' = 1$. By the result of Case (a) we have (8.17) for Case (d). The proof of Theorem 8.1 is now complete.

Theorem 8.1 will be used in Chapter 15 in the study of semi-iterative methods based on the symmetric SOR method and related methods.

We now consider the MSOR method with varying relaxation factors where one of the relaxation factors is unity. Thus as our basic method we let

$$\mathscr{G}_m = \prod_{k=m}^{1} \mathscr{L}_{\omega_k, \omega_k'} \tag{8.38}$$

where for some k either $\omega_k = 1$ or $\omega_k' = 1$. Since A is positive definite all eigenvalues of B are real and since det $\mathscr{G}_m = 0$, all eigenvalues of $\mathscr{G}_m$ are real. Moreover, for any polynomial $P_m(x)$ of degree m or less we have

$$\bar{S}(P_n(\mathscr{G}_m)) \geq (\omega_b - 1)^m \tag{8.39}$$

(See Exercise 4.) Thus, the optimum semi-iterative method based on the MSOR method is no better, as far as the spectral radius is concerned, than the ordinary SOR method with ω fixed and equal to ω_b.

11.9. COMPARISON OF NORMS

We now compare the norms of some of the above semi-iterative methods with the norms of certain variants of the MSOR method which were considered in Chapter 10. We shall assume that A is positive definite and has the form (8-1.2).

The semi-iterative methods whose norms we consider are the J–SI, RF–SI, GS–SI, GS–SSI, and CCSI methods. It is evident that the J–SI and RF–SI methods are about one-half as fast as the GS–SSI method. Moreover, we have already seen that the norms of the GS–SI method are somewhat larger than those of the GS–SSI method. Hence, the GS–SSI method and the CCSI method are the best as far as the $D^{1/2}$-norm and $A^{1/2}$-norm are concerned. We have already studied the norms of the CCSI method in Section 10.7.

We first note that we can extend Theorem 10-7.1 to include the cases $J(m) = F_6(m)$ and $J(m) = H_6(m)$, defined by (6.36) and (6.37), respectively. Indeed, we let

$$\begin{aligned} \varrho_0 &= (\bar{\mu}^2 + \bar{\mu}^4)^{1/2}, \qquad m_0 = 1 \qquad \text{for} \quad F_6(m) \\ \varrho_0 &= \bar{\mu}, \qquad\qquad\qquad m_0 = 1 \qquad \text{for} \quad H_6(m) \end{aligned} \tag{9.1}$$

This follows since $F_6(m)$ and $H_6(m)$ are decreasing functions of m for $m \geq 1$ and since $F_6(1) = (\bar{\mu}^2 + \bar{\mu}^4)^{1/2}$ and $H_6(m) = \bar{\mu}$.

We can extend Theorem 10-7.2 if we let

$$\begin{aligned} \theta(\varrho, r) &= \log(2(2)^{1/2}\gamma/\varrho), \qquad \gamma = \bar{\mu}(1 + \bar{\mu}^2)^{1/2}/(2^{1/2}r) \\ \theta(\varrho, 1) &= \log(2(2)^{1/2}/\varrho), \qquad \Gamma(z) = 2(2)^{1/2}e^{-z}(1 + e^{-2z})^{-1} = f_6(z) \end{aligned} \tag{9.2}$$

for $F_6(m)$ and

$$\begin{aligned} \theta(\varrho, r) &= \log 2\gamma/\varrho, \qquad \gamma = 2r^{-1/2}/(1 + r) \\ \theta(\varrho, 1) &= \log 2/\varrho, \qquad \Gamma(z) = 2e^{-z}(1 + e^{-2z})^{-1} = h_6(z) \end{aligned} \tag{9.3}$$

for $H_6(m)$. The details are left to the reader.

Young and Kincaid [1969] compared the GS–SSI and CCSI methods with respect to the $D^{1/2}$-norm and the $A^{1/2}$-norm. For the $D^{1/2}$-norm it can be shown that $F_6(m) \geq F_3(m)$ for $m \geq 1$ and that

$$\lim_{m\to\infty} [F_6(m)/F_3(m)] = (\bar{\mu}^2 + \bar{\mu}^4)^{1/2}(r + r^2)^{-1/2} < 1$$

and hence the CCSI method is best with respect to the $D^{1/2}$-norm. On

TABLE 9.1. Comparison of Exact and Approximate Theoretical Values of $y = m \log r^{-1}$

		$\varrho = 10^{-3}$				$\varrho = 10^{-6}$				$\varrho = 10^{-9}$			
		$D^{1/2}$		$A^{1/2}$		$D^{1/2}$		$A^{1/2}$		$D^{1/2}$		$A^{1/2}$	
r	Method	Exact	Approx.	Exact	Approx.	Exact	Approx.	Exact	Approx.	Exact	Approx.	Exact	Approx.
0.5	GS–SSI	8.553	8.553	8.235	8.235	15.461	15.461	15.143	15.143	22.369	22.369	22.051	22.051
	CCSI	8.150	8.150	8.150	8.150	15.058	15.058	15.058	15.058	21.966	21.966	21.966	21.966
0.6	GS–SSI	8.410	8.410	8.079	8.079	15.318	15.318	14.987	14.987	22.226	22.226	21.895	21.895
	CCSI	8.091	8.091	8.091	8.091	14.999	14.999	14.999	14.999	21.907	21.907	21.907	21.907
0.7	GS–SSI	8.280	8.280	7.942	7.942	15.188	15.188	14.849	14.849	22.096	22.096	21.757	21.757
	CCSI	8.044	8.044	8.044	8.044	14.952	14.952	14.952	14.952	21.860	21.860	21.860	21.860
0.8	GS–SSI	8.161	8.161	7.818	7.818	15.069	15.069	14.726	14.726	21.977	21.977	21.633	21.633
	CCSI	8.006	8.006	8.006	8.006	14.914	14.914	14.914	14.914	21.822	21.822	21.822	21.822
0.9	GS–SSI	8.051	8.051	7.705	7.705	14.959	14.959	14.613	14.613	21.866	21.866	21.520	21.520
	CCSI	7.975	7.975	7.975	7.975	14.882	14.882	14.882	14.882	21.790	21.790	21.790	21.790
1.0	GS–SSI	7.947	7.947	7.601	7.601	14.855	14.855	14.509	14.509	21.763	21.763	21.416	21.416
	CCSI	7.947	7.947	7.947	7.947	14.855	14.855	14.855	14.855	21.763	21.763	21.763	21.763

the other hand, for the $A^{1/2}$-norm we have $\lim_{m\to\infty} [H_6(m)/H_3(m)] = 2(1 + r)^{-3/2}$ which is less than unity for $r > 4^{1/3} - 1 \doteq 0.587$. Thus for the $A^{1/2}$-norm the GS–SSI method is slightly better.

To illustrate the above conclusions, in Table 9.1 we give the exact theoretical and the approximate theoretical values of $y = m \log r^{-1}$ both for the GS–SSI method and for the CCSI method. These results confirm the validity of the approximate formulas and also show that the GS–SSI method is slightly better than the CCSI method with the $A^{1/2}$-norm but slightly worse with the $D^{1/2}$-norm.

TABLE 9.2. Comparison of Observed and Theoretical Numbers of Iterations

		$\varrho = 10^{-3}$			$\varrho = 10^{-6}$			$\varrho = 10^{-9}$		
		m_O	m_T	m_A	m_O	m_T	m_A	m_O	m_T	m_A
CCSI	$D^{1/2}$	26	25.5	25.5	48	47.4	47.4	70	69.3	69.3
	$A^{1/2}$	20	25.5	25.5	42	47.4	47.4	64	69.3	69.3
GS–SSI	$D^{1/2}$	27	26.1	26.1	49	48.0	48.0	70	69.9	69.9
	$A^{1/2}$	20	25.1	25.1	42	46.9	46.9	63	68.9	68.9
GS–SI	$D^{1/2}$	26	36	—	46	59	—	68	82	—
	$A^{1/2}$	25	36	—	46	59	—	68	82	—

In Table 9.2 we compare the observed and theoretical numbers of iterations for the CCSI method, GS–SI method, and the GS–SSI method for the problem considered in Section 7.7. Results for other methods and definitions of m_0, m_T, and m_A have already been given in Table 10-7.2. It can be seen that while the GS–SI method requires the fewest iterations based on the $D^{1/2}$-norm, the theoretical values are considerably larger than for the other methods. This is due to the choice of starting vector. For some starting vector the GS–SI method would converge more slowly than the other methods.

SUPPLEMENTARY DISCUSSION

Flanders and Shortley [1950] used Chebyshev polynomials to accelerate the convergence of an iterative method for solving a homogeneous linear system with a singular matrix arising in connection with a matrix eigenvalue problem. The idea of using Chebyshev polynomials for Richardson's method and for other iterative methods was used by Lanczos [1952], Shortley [1953], Young [1954a, 1956], Sheldon [1955], Stiefel [1956], Varga [1957], Blair *et al.* [1959], and by many others. The term "semi-iterative methods" was first used by Varga [1957]. Frank [1960] applied the three-term recurrence relation for Chebyshev polynomials to the J–SI method obtaining thereby a scheme less subject to rounding errors (see also Blair *et al.*, 1959).

Section 11.6. Varga [1957] considered the GS–SI method and showed that, as far as the spectral radius is concerned, it converges approximately as fast as the SOR method. However, as observed by Tee [1963] and by Young and Kincaid, with the natural ordering the GS–SI method may converge very slowly indeed.

The analysis of the $D^{1/2}$-norms of the GS–SI and GS–SSI methods was given by Sheldon [1959]. The analysis of the $A^{1/2}$-norms was given by Young and Kincaid [1969].

Section 11.7. Wrigley [1962] considered the use of semi-iterative methods when the eigenvalues of the basic iterative method are complex. The use of Chebyshev polynomials was shown to be optimum if all eigenvalues of the basic method are known to lie in one of a class of ellipses in the complex plane. The results are analogous to those of Section 6.4. Wrigley gave a method for determining the optimum iteration parameters but concluded that it is "too complicated to be worthwhile" if the system is to be solved only once. However, he indicated that in cases where many systems are to be solved with the same matrix, substantial economies can often be effected by careful choice of the iteration parameters.

The use of semi-iterative methods with the SOR method was considered by Tee [1963], who concluded that it would not be effective.

Section 11.9. Most of the material in this section was given by Young and Kincaid [1969].

EXERCISES

Section 11.1

1. Verify that the sequence $1, -1, 1, -1, \ldots$ is Cesàro (C.1) summable.

2. Verify that the Cesàro (C.1) summability method is regular.

3. Give an example to show that even if $\sum_{k=0}^{n} \alpha_{n,k} = 1$, $n = 0, 1, 2, \ldots$, the summability method defined by

$$y_n = \sum_{k=0}^{n} \alpha_{n,k} x_k, \qquad n = 0, 1, 2, \ldots$$

is not necessarily regular.

Section 11.2

1. Develop a convergent semi-iterative method based on the linear stationary iterative method

$$u^{(n+1)} = Gu^{(n)} + k$$

which is completely consistent with $Au = b$ and where the eigenvalues μ of G lie in the ranges $-2 \le \mu \le 0.9$, $1.2 \le \mu \le 3$.

2. Consider the basic iterative method defined by $u^{(n+1)} = Gu^{(n)} + k$ for solving the linear system $Au = b$, where A is a nonsingular matrix and $k = (I - G)A^{-1}b$. Assume that all eigenvalues of G are real and lie in the interval $-2.0 \le \mu \le 0.5$.

(a) Show that the basic method is completely consistent. Is it convergent?

(b) Show how you would compute $v^{(1)}$, $v^{(2)}$, and $v^{(3)}$, the first three iterants of the optimum semi-iterative method with respect to the basic iterative method.

(c) Compute $\alpha_{n,k}$ for $n = 0, 1, 2$; $k = 0, 1, 2, \ldots, n$.

(d) If $P_n(x) = \sum_{k=0}^{n} \alpha_{n,k} x^k$ find $\bar{S}(P_2(G))$.

3. Carry out the previous exercise if the eigenvalues of G lie in the range $1.5 \le \mu \le 4$.

Section 11.3

1. Carry out four iterations of the J–SI, JOR–SI (with $\omega = 1.5$), and RF–SI methods for the system

$$\begin{pmatrix} 4 & 0 & -1 & -1 \\ 0 & 4 & -1 & -1 \\ -1 & -1 & 4 & 0 \\ -1 & -1 & 0 & 4 \end{pmatrix} \begin{pmatrix} u_1 \\ u_2 \\ u_3 \\ u_4 \end{pmatrix} = \begin{pmatrix} 0 \\ 1000 \\ 0 \\ 1000 \end{pmatrix}$$

using $u^{(0)} = (0, 0, 0, 0)^T$. For the J–SI and RF–SI methods, what is the average rate of convergence after two iterations?

2. Carry out two iterations of the RF–SI method and the J–SI method for the system

$$\begin{pmatrix} 1 & 0.9 & 0.9 \\ 0.9 & 1 & 0.9 \\ 0.9 & 0.9 & 1 \end{pmatrix} \begin{pmatrix} u_1 \\ u_2 \\ u_3 \end{pmatrix} = \begin{pmatrix} 1 \\ 0 \\ 0 \end{pmatrix}$$

with the starting vector $u^{(0)} = (1, 0, 1)^T$. What is the average rate of convergence for the first two iterations?

3. Compute $\bar{S}(\mathscr{B}_n)$ and $\bar{S}(\mathscr{Q}_n)$ for $n = 5$ and for the system

$$\begin{pmatrix} 4 & -1 \\ -1 & 1 \end{pmatrix} \begin{pmatrix} u_1 \\ u_2 \end{pmatrix} = \begin{pmatrix} 0 \\ 1 \end{pmatrix}$$

4. For the model problem with Laplace's equation and $h = \frac{1}{20}$, find the first five of the ϱ_i as defined by (3.6) for the J–SI method.

Section 11.4

1. Determine the $\gamma_k^{(n)}$ of (4.8) in terms of p and β_k for the case $n = 2$.

2. For the problem of Exercise 1, Section 11.3, carry out four iterations using Richardson's method with the optimum $\beta_k^{(m)}$ based on $m = 4$. Compare the results with that obtained by the RF method for the third iteration and for the fourth iteration.

3. For the model problem with Laplace's equation and $h = \frac{1}{20}$, find the $\beta_k^{(m)}$ as given by (4.23) for $m = 5$, $k = 0, 1, 2, 3, 4, 5$.

4. Solve the model problem with Laplace's equation and $h = \frac{1}{20}$ with the RF–SI method. Assume $u = 0$ on all sides and starting values unity. Iterate until $\max | u_i^{(n)} | \leq 10^{-6}$. Then attempt to applv Richardson's method with the m chosen as the number of iterations required for the RF method.

Section 11.5

1. Compute two iterations of the CCSI method for the system given in Exercise 1, Section 11.3. Compare with the results obtained using the J–SI method.

Section 11.6

1. Carry out two iterations using the GS–SI method and the GS–SSI method for the system

$$\begin{pmatrix} 4 & -1 & 0 \\ 1 & 4 & -1 \\ 0 & -1 & 4 \end{pmatrix} \begin{pmatrix} u_1 \\ u_2 \\ u_3 \end{pmatrix} = \begin{pmatrix} 2000 \\ 0 \\ 2000 \end{pmatrix}$$

Use zero starting values.

2. For the model problem with Laplace's equation with $h = \frac{1}{20}$ and with the red–black ordering, determine $\bar{S}(\mathscr{D}_m)$, $\| \overline{\mathscr{D}_m} \|_{D^{1/2}}$ and $\| \overline{\mathscr{D}_m} \|_{A^{1/2}}$ for $m = 20$.

3. In the problem of Exercise 2, carry out the GS–SSI method until $\max | u_i^{(n)} | \leq 10^{-6}$. Assume zero boundary values and starting values of unity.

4. For the model problem with Laplace's equation and $h = \frac{1}{20}$ and the natural ordering, estimate the number of iterations needed using the GS–SI method to reduce the $D^{1/2}$-norm of the error to 10^{-6} of its original value. How many would be required with the red–black ordering?

Section 11.7

1. Work out details of the proof of Theorem 7.1 for the case $m = 1$.

2. Give a direct proof of Theorem 7.1 for the case $m = 2$.

Section 11.8

1. Let A be a positive definite matrix of the form (8-1.2) and let $\bar{\mu} = S(B) = 0.8$, where $B = I - (\text{diag } A)^{-1}A$. Compute the region in the (ω, ω') plane which the roots of (8.1) are real for all μ in the interval $-\bar{\mu} \leq \mu \leq \bar{\mu}$. Evaluate $\sigma(\omega, \omega')$ as defined in (8.16) for the following values of ω and ω': $(\frac{1}{2}, \frac{1}{2})$, $(\frac{3}{4}, \frac{1}{2})$, $(1, 1)$, $(2, 1)$, $(4, 12)$, $(4, \frac{1}{2})$, $(4, 1)$.

2. In the previous exercise show that for $\omega = 2$, $\omega' = 1.5$, there is a value of μ in the range $-\bar{\mu} \leq \mu \leq \bar{\mu}$ such that the roots of (8.1) are complex.

3. In the problem of Exercise 1, find the optimum semi-iterative method based on the MSOR method for $\omega = 1$, $\omega' = 1$, and also for $\omega = 2$, $\omega' = 1$.

4. Let A be a positive definite matrix of the form (7-1). Show that if $\mathscr{G}_m$ is given by

$$\mathscr{G}_m = \prod_{k=m}^{1} \mathscr{L}_{\omega_k, \omega_k'}$$

where $(1 - \omega_k)(1 - \omega_k') = 0$ for some k, and if $P_n(x)$ is any polynomial of degree n or less such that $P_n(1) = 1$, then

$$\bar{S}(P_n(\mathscr{G}_m)) \geq (\omega_b - 1)^{mn}$$

Section 11.9

1. Verify the exact and approximate theoretical values of $m \log r^{-1}$ for the GS–SSI and CCSI methods given in Table 9.1 for the case $r = 0.9$, $\varrho = 10^{-6}$.

Chapter 12 / EXTENSIONS OF THE SOR THEORY: STIELTJES MATRICES

In Chapter 6 we showed that if A is a positive definite consistently ordered matrix, then a substantial improvement can be obtained using the SOR method, with a suitable relaxation factor, as compared with the GS method. We now seek to relax the conditions on A.

In Section 12.1 we shall show that there is a need for some restrictions on A in order that the SOR theory will hold. We shall give examples involving a consistently ordered matrix, a positive definite matrix, and an L-matrix where the SOR theory does not hold even approximately. We shall then show in Section 12.2 that if A is a Stieltjes matrix, then the SOR theory does hold to a large extent. On the other hand, for a given value of the spectral radius of the matrix B corresponding to the J method the best convergence is attained if A is consistently ordered.

12.1. THE NEED FOR SOME RESTRICTIONS ON A

The main result of the SOR theory which we would like to be able to extend is the following: If A is a consistently ordered positive definite matrix, then for suitable ω, the ratio $R(\mathscr{L}_\omega)/[R(\mathscr{L})]^{1/2}$ is approximately equal to 2 provided $R(B)$, the rate of convergence of the J method, is sufficiently small. We would be satisfied if there exists a positive constant k, independent of A, such that if $R(\mathscr{L})$ is sufficiently small then, for suitable ω, $R(\mathscr{L}_\omega)/[R(\mathscr{L})]^{1/2} \geq k$. In particular, if we have a family of matrices depending on a parameter, say a, and if $S(\mathscr{L}(a)) \to 1$ as $a \to a_0$, then we would hope that

$$\lim_{a \to a_0} R(\mathscr{L}_\omega(a))/[R(\mathscr{L}(a))]^{1/2} = k \tag{1.1}$$

It might happen that the J method would not even converge. For example, if A is positive definite, then the J method may not converge even though the GS method and the SOR method converge, the latter if $0 < \omega < 2$.

In order that the GS and SOR methods are defined, it is necessary that the diagonal elements of A do not vanish, and we shall assume this throughout.

We now consider a family of consistently ordered matrices where the theory does not hold. Let

$$A = \begin{pmatrix} 1 & -a & 0 & 0 \\ -a & 1 & 0 & 0 \\ 0 & 0 & 1 & -a \\ 0 & 0 & a & 1 \end{pmatrix} \tag{1.2}$$

where $a > 0$. Evidently A is consistently ordered and B has eigenvalues $a, -a, ia, -ia$. By (6-4.17) the optimum value of ω is unity and

$$S(\mathscr{L}) = a^2 = S(B)^2 \tag{1.3}$$

Hence the SOR method is no more effective than the GS method in this case, and the SOR theory does not hold.

Next, we give an example of a family of positive definite matrices where the theory does not apply. If

$$A = \begin{pmatrix} 1 & a & a \\ a & 1 & a \\ a & a & 1 \end{pmatrix} \tag{1.4}$$

and if $-\frac{1}{2} < a < 1$, then A is positive definite. The eigenvalues of A are $1 - a$, $1 - a$, $1 + 2a$, and the eigenvalues of B are $a, a, -2a$. If $\frac{1}{2} \le a < 1$, then the J method does not converge. On the other hand, the GS method does converge and we might hope that (1.1) holds. However, the results of Table 11-3.1 indicate that, even using $\bar{\omega} = \bar{\omega}(a)$, the empirically determined best ω, in each case, $R(\mathscr{L}_{\bar{\omega}}(a))/R(\mathscr{L}(a))$ converges to a value around 1.15 as $a \to 1-$. Thus the SOR theory does not hold in this case.

If, on the other hand, $-\frac{1}{2} < a < 0$ then as $a \to -\frac{1}{2}+$, the theory holds quite well. Indeed, if $a < 0$ we have

$$\begin{aligned} S(B) &= 2\,|\,a\,| \\ S(\mathscr{L}) &= \{3a^2 + |\,a\,|^3 + [|\,a\,|^3(1 + |\,a\,|]^2(4 + |\,a\,|))^{1/2}\}/2 \end{aligned} \tag{1.5}$$

We let ω_b be determined by

$$\omega_b = 2/\{1 + [1 - S(B)^2]^{1/2}\} \tag{1.6}$$

and let $\bar{\omega}$ be the value of ω which minimizes $S(\mathscr{L}_\omega)$, based on numerical results. As shown in Table 1.1, as $a \to -\frac{1}{2}+$, ω_b and $\bar{\omega}$ are very nearly equal and the ratio $R(\mathscr{L}_{\bar{\omega}})/[R(\mathscr{L})]^{1/2}$ is converging to a limiting value of approximately 1.5.

TABLE 1.1 Comparison of the J,GS, and SOR Methods

a	-0.45	-0.475	-0.495
$S(B)$	0.90000	0.95000	0.99000
$S(B)^2$	0.81000	0.90250	0.98010
$S(\mathscr{L})$	0.81099	0.90276	0.98011
$R(\mathscr{L})$	0.20950	0.10230	0.02009
ω_b	1.39286	1.52410	1.75274
$\bar{\omega}$	1.39297	1.52422	1.75273
$S(\mathscr{L}_{\bar{\omega}})$	0.70053	0.48439	0.21299
$R(\mathscr{L}_{\bar{\omega}})/[R(\mathscr{L})]^{1/2}$	1.530	1.541	1.502
$R(\mathscr{L}_{\bar{\omega}})/R(\mathscr{L})$	3.34	4.73	10.60

In the previous example the SOR theory is quite well satisfied when a is negative, i.e., when A is an L-matrix. In order that $S(\mathscr{L}) < 1$, by Theorem 4-5.1, A must be an M-matrix. We now give an example, considered by Kahan [1958] and by Varga [1962], which shows that the SOR theory may not hold even though A is an M-matrix. Thus if

$$A = \begin{pmatrix} 1 & -a & 0 \\ 0 & 1 & -a \\ -a & 0 & 1 \end{pmatrix} \tag{1.7}$$

and if $0 \leq a < 1$, then A is an M-matrix. The eigenvalues of B are a, $ae^{2\pi i/3}$, $ae^{4\pi i/3}$; hence

$$\bar{\mu} = S(B) = a \tag{1.8}$$

We shall show in Chapter 13 that the eigenvalues of $\mathscr{L}_\omega$ are related to those of B by

$$(\lambda + \omega - 1)^3 = \omega^3\mu^3\lambda \tag{1.9}$$

Hence if $\omega = 1$, we have $\lambda = 0$, $a^{3/2}$, $-a^{3/2}$ and

$$S(\mathscr{L}) = a^{3/2} \tag{1.10}$$

We seek to show that $S(\mathscr{L}_\omega)$ is minimized for $\omega = 1$. By Theorem 4-5.7, $S(\mathscr{L}_\omega)$ is a nonincreasing function of ω as ω increases from $\omega = 0$ to $\omega = 1$. Moreover, if we let

$$P(\lambda) = \lambda + \omega - 1 - \lambda^{1/3}\omega a \tag{1.11}$$

we have for $\omega \geq 1$,

$$P(-a^{3/2}) = (\omega - 1)(1 + a^{3/2}) \geq 0 \tag{1.12}$$

Since $P(\lambda) \to -\infty$ as $\lambda \to -\infty$ it follows that $P(\lambda)$ has a zero less than $-a^{3/2}$ and hence if $\omega \neq 1$,

$$S(\mathscr{L}_\omega) > a^{3/2} = S(\mathscr{L}) \tag{1.13}$$

Thus, although the GS method is better than the J method, nevertheless, no improvement can be obtained in using the SOR method with $\omega \neq 1$ even when the GS method is very slowly convergent, as is the case when a is very close to unity. Thus the SOR theory does not hold for this case.

The matrix A is actually consistently ordered in a generalized sense, as we shall show in Chapter 13. For such matrices there exists a relation between the eigenvalues of $\mathscr{L}_\omega$ and those of B similar to that derived in Chapter 5. However, the SOR method is not always effective. In the case of (1.7), we can permute the rows and corresponding columns to obtain the matrix

$$A = \begin{pmatrix} 1 & 0 & -a \\ -a & 1 & 0 \\ 0 & -a & 1 \end{pmatrix} \tag{1.14}$$

We can then replace (1.9) by

$$(\lambda + \omega - 1)^3 = \omega^3\mu^3\lambda^2 \tag{1.15}$$

Because of this relation, as will be shown in Chapter 13, a substantial improvement over the GS method can be achieved by using the SOR method with a suitable ω.

We have thus seen that if we assume that A has nonvanishing diagonal elements, then there are cases where the SOR theory does not hold and where A satisfies one of the following conditions:

(a) A is consistenly ordered,
(b) A is positive definite,
(c) A is an L-matrix.

In Chapter 6 we showed that the theory holds exactly if (a) and (b) hold. In the next section we show that the theory holds approximately if (b) and (c) hold, i.e., if A is a Stieltjes matrix.

12.2. STIELTJES MATRICES

We now prove

Theorem 2.1. If A is a Stieltjes matrix and if ω_b is given by

$$\omega_b = \frac{2}{1 + (1 - \bar{\mu}^2)^{1/2}} = 1 + \left(\frac{\bar{\mu}}{1 + (1 - \bar{\mu}^2)^{1/2}}\right)^2 \tag{2.1}$$

where $\bar{\mu} = S(B)$ and $B = I - (\operatorname{diag} A)^{-1}A$, then

$$\omega_b - 1 \leq S(\mathscr{L}_{\omega_b}) \leq (\omega_b - 1)^{1/2} \tag{2.2}$$

Proof. By Theorems 2-7.3 and 4-5.1 we have $\bar{\mu} < 1$. The first inequality of (2.2) follows from Theorem 4-1.2. By Theorem 4-3.5 it is sufficient to consider the case where $A = I - L - U$, where L and U are nonnegative matrices which are strictly lower triangular and strictly upper triangular, respectively, and where $L = U^{\mathrm{T}}$.

If λ is an eigenvalue of $\mathscr{L}_{\omega_b}$, then for some vector v, not necessarily real, with $(v, v) = 1$ we have

$$\mathscr{L}_{\omega_b} v = (I - \omega_b L)^{-1}(\omega_b U + (1 - \omega_b)I)v = \lambda v \tag{2.3}$$

and

$$(\omega_b U + (1 - \omega_b)I)v = \lambda(I - \omega_b L)v \tag{2.4}$$

Taking the inner product of both sides with v we obtain

$$\omega_b z^* + 1 - \omega_b = \lambda - \lambda\omega_b z \tag{2.5}$$

or

$$\lambda = (\omega_b z^* + 1 - \omega_b)/(1 - \omega_b z) \tag{2.6}$$

where

$$z = (v, Lv), \qquad z^* = (v, Lv)^* = (Lv, v) = (v, Uv) \tag{2.7}$$

If we let $z = re^{i\theta}$, we have

$$\begin{aligned} |\lambda|^2 &= \frac{(1-\omega_b)^2 + 2\omega_b(1-\omega_b)r\cos\theta + \omega_b^2 r^2}{1 - 2r\omega_b\cos\theta + \omega_b^2 r^2} \\ &= 1 - \frac{\omega_b(2-\omega_b)(1-2r\cos\theta)}{1 - 2r\omega_b\cos\theta + \omega_b^2 r^2} \end{aligned} \tag{2.8}$$

Since $L \geq 0$ we have

$$r = |z| = |(v, Lv)| \leq (|v|, L|v|) \tag{2.9}$$

and since $U^{\mathrm{T}} = L$,

$$(|v|, L|v|) = (L|v|, |v|) = (|v|, U|v|) \tag{2.10}$$

Therefore, by Theorem 2-2.2, we have

$$\begin{aligned} 2(|v|, L|v|) &= (|v|, L|v|) + (|v|, U|v|) \\ &= (|v|, B|v|) \leq \bar{\mu}(|v|, |v|) = \bar{\mu}(v, v) = \bar{\mu} \end{aligned}$$

and

$$r \leq \bar{\mu}/2 \tag{2.11}$$

Since $1 - 2r\cos\theta \geq 1 - 2r \geq 1 - \bar{\mu} > 0$, it follows from (2.8) that $|\lambda|^2 < 1$. From (2.8) we have

$$\frac{\partial|\lambda|^2}{\partial\theta} = \frac{-\omega_b(2-\omega_b)(2r\sin\theta)[\omega_b^2 r^2 - (\omega_b - 1)]}{(1 - 2\omega_b r\cos\theta + \omega_b^2 r^2)^2} \tag{2.12}$$

But by (2.11) and since $\omega_b^2\bar{\mu}^2 = 4(\omega_b - 1)$ by (2.1), it follows that $\omega_b^2 r^2 - (\omega_b - 1) \leq 0$. Therefore, for $0 \leq \theta \leq \pi$ the derivative is nonnegative and for $\pi \leq \theta \leq 2\pi$ the derivative is nonpositive. Thus the maximum value occurs when $\theta = \pi$. In addition, we have

$$\begin{aligned} \partial|\lambda|^2/\partial r &= [2\omega_b(2-\omega_b)/(1 - 2\omega_b r\cos\theta + \omega_b^2 r^2)^2] \\ &\quad \times [\omega_b^2 r + \cos\theta(1 - \omega_b - \omega_b^2 r^2)] \end{aligned} \tag{2.13}$$

which is positive if $\cos\theta < 0$, and in particular is positive if $\theta = \pi$. Thus the largest value of $|\lambda|^2$ as a function of z in the circle $|z| \leq \bar{\mu}/2$ occurs when $z = -\bar{\mu}/2$. The corresponding value of λ is, by (2.6)

$$\lambda = (-\omega_b\bar{\mu}/2 + 1 - \omega_b)/(1 + \omega_b\bar{\mu}/2) = -(\omega_b - 1)^{1/2} \tag{2.14}$$

since $\omega_b{}^2\bar{\mu}^2 = 4(\omega_b - 1)$. Therefore,

$$S(\mathscr{L}_{\omega_b}) \leq |\lambda| \leq (\omega_b - 1)^{1/2} \tag{2.15}$$

and the theorem follows.

Theorem 2.2. Under the hypotheses of Theorem 2.1

$$\bar{\mu}^2 \leq S(\mathscr{L}) \leq \bar{\mu}/(2 - \bar{\mu}) \tag{2.16}$$

$$\omega_b - 1 \leq \min_{\omega} S(\mathscr{L}_\omega) \leq (\omega_b - 1)^{1/2} \tag{2.17}$$

Proof. The argument depends on the use of several lemmas which we shall now prove.

Lemma 2.3. For any nonnegative matrix M and for any positive constant α,

$$\lim_{n\to\infty} [\operatorname{trace}(M + \alpha I)^n]^{1/n} = S(M) + \alpha \tag{2.18}$$

Proof. By Theorem 2-1.17, $\lambda_1 = S(M + \alpha I) = S(M) + \alpha$ is an eigenvalue of $M + \alpha I$. Since the trace of $M + \alpha I$ equals the sum of its eigenvalues $\lambda_1, \lambda_2, \ldots, \lambda_N$ we have

$$\operatorname{trace}(M + \alpha I)^n = \sum_{i=1}^{N} \lambda_i{}^n = \lambda_1{}^n\Big\{k + \sum_{\lambda_i \neq \lambda_1} (\lambda_i/\lambda_1)^n\Big\} \tag{2.19}$$

where N is the order of M and k is the multiplicity of λ_1. Evidently $|\lambda_i| < |\lambda_1|$ unless $\lambda_i = \lambda_1$ and hence

$$\lim_{n\to\infty} \Big\{\lambda_1{}^n\Big[k + \sum_{\lambda_i \neq \lambda_1} (\lambda_i/\lambda_1)^n\Big]\Big\}^{1/n} = \lambda_1 = S(M) + \alpha \tag{2.20}$$

and the lemma follows.

Lemma 2.4. If $L \geq 0$ and $U \geq 0$ and if $L^{\mathrm{T}} = U$ then $S(yL + y^{-1}U)$ is a nonincreasing function of the real variable y for $0 < y \leq 1$.

Proof. For any $\alpha > 0$ let us consider the trace of C^n where $C = yL + y^{-1}U + \alpha I$. For any real matrix $C = (c_{i,j})$ each element of the trace of $C^n = (c_{i,j}^{(n)})$ has the form

$$c_{i,i}^{(n)} = \sum_{i_1, i_2, \cdots, i_{n-1}} c_{i,i_1} c_{i_1,i_2} \cdots c_{i_{n-1},i}$$

where in the summation the integers $i_1, i_2, \ldots, i_{n-1}$ assume all values $1, 2, \ldots, N$. Thus, for instance, if $N = 2$, we have

$$C^2 = \begin{pmatrix} c_{1,1}c_{1,1} + c_{1,2}c_{2,1} & c_{1,1}c_{1,2} + c_{1,2}c_{2,2} \\ c_{2,1}c_{1,1} + c_{2,2}c_{2,1} & c_{2,1}c_{1,2} + c_{2,2}c_{2,2} \end{pmatrix}$$

$$= \begin{pmatrix} \sum_{i_1=1}^{2} c_{1,i_1}c_{i_1,1} & \sum_{i_1=1}^{2} c_{1,i_1}c_{i_1,2} \\ \sum_{i_1=1}^{2} c_{2,i_1}c_{i_1,1} & \sum_{i_1=1}^{2} c_{2,i_1}c_{i_1,2} \end{pmatrix}$$

and

$$c_{1,1}^{(3)} = \sum_{i_1,i_2=1}^{2} c_{1,i_1}c_{i_1,i_2}c_{i_2,1}, \qquad c_{2,2}^{(3)} = \sum_{i_1,i_2=1}^{2} c_{2,i_1}c_{i_1,i_2}c_{i_2,2}$$

If $C = yL + y^{-1}U + \alpha I$ and $E = L + U + \alpha I$, then

$$c_{i,i_1}c_{i_1,i_2} \cdots c_{i_{n-1},i} = y^p e_{i,i_1}e_{i_1,i_2} \cdots e_{i_{n-1},i}$$

where p equals the number of factors $c_{r,s}$ such that $c_{r,s}$ is below the main diagonal (i.e., such that $r > s$) minus the number of factors above the main diagonal. But corresponding to this term is another term of the form

$$c_{i,i_{n-1}}c_{i_{n-1},i_{n-2}} \cdots c_{i_2,i_1}c_{i_1,i} = y^{-p} e_{i,i_{n-1}}e_{i_{n-1},i_{n-2}} \cdots e_{i_2,i_1}e_{i_1,i}$$
$$= y^{-p} e_{i,i_1}e_{i_1,i_2} \cdots e_{i_{n-1}}e_i$$

by the symmetry of E. Thus the sum of these two terms in the expression for $c_{i,i}^{(n)}$ equals $y^p + y^{-p}$ times the sum of the corresponding terms in the expression for $e_{i,i}^{(n)}$, where $E^n = (e_{i,j}^{(n)})$. Since $y^p + y^{-p}$ is a nonincreasing function of y for $0 < y \leq 1$ and since $L \geq 0$ and $U \geq 0$, it follows that

$$\text{trace}(yL + y^{-1}U + \alpha I)^n$$

is a nonincreasing function of y for $0 < y \leq 1$. Hence, by Lemma 2.3 $S(yL + y^{-1}U) + \alpha$ is a nonincreasing function of y for $0 < y \leq 1$ and for any $\alpha > 0$. Therefore, by the continuity of $S(yL + y^{-1}U + \alpha I)$ as a function of α, $S(yL + y^{-1}U)$ is a nonincreasing function of y for $0 < y \leq 1$, and the lemma is proved.

Lemma 2.5. If $L, U \geq 0$ and $L^T = U$, then for $0 < y \leq 1$ we have

$$\bar{\mu} \leq S(yL + y^{-1}U) \leq \tfrac{1}{2}(y + y^{-1})\bar{\mu} \tag{2.21}$$

where $\bar{\mu} = S(L + U)$.

Proof. By Lemma 2.4, $S(yL + y^{-1}U) \geq S(L + U) = \bar{\mu}$. On the other hand, we have

$$yL + y^{-1}U = \tfrac{1}{2}(y + y^{-1})(L + U) + \tfrac{1}{2}(y - y^{-1})(L - U) \tag{2.22}$$

Since $yL + y^{-1}U$ is a nonnegative matrix, then, by Theorem 2-1.17 there exists a vector $v \geq 0$ such that $(yL + y^{-1}U)v = S(yL + y^{-1}U)v$. We normalize v so that $(v, v) = 1$. Clearly, $(v, (L - U)v) = (v, Lv) - (v, Uv) = 0$, since $L = U^T$ and since L, U, and v are real. Moreover, by Theorem 2-2.2 it follows that $(v, (L + U)v) \leq S(B) = \bar{\mu}$. Hence, from (2.22) we have

$$\begin{aligned} S(yL + y^{-1}U) &= (v, (yL + y^{-1}U)v) = \tfrac{1}{2}(y + y^{-1})(v, (L + U)v) \\ &\leq \tfrac{1}{2}(y + y^{-1})\bar{\mu} \end{aligned} \tag{2.23}$$

and the lemma follows.

To prove (2.16) we recall that, by Theorem 4-5.4, if $\bar{\lambda} = S(\mathscr{L})$, then since A is an L-matrix

$$S(\bar{\lambda}L + U) = \bar{\lambda} \tag{2.24}$$

Hence unless $\bar{\lambda} = 0$ we have

$$\bar{\lambda}^{1/2} = S(yL + y^{-1}U) \tag{2.25}$$

where $y = \bar{\lambda}^{1/2} < 1$. By Lemma 2.4 we have $S(yL + y^{-1}U) \geq S(L + U) = S(B) = \bar{\mu}$. Therefore $\bar{\lambda}^{1/2} \geq \bar{\mu}$ and $\bar{\lambda} \geq \bar{\mu}^2$.

On the other hand, by Lemma 2.5 and (2.25) we have

$$\bar{\lambda}^{1/2} = S(yL + y^{-1}U) \leq \tfrac{1}{2}(y + y^{-1})\bar{\mu} = \tfrac{1}{2}(\bar{\lambda}^{-1/2} + \bar{\lambda}^{1/2})\bar{\mu} \tag{2.26}$$

and

$$\bar{\lambda} \leq \bar{\mu}/(2 - \bar{\mu}) \tag{2.27}$$

The right inequality of (2.17) follows from Theorem 2.1. Since $S(\mathscr{L}_\omega) \geq |\omega - 1|$ by Theorem 4-1.2, we can prove the left inequality if we can show that for any ω such that $0 < \omega < \omega_b$, there is a real eigenvalue of $\mathscr{L}_\omega$ which is at least as large as $\omega_b - 1$. Evidently if $\omega \neq 0$, then λ is an eigenvalue of $\mathscr{L}_\omega$ if and only if $\alpha = (\lambda + \omega - 1)/\omega$ is an eigenvalue of $\lambda L + U$. By Lemma 2.5, since $L^{\mathrm{T}} = U$ it follows that for any λ in the range $0 < \lambda \leq 1$ we have

$$\lambda^{1/2}\bar{\mu} \leq S(\lambda L + U) \tag{2.28}$$

Consider now the function

$$\phi(\lambda) = S(\lambda L + U) - \frac{\lambda + \omega - 1}{\omega} \tag{2.29}$$

Clearly, $\phi(\lambda)$ is a continuous function and $\phi(1) = \bar{\mu} - 1 < 0$. We seek to show that if $\omega \leq \omega_b$, then $\phi(\omega_b - 1) \geq 0$. From this it will follow that there must be a root λ_1 of $\phi(\lambda) = 0$ in the closed interval $\omega_b - 1 \leq \lambda \leq 1$. But λ_1 will be an eigenvalue of $\mathscr{L}_\omega$ since we shall have $(\lambda_1 + \omega - 1)/\omega = S(\lambda_1 L + U)$ and hence, by Theorem 2-1.17, $(\lambda_1 + \omega - 1)/\omega$ will be an eigenvalue of $\lambda_1 L + U$. It will then follow that $S(\mathscr{L}_\omega) \geq \omega_b - 1$.

To show that $\phi(\omega_b - 1) \geq 0$ if $\omega \leq \omega_b$ we have from (2.28) and (2.29)

$$\begin{aligned}
\phi(\omega_b - 1) &\geq (\omega_b - 1)^{1/2}\bar{\mu} + \frac{2 - \omega_b}{\omega} - 1 \\
&\geq (\omega_b - 1)^{1/2}\bar{\mu} + \frac{2 - \omega_b}{\omega_b} - 1 \\
&= (\omega_b - 1)^{1/2}\bar{\mu} + \frac{2(1 - \omega_b)}{\omega_b} \\
&= (\omega_b - 1)^{1/2}\left(\bar{\mu} - \frac{2(\omega_b - 1)^{1/2}}{\omega_b}\right) = 0
\end{aligned} \tag{2.30}$$

since $\omega \leq \omega_b$ and $\omega_b\bar{\mu} = 2(\omega_b - 1)^{1/2}$. This completes the proof of Theorem 2.2.

We note that according to Theorem 2.2, if one has a matrix with Property A but which is not consistently ordered, then the convergence of the Gauss–Seidel method is less rapid than it would be if one had permuted the rows and columns of A to make it consistently ordered. For $\bar{\mu}$ very close to one, however, the difference is very slight. Indeed, let $\varepsilon = 1 - \bar{\mu}$. We have from (2.16)

$$\varepsilon(2-\varepsilon) \geq 1 - S(\mathscr{L}) \geq \frac{2\varepsilon}{1+\varepsilon} = \varepsilon(2-\varepsilon)\left(\frac{2}{2+\varepsilon(1-\varepsilon)}\right)$$
$$= \varepsilon(2-\varepsilon)\left(1 - \frac{\varepsilon(1-\varepsilon)}{2+\varepsilon(1-\varepsilon)}\right) \tag{2.31}$$

Since $2/[2+\varepsilon(1-\varepsilon)]$ is close to one for small ε, the loss in using a nonconsistently ordered matrix is small.

Let us now discuss the choice of ω_b. By Theorem 2.1 we can obtain a satisfactory value of ω_b if we can obtain a good estimate of $\bar{\mu}$. Also, by Theorem 2.2 we can estimate $\bar{\mu}$ closely if we can find $S(\mathscr{L})$. We may use any of the techniques described in Chapter 6 to estimate these quantities. However, the determination of $S(\mathscr{L})$ by the iterative method described in Section 6.6 may not be satisfactory unless all eigenvalues of $\mathscr{L}$ are real. The a priori methods, both exact and analytic, described in Sections 6.8 and 6.9 may often be used to find $\bar{\mu}$. (See Exercise 5.)

SUPPLEMENTARY DISCUSSION

Section 12.2. The material in this section is based on the work of Kahan [1958] and Varga [1959a]. The proof of Theorem 2.1 is based on that of Wachspress [1966]. The proof of Theorem 2.2 is based on that of Varga [1959a]. The technique can also be used to give a concise proof of Theorem 2.1 (see Varga, 1962).

The result (2.16) confirms a conjecture of Young [1950] that, under certain conditions, if A has Property A, then $S(\mathscr{L}) \geq \bar{\mu}^2$ whether or not A is consistently ordered.

Varga [1959a] showed that under certain conditions $S(\mathscr{L}_{\omega_b}) > \omega_b - 1$ and $\min_\omega S(\mathscr{L}_\omega) > \omega_b - 1$ unless A is consistently ordered. This implies, in particular, that under the given conditions if A has Property A, then it is best to permute the rows and columns of A so that the resulting matrix is consistently ordered.

EXERCISES

Section 12.1

1. Show that if $\bar{\mu}^2 \geq 1$ and if ω_b satisfies $\omega_b^2\bar{\mu}^2 = 4(\omega_b - 1)$, then $|\omega_b - 1| = 1$.

2. For the matrix (1.4) with $a = -0.45$, find $S(B)$, $S(\mathscr{L})$, ω_b and $S(\mathscr{L}_{\omega_b})$.

3. Show that if A is given by (1.4) and if $0 \leq a < 1$, then $S(\mathscr{L}) = a^{3/2}$. Find $S(\mathscr{L})$ if $-\frac{1}{2} < a \leq 0$. In particular, find $S(\mathscr{L})$ for $a = \frac{1}{2}, -\frac{1}{2}$.

4. Show that the matrix A given by (1.7) is an M-matrix if $0 \leq a < 1$.

5. Find the value of ω which minimizes $S(\mathscr{L}_\omega)$ for the matrix (1.14) with $a = 0.45$.

Section 12.2

1. Show that if A is a consistently ordered matrix with positive diagonal elements, then for any $\lambda > 0$

$$S(\lambda L + U) = \lambda^{1/2}\bar{\mu}$$

where $\bar{\mu} = S(B)$ and $B = I - (\operatorname{diag} A)^{-1}A = L + U$ where L and U are strictly lower and strictly upper triangular matrices, respectively. Illustrate the above result for the case

$$A = \begin{pmatrix} 1 & -\alpha & 0 \\ -\alpha & 1 & -\alpha \\ 0 & -\alpha & 1 \end{pmatrix}$$

where $0 < \alpha < \frac{1}{2}$.

2. Consider a linear system involving the matrix (1.4).

(a) Show that if $0 < -a < \frac{1}{2}$, then for $\lambda > 0$, $S(\lambda L + U) \geq \lambda^{1/2}\bar{\mu}$ where $\bar{\mu} = S(B)$.
(*Hint*: Show $\det(\lambda L + U - kI)$ has a root $k \geq \lambda^{1/2}\bar{\mu}$.)

(b) Compute $\bar{\mu}$, $\omega_b = 2(1 + [1 - \bar{\mu}^2]^{1/2})^{-1}$ if $a = -0.40$. Give bounds on $S(\mathscr{L}_{\omega_b})$, $S(\mathscr{L})$, and $\min_\omega S(\mathscr{L}_\omega) = S(\mathscr{L}_{\omega^*})$.

(c) Compute the exact values of $S(\mathscr{L}_{\omega_b})$, ω^*, and $S(\mathscr{L}_{\omega^*})$.

3. Graph the functions $g_1(\lambda) = \lambda^{1/2}\bar{\mu}$ and $g_2(\lambda) = (\lambda + \omega - 1)/\omega$ where $0 < \bar{\mu} < 1$, $1 < \omega < 2$. Show that the curves intersect if $\omega \leq \omega_b = 2/(1 + [1 - \bar{\mu}^2]^{1/2})$ and have one intersection if $\omega = \omega_b$. Let $\bar{\lambda}$ be the largest value of λ corresponding to one of the intersections. For what value of ω will $\bar{\lambda}$ be minimized? If $S(\lambda L + U) \geq \lambda^{1/2}\bar{\mu}$ what does this tell us about $\bar{\lambda}$ if $g_1(\lambda) = S(\lambda L + U)$? Since $S(\mathscr{L}_\omega)$ is not less than $\bar{\lambda}$, what can be said about $\min_\omega S(\mathscr{L}_\omega)$?

4. Find the eigenvalues of B for the nine-point model problem. What value of ω would you use for the case $h = \frac{1}{20}$? Give bounds for $S(\mathscr{L}_\omega)$ for your choice of ω.

5. Use the method of separation of variables to find $S(\mathscr{L})$ for the problem described in the previous exercise. Compare $S(\mathscr{L})$ and $S(B)^2$.

6. Verify that if $M \geq 0$ and $\alpha > 0$, then

$$\lim_{n\to\infty} (\text{trace}(M + \alpha I)^n)^{1/n} = S(M) + \alpha$$

for the case

$$M = \begin{pmatrix} 0 & 2 \\ 2 & 0 \end{pmatrix}$$

Does the limit of $(\text{trace } M^n)^{1/n}$ exist?

7. Prove that if A is a Stieltjes matrix, then $S(\mathscr{L}) = 0$ if and only if $S(B) = 0$. Thus, prove Theorem 2.2 for the case where $S(\mathscr{L}) = 0$.

Chapter 13 / GENERALIZED CONSISTENTLY ORDERED MATRICES

13.1. INTRODUCTION

In Chapter 5 we showed that if A is a consistently ordered matrix with nonvanishing diagonal elements, then the eigenvalues λ of $\mathscr{L}_\omega$ and the eigenvalues μ of B satisfy the relation

$$(\lambda + \omega - 1)^2 = \omega^2 \mu^2 \lambda \tag{1.1}$$

In this chapter for any positive integers q and r we define a class of *(q, r)-consistently ordered matrices* (*CO(q, r)-matrices*) and show that for any such matrix with nonvanishing diagonal elements the relation (1.1) can be replaced by

$$(\lambda + \omega - 1)^p = \omega^p \mu^p \lambda^r \tag{1.2}$$

where

$$p = q + r \tag{1.3}$$

If $q = 1$ and if the eigenvalues of B satisfy certain conditions, then for suitable ω, the SOR method converges substantially faster than the GS method. Moreover, given any CO(q, r)-matrix one can permute the rows and corresponding columns of A to obtain a CO$(1, r')$-matrix for some r'. We also define for any positive integers q and r the class of matrices with Property $A_{q,r}$ and show that given any such matrix one can permute the rows and columns of A to obtain a CO(q, r)-matrix. Thus, if A has Property $A_{q,r}$ for some q and r one can often make effective use of the SOR method if one first permutes the rows and columns of A to obtain a CO($1, r'$)-matrix.

In Section 13.4 and 13.5 we shall define the class of *generalized* (q, r)-*consistently ordered matrices* (GCO(q, r)-*matrices*) and show that every such matrix which is an irreducible L-matrix is also a CO(q, r)-matrix. In Section 13.6 we shall consider various canonical forms which can be obtained with a CO(q, r)-matrix, or with a matrix with Property $A_{q,r}$, by certain permutations of the rows and columns of A. Thus we extend the results of Chapter 5 where we showed that any consistently ordered matrix can be reduced to a T-matrix by a nonmigratory permutation while every matrix with Property A is either a diagonal matrix or can be reduced to a matrix of the form (2-6.1) by a more general permutation. We shall also describe a systematic procedure for determining whether a matrix has Property $A_{q,r}$ and, if so, whether it is a CO(q, r)-matrix. A comparison of our discussion with that of others will be given in Section 13.7.

13.2. CO(q, r)-MATRICES, PROPERTY $A_{q,r}$, AND ORDERING VECTORS

Analogous to Definition 5-3.2, we have

Definition 2.1. For given positive integers q and r, the matrix A of order N is a *(q, r)-consistently-ordered matrix* (*a* CO(q, r)-*matrix*) if for some t, there exist disjoint subsets $S_1, S_2, \ldots, S_t$ of $W = \{1, 2, \ldots, N\}$ such that $\sum_{k=1}^{t} S_k = W$ and such that: if $a_{i,j} \neq 0$ and $i < j$, then $i \in S_1 + S_2 + \cdots + S_{t-r}$ and $j \in S_{k+r}$, where S_k is the subset containing i; if $a_{i,j} \neq 0$ and $i > j$, then $i \in S_{q+1} + S_{q+2} + \cdots + S_t$ and $j \in S_{k-q}$ where S_k is the subset containing i.

We remark that some, but not all, of the S_k may be empty. Evidently a CO(1, 1)-matrix is a consistently ordered matrix in the sense of Definition 5-3.2 and vice versa.

Analogous to Definition 2-6.1, we have

Definition 2.2. For given positive integers q and r, a matrix A of order N has *Property* $A_{q,r}$ if there exist $p = q + r$ disjoint subsets $S_1, S_2, \ldots, S_p$ of $W = \{1, 2, \ldots, N\}$ such that $\sum_{k=1}^{p} S_k = W$ and such that if $a_{i,j} \neq 0$ and $i \neq j$, then either $i \in S_1 + S_2 + \cdots + S_q$ and $j \in S_{k+r}$, where S_k is the set containing i or else $i \in S_{q+1} + S_{q+2} + \cdots + S_p$ and $j \in S_{k-q}$ where S_k is the set containing i.

We remark that some of the S_k may be empty.

Definition 2.3. For a given integer $p \geq 2$, a matrix A of order N has *Property* A_p if A has Property $\mathrm{A}_{1,p-1}$.

Evidently Property A_2 is equivalent to Property A. Later we shall prove that A has Property A_p if and only if A has Property $\mathrm{A}_{q,r}$ for all pairs of positive integers q and r such that $q + r = p$.

In order to provide an effective means for determining if a matrix has Property $\mathrm{A}_{q,r}$ for some $q \geq 1$, $r \geq 1$ and/or if it is a CO(q, r)-matrix, we introduce the notion of (q, r)-ordering vectors.

Definition 2.4. The vector $\gamma = (\gamma_1, \gamma_2, \ldots, \gamma_N)^{\mathrm{T}}$ where $\gamma_1, \gamma_2, \ldots, \gamma_N$ are integers, is a *(q, r)-ordering vector* for the matrix A if for any i and j such that $a_{i,j} \neq 0$ and $i \neq j$ we have

$$\gamma_j - \gamma_i = r \qquad \text{or} \qquad -q \tag{2.1}$$

Definition 2.5. The (q, r)-ordering vector γ is a *compatible (q, r)-ordering vector* if for any i and j such that $a_{i,j} \neq 0$ we have

$$\gamma_j - \gamma_i = r \qquad \text{if} \quad i < j \tag{2.2}$$

and

$$\gamma_j - \gamma_i = -q \qquad \text{if} \quad i > j \tag{2.3}$$

Theorem 2.1. A matrix A is a CO(q, r)-matrix if and only if there exists a compatible (q, r)-ordering vector for A.

The proof is similar to that of Theorem 5-3.2, and will be left as an exercise.

As an example, consider the matrix (Broyden, 1968)

$$A = \begin{bmatrix} 0 & 0 & x & 0 & 0 & 0 & 0 \\ 0 & 0 & 0 & 0 & x & 0 & 0 \\ 0 & 0 & 0 & x & 0 & 0 & 0 \\ 0 & x & 0 & 0 & 0 & 0 & x \\ x & 0 & 0 & 0 & 0 & x & 0 \\ 0 & 0 & x & 0 & 0 & 0 & 0 \\ 0 & 0 & 0 & 0 & x & 0 & 0 \end{bmatrix} \tag{2.4}$$

One can easily verify that the vector $\gamma = (1, 2, 3, 5, 4, 6, 7)^T$ is a compatible (3, 2)-ordering vector. Hence, A is a CO(3, 2)-matrix.

Theorem 2.2. If for given positive integers q and r the matrix A has Property $A_{q,r}$, then there exists a (q, r)-ordering vector γ for A whose components lie in the range $1 \leq \gamma_i \leq q + r$ and there exists a permutation matrix P such that $P^{-1}AP$ is a CO(q, r)-matrix. On the other hand, if there exists a (q, r)-ordering vector or if there exists a permutation matrix P such that $P^{-1}AP$ is a CO(q, r)-matrix, then A has Property $A_{q,r}$.

Proof. If A has Property $A_{q,r}$, then there exist sets $S_1, S_2, \ldots, S_p$, where $p = q + r$, satisfying the conditions of Definition 2.2. If we let $\gamma_i = k$ for $i \in S_k$, then we have the desired (q, r)-ordering vector γ. If we permute the rows and columns of A so that those corresponding to S_1 come first, those corresponding to S_2 come next, etc., then the sets $S_1, S_2, \ldots$ satisfy the conditions of Definition 2.1 for the new matrix. Hence, for some permutation matrix P, the matrix $P^{-1}AP$ is a CO(q, r)-matrix.

If a (q, r)-ordering vector γ exists, let S_1 be the set of all i such that $\gamma_i = \alpha = \min(\gamma_1, \gamma_2, \ldots, \gamma_N)$, S_2 be the set of all i such that $\gamma_i = \alpha + 1$, etc. Let $S_1' = S_1 + S_{p+1} + S_{2p+1} + \cdots$, $S_2' = S_2 + S_{p+2} + S_{2p+2} + \cdots, \ldots, S_p' = S_p + S_{2p} + S_{3p} + \cdots$. The reader should verify that the sets $S_1', S_2', \ldots, S_p'$ satisfy the conditions of Definition 2.2; hence A has Property $A_{q,r}$.

It is evident from Definition 2.2 that A has Property $A_{q,r}$ if and only if $P^{-1}AP$ has Property $A_{q,r}$ for all permutation matrices P. Thus, if $P^{-1}AP = A'$ is a CO(q, r)-matrix, then in order to show that A has Property $A_{q,r}$ it is sufficient to show that A' has Property $A_{q,r}$. But by Theorem 2.1 there exists a compatible (q, r)-ordering vector for A', and, as in the previous paragraph, A' has Property $A_{q,r}$. This completes the proof of Theorem 2.2.

Theorem 2.3. If A is a CO(q, r)-matrix, then, for any positive integer k, A is a CO(kq, kr)-matrix. Moreover, A is a CO($q/d, r/d$)-matrix where d is any positive common factor of q and r.

Proof. Evidently, if γ is a compatible (q, r)-ordering vector, then $k\gamma$ is a compatible (kq, kr)-ordering vector; hence A is a CO(kq, kr)-matrix. Let $S_1, S_2, \ldots, S_t$ be the sets of Definition 2.1. If d is any positive common factor of q and r, let $S_1' = S_1 + S_2 + \cdots + S_d$, $S_2' = S_{d+1} + S_{d+2} + \cdots + S_{2d}$, etc. Evidently, the sets $S_1', S_2', \ldots$

satisfy the conditions of Definition 2.1 with q and r replaced by q/d and r/d, respectively. Hence A is a $\mathrm{CO}(q/d, r/d)$-matrix.

Theorem 2.4. If A has Property $\mathrm{A}_{q,r}$, then A has Property $\mathrm{A}_{kq,kr}$ for any positive integer k, and A has Property $\mathrm{A}_{q/d,r/d}$ for any common factor d of q and r. If A has Property A_p and if d is any positive divisor of p such that $d < p$, then A has Property $\mathrm{A}_{p/d}$.

Proof. If A has Property $\mathrm{A}_{q,r}$, then there exists a (q, r)-ordering vector γ. Evidently $k\gamma$ is a (kq, kr)-ordering vector for A; hence A has Property $\mathrm{A}_{kq,kr}$.

Evidently $A' = P^{-1}AP$ is a $\mathrm{CO}(q, r)$-matrix for some permutation matrix P by Theorem 2.2. By Theorem 2.3, if q is a positive common factor of q and r, A' is a $\mathrm{CO}(q/d, r/d)$-matrix. Hence A has Property $\mathrm{A}_{q/d,r/d}$, by Theorem 2.2.

Suppose now that A has Property A_p and that d is any positive divisor of p such that $d < p$. Let $S_1, S_2, \ldots, S_p$ be any sets satisfying the conditions of Definition 2.2 for $q = 1$, $r = p - 1$. Let $p' = p/d$ and $S_1'' = S_1 + S_{p'+1} + \cdots + S_{(d-1)p'+1}$, $S_2'' = S_2 + S_{p'+2} + \cdots + S_{(d-1)p'+2}$, $\ldots$, $S_{p'}'' = S_{p'} + S_{2p'} + \cdots + S_{dp'}$. The reader should verify that the sets $S_1'', S_2'', \ldots, S_p''$ satisfy the conditions of Definition 2.2 for $q = 1$, $r = p' - 1$; hence A has Property $\mathrm{A}_{1,p'-1}$, i.e., Property $\mathrm{A}_{p'}$. This completes the proof of Theorem 2.4.

The following theorem is an extension of a result of Varga [1962, p. 109, Exercise 2]:

Theorem 2.5. If A has Property A_p, then A has Property $\mathrm{A}_{q,r}$ for any positive integers q and r such that $q + r = p$.

Proof. Let $S_1, S_2, \ldots, S_p$ be sets satisfying the conditions of Definition 2.2 for $q = 1$, $r = p - 1$. Suppose first that q and r are relatively prime, and consider the function σ defined by

$$\sigma(1) = 1,\ \sigma(2) = \phi(1+q),\ \sigma(3) = \phi(1+2q), \ldots, \sigma(p) = \phi(1+(p-1)q) \tag{2.5}$$

where $\phi(k)$ is an integer such that $1 \leq \phi(k) \leq p$ and $\phi(k) \equiv k \pmod{p}$. Clearly, the numbers $\sigma(1), \sigma(2), \ldots, \sigma(p)$ are all different since p and q are relatively prime. Hence σ defines a permutation on the integers $1, 2, \ldots, p$.

Let us define the sets $T_1, T_2, \ldots, T_p$ by $T_1 = S_1$, $T_2 = S_{\sigma^{-1}(2)}$, $\ldots$,

$T_p = S_{\sigma^{-1}(p)}$. We show that if $i \in T_k$, $a_{i,j} \neq 0$ and $j \neq i$, then $j \in T_{\phi(k+r)}$. This will clearly suffice to show that the sets $T_1, T_2, \ldots, T_p$ satisfy the conditions of Definition 2.2 for q and r. If $i \in T_k$, then $i \in S_{\sigma^{-1}(k)}$. Hence $j \in S_{\phi(\sigma^{-1}(k)+p-1)} = T_{\sigma[\phi(\sigma^{-1}(k)+p-1)]}$. It remains to show that

$$\sigma[\phi(\sigma^{-1}(k) + p - 1)] = \phi(k + r) \tag{2.6}$$

for all k. This is equivalent to showing that

$$\phi\{1 + [\phi(\sigma^{-1}(k) + p - 1) - 1]q\} = \phi(k + r) \tag{2.7}$$

or

$$1 + [\phi(\sigma^{-1}(k) + p - 1) - 1]q = k + r + np \tag{2.8}$$

for some integer n. This is equivalent to

$$[(\sigma^{-1}(k) + p - 1) + mp - 1]q = k + r + np - 1 \tag{2.9}$$

for some integers m and n, or

$$\sigma^{-1}(k) = 2 - (m + 1)p + \frac{k + r + np - 1}{q} \tag{2.10}$$

But since $q + r = p$, we have, by (2.5),

$$\begin{aligned}\sigma&\left[2 - (m + 1)p + \frac{k + r + np - 1}{q}\right] \\ &= \phi(1 + (1 - (m + 1)p)q + k + r + np - 1) \\ &= \phi(k + (n + 1)p - (m + 1)pq = k \end{aligned} \tag{2.11}$$

Therefore (2.10) holds, and hence A has Property $A_{q,r}$.

As an example, consider the case $p = 5$, $q = 2$. Assume we have sets $S_1, S_2, \ldots, S_5$ such that if $a_{i,j} \neq 0$ and $i \neq j$, then $j \in S_{\phi(k+p-1)}$ if $i \in S_k$. The permutation σ defined by (2.5) is given by $\sigma(1) = 1$, $\sigma(2) = 3$, $\sigma(3) = 5$, $\sigma(4) = 2$, $\sigma(5) = 4$. Therefore, we have

$$T_1 = S_1, \qquad T_2 = S_4, \qquad T_3 = S_2, \qquad T_4 = S_5, \qquad T_5 = S_3$$

If $a_{i,j} \neq 0$, $i \neq j$, and $i \in T_1$, then $i \in S_1$ and $j \in S_5$. Hence $j \in T_4$. If $i \in T_2$, then $i \in S_4$ and $j \in S_3 = T_5$. Similarly, if $i \in T_3$, then $j \in T_1$, if $i \in T_4$, $j \in T_2$, and if $i \in T_5$, $j \in T_3$. Thus if $a_{i,j} \neq 0$ and $i \neq j$, then $j \in T_{\phi(k+p-q)}$ if $i \in T_k$. Thus A has Property $A_{2,3}$. A (2, 3)-ordering vector is given by $\gamma_i = 1$ if $i \in T_1$, $\gamma_i = 2$ if $i \in T_2$, etc.

Suppose now that q and r are not relatively prime. If d is the highest common factor of q and r, then d divides p. Hence, by Theorem 2.4, A has Property $A_{p'}$ where $p' = p/d$. Thus A has Property $A_{q',r'}$ where $q' = q/d$, $r' = r/d$, since q' and r' are relatively prime and $q' + r' = p'$. But by Theorem 2.4, it follows that A has Property $A_{dq',dr'}$, and hence Property $A_{q,r}$. This completes the proof of Theorem 2.5.

Theorem 2.6. The matrix A has Property $A_{q,r}$ if and only if A has Property $A_{p'}$ where $p' = (q + r)/d$ and d is the greatest common factor of q and r.

Proof. If A has Property $A_{q,r}$, then, by Theorem 2.4, A has Property $A_{q',r'}$ where $q' = q/d$, $r' = r/d$. Let $S_1, S_2, \ldots, S_{p'}$ be the sets satisfying the conditions of Definition 2.2 for Property $A_{q',r'}$. Since q' and r' are relatively prime, we can use the inverse of the permutation (2.5) to obtain sets $T_1, T_2, \ldots, T_{p'}$, which satisfy the conditions of Definition 2.2 for $q = 1$, $r = p' - 1$. This proves that A has Property A_p'.

Suppose now that A has Property A_p' and that q and r are integers such that $(q + r)/d = p'$, where d is the greatest common factor of q and r. Then A has Property $A_{1,(q+r)/d-1}$ and, by Theorem 2.5, A also has Property $A_{q/d,r/d}$. Hence, by Theorem 2.4, A has Property $A_{q,r}$.

Theorem 2.7. Let A be a matrix with Property A_p and let q and r be any two positive integers such that $q + r = p$. If $\hat{q} = kq/s$ and $\hat{r} = kr/s$ for some integer k and for some common factor s of q and r, then A has Property $A_{\hat{q},\hat{r}}$.

Proof. By Theorem 2.5 A has Property $A_{q,r}$, and by Theorem 2.4 A has Property $A_{q/s,r/s}$. Hence A has Property $A_{\hat{q},\hat{r}}$ for $\hat{q} = kq/s$, $\hat{r} = kq/r$ and for any integer k.

Using Theorems 2.6 and 2.7, we can find values of $\hat{q}$ and $\hat{r}$ such that A has Property $A_{\hat{q},\hat{r}}$, given that A has Property $A_{q,r}$. For example, if A has Property $A_{2,6}$, then by Theorem 2.6, A has Property A_4. Consequently, A has Property $A_{\hat{q},\hat{r}}$ for the following sets of values of $(\hat{q}, \hat{r})$

$$\begin{array}{lllll} (3, 1), & (6, 2), & (9, 3), & (12, 4), & \cdots \\ (1, 1), & (2, 2), & (3, 3), & (4, 4), & \ldots \\ (1, 3), & (2, 6), & (3, 9), & (4, 12), & \ldots \end{array}$$

It would be possible to develop the theory using Property A_p but avoiding Property $A_{q,r}$. However, it appears that the use of Property

$A_{q,r}$ significantly simplifies the discussion. One could eliminate the use of "Property $A_{q,r}$" by replacing it at each stage by the requirement that "for some permutation matrix P the matrix $P^{-1}AP$ is a CO(q, r)-matrix." One would define Property A_p by Definition 2.2 with $q = 1, r = p - 1$, and would then prove that if A is permutationally similar to a CO(q, r)-matrix, then the sets $S_1, S_2, \ldots, S_p$ specified in Definition 2.2 exist.

An important property of CO(q, r)-matrices is illustrated by the following theorem, which is a generalization of Theorem 5-3.3.

Theorem 2.8. If A is a CO(q, r)-matrix, then

$$\Delta = \det(\alpha^q C_L + \alpha^{-r} C_U - kD) \tag{2.12}$$

is independent of α for all $\alpha \neq 0$ and for all k. Here $D = \operatorname{diag} A$ and C_L and C_U are strictly lower and strictly upper triangular matrices, respectively, such that $A = D - C_L - C_U$.

Proof. The general term of Δ is

$$t(\sigma) = \pm \prod_{i=1}^{N} a_{i,\sigma(i)} \alpha^{qn_L - rn_U} k^{N-(n_L+n_U)} \tag{2.13}$$

where σ is a permutation defined on the integers $1, 2, \ldots, N$. Here n_L and n_U are, respectively, the number of values of i such that $i > \sigma(i)$ and such that $i < \sigma(i)$. Let γ be a compatible (q, r)-ordering vector for A. Evidently, we have

$$qn_L = \sum_{\substack{i=1 \\ \sigma(i)<i}}^{N} (\gamma_i - \gamma_{\sigma(i)}), \qquad rn_U = \sum_{\substack{i=1 \\ \sigma(i)>i}}^{N} (\gamma_{\sigma(i)} - \gamma_i) \tag{2.14}$$

since $\sigma(i) > i$ and $a_{i,\sigma(i)} \neq 0$ if and only if $\gamma_{\sigma(i)} > \gamma_i$, and $\sigma(i) < i$ and $a_{i,\sigma(i)} \neq 0$ if and only if $\gamma\sigma_{(i)} < \gamma_i$. However, since σ is a permutation, we have

$$qn_L - rn_U = \sum_{\substack{i=1 \\ \sigma(i)\neq i}}^{N} (\gamma_i - \gamma_{\sigma(i)}) = \sum_{\substack{i=1 \\ \sigma(i)\neq i}}^{N} \gamma_i - \sum_{\substack{i=1 \\ \sigma(i)\neq i}}^{N} \gamma_{\sigma(i)} = 0 \tag{2.15}$$

Therefore, $t(\sigma)$ is independent of α and the theorem follows.

The preceding theorem can be applied to study the eigenvalues of the SOR method.

Theorem 2.9. Let A be a CO(q, r)-matrix with nonvanishing diagonal elements and let $B = I - (\operatorname{diag} A)^{-1}A$ and $\mathscr{L}_\omega = (I - \omega L)^{-1}$

$\times(\omega U + (1 - \omega)I)$, where L and U are strictly lower and strictly upper triangular matrices, respectively, such that $B = L + U$.

(a) If μ is any eigenvalue of B, then $\theta\mu$ is also an eigenvalue of B where θ is any p'th root of unity where

$$p' = (q + r)/d \tag{2.16}$$

and where d is the greatest common divisor of q and r.

(b) If λ satisfies

$$(\lambda + \omega - 1)^p = \omega^p \mu^p \lambda^r \tag{2.17}$$

where $p = q + r$, for some eigenvalue μ of B, then λ is an eigenvalue of $\mathscr{L}_\omega$.

(c) If λ is an eigenvalue of $\mathscr{L}_\omega$, then there exists an eigenvalue μ of B such that (2.17) holds.

Proof. If μ is an eigenvalue of B, then $\det(L + U - \mu I) = 0$. By Theorem 2.8 we have

$$\det(L + U - \mu I) = \det(\theta^{q/d} L + \theta^{-r/d} U - \mu I) = 0 \tag{2.18}$$

Hence, since $\theta^{(q+r)/d} = \theta^{p'} = 1$, it follows that

$$\det(L + U - \theta^{-q/d}\mu I) = 0 \tag{2.19}$$

and $\theta^{-q/d}\mu$ is an eigenvalue of B. Also, by Theorem 2.8 we have

$$\det((\theta^{-1})^{q/d} + (\theta^{-1})^{-r/d} - \mu I) = 0 \tag{2.20}$$

and

$$\det(L + U - \theta^{q/d}\mu I) = 0 \tag{2.21}$$

so that $\theta^{q/d}\mu$ is also an eigenvalue of B. Similarly, we can show that $\theta^{kq/d}\mu$ is an eigenvalue of B for $k = 0, \pm 1, \pm 2, \ldots$. Since q/d and r/d are relatively prime, it follows that q/d and p' are relatively prime. Hence, by a theorem of number theory there exist positive integers s and t such that

$$s(q/d) + tp' = 1 \tag{2.22}$$

Therefore

$$\theta^{sq/d}\mu = \theta^{1-tp'}\mu = \theta\mu \tag{2.23}$$

is an eigenvalue of B, and (a) follows.

As in the proof of Theorem 5-2.2, the characteristic equation of $\mathscr{L}_\omega$ is

$$\det(\omega U + \lambda\omega L - (\lambda + \omega - 1)I) = 0 \tag{2.24}$$

If $\lambda = 0$ satisfies (2.17), then $\omega = 1$ and (2.24) is satisfied. Thus λ is an eigenvalue of $\mathscr{L}_\omega$. If $\lambda \neq 0$, we have

$$\det\left(\lambda^{1-(r/p)}L + \lambda^{-r/p}U - \frac{\lambda + \omega - 1}{\omega\lambda^{r/p}} I\right) = 0 \tag{2.25}$$

or, by Theorem 2.8

$$\det\left(L + U - \frac{\lambda + \omega - 1}{\omega\lambda^{r/p}} I\right) = 0 \tag{2.26}$$

If (2.17) holds for some eigenvalue μ of B, then (2.26) holds; therefore, (2.24) holds and λ is an eigenvalue of $\mathscr{L}_\omega$. This proves (b).

If $\lambda = 0$ is an eigenvalue of $\mathscr{L}_\omega$, then $\omega = 1$ since $\det(\mathscr{L}_\omega) = (1-\omega)^N$. Moreover, (2.17) holds for any eigenvalue μ of B. If $\lambda \neq 0$ and λ is an eigenvalue of $\mathscr{L}_\omega$, then (2.26) holds. Hence $(\lambda + \omega - 1)/(\omega\lambda^{r/p})$ is an eigenvalue of B. Therefore (c) holds and the proof of Theorem 2.9 is complete.

13.3. DETERMINATION OF THE OPTIMUM RELAXATION FACTOR

We now consider the application of the results of the previous section to the determination of the optimum value of ω for use with the SOR method when A is a CO(q, r)-matrix. Varga [1959] gave a complete analysis for the case $q = 1$ assuming that the eigenvalues of B^{q+r} are real and nonnegative and that $S(B) < 1$. Nichols and Fox [1969] considered the case of more general q and showed that the SOR method is not effective if $q > 1$. Thus, if one has a CO(q, r)-matrix, or more generally, a matrix with Property $A_{q,r}$, one should permute the rows and columns of A as described in the previous section in order to obtain a CO$(1, r')$-matrix where $r' = p' - 1$ and $p' = p/d$ where d is the greatest common divisor of q and r.

We now prove

Theorem 3.1. Let A be a CO(q, r)-matrix with nonvanishing diagonal elements such that all eigenvalues of B^{q+r} are real and nonnegative and $\bar{\mu} = S(B) < 1$.

(a) If $q = 1$ and $r = p - 1$ and if ω_c is the smallest positive root of

$$(p - 1)^{p-1}(\omega_c \bar{\mu})^p = p^p(\omega_c - 1) \tag{3.1}$$

then

$$\bar{S}(\mathscr{L}_{\omega_c}) = S(\mathscr{L}_{\omega_c}) = (p - 1)(\omega_c - 1) \tag{3.2}$$

and if $\omega \neq \omega_c$, then

$$\bar{S}(\mathscr{L}_\omega) > S(\mathscr{L}_{\omega_c}) \tag{3.3}$$

(b) If $q > 1$ and $r = p - q$, then

$$\bar{S}(\mathscr{L}) = S(\mathscr{L}) = \bar{\mu}^{p/q} \tag{3.4}$$

and if $\omega \neq 1$ and $0 < \omega < 1$, then

$$\bar{S}(\mathscr{L}_\omega) = S(\mathscr{L}_\omega) > S(\mathscr{L}) \tag{3.5}$$

Proof. We give a proof only for the case $p = 3$. A proof of (a) for the general case can be found in Varga [1959].† A proof of both (a) and (b) is given by Nichols and Fox [1969].

To prove (a) for the case $p = 3$, we let $q = 1$, $r = 2$. By Theorem 4-1.2 and the fact that $S(\mathscr{L}) < 1$, as we shall show, we can assume $0 < \omega < 2$. We write (2.17) in the form

$$P(y) = y^3 - by^2 + c = 0 \tag{3.6}$$

where

$$y = \lambda^{1/3}, \qquad b = \omega\mu, \qquad c = \omega - 1 \tag{3.7}$$

If $c < 0$ and $b > 0$, i.e., if $0 < \omega < 1$ and $\mu > 0$, then there is precisely one positive root by Descartes' rule of signs. This positive root is an increasing function of μ for μ real and for given ω and a decreasing function of ω for given μ. To prove the second statement we have

$$dy/d\omega = -(1 - \mu y^2)/(3y^2 - 2by) \tag{3.8}$$

Evidently $P(1) = 1 + c - b = \omega(1 - \mu)$ so that if $\omega > 0$ and $\mu < 1$, we have $P(1) > 0$. Also $P(0) = \omega - 1 < 0$ if $\omega < 1$. Hence the positive root of (3.6) is less than unity. Since $P(b) = c < 0$ the positive root is greater than b and hence the derivative in (3.8) is negative.

† Varga was also aware of the conclusion (b). (See Exercises 1 and 2 of Varga, 1962, p. 109.)

If μ is a complex eigenvalue of B, then $\mu\gamma$ is a real eigenvalue of B where γ is some complex cube root of unity. If y is a root of (3.6), then γy is a complex root of

$$y^3 - \gamma\mu b y^2 + c = 0 \tag{3.9}$$

Since for $\mu > 0$ and $0 < \omega \leq 1$ the positive root of (3.6) exceeds the modulus of either complex root it follows that

$$S(\mathscr{L}) \leq S(\mathscr{L}_\omega) < 1, \qquad 0 < \omega \leq 1 \tag{3.10}$$

The fact that $S(\mathscr{L}) < 1$ follows since the roots of (3.6) are μ, 0, 0 if $\omega = 1$ and hence

$$S(\mathscr{L}) = \bar{\mu}^3 \tag{3.11}$$

If $\omega > 1$ and $\mu > 0$, then $b > 0$, $c > 0$, and by Descartes' rule of signs (3.6) has one negative root and either two positive roots or two complex roots. There will be two positive roots if

$$c \leq 4b^3/27 \tag{3.12}$$

If (3.12) holds, the one positive root will be less than $2b/3$ and the other will be greater than $2b/3$. Moreover, by (3.8) both the largest positive root of (3.6) and the negative root are decreasing functions of ω. Since the roots of (3.6) are μ, 0, 0 when $\omega = 1$, it follows that the largest root is less than unity for $1 \leq \omega < 2$, provided (3.12) holds

$$P(-1) = -[2 - \omega(1 - \mu)] < 0 \tag{3.13}$$

Hence all roots of (3.6) are greater than -1 and hence $|y| < 1$ for any root y of (3.6) provided (3.12) holds.

If $p = 3$, then (3.1) becomes

$$4(\omega_c\bar{\mu})^3 = 27(\omega_c - 1) \tag{3.14}$$

By Descartes' rule (3.14) has either two positive real roots or none. Evidently $Q(\omega) = 4\omega^3\bar{\mu}^3 - 27\omega + 27$ has a minimum value for $\omega = 3/(2\bar{\mu}^{3/2})$. Moreover,

$$Q\left(\frac{3}{2\bar{\mu}^{3/2}}\right) = 27\left(1 - \frac{1}{\bar{\mu}^{3/2}}\right) < 0 \tag{3.15}$$

so that there are indeed two positive roots. If ω_c is the smaller root and moreover, if $1 \leq \omega \leq \omega_c$, then

$$\omega - 1 \leq 4(\omega\bar{\mu})^3/27 \tag{3.16}$$

If $\omega = \omega_c$ and $\mu = \bar{\mu}$, then the roots of (3.6) are

$$2\omega_c\bar{\mu}/3, \qquad 2\omega_c\bar{\mu}/3, \qquad y_3 \tag{3.17}$$

Since $y_3 + 4\omega_c\bar{\mu}/3 = \omega_c\bar{\mu}$, it follows that

$$\begin{aligned} P(-2\omega_c\bar{\mu}/3) &\leq -16\omega_c{}^3\bar{\mu}^3/27 \\ y_3 &= -\omega_c\bar{\mu}/3 \end{aligned} \tag{3.18}$$

Thus the root radius of (3.6) is $2\omega_c\bar{\mu}/3$, and (3.2) follows.

Let us now consider the case where (3.12) does not hold. If $y = \varrho e^{i\theta}$ is a root of (3.6), then we have

$$\begin{aligned} \varrho \cos \theta + c\varrho^{-2} \cos 2\theta &= b \\ \varrho \sin \theta - c\varrho^{-2} \sin 2\theta &= 0 \end{aligned} \tag{3.19}$$

If $\varrho^3 \leq 2c$, then the second of the above equations is satisfied by

$$\cos \theta = \varrho^3/(2c)$$

as well as by $\theta = 0, \pi$. Since (3.12) does not hold there are no roots corresponding to $\theta = 0$. Substituting in the first equation of (3.19) we obtain

$$R(\varrho^2) = \varrho^6 - bc\varrho^2 - c^2 = 0 \tag{3.20}$$

By Descartes' rule there is precisely one positive root of (3.20). Clearly, this root increases if b or c increases. Thus, it increases if μ or if ω increases.

If $\omega = \omega_c$, then

$$\varrho = [2(\omega_c - 1)]^{1/3} \tag{3.21}$$

If $\omega > \omega_c$, and if $\varrho = [2(\omega_c - 1)]^{1/3}$, then $\varrho^3 < 2c$. If the root radius of (3.6) were ϱ, then we would have

$$R(\varrho^2) = \varrho^6 - bc\varrho^2 - c^2 = 0 \tag{3.22}$$

But,

$$R(\varrho^2) < \varrho^6 - \omega_c\bar{\mu}(\omega_c - 1)\varrho^2 - (\omega_c - 1)^2 = 0 \tag{3.23}$$

Hence there is a root of (3.20) greater than ϱ if $\mu = \bar{\mu}$. This proves (3.3) for the case $p = 3$.

To prove (b) for the case $p = 3$, $q = 2$, we first note that as before, $S(\mathscr{L}_\omega)$ is a decreasing function of ω for $0 < \omega \leq 1$. Moreover,

$$S(\mathscr{L}) = \bar{\mu}^{3/2} \tag{3.24}$$

We consider the characteristic equation

$$P_2(y) = y^3 - \omega\bar{\mu}y + \omega - 1 = 0 \tag{3.25}$$

If $\omega > 1$, we have

$$P_2(-\bar{\mu}^{3/2}) = -\bar{\mu}^{9/2} + \omega\bar{\mu}^{5/2} + \omega - 1 \geq \bar{\mu}^{5/2}(1 - \bar{\mu}^2) > 0 \tag{3.26}$$

Hence there is a root of (3.25) less than $-\bar{\mu}^{3/2}$ so that $S(\mathscr{L}_\omega) > \bar{\mu}^{3/2}$. Thus we have proved Theorem 3.1 for the case $p = 3$. For the more general case, if $q \neq 1$, then when $\omega = 1$, there are q roots of

$$y^p - \omega\bar{\mu}y^r + \omega - 1 = 0 \tag{3.27}$$

with modulus $\bar{\mu}^{1/q}$ including $y = \bar{\mu}^{1/q}$. The rest of the roots vanish. As ω increases, the root $\bar{\mu}^{1/q}$ decreases, but some of the other roots increase in modulus. Thus the minimum value of $S(\mathscr{L})$ is attained when $\omega = 1$. If $q = 1$, on the other hand, then all roots vanish except $y = \bar{\mu}$. As ω increases, this root decreases while the others move out from the origin until for $\omega = \omega_c$ we have a double root. Further increases in ω yield complex roots with increasing modulus.

Let us now investigate the behavior of $R(\mathscr{L}_{\omega_c})$ for $\bar{\mu}$ close to unity. If we let

$$\bar{\mu} = 1 - \varepsilon, \qquad S(\mathscr{L}_{\omega_c}) = 1 - \delta \tag{3.28}$$

then we have

$$\begin{aligned} \delta &= 1 - (p - 1)(\omega_c - 1) \\ \omega_c - 1 &= (1 - \delta)/(p - 1) \\ \omega_r &= (p - \delta)/(p - 1) \end{aligned} \tag{3.29}$$

Therefore by (3.1) we have

$$(p - \delta)(1 - \varepsilon) = p(1 - \delta)^{1/p} \tag{3.30}$$

and

$$\varepsilon = 1 - [p(1 - \delta)^{1/p}/(p - \delta)] = (p - 1)\delta^2/(2p^2) + O(\delta^3) \tag{3.31}$$

so that for sufficiently small ε

$$\delta = [2p/(p-1)]^{1/2}(p\varepsilon)^{1/2} + O(\varepsilon) \tag{3.32}$$

Therefore,

$$\begin{aligned} R(\mathscr{L}_{\omega_c}) &= -\log(1-\delta) \sim \delta \sim [2p/(p-1)]^{1/2}(p\varepsilon)^{1/2} \\ R(\mathscr{L}) &= -p\log(1-\varepsilon) \sim p\varepsilon \end{aligned} \tag{3.33}$$

hence for $\bar{\mu}$ near 1 we have

$$R(\mathscr{L}_{\omega_c})/[R(\mathscr{L})]^{1/2} \sim [2p/(p-1)]^{1/2} \tag{3.34}$$

Moreover,

$$R(\mathscr{L}_{\omega_c}) \sim [2p^2/(p-1)]^{1/2}\varepsilon^{1/2} \tag{3.35}$$

and

$$R(\mathscr{L}_{\omega_c})/[R(B)]^{1/2} \sim [2p^2/(p-1)]^{1/2} \tag{3.36}$$

Since $2p^2/(p-1)$ is an increasing function of p for $p \geq 2$, it follows that the larger p, the faster the convergence, for a given $\bar{\mu}$. On the other hand, as shown by Varga [1962], the maximum of the ratio (3.34) is achieved for $p = 2$, since $2p/(p-1)$ is a decreasing function of p.

Suppose one has a matrix A with Property A_4. One can permute the rows and columns of A and obtain either a CO(1, 1)– or a CO(1, 3)-matrix. If all eigenvalues of B^2 are real and positive, then so are all those of B^4, but not necessarily conversely. Also, by (3.36) the ratios $R(\mathscr{L}_{\omega_c})/[R(B)]^{1/2}$ will be approximately $8^{1/2} \doteq 2.83$ and $(32/3)^{1/2} \doteq 3.27$, respectively, so that the use of the CO(1, 3)-matrix is slightly better.

13.4. GENERALIZED CONSISTENTLY ORDERED MATRICES

The conclusion of Theorem 2.9 suggests the following generalization of the concept of CO(q, r)-matrices.

Definition 4.1. A matrix A is a *generalized* (q, r)*-consistently ordered matrix* (a GCO(q, r)-matrix) if

$$\Delta = \det(\alpha^q C_L + \alpha^{-r} C_U - kD) \tag{4.1}$$

is independent of α for all $\alpha \neq 0$ and for all k. Here, D, C_L, and C_U are as defined in Theorem 2.8.

Evidently, by Theorem 2.8, every CO(q, r)-matrix is a GCO(q, r)-matrix. However, not every GCO(q, r)-matrix is a CO(q, r)-matrix. For example, if

$$A = \begin{pmatrix} 1 & 0 & 0 \\ 1 & 1 & 0 \\ 1 & 1 & 1 \end{pmatrix} \tag{4.2}$$

it is clear that A is a GCO(1, 1)-matrix. However, A does not have Property A and hence is not a CO(1, 1)-matrix.

Even if A is assumed to have Property $A_{q,r}$ in addition to being a GCO(q, r)-matrix, A need not be a CO(q, r)-matrix. For example, let A be given by

$$A = \begin{pmatrix} I & -F & 0 & -G \\ -H & I & -F & 0 \\ 0 & -H & I & -F \\ 0 & 0 & -H & I \end{pmatrix} \tag{4.3}$$

where

$$I = \begin{pmatrix} 1 & 0 \\ 0 & 1 \end{pmatrix}, \quad F = \begin{pmatrix} 1 & \frac{1}{2} \\ \frac{1}{2} & 1 \end{pmatrix}, \quad G = \begin{pmatrix} 1 & -1 \\ -1 & 1 \end{pmatrix}, \quad H = \begin{pmatrix} \frac{1}{2} & \frac{1}{2} \\ \frac{1}{2} & \frac{1}{2} \end{pmatrix} \tag{4.4}$$

(See Broyden, 1968.) It is easy to show that A has Property $A_{1,1}$ but is not consistently ordered. On the other hand, one can easily show that A is a GCO(1, 1)-matrix.

In the next section we show that if A is an irreducible L-matrix which is a GCO(q, r)-matrix, then A is a CO(q, r)-matrix.

13.5. RELATION BETWEEN GCO(q, r)-MATRICES AND CO(q, r)-MATRICES

In order to study the relation between CGO(q, r)-matrices and CO(q, r)-matrices, we shall use methods of graph theory (see Section 2.5). For a given matrix A, let us define a *chain* as a set of positive integers $i_1, i_2, \ldots, i_s$ where $s \geq 2$ such that

$$a_{i_1, i_2} a_{i_2, i_3} \cdots a_{i_{s-1}, i_s} \neq 0 \tag{5.1}$$

A chain is *simple* if no element appears more than once except possibly i_1 which may equal i_s. If $i_1 = i_s$ the chain is *closed*. The *length* of the chain is $s - 1$.

From Theorem 2-5.2 we have

Lemma 5.1. If A is an irreducible matrix of order N, then for any i and j such that $1 \leq i \leq N$, $1 \leq j \leq N$ and $i \neq j$ there is a chain from i to j.

For any pair of positive integers q and r let the *increment* of a chain be defined as follows. The increment of the chain i, j is $-q$ if $i > j$, r if $i < j$, and 0 if $i = j$. The increment of a chain $i_1, i_2, \ldots, i_s$ is the sum of the increments of the chains i_j, i_{j+1}, $j = 1, 2, \ldots, s - 1$.

Lemma 5.2. If A is an irreducible matrix such that the increment around any simple closed chain is zero, then for any i and j the increment in any chain from i to j is the same for all chains joining i to j.

Proof. Consider any two chains C and C' joining i to j. Since A is irreducible, there is a chain C'' from j to i. The lemma will follow if we can show that the increment of $C + C''$ and the increment of $C' + C''$ both vanish. This is clearly equivalent to showing that the increment around any closed chain vanishes.

Let C: $i_1, i_2, \ldots, i_{s-1}, i_s$, where $i_s = i_1$, be any closed chain. We construct a simple closed chain C' with the same increment as follows. Let i_1 be the first element of C'. If $s = 2$, then $i_2 = i_1$ and $C' = \{i_1, i_1\}$. Since C' is a simple closed chain, its increment, and hence that of C, vanishes. If $s > 2$ and $i_2 = i_1$, then the chain i_1, i_1 has increment zero and we can discard i_2 without changing the increment of C. If $i_2 \neq i_1$, then we retain i_2. Next, we consider i_3. If $s = 3$, then $i_3 = i_1$. Our chain C' is either i_1, i_2, i_1 or, if i_2 was discarded, i_1, i_1. In either case C' is simple and has vanishing increment. Therefore C has a vanishing increment. Suppose now that for some $k < s$ we have constructed a simple chain from i_1 to i_k, perhaps by discarding some of the elements of the original chain C. If $k + 1 = s$ and if $i_k = i_1$, then we can discard $i_{k+1} = i_1$ since the simple closed chain i_1, i_1 has zero increment. If $k + 1 = s$ and $i_k \neq i_1$, then we retain $i_{k+1} = i_1$ and have a simple closed chain whose increment vanishes. If $k + 1 < s$, and if i_{k+1} has appeared previously, then the chain joining the two occurrences of i_{k+1} is a simple closed chain. This simple closed chain has increment zero. We can discard all the elements of this simple closed chain except for one of the i_{k+1}'s. Continuing this process we obtain a simple closed chain C' with the same increment as C. Since the increment of C' vanishes, it follows that the increment of C vanishes. This completes the proof of Lemma 5.2.

Lemma 5.3. Let A be an irreducible matrix of order N such that the increment around any simple closed chain is zero. Then there exist integers $\gamma_1, \gamma_2, \ldots, \gamma_N$ such that if $a_{i,j} \neq 0$ and $i \neq j$, then

$$\begin{aligned} \gamma_j - \gamma_i &= -q \qquad \text{if} \quad i > j \\ \gamma_j - \gamma_i &= r \qquad \text{if} \quad i < j \end{aligned} \tag{5.2}$$

Proof. Let $\gamma_1 = 0$. For each j let $\gamma_j = \Delta(1, j, C)$ where C is some chain joining 1 to j, and where $\Delta(1, j, C)$ is the increment from 1 to j along the chain C. By Lemma 5.1, there is such a chain since A is irreducible. By Lemma 5.2, the increment is the same for all chains joining i and j. Hence the above construction gives a single-valued function γ_i of i.

Suppose now that $a_{i,j} \neq 0$ and $i > j$. Since A is irreducible, there is a chain C joining 1 to i. Let C'' be the chain formed by adding the segment C': $\{i, j\}$, to C. Then $\Delta(1, j, C'') = \Delta(1, i, C) + \Delta(i, j, C')$ $= \Delta(1, i, C) - q$. But

$$\begin{aligned} \gamma_i &= \Delta(1, i, C) \\ \gamma_j &= \Delta(1, j, C'') = \Delta(1, i, C) - q = \gamma_i - q \end{aligned}$$

Similarly, if $a_{i,j} \neq 0$ and $i < j$, we can show that $\gamma_j = \gamma_i + r$. This completes the proof of Lemma 5.3.

Lemma 5.4. If A is an L-matrix which is also a GCO(q, r)-matrix, then the increment around any simple closed chain vanishes.

Proof. Since A is a GCO(q, r)-matrix, the determinant $\det(\alpha^q C_L + \alpha^{-r} C_U - kD)$ is independent of α for all $\alpha \neq 0$ and for all k. Since A is an L-matrix, the diagonal elements of A are positive. We use induction to show that the increment for any simple closed chain of length n vanishes. If $n = 1$, then the increment of the chain i, i is zero by definition. Suppose now that all simple closed chains of length n or less have zero increment. Consider all terms of $\det(\alpha^q C_L + \alpha^{-r} C_U - kD)$ which have precisely $N - (n + 1)$ factors from the main diagonal. As in Section 2.1 any such term corresponds to a permutation which can be represented by $N - (n + 1)$ cycles of length one plus one or more other cycles of length at least two and not more than $n + 1$. By the induction hypothesis, all terms not involving a cycle of length $n + 1$ are independent of α since the increments around the simple closed chains corresponding to the cycles of length n or less vanish. Since $\det(\alpha^q C_L + \alpha^{-r} C_U - kD)$ is a polynomial in k which is independent of α for all $\alpha \neq 0$ and for all k, the coefficient of each power of k must be independent of α. The coefficient of $k^{N-(n+1)}$ is the sum of the terms having exactly $N-(n+1)$ factors from the main diagonal. For the sum of these terms to be independent of α it is necessary that the sum of all terms corresponding to $(n+1)$-cycles vanish. But if $\alpha > 0$, all such terms have the same sign. Thus, if $\alpha > 0$ each element of $\alpha^q C_L + \alpha^{-r} C_U - kD$ which is not on the main diagonal is positive. The terms in question are all of one sign if $n+1$ is odd and of the other sign if $n+1$ is even. In the former case the associated permutation has an even number, namely zero, of even cycles; in the latter case there are an odd number, namely one, of even cycles (see Section 2.1). Since each term has the same sign, their sum will be independent of α for all $\alpha \neq 0$ if and only if each term corresponding to a cycle of length $n+1$ is independent of α. If there were a simple closed chain, of length $n+1$ with a nonzero increment, then since the diagonal elements of A do not vanish, the term associated with the corresponding $(n+1)$-cycle plus $N-(n+1)$ one-cycles would not vanish and the determinant would not be independent of α. Consequently, the increment around any simple closed chain of length $n+1$ vanishes. The lemma now follows by induction.

By Lemma 5.3 and 5.4 and by Theorem 2.1 we have

Theorem 5.5. If A is an irreducible GCO(q, r)-matrix which is an L-matrix, then A is a CO(q, r)-matrix.

13.6. COMPUTATIONAL PROCEDURES: CANONICAL FORMS

Given a matrix A and positive integers q and r we can use procedures analogous to those of Section 5.7 to determine whether or not A has Property $A_{q,r}$ and, if so, whether it is a CO(q, r)-matrix. To test whether a matrix A has Property $A_{q,r}$ it is sufficient by Theorems 2.2 and 2.4 to see whether A has Property $A_{q',r'}$, where $q' = q/d$, $r' = r/d$ and d is the highest common factor of q and r. Moreover, by Theorem 2.2 it is sufficient to seek an ordering vector each of whose components lies between 1 and p', where $p' = q' + r'$.

If a (q, r)-ordering vector is found to exist, then we seek a compatible ordering vector in a manner similar to that of Section 5.7. If such a vector exists, then A is a CO(q, r)-matrix. Otherwise, we can permute the rows and columns of A to obtain a CO(q, r)-matrix.

As we have seen in Section 13.3, the SOR method is effective only if

$q = 1$. Consequently, we should test only those values of q and r with $q = 1$. (Actually, if A has Property $A_{q,r}$ for some q and r, then A has Property $A_{1,r'}$ for some r'.)

One might first test whether A is an irreducible L-matrix. If such is the case, then in order that $P^{-1}AP$ should be a GCO(q, r)-matrix for some permutation matrix P, A must have Property $A_{q,r}$. Thus, if A does not have Property $A_{q,r}$, then A is not a GCO(q, r)-matrix.

As shown in Chapter 5, given a CO(1, 1)-matrix A, one can find a nonmigratory permutation P such that $P^{-1}AP$ is a T-matrix. Similarly, given a matrix A with Property $A_{1,1}$, one can find a permutation matrix Q such that $Q^{-1}AQ$ is a T-matrix with either one or two diagonal blocks. A natural generalization of a T-matrix is a $T(q, r)$-matrix defined as follows:

Definition 6.1. Given the positive integers q, r, and t, the matrix A is a *$T(q, r, t)$-matrix* if it can be partitioned into the $t \times t$ block form $A = (A_{i,j})$ where, for each i, $A_{i,i} = D_i$ is a square diagonal matrix and where all other blocks vanish except possibly for the blocks $A_{i,i+r}$, $i = 1, 2, \ldots, t - r$, and $A_{i,i-q}$, $i = q + 1, q + 2, \ldots, t$. If A is a $T(q, r, t)$-matrix for some t, then A is a *$T(q, r)$-matrix*.

An example of a $T(2, 3, 8)$-matrix is

$$A = \begin{pmatrix} D_1 & 0 & 0 & H_1 & 0 & 0 & 0 & 0 \\ 0 & D_2 & 0 & 0 & H_2 & 0 & 0 & 0 \\ K_1 & 0 & D_3 & 0 & 0 & H_3 & 0 & 0 \\ 0 & K_2 & 0 & D_4 & 0 & 0 & H_4 & 0 \\ 0 & 0 & K_3 & 0 & D_5 & 0 & 0 & H_5 \\ 0 & 0 & 0 & K_4 & 0 & D_6 & 0 & 0 \\ 0 & 0 & 0 & 0 & K_5 & 0 & D_7 & 0 \\ 0 & 0 & 0 & 0 & 0 & K_6 & 0 & D_8 \end{pmatrix} \tag{6.1}$$

where the D_i are square diagonal matrices.

It turns out that not every CO(q, r)-matrix can be transformed into a $T(q, r)$-matrix by a nonmigratory permutation, Moreover, not every matrix with Property $A_{q,r}$ can be transformed into a $T(q, r)$-matrix by a general permutation. However, one can obtain a *quasi-$T(q, r)$-matrix* from a CO(q, r)-matrix by a nonmigratory permutation and from a matrix with Property $A_{q,r}$ by a general permutation. A quasi-$T(q, r)$-matrix can be obtained from a $T(q, r)$-matrix by deletion of certain of the rows and the corresponding columns of blocks.

An example of a quasi-$T(2, 3)$-matrix is the matrix

$$\tilde{A} = \begin{bmatrix} D_1 & 0 & 0 & H_1 & 0 & 0 & 0 \\ 0 & D_2 & 0 & 0 & H_2 & 0 & 0 \\ K_1 & 0 & D_3 & 0 & 0 & 0 & 0 \\ 0 & K_2 & 0 & D_4 & 0 & H_4 & 0 \\ 0 & 0 & K_3 & 0 & D_5 & 0 & H_5 \\ 0 & 0 & 0 & 0 & K_5 & D_7 & 0 \\ 0 & 0 & 0 & 0 & 0 & 0 & D_8 \end{bmatrix} \tag{6.2}$$

which is obtained from (6.1) by deleting the sixth row and column of blocks.

Under certain conditions, a CO(q, r)-matrix can be transformed to a $T(q, r)$-matrix by a nonmigratory permutation. Similarly, it is sometimes possible to transform a matrix with Property $A_{q,r}$ into a $T(q, r)$-matrix with at most $(q + r)^2$ blocks by a general permutation. In such cases the $T(q, r)$-matrix is said to be a *canonical form*. In other cases we may have to settle for a quasi-$T(q, r)$-matrix as a canonical form.

Let us introduce the following definitions:

Definition 6.2. A matrix A is a *strongly (q, r)-consistently ordered matrix* (a CO*(q, r)-matrix) if it is a CO(q, r)-matrix and there exist nonempty sets $S_1, S_2, \ldots, S_t$ satisfying the conditions of Definition 2.1.

Definition 6.3. A (q, r)-ordering vector γ is *complete* if for each integer k such that $\min_j \gamma_j \le k \le \max_j \gamma_j$ there exists i such that $\gamma_i = k$.

Definition 6.4. A matrix A has *strong Property* $A_{q,r}$ (Property $A^*_{q,r}$) if it has Property $A_{q,r}$ and if for some $t \le p$, where $p = q + r$, there exist nonempty sets $S_1, S_2, \ldots, S_t$ satisfying the conditions of Definition 2.2. If A has Property $A^*_{1,p-1}$, then A has Property A_p^*.

We state without proof the following theorems:

Theorem 6.1. A is a CO*(q, r)-matrix if and only if one and hence all of the following conditions hold:

(a) There exists a nonmigratory permutation matrix P such that $P^{-1}AP$ is a $T(q, r)$-matrix.

(b) There exists a complete and compatible (q, r)-ordering vector for A.

Theorem 6.2. If A is an irreducible CO(q, r)-matrix, then there exists a unique compatible ordering vector γ such that $\gamma_1 = 1$. If A is also a CO*(q, r)-matrix, then γ is complete.

Theorem 6.3. If A is a CO*(q, r)-matrix, then A is a CO*(q', r')-matrix where $q' = q/d$, $r' = r/d$ and d is any common factor of q and r.

Theorem 6.4. If A has Property $\mathrm{A}^*_{q,r}$, then there exists a complete (q, r)-ordering vector γ for A whose components lie in the range $1 \le \gamma_i \le q + r$ and there exists a permutation matrix P such that $P^{-1}AP$ is a $T(q, r)$-matrix. On the other hand, if there exists a complete (q, r)-ordering vector or if there exists a permutation matrix P such that $P^{-1}AP$ is a $T(q, r)$-matrix, then A has Property $\mathrm{A}^*_{q,r}$.

Theorem 6.5. If A has Property $\mathrm{A}^*_{q,r}$, then A has Property $\mathrm{A}^*_{q/d,r/d}$ for any common factor d of q and r. If A has Property $\mathrm{A}_p{}^*$, then A has Property $\mathrm{A}^*_{p/d}$ where d is any divisor of p such that $d < p$ and A has Property $\mathrm{A}^*_{q,r}$ for any pair of relatively prime integers such that $q + r = p$.

Theorem 6.6. If A has Property $\mathrm{A}^*_{q,r}$, then A has Property $\mathrm{A}^*_{p'}$ where $p' = (q + r)/d$ and where d is the greatest common factor of q and r.

Theorem 6.7. Let A be an irreducible matrix or a matrix with nonvanishing diagonal elements such that $S(B) \neq 0$, where $B = I - (\operatorname{diag} A)^{-1}A$.

(a) If A has Property A_p, then A has Property $\mathrm{A}_p{}^*$ and A is permutationally similar to a $T(1, p - 1, p)$-matrix.

(b) If A has Property $\mathrm{A}_{q,r}$, then A has Property $\mathrm{A}^*_{q',r'}$ where $q' = q/d$, $r' = r/d$, and d is the greatest common factor of q and r.

Theorem 6.8. If A has Property A_p, then, for some $p' \le p$, A has Property $\mathrm{A}^*_{p'}$.

An example of a CO$(1, 2)$-matrix which is not a CO*(q, r)-matrix is given by

$$A = \begin{pmatrix} 0 & 0 & 1 & 0 \\ 1 & 0 & 0 & 0 \\ 0 & 1 & 0 & 1 \\ 0 & 0 & 0 & 0 \end{pmatrix} \tag{6.3}$$

The reader should verify that A is a quasi-$T(1, 2)$-matrix but that there does not exist a nonmigratory permutation matrix P such that $P^{-1}AP$ is a $T(1, 2)$-matrix. On the other hand, A has Property $A_{1,2}$ and hence has Property $A^*_{p'}$ for some $p' \leq 3$. In fact, in this case A actually has Property A_3^*. Hence, one can obtain a $T(1, 2)$-matrix by a migratory permutation.

We now give an example of an irreducible CO(2, 3)-matrix which is not a CO*(2, 3)-matrix. Let

$$A = \begin{pmatrix} 0 & 0 & 1 & 0 & 0 \\ 1 & 0 & 0 & 0 & 0 \\ 0 & 0 & 0 & 0 & 1 \\ 0 & 1 & 0 & 0 & 0 \\ 0 & 0 & 0 & 1 & 0 \end{pmatrix} \tag{6.4}$$

The unique compatible ordering vector such that $\gamma_1 = 1$ is $\gamma = (1, 3, 4, 5, 7)$ which is not complete. Therefore by Theorem 6.1, A is not a CO*(q, r)-matrix. On the other hand, by Theorem 6.7, A has Property $A^*_{2,3}$.

Given a $T(q, r)$-matrix we can permute the rows and columns of blocks to obtain at most p^2 blocks, where $p = q + r$. Thus, for instance, for the $T(1, 2)$-matrix

$$A = \begin{pmatrix} 0 & 0 & H_1 & 0 \\ K_1 & 0 & 0 & H_2 \\ 0 & K_2 & 0 & 0 \\ 0 & 0 & K_3 & 0 \end{pmatrix} \tag{6.5}$$

we can permute the rows and columns of blocks to obtain the form

$$A = \left(\begin{array}{cc|c|c} 0 & 0 & 0 & H_1 \\ 0 & 0 & 0 & K_3 \\ \hline K_1 & H_2 & 0 & 0 \\ \hline 0 & 0 & K_2 & 0 \end{array}\right) \tag{6.6}$$

which is a $T(1, 2)$-matrix with nine blocks.

Given a matrix A and the integers q and r we seek to determine two canonical forms: one under nonmigratory permutations, and the other under general permutations. For the first canonical form if we can construct a complete compatible ordering vector γ, then our canonical form is a $T(q, r)$-matrix. If a compatible ordering vector exists but not a complete and compatible ordering vector, then our canonical form is a quasi-$T(q, r)$-matrix. If A is irreducible, then by Theorem 6.2 we need only test one compatible ordering vector for completeness.

For the second canonical form we first see whether a (q, r)-ordering vector can be constructed. If a complete (q, r)-ordering vector exists, then our canonical form is a $T(q, r)$-matrix with at most $(q + r)^2$ blocks. If an ordering vector exists but no complete ordering vector, then we settle for a quasi-$T(q, r)$-matrix with at most $(q + r)^2$ blocks. Alternatively, we can obtain a $T(1, p', 1)$-matrix for some $p' < q + r$.

The number of practical cases where matrices with Property A_p arise for $p > 2$ appears to be rather small. One such case is the four-point discrete analog of Poisson's equation

$$u_{xx} + u_{yy} = G(x, y) \tag{6.7}$$

in a triangular mesh. For example, suppose we wish to solve the Dirichlet problem in the equilateral triangle of side-length unity and we superimpose a triangular mesh of length $h = 1/M$. For each mesh point (x, y) we replace Poisson's equation by the difference equation

$$\frac{4}{3h^2}\left\{u(x + h, y) + u\left(x - \frac{h}{2}, y + \frac{3^{1/2}h}{2}\right)\right.$$
$$\left. + u\left(x - \frac{h}{2}, y - \frac{3^{1/2}h}{2}\right) - 3u(x, y)\right\} = G(x, y) \tag{6.8}$$

As can be seen by examining the situation shown in Figure 6.1, for the case $h = \frac{1}{6}$, there is a natural grouping of the interior mesh points into

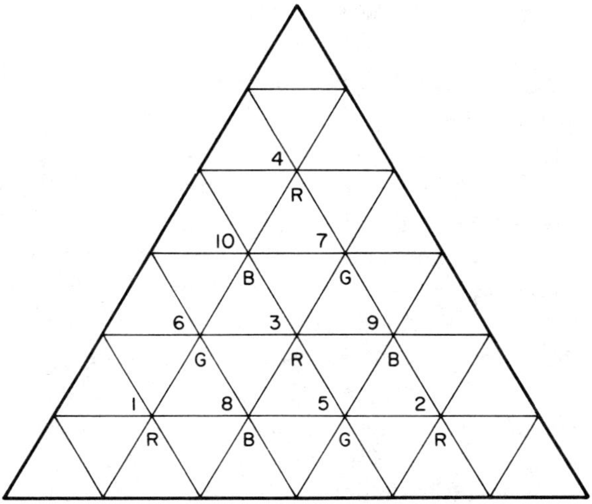

Figure 6.1.

three classes, red (R), black (B), and green (G). The equations for the red points involve only black points, the equations for black points involve only green points, and the equations for green points involve only red points. If we number the points as shown with the red points first, then the green, and then the black, we obtain a $T(1, 2, 3)$-matrix. In fact, for the particular case we have, after multiplying the system through by $-3h^2/4$, the system $Au = b$, where A has the form

$$A = \left(\begin{array}{cccc|ccc|ccc} 3 & 0 & 0 & 0 & 0 & 0 & 0 & -1 & 0 & 0 \\ 0 & 3 & 0 & 0 & 0 & 0 & 0 & 0 & -1 & 0 \\ 0 & 0 & 3 & 0 & 0 & 0 & 0 & -1 & -1 & -1 \\ 0 & 0 & 0 & 3 & 0 & 0 & 0 & 0 & 0 & -1 \\ \hline 0 & -1 & -1 & 0 & 3 & 0 & 0 & 0 & 0 & 0 \\ -1 & 0 & -1 & 0 & 0 & 3 & 0 & 0 & 0 & 0 \\ 0 & 0 & -1 & -1 & 0 & 0 & 3 & 0 & 0 & 0 \\ \hline 0 & 0 & 0 & 0 & -1 & -1 & 0 & 3 & 0 & 0 \\ 0 & 0 & 0 & 0 & -1 & 0 & -1 & 0 & 3 & 0 \\ 0 & 0 & 0 & 0 & 0 & -1 & -1 & 0 & 0 & 3 \end{array}\right) \qquad (6.9)$$

Another application of matrices with Property A_p with $p > 2$ is given by Tee [1964] and involves the solution of a parabolic problem with periodic boundary conditions.

13.7. RELATION TO OTHER WORK

We have already discussed in Section 5.8 the relation between various classes of matrices considered by Varga and matrices with Property A_2* and CO*(1, 1)-matrices. We now extend the discussion to include the more general classes of matrices considered in this chapter.

As defined by Varga, a matrix is *p-cyclic*, for $p \geq 2$, if it has nonvanishing diagonal elements and if $B = I - (\operatorname{diag} A)^{-1}A$ is permutationally similar to a $T(1, p - 1, p)$-matrix. Thus if A is p-cyclic, then A has Property A_p*, and hence Property A_p, but not necessarily conversely, since if A has Property A_p*, A may be similar to a $T(1, p - 1, t)$-matrix with $t < p$. However, from Theorem 6.7, it follows that if A is an irreducible matrix with nonvanishing diagonal elements which has Property A_p*, then A is a p-cyclic matrix. Thus, although Property A_p is somewhat more general than Property A_p* and p-cyclic, the difference between Property A_p* and p-cyclic is very minor indeed.

Verner and Bernal [1968] defined a GCO(q, r)-matrix A as one with nonvanishing diagonal elements such that the eigenvalues of

$$B(\alpha) = \alpha^q L + \alpha^{-r} U \tag{7.1}$$

are independent of α for all $\alpha \neq 0$. Here $B = I - (\operatorname{diag} A)^{-1}A$ and L and U are strictly lower and strictly upper triangular matrices, respectively, such that $L + U = B$. It is easy to show that if the diagonal elements of A do not vanish, then our definition is equivalent to that of Verner and Bernal. Thus, our definition is very slightly more general. If A is an irreducible L-matrix, then if A is a GCO(q, r)-matrix in the sense of Verner and Bernal, and hence in our sense, A is a CO(q, r)-matrix.

Varga gave a definition of $(1, p - 1)$-consistently-ordered-matrices which is the same as that of Verner and Bernal except for the additional requirement that A be p-cyclic. An irreducible L-matrix which is a GCO$(1, p - 1)$-matrix in the sense of Varga, and hence in the sense of Verner and Bernal, is a CO$(1, p - 1)$-matrix. On the other hand, if A is a CO$(1, p - 1)$-matrix with nonvanishing diagonal elements, then A is a GCO$(1, p - 1)$-matrix in the sense of Verner and Bernal. Moreover, A has Property A_p. If A is irreducible, then, by Theorem 6.7, A has Property A_p^* and is permutationally similar to a $T(1, p - 1, p)$-matrix. Hence A is p-cyclic and satisfies all the requirements of Varga's definition.

The definitions of p-cyclic matrices and generalized consistently ordered matrices given by Varga, and by Verner and Bernal, are applicable to the case where $B = I - D^{-1}A$ where D is a nonsingular matrix which is not necessarily diag A. In fact, D need not be a diagonal matrix. If D is a block diagonal matrix, B corresponds to the block Jacobi method. We shall consider block and group methods in Chapter 14.

SUPPLEMENTARY DISCUSSION

Much of the material in this chapter is given in Young [1970]. It is based on earlier work of Varga [1959, 1962], Kjellberg [1961], Broyden [1964, 1968], and Verner and Bernal [1968]. The results of Section 13.5, in particular, were proved by Broyden [1968] using a different method. The introduction of Property $A_{q,r}$ and the relation to (q, r)-consistently-ordered-matrices is believed to be new.

EXERCISES

Section 13.1

1. Solve (1.2) for λ if $\mu = 0.8$ and $\omega = 0.8, 1.0, 1.2, 1.4$ for the cases $p = 3$, $r = 2$; $p = 3$, $r = 1$; $p = 1$, $r = 1$.

Section 13.2

1. Prove Theorem 2.1.

2. Verify the last part of the proof of Theorem 2.4.

3. If A has Property $A_{6,10}$, find all $\hat{q}$, $\hat{r}$ such that A has Property $A_{\hat{q},\hat{r}}$. For what values of p does A have Property A_p?

4. Find a migratory permutation matrix P such that $P^{-1}AP$ is a CO(2, 1)-matrix where A is given by (6.9).

5. Show that the matrix A obtained in the same way as (6.9) is obtained except with the numbering of the points (see Figure 6.1) given by the accompanying figure has Property A_3. Is it a CO(2, 1)-matrix or a CO(1, 2)-matrix?

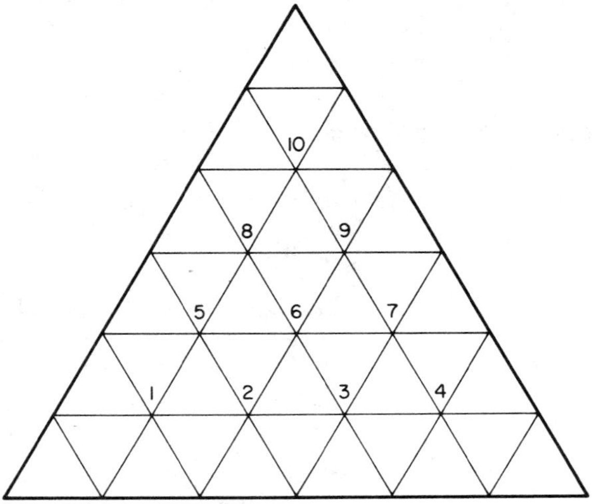

6. Let A be a matrix such that for some permutation matrix P the matrix $P^{-1}AP$ is a CO(3, 12)-matrix. Show that A has Property $A_{9,6}$.

7. Determine the permutation σ which will transform a CO(1, 6)-matrix into a CO(4, 3)-matrix. Perform a suitable permutation on the matrix given below to obtain a CO(4, 3)-matrix.

$$A = \begin{pmatrix} 0 & 0 & 0 & 0 & 0 & 0 & 1 \\ 1 & 0 & 0 & 0 & 0 & 0 & 0 \\ 0 & 1 & 0 & 0 & 0 & 0 & 0 \\ 0 & 0 & 1 & 0 & 0 & 0 & 0 \\ 0 & 0 & 0 & 1 & 0 & 0 & 0 \\ 0 & 0 & 0 & 0 & 1 & 0 & 0 \\ 0 & 0 & 0 & 0 & 0 & 1 & 0 \end{pmatrix}$$

8. Verify that $\gamma = (1, 2, 3, 5, 4, 6, 7)^{\mathrm{T}}$ is a compatible (3, 2)-ordering vector for the matrix A given by (2.4).

9. If A has Property $\mathrm{A}_{4,2}$, find the set of all (q, r) such that A has Property $\mathrm{A}_{q,r}$. Does A have Property $\mathrm{A}_{3,3}$ and/or Property $\mathrm{A}_{1,1}$? Give an example.

10. If A has Property $\mathrm{A}_{4,2}$, what should be done in order to achieve maximum effectiveness with the SOR method assuming that all eigenvalues of B^3 are real and nonnegative?

11. Verify that the matrix

$$A = \begin{pmatrix} 4 & 0 & 0 & -1 \\ -1 & 4 & 0 & 0 \\ 0 & -1 & 4 & 0 \\ 0 & 0 & -1 & 4 \end{pmatrix}$$

has Property $\mathrm{A}_{2,2}$ and Property $\mathrm{A}_{1,1}$. Determine a migratory permutation matrix P so that $P^{-1}AP$ is a CO(2, 2)-matrix.

12. Show that the matrix

$$A = \begin{pmatrix} 3 & -1 & 0 & 0 & 0 & 0 & 0 & 0 \\ 0 & 3 & -1 & 0 & 0 & 0 & 0 & -1 \\ 0 & 0 & 3 & -1 & 0 & 0 & -1 & 0 \\ 0 & 0 & 0 & 3 & 0 & -1 & 0 & 0 \\ 0 & 0 & 0 & -1 & 3 & 0 & 0 & 0 \\ 0 & 0 & -1 & 0 & -1 & 3 & 0 & 0 \\ 0 & -1 & 0 & 0 & 0 & -1 & 3 & 0 \\ -1 & 0 & 0 & 0 & 0 & 0 & 0 & 3 \end{pmatrix}$$

has Property $\mathrm{A}_{2,1}$ and Property $\mathrm{A}_{1,2}$ but is not a CO(1, 2)-matrix or a

CO(2, 1)-matrix. Find migratory permutation matrices P and Q such that $P^{-1}AP$ is a CO(1, 2)-matrix and $Q^{-1}AQ$ is a CO(2, 1)-matrix.

13. Develop the theory of Section 13.2 without the use of Property $A_{q,r}$, i.e., using only Property A_p.

Section 13.3

1. For the matrix

$$A = \begin{pmatrix} 3 & 0 & -1 \\ -1 & 3 & 0 \\ 0 & -1 & 3 \end{pmatrix}$$

determine ω_c and the behavior of the eigenvalues of the SOR method for various values of ω. Do the same for

$$A_1 = \begin{pmatrix} 3 & -1 & 0 \\ 0 & 3 & -1 \\ -1 & 0 & 3 \end{pmatrix}$$

2. Consider the following matrix:

$$A = \begin{pmatrix} 1 & 0 & 0 & -a \\ -a & 1 & 0 & 0 \\ 0 & -a & 1 & 0 \\ 0 & 0 & -a & 1 \end{pmatrix}$$

Show that A has Property A_4 and also Property A_2. If $a = 0.99$, find the optimum ω to use when A is ordered to be a CO(1, 3)-matrix and also a CO(1, 1)-matrix. Also, find $S(\mathscr{L}_\omega)$ in each case. Also, verify (3.33) in each case.

3. Suppose A is a CO(1, 3)-matrix such that A has nonvanishing diagonal elements, the eigenvalues of B^4 are real and nonnegative, and $\bar{\mu} = S(B) = 0.9$. Find the optimum ω, say ω_c, and $S(\mathscr{L}_{\omega_c})$. What would be the best ω if A were a CO(3, 1)-matrix and what would $S(\mathscr{L}_\omega)$ be?

4. Verify that the positive root of (3.20) increases as b increases and as c increases.

5. Prove that if $b > 0$, $c > 0$, then the cubic equation (3.6) has three real roots if (3.12) holds.

Section 13.4

1. Let A be a matrix with nonvanishing diagonal elements and let $B = I - (\text{diag } A)^{-1}A$, $L + U = B$ where L and U are strictly lower and strictly upper triangular matrices, respectively. Show that the eigenvalues of

$$B(\alpha) = \alpha^q L + \alpha^{-r} U$$

are independent of α for all $\alpha \neq 0$ if and only if A is a GCO(q, r)-matrix.

2. Verify that the matrix A given by (4.3) is a GCO(1, 1)-matrix.

3. Prove directly that if A is an irreducible GCO(1, 1)-matrix of order 3, then A is a CO*(1, 1)-matrix.

4. Prove directly if A is an irreducible GCO(1, 1)-matrix which is also an L-matrix of order 4, then A is a CO*(1, 1)-matrix.

Section 13.5

1. Carry out the argument used to prove Theorem 5.5 in detail for the case $N = 5$.

Section 13.6

1. Show that the matrix

$$A = \begin{pmatrix} 0 & 0 & 1 \\ 1 & 0 & 0 \\ 0 & 0 & 0 \end{pmatrix}$$

is a CO(q, r)-matrix for any positive integers q and r. Find all q and r such that A is a CO*(q, r)-matrix and such that A has Property $A^*_{q,r}$.

2. Show that every CO(1, 1)-matrix is a CO*(1, 1)-matrix and every matrix with Property A_2 has Property A_2^*.

3. Show that the matrix

$$A = \begin{pmatrix} 0 & 0 & 1 & 0 \\ 1 & 0 & 0 & 0 \\ 0 & 1 & 0 & 1 \\ 0 & 0 & 0 & 0 \end{pmatrix}$$

is a CO(1, 1)-matrix, but that there does not exist a nonmigratory permutation matrix P such that $P^{-1}AP$ is a T(1, 2)-matrix. Does there exist a migratory permutation matrix Q such that $Q^{-1}AQ$ is a T(1, 2)-matrix? Show that A is a quasi-T(1, 2)-matrix.

Chapter 14 / GROUP ITERATIVE METHODS

Each iteration of the J method can be considered as the solution of N systems of equations, each system having one equation and one unknown. Thus for the ith equation one treats u_i as unknown and the u_j, for $j \neq i$, as known. This can also be done for the GS method except that for the known values of u_j one uses the latest available values. A similar interpretation can be given for the SOR method if we write the equations in the form

$$\frac{a_{i,i}}{\omega} u_i + \left(1 - \frac{1}{\omega}\right) a_{i,i} u_i + \sum_{\substack{j=1 \\ j \neq i}}^{N} a_{i,j} u_j = b_i, \qquad i = 1, 2, \ldots, N \tag{1}$$

and if one treats the first term as unknown and all other terms as known.

With *group iterative methods*, one first assigns the equations to subsets, or groups, such that each equation belongs to one and only one group. One then solves the groups of equations for the corresponding unknowns u_i, treating the other values of u_j as known. A special case of a grouping is a *partitioning*. Here for some integers $n_1, n_2, \ldots, n_q$ such that $1 \leq n_1 < n_2 < \cdots < n_q = N$ the equations for $i = 1, 2, \ldots, n_1$ belong to the first group, those for $i = n_1 + 1, n_1 + 2, \ldots, n_2$ belong to the second group, etc. Methods based on partitionings are usually known as *block methods*.

In the case of the discrete Dirichlet problem one could use a grouping involving all equations associated with points of a given row. If the mesh points are numbered in the natural ordering, then such a grouping would correspond to a partitioning. However, if each group involves equations for a given column, then the grouping would in general not correspond to a partitioning.

Normally, one uses a direct method to solve the subsystems involved in carrying out a group iteration method. This is usually feasible since the subsystems are much smaller than the original system. Also, in many cases the matrix of the smaller system has properties which make its solution particularly convenient. For example, in many cases the submatrix is tri-diagonal and the number of multiplications needed for a direct solution is proportional to m, the order of the submatrix, instead of to m^3 as would be needed with the Gauss elimination method in the general case.

We shall define several group iterative methods in Section 14.1 and in Section 14.2, we shall show they can be effectively carried out. Convergence properties will be studied in Section 14.3. Applications to problems arising from partial differential equations will be considered in Section 14.4. In Section 14.5, we shall give a comparison of point and group iterative methods and shall show that many of the results of Chapters 7–13 can be extended to group methods.

14.1. CONSTRUCTION OF GROUP ITERATIVE METHODS

The first step in constructing a group iterative method is to divide the integers $1, 2, \ldots, N$ into groups such that each such integer belongs to one and only one group. We do not assume that the groups consist of consecutive integers but we shall assume that the groups are ordered. We define

Definition 1.1. An *ordered grouping* π of $W = \{1, 2, \ldots, N\}$ is a subdivision of W into disjoint subsets $R_1, R_2, \ldots, R_q$ such that $R_1 + R_2 + \cdots + R_q = W$. Two ordered groupings π and π' defined by $R_1, R_2, \ldots, R_q$ and $R_1', R_2', \ldots, R_{q'}'$, respectively, are identical if $q = q'$ and if $R_1 = R_1', R_2 = R_2', \ldots, R_q = R_{q}'$. Examples of ordered groupings for the case $N = 4$ are

$$
\begin{array}{lllll}
\pi_0\colon & R_1 = \{1\}, & R_2 = \{2\}, & R_3 = \{3\}, & R_4 = \{4\} \\
\pi_1\colon & R_1 = \{1, 2\}, & R_2 = \{3, 4\} & & \\
\pi_2\colon & R_1 = \{2\}, & R_2 = \{1, 3\}, & R_3 = \{4\} & \\
\pi_3\colon & R_1 = \{4\}, & R_2 = \{3, 1\}, & R_3 = \{2\} & \\
\pi_4\colon & R_1 = \{2, 1\}, & R_2 = \{4, 3\} & &
\end{array}
$$

Evidently π_0, π_1, and π_4 define partitionings. Moreover, $\pi_1 = \pi_4$, but $\pi_2 \neq \pi_3$.

We shall refer to the subsets $R_1, R_2, \ldots, R_q$ as *groups* even though they are not groups in the usual mathematical sense. We let π_0 denote the ordered grouping defined by $R_k = \{k\}$, $k = 1, 2, \ldots, N$.

Given a matrix A and an ordered grouping π we define the submatrices $A_{r,s}$ for $r, s = 1, 2, \ldots, q$ as follows: $A_{r,s}$ is formed from A by deleting all rows except those corresponding to R_r and all columns except those corresponding to R_s. Given a column vector u we define column vectors $U_1, U_2, \ldots, U_q$ where U_r is formed from u by deleting all elements of u except those corresponding to R_r. Similarly, we define column vectors $B_1, B_2, \ldots, B_q$ given the column vector b. Evidently, the system $Au = b$ can be written in the equivalent form

$$\sum_{s=1}^{q} A_{r,s} U_s = B_r, \qquad r = 1, 2, \ldots, q \tag{1.1}$$

For example, if $N = 3$ and π is defined by $R_1 = \{1, 3\}$, $R_2 = \{2\}$, we have

$$\begin{aligned} A_{1,1} &= \begin{pmatrix} a_{1,1} & a_{1,3} \\ a_{3,1} & a_{3,3} \end{pmatrix}, & A_{1,2} &= \begin{pmatrix} a_{1,2} \\ a_{3,2} \end{pmatrix}, & U_1 &= \begin{pmatrix} u_1 \\ u_3 \end{pmatrix}, & B_1 &= \begin{pmatrix} b_1 \\ b_3 \end{pmatrix} \\ A_{2,1} &= (a_{2,1} \quad a_{2,3}), & A_{2,2} &= (a_{2,2}), & U_2 &= (u_2), & B_2 &= (b_2) \end{aligned} \tag{1.2}$$

so that (1.1) becomes

$$\begin{aligned} \begin{pmatrix} a_{1,1} & a_{1,3} \\ a_{3,1} & a_{3,3} \end{pmatrix} \begin{pmatrix} u_1 \\ u_3 \end{pmatrix} + \begin{pmatrix} a_{1,2} \\ a_{3,2} \end{pmatrix} (u_2) &= \begin{pmatrix} b_1 \\ b_3 \end{pmatrix} \\ (a_{2,1} \quad a_{2,3}) \begin{pmatrix} u_1 \\ u_3 \end{pmatrix} + (a_{2,2})(u_2) &= (b_2) \end{aligned} \tag{1.3}$$

which is equivalent to the original system.

We assume that $A_{r,r}$ is nonsingular for all r. The group J method is defined by

$$A_{r,r} U_r^{(n+1)} + \sum_{\substack{s=1 \\ s \neq r}}^{q} A_{r,s} U_s^{(n)} = B_r, \qquad r = 1, 2, \ldots, q \tag{1.4}$$

or, equivalently

$$U_r^{(n+1)} = \sum_{\substack{s=1 \\ s \neq r}}^{q} B_{r,s} U_s^{(n)} + C_r, \qquad r = 1, 2, \ldots, q \tag{1.5}$$

where

$$B_{r,s} = \begin{cases} -A_{r,r}^{-1}A_{r,s} & \text{if } r \neq s \\ 0 & \text{if } r = s \end{cases} \tag{1.6}$$

and

$$C_r = A_{r,r}^{-1}B_r \tag{1.7}$$

Evidently, we may write (1.5) in the matrix form

$$u^{(n+1)} = B^{(\pi)}u^{(n)} + c^{(\pi)} \tag{1.8}$$

where

$$\begin{aligned} B^{(\pi)} &= (D^{(\pi)})^{-1}C^{(\pi)} \\ c^{(\pi)} &= (D^{(\pi)})^{-1}b \end{aligned} \tag{1.9}$$

$$C^{(\pi)} = D^{(\pi)} - A \tag{1.10}$$

$$D^{(\pi)} = \operatorname{diag}_\pi A \tag{1.11}$$

Here $\operatorname{diag}_\pi A$ is the matrix formed from A by replacing by zeros all $a_{i,j}$ unless i and j belong to the same group.

For the group GS method we have

$$A_{r,r}U_r^{(n+1)} + \sum_{s=1}^{r-1} A_{r,s}U_s^{(n+1)} + \sum_{s=r+1}^{q} A_{r,s}U_s^{(n)} = B_r, \quad r = 1, 2, \ldots, q \tag{1.12}$$

or equivalently

$$U_r^{(n+1)} = \sum_{s=1}^{r-1} B_{r,s}U_s^{(n+1)} + \sum_{s=r+1}^{q} B_{r,s}U_s^{(n)} + C_r \tag{1.13}$$

The group GS method can also be written in the matrix form

$$u^{(n+1)} = \mathscr{L}^{(\pi)}u^{(n)} + (I - L^{(\pi)})^{-1}c^{(\pi)} \tag{1.14}$$

where

$$\mathscr{L}^{(\pi)} = (I - L^{(\pi)})^{-1}U^{(\pi)} \tag{1.15}$$

Here

$$L^{(\pi)} = (D^{(\pi)})^{-1}C_L^{(\pi)}, \qquad U^{(\pi)} = (D^{(\pi)})^{-1}C_U^{(\pi)} \tag{1.16}$$

and $C_L^{(\pi)}$ and $C_U^{(\pi)}$ are formed from A by replacing all elements of A by zero except those $a_{i,j}$ such that i and j belong to different groups and such that the group containing i comes after and before, respectively, the group containing j.

The matrix forms for the group JOR and group SOR methods are, respectively,

$$u^{(n+1)} = B_\omega^{(\pi)} u^{(n)} + \omega c^{(\pi)} \tag{1.17}$$

and

$$u^{(n+1)} = \mathscr{L}_\omega^{(\pi)} u^{(n)} + (I - \omega L^{(\pi)})^{-1} \omega c^{(\pi)} \tag{1.18}$$

where

$$B_\omega^{(\pi)} = \omega B^{(\pi)} + (1 - \omega) I \tag{1.19}$$

$$\mathscr{L}_\omega^{(\pi)} = (I - \omega L^{(\pi)})^{-1} (\omega U^{(\pi)} + (1 - \omega) I) \tag{1.20}$$

For actual computation with the group SOR method one would solve the system

$$(D^{(\pi)} - \omega C_L^{(\pi)}) u^{(n+1)} - (\omega C_U^{(\pi)} + (1 - \omega) I) u^{(n)} = \omega b \tag{1.21}$$

for $u^{(n+1)}$. This is equivalent to solving

$$\begin{aligned} A_{r,r} U_r^{(n+1)} &+ \omega \left\{ \sum_{s=1}^{r-1} A_{r,s} U_s^{(n+1)} + \sum_{s=r+1}^{q} A_{r,s} U_s^{(n)} \right\} \\ &+ (\omega - 1) A_{r,r} U_r^{(n)} = \omega B_r, \qquad r = 1, 2, \ldots, q \end{aligned} \tag{1.22}$$

successively for $U_1^{(n+1)}, U_2^{(n+1)}, \ldots, U_q^{(n+1)}$.

As an example, let us consider the matrix

$$A = \begin{pmatrix} 4 & -1 & 0 & -1 & 0 \\ -1 & 4 & -1 & 0 & -1 \\ 0 & -1 & 4 & 0 & 0 \\ -1 & 0 & 0 & 4 & -1 \\ 0 & -1 & 0 & -1 & 4 \end{pmatrix} \tag{1.23}$$

arising from the solution of the discrete Dirichlet problem for the region shown in Figure 1.1. Let us first consider the grouping by rows corresponding to the partition

$$\pi_1: \quad R_1 = \{1, 2, 3\}, \qquad R_2 = \{4, 5\}$$

In this case of π_1 the matrices $A_{r,s}$ are given by

$$\begin{aligned} A_{1,1} &= \begin{pmatrix} 4 & -1 & 0 \\ -1 & 4 & -1 \\ 0 & -1 & 4 \end{pmatrix}, \qquad & A_{1,2} &= \begin{pmatrix} -1 & 0 \\ 0 & -1 \\ 0 & 0 \end{pmatrix} \\ A_{2,1} &= \begin{pmatrix} -1 & 0 & 0 \\ 0 & -1 & 0 \end{pmatrix}, \qquad & A_{2,2} &= \begin{pmatrix} 4 & -1 \\ -1 & 4 \end{pmatrix} \end{aligned} \tag{1.24}$$

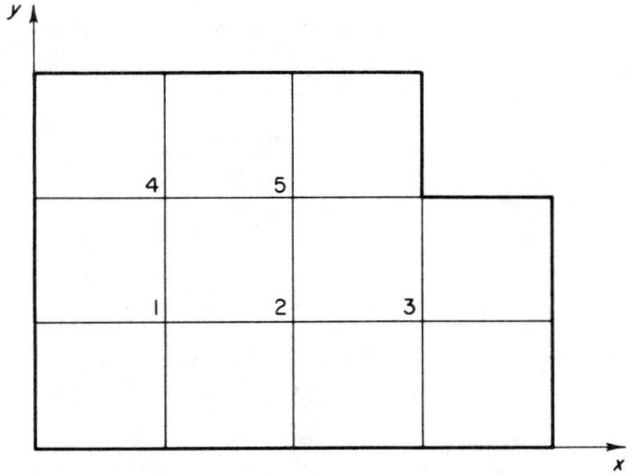

Figure 1.1.

The reader should verify that

$$D^{(\pi_1)} = \begin{pmatrix} 4 & -1 & 0 & 0 & 0 \\ -1 & 4 & -1 & 0 & 0 \\ 0 & -1 & 4 & 0 & 0 \\ 0 & 0 & 0 & 4 & -1 \\ 0 & 0 & 0 & -1 & 4 \end{pmatrix}, \quad C_L^{(\pi_1)} = \begin{pmatrix} 0 & 0 & 0 & 0 & 0 \\ 0 & 0 & 0 & 0 & 0 \\ 0 & 0 & 0 & 0 & 0 \\ 1 & 0 & 0 & 0 & 0 \\ 0 & 1 & 0 & 0 & 0 \end{pmatrix} \tag{1.25}$$

$$C_U^{(\pi_1)} = \begin{pmatrix} 0 & 0 & 0 & 1 & 0 \\ 0 & 0 & 0 & 0 & 1 \\ 0 & 0 & 0 & 0 & 0 \\ 0 & 0 & 0 & 0 & 0 \\ 0 & 0 & 0 & 0 & 0 \end{pmatrix}, \qquad (D^{(\pi_1)})^{-1} = \begin{pmatrix} \frac{15}{56} & \frac{4}{56} & \frac{1}{56} & 0 & 0 \\ \frac{4}{56} & \frac{16}{56} & \frac{4}{56} & 0 & 0 \\ \frac{1}{56} & \frac{4}{56} & \frac{15}{56} & 0 & 0 \\ 0 & 0 & 0 & \frac{4}{15} & \frac{1}{15} \\ 0 & 0 & 0 & \frac{1}{15} & \frac{4}{15} \end{pmatrix}$$

and

$$B^{(\pi_1)} = \begin{pmatrix} 0 & 0 & 0 & \frac{15}{56} & \frac{4}{56} \\ 0 & 0 & 0 & \frac{4}{56} & \frac{16}{56} \\ 0 & 0 & 0 & \frac{1}{56} & \frac{4}{56} \\ \frac{4}{15} & \frac{1}{15} & 0 & 0 & 0 \\ \frac{1}{15} & \frac{4}{15} & 0 & 0 & 0 \end{pmatrix}, \qquad L^{(\pi_1)} = \begin{pmatrix} 0 & 0 & 0 & 0 & 0 \\ 0 & 0 & 0 & 0 & 0 \\ 0 & 0 & 0 & 0 & 0 \\ \frac{4}{15} & \frac{1}{15} & 0 & 0 & 0 \\ \frac{1}{15} & \frac{4}{15} & 0 & 0 & 0 \end{pmatrix} \tag{1.26}$$

$$U^{(\pi_1)} = \begin{pmatrix} 0 & 0 & 0 & \frac{15}{56} & \frac{4}{56} \\ 0 & 0 & 0 & \frac{4}{56} & \frac{16}{56} \\ 0 & 0 & 0 & \frac{1}{56} & \frac{4}{56} \\ 0 & 0 & 0 & 0 & 0 \\ 0 & 0 & 0 & 0 & 0 \end{pmatrix}$$

Next, let us consider the grouping by columns corresponding to the ordered grouping

$$\pi_2: \quad R_1 = \{1, 4\}, \qquad R_2 = \{2, 5\}, \qquad R_3 = \{3\}$$

In this case, the matrices $A_{r,s}$ are given by

$$\begin{aligned}
&A_{1,1} = \begin{pmatrix} 4 & -1 \\ -1 & 4 \end{pmatrix}, \quad && A_{1,2} = \begin{pmatrix} -1 & 0 \\ 0 & -1 \end{pmatrix}, \quad && A_{1,3} = \begin{pmatrix} 0 \\ 0 \end{pmatrix} \\
&A_{2,1} = \begin{pmatrix} -1 & 0 \\ 0 & -1 \end{pmatrix}, && A_{2,2} = \begin{pmatrix} 4 & -1 \\ -1 & 4 \end{pmatrix}, && A_{2,3} = \begin{pmatrix} -1 \\ 0 \end{pmatrix} \\
&A_{3,1} = (0 \quad 0), && A_{3,2} = (-1 \quad 0), && A_{3,3} = (4)
\end{aligned} \tag{1.27}$$

Here we have

$$D^{(\pi_2)} = \begin{pmatrix} 4 & 0 & 0 & -1 & 0 \\ 0 & 4 & 0 & 0 & -1 \\ 0 & 0 & 4 & 0 & 0 \\ -1 & 0 & 0 & 4 & 0 \\ 0 & -1 & 0 & 0 & 4 \end{pmatrix}, \qquad C_L^{(\pi_2)} = \begin{pmatrix} 0 & 0 & 0 & 0 & 0 \\ 1 & 0 & 0 & 0 & 0 \\ 0 & 1 & 0 & 0 & 0 \\ 0 & 0 & 0 & 0 & 0 \\ 0 & 0 & 0 & 1 & 0 \end{pmatrix}$$

$$C_U^{(\pi_2)} = \begin{pmatrix} 0 & 1 & 0 & 0 & 0 \\ 0 & 0 & 1 & 0 & 0 \\ 0 & 0 & 0 & 0 & 0 \\ 0 & 0 & 0 & 0 & 1 \\ 0 & 0 & 0 & 0 & 0 \end{pmatrix} \tag{1.28}$$

and

$$B^{(\pi_2)} = \begin{pmatrix} 0 & \frac{4}{15} & 0 & 0 & \frac{1}{15} \\ \frac{4}{15} & 0 & \frac{4}{15} & \frac{1}{15} & 0 \\ 0 & \frac{1}{4} & 0 & 0 & 0 \\ 0 & \frac{1}{15} & 0 & 0 & \frac{4}{15} \\ \frac{1}{15} & 0 & \frac{1}{15} & \frac{4}{15} & 0 \end{pmatrix}, \qquad L^{(\pi_2)} = \begin{pmatrix} 0 & 0 & 0 & 0 & 0 \\ \frac{4}{15} & 0 & 0 & \frac{1}{15} & 0 \\ 0 & \frac{1}{4} & 0 & 0 & 0 \\ 0 & 0 & 0 & 0 & 0 \\ \frac{1}{15} & 0 & 0 & \frac{4}{15} & 0 \end{pmatrix}$$

$$U^{(\pi_2)} = \begin{pmatrix} 0 & \frac{4}{15} & 0 & 0 & \frac{1}{15} \\ 0 & 0 & \frac{4}{15} & 0 & 0 \\ 0 & 0 & 0 & 0 & 0 \\ 0 & \frac{1}{15} & 0 & 0 & \frac{4}{15} \\ 0 & 0 & \frac{1}{15} & 0 & 0 \end{pmatrix} \tag{1.29}$$

Thus in the case of π_2 the matrices $L^{(\pi_2)}$, $U^{(\pi_2)}$, etc. are not strictly triangular matrices as would be the case if π_2 were a partitioning. However, we emphasize again that to actually carry out the iteration process one should use (1.4), (1.12), (1.22), etc. In the next section we shall show how this can be done effectively if the $A_{r,r}$ are tri-diagonal matrices.

14.2. SOLUTION OF A LINEAR SYSTEM WITH A TRI-DIAGONAL MATRIX

A basic step in carrying out group iteration methods is the determination of a vector w which satisfies the system

$$Qw = v \tag{2.1}$$

where Q is a given matrix and v is a given vector. We now describe a procedure for doing this in the case where Q is a tri-diagonal matrix. This procedure, which is a variant of the Gauss elimination method, was used by Thomas [1949] in connection with the solution of partial difference equations.

Let us write the system (2.1) in the form

$$\begin{pmatrix} B_1 & C_1 & 0 & 0 & \cdots & 0 & 0 & 0 \\ A_2 & B_2 & C_2 & 0 & \cdots & 0 & 0 & 0 \\ 0 & A_3 & B_3 & C_3 & \cdots & 0 & 0 & 0 \\ \vdots & \vdots & \vdots & \vdots & \cdots & \vdots & \vdots & \vdots \\ 0 & 0 & 0 & 0 & \cdots & A_{M-1} & B_{M-1} & C_{M-1} \\ 0 & 0 & 0 & 0 & \cdots & 0 & A_M & B_M \end{pmatrix} \begin{pmatrix} w_1 \\ w_2 \\ w_3 \\ \vdots \\ w_{M-1} \\ w_M \end{pmatrix} = \begin{pmatrix} D_1 \\ D_2 \\ D_3 \\ \vdots \\ D_{M-1} \\ D_M \end{pmatrix} \tag{2.2}$$

We apply the Gauss elimination method as follows. Divide the first equation by B_1 and then subtract A_2 times the first equation from the second. Next, divide the second equation by B_2', the value of B_2 as modified, and subtract A_3 times the modified second row from the third. Continuing in this way we get an equivalent linear system with an upper triangular matrix which can be solved by back substitution.

It is easily verified that the w_i determined in this way are given by

$$\begin{aligned} b_1 &= \frac{C_1}{B_1}, & b_i &= \frac{C_i}{B_i - A_i b_{i-1}}, & i &= 2, 3, \ldots, M-1 \\ q_1 &= \frac{D_1}{B_1}, & q_i &= \frac{D_i - A_i q_{i-1}}{B_i - A_i b_{i-1}}, & i &= 2, 3, \ldots, M \\ w_M &= q_M, & w_i &= q_i - b_i w_{i+1}, & i &= M-1, M-2, \ldots, 1 \end{aligned} \tag{2.3}$$

We show that the above process can be carried out if the following conditions are satisfied

$$\begin{aligned} &B_i > 0, \qquad i = 1, 2, \ldots, M \\ &B_1 > | C_1 | \\ &B_i \geq | C_i | + | A_i |, \quad C_i \neq 0, \quad A_i \neq 0, \quad i = 2, 3, \ldots, M-1 \\ &B_M \geq | A_M | \end{aligned} \tag{2.4}$$

It is only necessary to show $| B_i - A_i b_{i-1} | > 0$ for $i = 2, 3, \ldots, M$. We first show that $| b_i | < 1$ for $i = 1, 2, \ldots, M-1$. Evidently, $| b_1 | = | C_1/B_1 | < 1$. Assume now that $| b_i | < 1$ for some $i \leq M-2$. To show that $| b_{i+1} | < 1$ we need only show that $| B_{i+1} - A_{i+1} b_i | > | C_{i+1} |$. But $| A_{i+1} b_i | < | A_{i+1} |$ since $| b_i | < 1$ and $A_{i+1} \neq 0$. Therefore, $B_{i+1} - A_{i+1} b_i > 0$. Consequently, $| B_{i+1} - A_{i+1} b_i | = B_{i+1} - A_{i+1} b_i > B_{i+1} - A_{i+1} \geq B_{i+1} - | A_{i+1} |$. But $B_{i+1} - | A_{i+1} | \geq | C_{i+1} |$ and hence $| b_{i+1} | < 1$. Thus, by induction we have $| b_i | < 1$, $i = 1, 2, \ldots, M-1$. Therefore, since $| A_i | \leq B_i$ it follows that $| B_i - A_i b_{i-1} | \geq B_i - | A_i | | b_{i-1} | > | B_i | - | A_i | \geq 0$ for $i = 2, 3, \ldots, M-1$. Also, $B_M - b_{M-1} A_M > 0$ if $A_M = 0$. Otherwise, $| B_M - b_{M-1} A_M | \geq B_M - | b_{M-1} | | A_M | > B_M - | A_M | \geq 0$. Thus $| B_i - b_{i-1} A_i | > 0$ for $i = 1, 2, \ldots, M$.

The work required to solve a linear system with a tri-diagonal matrix is very much less than it would be if Q were a full matrix. On the other hand, approximately four times as much work is required as in the case where Q is diagonal. We now describe a procedure developed by Cuthill and Varga [1959], which, in the case Q is positive definite, enables one to solve (2.1) with essentially the same amount of work as though Q were diagonal. There are some "overhead" operations required but these are usually not significant since one will normally solve many systems with the same matrix Q.

We illustrate the procedure by considering the case

$$M = \begin{pmatrix} b_1 & c_1 & 0 \\ c_1 & b_2 & c_2 \\ 0 & c_2 & b_3 \end{pmatrix}, \qquad v = \begin{pmatrix} v_1 \\ v_2 \\ v_3 \end{pmatrix} \tag{2.5}$$

We seek a diagonal matrix D and an upper bi-diagonal matrix T of the forms

$$D = \begin{pmatrix} d_1 & 0 & 0 \\ 0 & d_2 & 0 \\ 0 & 0 & d_3 \end{pmatrix}, \qquad T = \begin{pmatrix} 1 & e_1 & 0 \\ 0 & 1 & e_2 \\ 0 & 0 & 1 \end{pmatrix} \tag{2.6}$$

such that

$$Q = DT^{\mathrm{T}}TD \tag{2.7}$$

The reader should show by induction that the d_k and e_k are given by

$$\begin{aligned} d_1 &= b_1^{1/2}, \qquad d_k = [b_k - (c_{k-1}/d_{k-1})^2]^{1/2}, \qquad k = 2, 3 \\ e_k &= c_k/(d_k d_{k+1}), \qquad k = 1, 2 \end{aligned} \tag{2.8}$$

He should also show that since M is positive definite, each d_k is real and positive; hence, the process can be carried out in terms of real operations. The proof depends on the use of a three-term recurrence relation involving the determinants

$$\det(b_1), \qquad \det\begin{pmatrix} b_1 & c_1 \\ c_1 & b_2 \end{pmatrix}, \qquad \det\begin{pmatrix} b_1 & c_1 & 0 \\ c_1 & b_2 & c_2 \\ 0 & c_2 & b_3 \end{pmatrix}$$

To solve the system (2.1) we use (2.8) obtaining

$$DT^{\mathrm{T}}TDw = v \tag{2.9}$$

or

$$T^{\mathrm{T}}T(Dw) = D^{-1}v \tag{2.9$'$}$$

If we let

$$z = Dw, \qquad y = Tz \tag{2.10}$$

we obtain

$$T^{\mathrm{T}}y = D^{-1}v = s \tag{2.11}$$

Solving the system

$$T^{\mathrm{T}}y = \begin{pmatrix} 1 & 0 & 0 \\ e_1 & 1 & 0 \\ 0 & e_2 & 1 \end{pmatrix} \begin{pmatrix} y_1 \\ y_2 \\ y_3 \end{pmatrix} = \begin{pmatrix} s_1 \\ s_2 \\ s_3 \end{pmatrix} = D^{-1}v \tag{2.12}$$

we obtain

$$\begin{aligned} y_1 &= s_1 \\ y_2 &= s_2 - e_1 y_1 \\ y_3 &= s_3 - e_2 y_2 \end{aligned} \tag{2.13}$$

Similarly, solving the system

$$Tz = \begin{pmatrix} 1 & e_1 & 0 \\ 0 & 1 & e_2 \\ 0 & 0 & 1 \end{pmatrix} \begin{pmatrix} z_1 \\ z_2 \\ z_3 \end{pmatrix} = \begin{pmatrix} y_1 \\ y_2 \\ y_3 \end{pmatrix} = y \tag{2.14}$$

we obtain

$$\begin{aligned} z_3 &= y_3 \\ z_2 &= y_2 - e_2 z_3 \\ z_1 &= y_1 - e_1 z_2 \end{aligned} \tag{2.15}$$

Finally, we solve $z = Dw$ for w which gives us $w = D^{-1}z$ and therefore

$$\begin{aligned} w_1 &= z_1/d_1 \\ w_2 &= z_2/d_2 \\ w_3 &= z_3/d_3 \end{aligned} \tag{2.16}$$

As an illustration, let us solve

$$\begin{pmatrix} 4 & -1 & 0 \\ -1 & 4 & -1 \\ 0 & -1 & 4 \end{pmatrix} \begin{pmatrix} w_1 \\ w_2 \\ w_3 \end{pmatrix} = \begin{pmatrix} v_1 \\ v_2 \\ v_3 \end{pmatrix} \tag{2.17}$$

by the normalized scheme. We obtain by (2.8)

$$\begin{aligned} &d_1 = 2, \qquad d_2 = \tfrac{1}{2}(15)^{1/2} \doteq 1.93649, \\ &d_3 = [4 - (4/15)]^{1/2} \doteq 1.93218 \\ &e_1 = -15^{-1/2} \doteq -0.2582, \qquad e_2 = -14^{-1/2} \doteq -0.2673 \end{aligned} \tag{2.18}$$

In practice, one usually works with the z_i rather than with the w_i. Only when the process has converged is w computed from the final value of z. In the case of group methods, if A is positive definite, then each $A_{r,r}$ is positive definite and we determine D_r and T_r as described above. Instead of seeking the U_r we seek

$$V_r = D_r U_r \tag{2.19}$$

Once the D_r and T_r have been computed, we can determine the components of V_r for each iteration by $2m$ multiplications where m is the order of D_r. The rates of convergence of the group J, GS, and SOR methods are the same as though we were iterating with the U_r. Moreover, the amount of work per iteration is roughly the same as that for the corresponding point methods. On the other hand, as we will show below, the rapidity of convergence of group methods may be considerably greater than for point methods.

14.3. CONVERGENCE ANALYSIS

In this section we study the convergence properties of the group J, GS, and SOR methods. We seek to derive a relation between the eigenvalues of $\mathscr{L}_\omega^{(\pi)}$ and $B^{(\pi)}$ similar to that obtained in Chapter 5 for $\mathscr{L}$ and B. To do this we shall generalize the concepts of Property A, consistently ordered matrices, ordering vectors, etc.

Given a matrix A and an ordered grouping π, with q groups, we define the $q \times q$ matrix $Z = (z_{r,s})$ by

$$z_{r,s} = \begin{cases} 0 & \text{if} \quad A_{r,s} = 0 \\ 1 & \text{if} \quad A_{r,s} \neq 0 \end{cases} \tag{3.1}$$

Here the matrices $A_{r,s}$ are as defined in Section 14.1.

Definition 3.1. The matrix A has Property $\mathrm{A}^{(\pi)}$ if Z has Property A.

Definition 3.2. The matrix A is a *π-consistently ordered matrix* (a π-CO-matrix) if Z is consistently ordered.

If $\pi = \pi_0$, then A is a consistently ordered matrix (a CO-matrix).

It is evident that given a matrix with Property $\mathrm{A}^{(\pi)}$ one can reorder the groups of π to obtain an ordered grouping π' such that A is π'-consistently ordered.

Let us now consider the matrix†

$$\hat{A}^{(\pi)} = P^{-1}AP = \begin{pmatrix} A_{1,1} & A_{1,2} & \cdots & A_{1,q} \\ A_{2,1} & A_{2,2} & \cdots & A_{2,q} \\ \vdots & \vdots & \cdots & \vdots \\ A_{q,1} & A_{q,2} & \cdots & A_{q,q} \end{pmatrix} \tag{3.2}$$

where $P = P^{(\pi)}$ is the permutation matrix associated with the permutation σ where

(a) $\sigma(i) < \sigma(j)$ or $\sigma(i) > \sigma(j)$ if the group containing i comes before or after, respectively, the group containing j,

(b) if i and j belong to the same group and $i \neq j$, then $\sigma(i) < \sigma(j)$ or $\sigma(i) > \sigma(j)$ if $i < j$ or $i > j$, respectively.

† In this chapter we shall use the notation $\hat{H}$ as indicated. Normally we let $\hat{H} = D^{-1/2}HD^{-1/2}$ where D is a certain diagonal matrix (see Chapter 4 and other previous chapters).

Thus, for example, if $N = 5$ and π is defined by $R_1 = \{1, 4\}$, $R_2 = \{2, 5\}$, $R_3 = \{3\}$, we let

$$\sigma(1) = 1, \qquad \sigma(4) = 2, \qquad \sigma(2) = 3, \qquad \sigma(5) = 4, \qquad \sigma(3) = 5$$

The same permutation would correspond to the ordered grouping defined by $R_1 = \{4, 1\}$, $R_2 = \{5, 2\}$, $R_3 = \{3\}$.

The proof of the following theorem is straightforward and is left as an exercise.

Theorem 3.1. If A has Property $\mathrm{A}^{(\pi)}$, then $C^{(\pi)}$ has Property A, and if $D^{(\pi)}$ is nonsingular, $B^{(\pi)}$ has Property A. If A is a π-CO-matrix, then $\hat{C}^{(\pi)}$ is a CO-matrix, and if $D^{(\pi)}$ is nonsingular, $\hat{B}^{(\pi)}$ is a CO-matrix.

As an example let us consider the matrix

$$A = \begin{pmatrix} 4 & -1 & 0 & -1 & 0 & 0 & 0 & 0 & 0 \\ -1 & 4 & -1 & 0 & -1 & 0 & 0 & 0 & 0 \\ 0 & -1 & 4 & 0 & 0 & -1 & 0 & 0 & 0 \\ -1 & 0 & 0 & 4 & -1 & 0 & -1 & 0 & 0 \\ 0 & -1 & 0 & -1 & 4 & -1 & 0 & -1 & 0 \\ 0 & 0 & -1 & 0 & -1 & 4 & 0 & 0 & -1 \\ 0 & 0 & 0 & -1 & 0 & 0 & 4 & -1 & 0 \\ 0 & 0 & 0 & 0 & -1 & 0 & -1 & 4 & -1 \\ 0 & 0 & 0 & 0 & 0 & -1 & 0 & -1 & 4 \end{pmatrix} \tag{3.3}$$

which corresponds to the model problem for the case $h = \frac{1}{4}$ with the points labeled as indicated in Figure 3.1. With the ordered grouping π defined by $R_1 = \{1, 2, 4, 5\}$, $R_2 = \{3, 6\}$, $R_3 = \{7, 8\}$, $R_4 = \{9\}$, we have $\sigma(i) = i$ for $i = 1, 2, 6, 7, 8, 9$, $\sigma(4) = 3$, $\sigma(5) = 4$, and $\sigma(3) = 5$. Therefore,

$$\hat{A}^{(\pi)} = \left(\begin{array}{cccc|cc|cc|c} 4 & -1 & -1 & 0 & 0 & 0 & 0 & 0 & 0 \\ -1 & 4 & 0 & -1 & -1 & 0 & 0 & 0 & 0 \\ -1 & 0 & 4 & -1 & 0 & 0 & -1 & 0 & 0 \\ 0 & -1 & -1 & 4 & 0 & -1 & 0 & -1 & 0 \\ \hline 0 & -1 & 0 & 0 & 4 & -1 & 0 & 0 & 0 \\ 0 & 0 & 0 & -1 & -1 & 4 & 0 & 0 & -1 \\ \hline 0 & 0 & -1 & 0 & 0 & 0 & 4 & -1 & 0 \\ 0 & 0 & 0 & -1 & 0 & 0 & -1 & 4 & -1 \\ \hline 0 & 0 & 0 & 0 & 0 & -1 & 0 & -1 & 4 \end{array}\right) \tag{3.4}$$

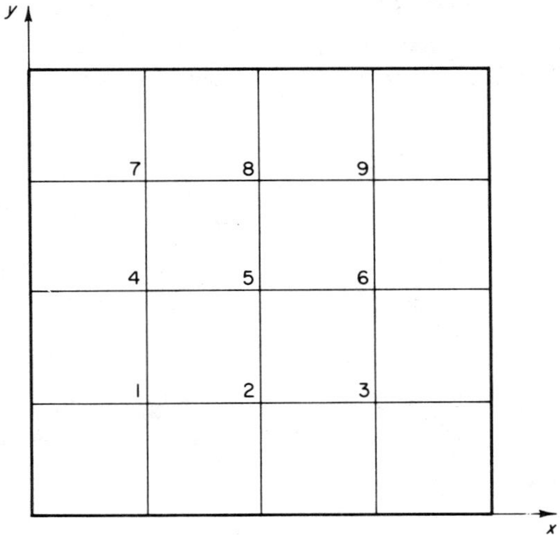

Figure 3.1.

The matrix Z is given by

$$Z = \begin{pmatrix} 1 & 1 & 1 & 0 \\ 1 & 1 & 0 & 1 \\ 1 & 0 & 1 & 1 \\ 0 & 1 & 1 & 1 \end{pmatrix} \tag{3.5}$$

Evidently the vector $\gamma = (1, 2, 2, 3)^{\mathrm{T}}$ is a compatible ordering vector for Z and hence A has Property $A^{(\pi)}$ and is a π-CO-matrix. The reader should verify that $\hat{C}^{(\pi)}$ is a CO-matrix with the compatible ordering vector $\gamma' = (1, 1, 1, 1, 2, 2, 2, 2, 3)^{\mathrm{T}}$. He should also verify that if π' is defined by $R_1 = \{1, 2, 4, 5\}$, $R_2 = \{3, 6\}$, $R_3 = \{9\}$, $R_4 = \{7, 8\}$, then the corresponding matrix

$$Z' = \begin{pmatrix} 1 & 1 & 0 & 1 \\ 1 & 1 & 1 & 0 \\ 0 & 1 & 1 & 1 \\ 1 & 0 & 1 & 1 \end{pmatrix} \tag{3.6}$$

has Property A but is not a CO-matrix. Thus, by Definition 3.1, A has Property $\mathrm{A}^{(\pi')}$. However, A is not a π'-CO-matrix.

Theorem 3.1 suggests two generalizations of Property $\mathrm{A}^{(\pi)}$ and also of π-CO-matrices. One generalization involves $C^{(\pi)}$ and $\hat{C}^{(\pi)}$, respectively, and the other involves $B^{(\pi)}$ and $\hat{B}^{(\pi)}$, respectively. We shall see that neither generalization includes the other. That the generalizations involving $C^{(\pi)}$ and $\hat{C}^{(\pi)}$ are not trivial can be seen from the following example involving

$$A = \left(\begin{array}{cc|cc|cc} 1 & 0 & 1 & 0 & 0 & 0 \\ 0 & 1 & 0 & 0 & 1 & 0 \\ \hline 1 & 0 & 1 & 0 & 0 & 0 \\ 0 & 0 & 0 & 1 & 0 & 1 \\ \hline 0 & 1 & 0 & 0 & 1 & 0 \\ 0 & 0 & 0 & 1 & 0 & 1 \end{array}\right) \tag{3.7}$$

with the ordered grouping π defined by $R_1 = \{1, 2\}$, $R_2 = \{3, 4\}$, $R_3 = \{5, 6\}$. Evidently A is not a π-CO-matrix nor does it have Property $\mathrm{A}^{(\pi)}$ since

$$Z = \begin{pmatrix} 1 & 1 & 1 \\ 1 & 1 & 1 \\ 1 & 1 & 1 \end{pmatrix} \tag{3.8}$$

does not have Property A. On the other hand, $C^{(\pi)} = \hat{C}^{(\pi)}$ is consistently ordered with the ordering vector $\gamma = (1, 1, 2, 1, 2, 2)^{\mathrm{T}}$.

A still further generalization of the concept of a π-consistently ordered matrix is suggested by the following theorem:

Theorem 3.2. Let A be a matrix and π an ordered grouping such that $\hat{C}^{(\pi)}$ is consistently ordered. Then

$$\Delta = \det(\alpha C_L^{(\pi)} + \alpha^{-1} C_U^{(\pi)} - k D^{(\pi)}) \tag{3.9}$$

is independent of α for all $\alpha \neq 0$ and for all k.

Proof. Evidently

$$\Delta = \det(\alpha \hat{C}_L^{(\pi)} + \alpha^{-1} \hat{C}_U^{(\pi)} - k \hat{D}^{(\pi)}) \tag{3.10}$$

As in the proof of Theorem 5-3.3 the general term of Δ is

$$t(\sigma) = \pm \prod_{i=1}^{N} \hat{a}_{i,\sigma(i)} \alpha^{n_L - n_U} k^{N - n_L - n_U} \tag{3.11}$$

where $\hat{A} = (\hat{a}_{i,j})$ and n_L and n_U denote, respectively, the number of factors $a_{i,\sigma(i)}$ such that i and j belong to different groups and the group containing i comes after and before, respectively, the group containing

$\sigma(i)$. Let γ be a compatible ordering vector for $\hat{C}^{(\pi)}$. Then

$$n_L = \sum_{i=1}^{N}{}' (\gamma_i - \gamma_{\sigma(i)}), \qquad n_U = \sum_{i=1}^{N}{}'' (\gamma_{\sigma(i)} - \gamma_i) \tag{3.12}$$

where the first and second sums are taken for all i such that the group containing i comes after or before, respectively, the group containing $\sigma(i)$. If we let T denote the set of all i such that i and $\sigma(i)$ belong to the same group, then we have

$$n_L - n_U = \sum_{\substack{i=1 \\ i \notin T}}^{N} (\gamma_i - \gamma_{\sigma(i)}) = \sum_{\substack{i=1 \\ i \notin T}}^{N} \gamma_i - \sum_{\substack{i=1 \\ i \notin T}}^{N} \gamma_{\sigma(i)} \tag{3.13}$$

But if $i \in T$, then $\sigma(i) \in T$ since $\sigma(\sigma(i)) = i$. Hence σ defines a permutation on the set of integers not in T and $n_L = n_U$. Thus $t(\sigma)$ is independent of α and the theorem now follows.

We now define

Definition 3.3. A matrix A is a *generalized π-consistently ordered matrix* (a π-GCO-matrix) if Δ given by (3.9) is independent of α for all $\alpha \neq 0$ and for all k.

From Theorems 3.1 and 3.2, we have

Theorem 3.3. If A is a π-CO-matrix, then A is a π-GCO-matrix. More generally, if $\hat{C}^{(\pi)}$ is a CO-matrix or if $D^{(\pi)}$ is nonsingular and $\hat{B}^{(\pi)}$ is a CO-matrix, then A is a π-GCO-matrix.

Definition 3.3 is closely related to a definition introduced by Varga [1962] but is slightly more general. In our notation, Varga's definition requires not only that Δ be independent of α for all $\alpha \neq 0$ and for all k, but also that $D^{(\pi)}$ be nonsingular and that $B^{(\pi)}$ have Property A.

If A has Property A, then, for any ordered grouping π, $C^{(\pi)}$ has Property A, even though A may not have Property $A^{(\pi)}$. Thus, a matrix may be a π-GCO-matrix in the sense of Definition 3.3 even though it is not a π-CO-matrix. This situation holds for the case

$$A = \left(\begin{array}{cc|cc|cc} 2 & -1 & 0 & -1 & -1 & -1 \\ -1 & 2 & 0 & -1 & -1 & -1 \\ \hline 0 & 0 & 2 & -1 & -1 & -1 \\ -1 & -1 & -1 & 2 & 0 & 0 \\ \hline -1 & -1 & -1 & 0 & 2 & -1 \\ -1 & -1 & -1 & 0 & -1 & 2 \end{array}\right) \tag{3.14}$$

with π determined by $R_1 = \{1, 2\}$, $R_2 = \{3, 4\}$, $R_3 = \{5, 6\}$. Evidently $\hat{C}^{(\pi)} = C^{(\pi)}$ is a CO-matrix with a compatible ordering vector $\gamma = (1, 1, 1, 2, 2, 2)^T$ and hence A is a π-GCO-matrix. On the other hand, A does not have Property $A^{(\pi)}$. Moreover, $\hat{B}^{(\pi)} = B^{(\pi)}$ does not have Property A since

$$\hat{B}^{(\pi)} = B^{(\pi)} = \tfrac{1}{3}\begin{pmatrix} 0 & 0 & 0 & 3 & 3 & 3 \\ 0 & 0 & 0 & 3 & 3 & 3 \\ 1 & 1 & 0 & 0 & 2 & 2 \\ 2 & 2 & 0 & 0 & 1 & 1 \\ 3 & 3 & 3 & 0 & 0 & 0 \\ 3 & 3 & 3 & 0 & 0 & 0 \end{pmatrix} \tag{3.15}$$

Thus A is not a π-GCO-matrix in the sense of Varga.

On the other hand, as shown by Varga [1962, Section 4.1], even though $\hat{B}^{(\pi)}$ is a CO-matrix, $\hat{C}^{(\pi)}$ need not be a CO-matrix. For example, if

$$A = \left(\begin{array}{cc|cc|cc} 2 & -1 & 0 & -1 & -1 & -1 \\ -1 & 2 & 0 & -1 & -1 & -1 \\ \hline -1 & 1 & 2 & -1 & 2 & -2 \\ 2 & -2 & -1 & 2 & -1 & 1 \\ \hline -1 & -1 & -1 & 0 & 2 & -1 \\ -1 & -1 & -1 & 0 & -1 & 2 \end{array}\right) \tag{3.16}$$

and if π is defined by $R_1 = \{1, 2\}$, $R_2 = \{3, 4\}$, $R_3 = \{5, 6\}$, then A does not have Property $A^{(\pi)}$ and is not a π-CO-matrix. Moreover $\hat{C}^{(\pi)} = C^{(\pi)}$ is not a CO-matrix. On the other hand,

$$B^{(\pi)} = \hat{B}^{(\pi)} = \left(\begin{array}{cc|cc|cc} 0 & 0 & 0 & 1 & 1 & 1 \\ 0 & 0 & 0 & 1 & 1 & 1 \\ \hline 0 & 0 & 0 & 0 & -1 & 1 \\ -1 & 1 & 0 & 0 & 0 & 0 \\ \hline 1 & 1 & 1 & 0 & 0 & 0 \\ 1 & 1 & 1 & 0 & 0 & 0 \end{array}\right) \tag{3.17}$$

and hence $\hat{B}^{(\pi)}$ is a CO-matrix with compatible ordering vector $\gamma = (1, 1, 1, 2, 2, 2)^T$, and A is a π-GCO-matrix. It should be noted, however, that if A is an M-matrix, then $\hat{B}^{(\pi)}$ is not a CO-matrix unless $\hat{C}^{(\pi)}$ is a CO-matrix. For, $D^{(\pi)}$ is an M-matrix by Theorem 4-5.8, and the diagonal elements of $D^{(\pi)}$ can easily be shown to be positive. Since

$(D^{(\pi)})^{-1} \geq 0$ it follows that every nonzero element of $\hat{C}^{(\pi)}$ corresponds to a nonzero element of $\hat{B}^{(\pi)}$.

The main result of this section is the following, which is a slight variant of a theorem of Arms *et al.* [1956]:

Theorem 3.4. If A is a π-GCO-matrix such that $D^{(\pi)}$ is nonsingular, then the conclusions of Theorem 5-3.4 are valid if we replace B by $B^{(\pi)}$ and $\mathscr{L}_\omega$ by $\mathscr{L}_\omega^{(\pi)}$.

The proof is similar to that of Theorem 5-3.4 and will be left as an exercise.

As noted above, given that A is a π-GCO-matrix we cannot conclude that A is a π-CO-matrix. We can, however, prove the following:

Theorem 3.5. If A is a π-GCO-matrix which is an irreducible M-matrix, then $\hat{B}^{(\pi)}$ and $\hat{C}^{(\pi)}$ are CO-matrices.

Proof. If A is a π-GCO-matrix, then Δ as given by (3.9) is independent of α for all $\alpha \neq 0$ and for all k. But

$$\Delta = \det \hat{D}^{(\pi)} \det[\alpha(\hat{D}^{(\pi)})^{-1}\hat{C}_L^{(\pi)} + \alpha^{-1}(\hat{D}^{(\pi)})^{-1}\hat{C}_U^{(\pi)} - kI] \quad (3.18)$$

Moreover, by Theorem 7-5.8, $D^{(\pi)}$ and $\hat{D}^{(\pi)}$ are M-matrices and, hence, $(\hat{D}^{(\pi)})^{-1} \geq 0$ and every diagonal element of $(\hat{D}^{(\pi)})^{-1}$ is positive. Thus, since $\hat{C}^{(\pi)} \geq 0$ it follows that

$$A' = I - (\hat{D}^{(\pi)})^{-1}\hat{C}_L^{(\pi)} - (\hat{D}^{(\pi)})^{-1}C_U^{(\pi)} \quad (3.19)$$

is an irreducible L-matrix. Therefore, by Theorem 13-5.5, A' is a CO-matrix. Hence $\hat{B}^{(\pi)}$ and $\hat{C}^{(\pi)}$ are CO-matrices and the theorem follows.

In spite of the fact that not every π-GCO-matrix is a π-CO-matrix, nevertheless the most important applications of group iterative methods involve π-CO-matrices. If A is positive definite, then we can obtain results analogous to those of Section 6.2. Thus since each $A_{r,r}$ is positive definite, it follows that $B^{(\pi)}$ is similar to the symmetric matrix

$$\tilde{B}^{(\pi)} = (D^{(\pi)})^{1/2}B^{(\pi)}(D^{(\pi)})^{-1/2} \quad (3.20)$$

Moreover, since $B^{(\pi)}$ has Property A, because

$$A = D^{(\pi)} - D^{(\pi)}B^{(\pi)} \quad (3.21)$$

and since

$$(D^{(\pi)})^{-1/2}A(D^{(\pi)})^{-1/2} = I - \tilde{B}^{(\pi)} \tag{3.22}$$

is positive definite, it follows that

$$\bar{\mu}^{(\pi)} = S(B^{(\pi)}) < 1 \tag{3.23}$$

Hence given $\bar{\mu}^{(\pi)}$ we can determine the optimum relaxation factor $\omega_b^{(\pi)}$ by

$$\omega_b^{(\pi)} = \frac{2}{1 + [1 - (\bar{\mu}^{(\pi)})^2]^{1/2}} \tag{3.24}$$

and we have

$$S(\mathscr{L}^{(\pi)}_{\omega_b^{(\pi)}}) = \omega_b^{(\pi)} - 1 \tag{3.25}$$

14.4. APPLICATIONS

The results of Section 14.3 can be applied directly to systems of linear equations arising from the five-point difference equations considered in Section 2.8. A frequently used choice of π is the grouping of lines of points. It is easy to show that with such a grouping of points, according to lines, A is a π-CO-matrix for all orderings of the lines.

We now seek to determine the eigenvalues of the line J method for the model problem. The method is similar to that used in Section 4.6. We assume that the lines are ordered with increasing y. By (4-6.4) the iterative method is defined by

$$\begin{aligned} u^{(n+1)}(x, y) = {} & \tfrac{1}{4}u^{(n+1)}(x + h, y) + \tfrac{1}{4}u^{(n+1)}(x - h, y) \\ & + \tfrac{1}{4}u^{(n)}(x, y + h) + \tfrac{1}{4}u^{(n)}(x, y - h) \end{aligned} \tag{4.1}$$

if $(x, y) \in R_h$. Evidently (4-6.7) becomes

$$\begin{aligned} \varepsilon^{(n+1)}(x, y) = {} & \tfrac{1}{4}\varepsilon^{(n+1)}(x + h, y) + \tfrac{1}{4}\varepsilon^{(n+1)}(x - h, y) \\ & + \tfrac{1}{4}\varepsilon^{(n)}(x, y + h) + \tfrac{1}{4}\varepsilon^{(n)}(x, y - h) \end{aligned} \tag{4.2}$$

To determine an eigenvalue μ and an eigenvector v of $B^{(\pi)}$ we seek a number μ and a function $v(x, y)$ which vanishes on S_h such that

$$\mu v(x, y) = \tfrac{1}{4}[\mu v(x+h, y) + \mu v(x - h, y) + v(x, y+h) + v(x, h - h)] \tag{4.3}$$

The above equation is obtained from (4.2) by replacing $\varepsilon^{(n+1)}(x, y)$ by

$\mu v(x, y)$ and $\varepsilon^{(n)}(x, y)$ by $v(x, y)$. Following the methods of Section 4.6, we can show that the eigenvalues are

$$\mu_{p,q} = \cos q\pi h/(2 - \cos p\pi h), \qquad p, q = 1, 2, \ldots, h^{-1} - 1 \tag{4.4}$$

and the corresponding eigenfunctions are

$$v_{p,q}(x, y) = \sin p\pi x \sin q\pi y \tag{4.5}$$

Therefore, we have

$$\bar{\mu}^{(\pi)} = S(B^{(\pi)}) = \cos \pi h/(2 - \cos \pi h) \sim 1 - \pi^2 h^2 \tag{4.6}$$

as compared with $S(B) = \cos \pi h \sim 1 - \frac{1}{2}\pi^2 h^2$ for the point J method. Thus the convergence rate of the line J method is approximately twice that of the J method for small h. Similarly, the line GS method converges approximately twice as fast as the line J method. By (3.25) and Theorem 6-3.1, the line SOR method, using the relaxation factor

$$\omega_b^{(\pi)} = \frac{2}{1 + [1 - S(B^{(\pi)})^2]^{1/2}} \tag{4.7}$$

converges approximately $2^{1/2}$ times as fast as the ordinary SOR method.

While the saving of a factor of $2^{1/2}$ is a relatively modest improvement as compared with the gain achieved in replacing the GS method by the SOR method, nevertheless, it is worthwhile if one has a great many problems to solve. It is, of course, necessary to use a scheme such as that described in Section 14.2 so that the amount of work per iteration will not be appreciably greater than for point iteration.

With two-line iterative methods one uses the ordered grouping involving π_2 where R_1 corresponds to the points of the first two rows of mesh points, R_2 corresponds to the third and fourth rows, etc. One can show that the matrix A is a π_2-CO-matrix; hence the same relation holds between the eigenvalues of $\mathcal{L}_\omega^{(\pi_2)}$ and $B^{(\pi_2)}$ that holds between $\mathcal{L}_\omega$ and B. Moreover, if A is positive definite, then $\bar{\mu}^{(\pi_2)} = S(B^{(\pi_2)}) < 1$, and we can determine the optimum ω by (4.7). Varga [1960] has shown that the rate of convergence of the two-line SOR method with $\omega_b^{(\pi_2)}$ is approximately twice that of $\mathcal{L}_{\omega_b}$. Parter [1961] showed that the k-line SOR method with optimum ω converges approximately $(2k)^{1/2}$ as fast as $\mathcal{L}_{\omega_b}$. Of course, one must solve more involved systems of equations at each iteration. However, Varga [1960] showed how by the use of a factorization technique, similar to that described in Section 14.2, if A

is positive definite, one can solve the five-diagonal system effectively, and can carry out the two-line SOR method with essentially the same amount of work per point as for point SOR.

One of the advantages in using group methods is that frequently A is a π-CO-matrix for some π even though A is not a CO-matrix. Thus, for example, for a nine-point difference equation involving the points $(x_0 + ph, y_0 + qh)$, $p, q = -1, 0, 1$ we do not have a consistently ordered matrix. However, with a grouping by lines (in any order) we do have a π-consistently ordered matrix. Similarly, with difference equations involving more mesh points one can frequently use two-line methods so that A is π-consistently ordered. Thus, for example, consider the usual 13-point difference equation corresponding to the bi-harmonic equation

$$\frac{\partial^4 u}{\partial x^4} + 2\frac{\partial^4 u}{\partial x^2\,\partial y^2} + \frac{\partial^4 u}{\partial y^4} = 0$$

which involves the points $(x_0 + ph, y_0 + qh)$ where: $p = 0$, $q = 0, \pm 1, \pm 2$; $q = 0$, $p = \pm 1, \pm 2$; $p = \pm 1$, $q = \pm 1$. The difference equation involves points on five lines. Nevertheless, if π corresponds to the two-line grouping, the resulting matrix is π-consistently ordered (see Heller, 1960 and Varga, 1960).

14.5. COMPARISON OF POINT AND GROUP ITERATIVE METHODS

One might expect that group iterative methods are always better than point methods. However, the following example given by Arms and Gates [1956] shows that this is not the case. Consider the matrix

$$A = \begin{pmatrix} 5 & 2 & 2 \\ 2 & 5 & 4 \\ 2 & 4 & 5 \end{pmatrix} \tag{5.1}$$

with the ordered grouping π defined by $R_1 = \{1, 2\}$, $R_2 = \{3\}$. One can easily verify that for the Gauss–Seidel method we have

$$S(\mathscr{L}) = [104 + (2816)^{1/2}]/250 \doteq 0.628$$

On the other hand, for the group GS method we obtain

$$S(\mathscr{L}^{(\pi)}) = 68/105 \doteq 0.648 > S(\mathscr{L})$$

Thus, in this case the group GS method is slower than the point GS method.

Let us now assume that A is an M-matrix. Evidently, for any ordered grouping π we have

$$C^{(\pi)}[A] \leq C[A] \tag{5.2}$$

and hence by Corollary 4-5.6, the group J method converges at least as rapidly as the J method.

Evidently the group GS method corresponds to the splitting

$$A = (D^{(\pi)} - C_L^{(\pi)}) - C_U^{(\pi)} \tag{5.3}$$

while the GS method corresponds to the splitting

$$A = (D - C_L) - C_U \tag{5.4}$$

Since $C_U^{(\pi)} \leq C_U$, it follows from Corollary 4-5.6 that $S(\mathscr{L}^{(\pi)}) \leq S(\mathscr{L})$.

Suppose now that A is a Stieltjes matrix and a π-GCO-matrix. As we have shown earlier $\bar{\mu}^{(\pi)} = S(B^{(\pi)}) < 1$ and $\omega_b^{(\pi)}$ and $S(\mathscr{L}_{\omega_b^{(\pi)}}^{(\pi)})$ are given by (3.24) and (3.25), respectively. Moreover, since $S(B^{(\pi)}) \leq S(B)$ we have

$$\omega_b^{(\pi)} \leq \omega_b \tag{5.5}$$

Also, by Theorem 12-2.2 we have

$$\min_{\omega} S(\mathscr{L}_\omega) \geq \omega_b - 1 \geq S(\mathscr{L}_{\omega_b^{(\pi)}}^{(\pi)}) \tag{5.6}$$

Thus the group SOR method with $\omega = \omega_b^{(\pi)}$ converges at least as fast as the point SOR method with any ω.

We note that most of the analysis of Chapters 7, 8, 10, and 11 can be applied to the case where A has the form

$$A = \begin{pmatrix} D_1 & H \\ K & D_2 \end{pmatrix} \tag{5.7}$$

where D_1 and D_2 are square nonsingular matrices. We let

$$B^{(\pi)} = \begin{pmatrix} 0 & -D_1^{-1}H \\ -D_2^{-1}K & 0 \end{pmatrix} = \begin{pmatrix} 0 & F \\ G & 0 \end{pmatrix} \tag{5.8}$$

which, if A is positive definite, is similar to the symmetric matrix

$$\tilde{B}^{(\pi)} = D^{1/2}B^{(\pi)}D^{-1/2} = \begin{pmatrix} 0 & \tilde{F} \\ \tilde{G} & 0 \end{pmatrix} \tag{5.9}$$

and

$$D^{(\pi)} = \begin{pmatrix} D_1 & 0 \\ 0 & D_2 \end{pmatrix}, \qquad \tilde{F} = D_1^{1/2} F D_2^{-1/2}, \qquad \tilde{G} = D_2^{1/2} G D_1^{-1/2}, \qquad \tilde{F}^{\mathrm{T}} = \tilde{G}$$

An analysis of the various norms can be carried out as in Chapter 7 with the $D^{1/2}$-norm corresponding to the matrix D which is no longer diagonal. For the $A^{1/2}$-norm we can show, as before, that we can without loss of generality assume that $D = I$. The analysis of the variants of the MSOR method given in Chapter 10 and much of the analysis of semi-iterative methods which relates to a matrix of the form (7-1) is applicable (e.g., Sections 11.6, 11.8, and 11.9). Other results (e.g., Section 11.3) are applicable if D is a block diagonal matrix.

We also note that much of the analysis of Chapter 13 is applicable to group methods. Thus we can define π — CO(q, r)-matrices, Property $\mathrm{A}_{q,r}^{(\pi)}$, π-(q, r)-ordering vectors, etc., as in Section 13.2, based on the matrix Z defined by (3.1). We can also define π-GCO(q, r)-matrices as in Section 13.3, and we can prove an analog of Theorem 3.5 to show that for every π-GCO(q, r)-matrix which is an irreducible M-matrix $\hat{B}^{(\pi)}$ and $\hat{C}^{(\pi)}$ are CO(q, r)-matrices.

SUPPLEMENTARY DISCUSSION

The idea of group iteration goes back to Southwell and his associates (see, for instance, Southwell, 1946). The term "group iteration" was used by Geiringer [1949]. Arms *et al.* [1956] and Friedman [1957] analyzed the convergence rates of block SOR methods. Subsequent work was done by Varga [1960], Parter [1959, 1961, 1965], Keller [1960], and others. The introduction of the procedure of Cuthill and Varga [1959] for efficiently carrying out the process numerically made the use of block methods practical.

Section 14.3. The definition of π-GCO-matrices is based on that given by Verner and Bernal [1968].

Section 14.4. The formula for $\bar{\mu}$ for the line Jacobi method for the model problem is due to Arms *et al.* [1956].

Section 14.5. The comparison between point and group methods is based on the work of Varga [1960] on splittings.

EXERCISES

Section 14.1

1. Show that each of the methods given in Section 14.1 is completely consistent for $\omega \neq 0$ if $D^{(\pi)}$ is nonsingular.

2. Carry out one iteration of the group J method and the group SOR method with $\omega = 1.5$ for the linear system

$$\begin{pmatrix} 4 & -1 & -1 \\ -1 & 4 & -1 \\ -1 & -1 & 4 \end{pmatrix} \begin{pmatrix} u_1 \\ u_2 \\ u_3 \end{pmatrix} = \begin{pmatrix} 1 \\ 2 \\ 3 \end{pmatrix}$$

with

$$u^{(0)} = \begin{pmatrix} 1 \\ 0 \\ 1 \end{pmatrix}$$

and the ordered grouping defined by $R_1 = \{1, 3\}$, $R_2 = \{2\}$. Determine $D^{(\pi)}$, $C^{(\pi)}$, $B^{(\pi)}$, $\mathscr{L}^{(\pi)}$, $\mathscr{L}_\omega^{(\pi)}$.

3. For the model problem for Laplace's equation and $h = \frac{1}{4}$ carry out two iterations of the group SOR method using groupings by rows with $\omega = 1.5$. Assume boundary unity for $y = 0$ and zero elsewhere. Use zero starting values, and the natural ordering.

Section 14.2

1. Derive (2.3) in the following manner: Assume that

$$w_i = \alpha_i w_{i+1} + \beta_i, \qquad i = 0, 1, \ldots, M - 1$$

Then show that the α_i and β_i satisfy a certain recursion relation. (Note that $\alpha_1 = -C_1/B_1$, $\beta_1 = D_1/B_1$.)

2. Consider the model problem with Laplace's equation and $h = \frac{1}{3}$. With the points labeled as indicated in the accompanying figure and the partition π determined by

$$R_1 = \{1, 3\}, \qquad R_2 = \{2, 4\}$$

carry out one iteration of the group SOR method with $\omega = 1.5$.

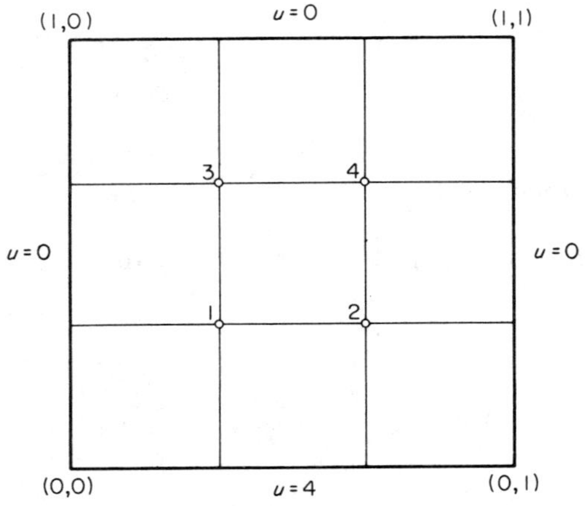

3. Prove (2.8) and show that each d_k is real and positive.

4. Give an example to show that the process (2.3) may fail if we omit the requirement that $A_i \neq 0$ in (2.4).

5. Prove that the process (2.3) can be carried out even with the A_i and C_i are complex provided (2.4) holds.

6. Can the process (2.3) always be carried out if we replace the conditions

$$B_1 > |C_1| \qquad \text{and} \qquad B_M \geq |A_M|$$

by

$$B_1 \geq |C_1| \qquad \text{and} \qquad B_M > |A_M|$$

respectively?

7. Apply the algorithm (2.3) for the system

$$\begin{aligned} 2u_1 - u_2 \qquad\quad &= 1 \\ -u_1 + 2u_2 - \quad u_3 &= 2 \\ - \quad u_2 + 2u_3 &= 3 \end{aligned}$$

8. Apply the Cuthill–Varga algorithm to the preceding example.

Section 14.3

1. Compute $D^{(\pi)}$, $C_L^{(\pi)}$, $C_U^{(\pi)}$, $C^{(\pi)}$, $B^{(\pi)}$, $L^{(\pi)}$, and $U^{(\pi)}$ as well as $\hat{A}^{(\pi)}$, $\hat{D}^{(\pi)}$, etc., for the matrix A of (1.23) with π defined by

$$R_1 = \{1, 4\}, \qquad R_2 = \{2, 3\}, \qquad R_3 = \{5\}$$

Determine whether or not A has Property $A^{(\pi)}$ and if so whether it is a π-CO-matrix. Also determine whether $C^{(\pi)}$ and/or $B^{(\pi)}$ have Property A and whether or not $\hat{C}^{(\pi)}$ and/or $\hat{B}^{(\pi)}$ are CO-matrices.

2. How many different ordered groupings are there for the case $N = 4$? If A is a matrix of order 4 with no vanishing elements, for how many ordered groupings does A have Property $\mathrm{A}^{(\pi)}$, and for how many is A a π-CO-matrix?

3. Give an elementary proof of the following theorem: If A is a π-CO-matrix such that $D^{(\pi)}$ is nonsingular, then the conclusion of Theorem 3.2 holds.

4. Show that if A is given by (3.16) with π as indicated, then $C^{(\pi)}$ is not a CO-matrix.

5. Show that A has Property $\mathrm{A}^{(\pi)}$ if and only if there exists an ordered grouping π', obtained from π by a reordering of the groups of π, such that A is as π'-CO-matrix.

6. Consider the discrete analog of the Dirichlet problem for the region shown in the accompanying figure with $h = 1$. With the indicated labeling of the points and the ordered grouping π determined by the sets

$$R_1 = \{1, 4\}, \qquad R_2 = \{2, 5\}, \qquad R_3 = \{3\}, \qquad R_4 = \{6\}, \qquad R_5 = \{7\}$$

find the matrices A and $\hat{A}^{(\pi)}$. Does the matrix A have Property $\mathrm{A}^{(\pi)}$?

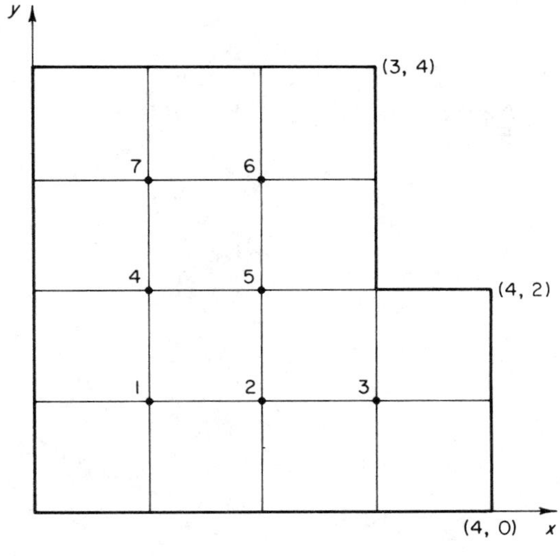

Is it a π-CO-matrix? If not, construct a new ordered grouping π' by reordering the R_i so that A is a π'-CO-matrix. Also, exhibit a compatible π'-ordering vector. Compute $B^{(\pi')}$ and show that it is a CO-matrix.

7. The matrix given by

$$A = \begin{pmatrix} 4 & -1 & -1 & 0 \\ -1 & 4 & 0 & -1 \\ -1 & 0 & 4 & -1 \\ 0 & -1 & -1 & 4 \end{pmatrix}$$

has Property A. Does there exist an ordered grouping π such that A does not have Property $\mathrm{A}^{(\pi)}$?

Section 14.4

1. Compute the eigenvalues of the line J method for the nine-point difference approximation involving the points $(x_0 + ph,\ y_0 + qh)$, p, $q = -1, 0, 1$ to Laplace's equation in the unit square. How does the convergence rate for the nine-point line J method compare to the convergence rate for the five-point line J method?

2. Verify that for the nine-point difference approximation involving the points $(x_0 + ph, y_0 + qh)$, p, $q = -1, 0, 1$ to Laplace's equation in the unit square with $h = \frac{1}{4}$ that the matrix A is not consistently ordered. Show that grouping by lines, the matrix $A^{(\pi)}$ is π-consistently ordered.

Section 14.5

1. For the matrix (5.1) with the ordered grouping π defined by $R_1 = \{1, 2\}$, $R_2 = \{3\}$ verify that

$$S(\mathscr{L}^{(\pi)}) > S(\mathscr{L})$$

Chapter 15 / SYMMETRIC SOR METHOD AND RELATED METHODS

15.1. INTRODUCTION

The *symmetric SOR method* (SSOR method) can be considered as two half iterations. The first half iteration is the same as the SOR method, while the second half iteration is the SOR method with the equations taken in reverse order. Thus we determine $u^{(n+1/2)}$ from $u^{(n)}$ by the (forward) SOR method

$$u^{(n+1/2)} = \mathscr{L}_\omega u^{(n)} + (I - \omega L)^{-1}\omega c \tag{1.1}$$

and $u^{(n+1)}$ from $u^{(n+1/2)}$ by the (backward) SOR method

$$u^{(n+1)} = \mathscr{U}_\omega u^{(n+1/2)} + (I - \omega U)^{-1}\omega c \tag{1.2}$$

where

$$\begin{aligned} \mathscr{L}_\omega &= (I - \omega L)^{-1}(\omega U + (1 - \omega)I) \\ \mathscr{U}_\omega &= (I - \omega U)^{-1}(\omega L + (1 - \omega)I) \end{aligned} \tag{1.3}$$

From (1.1) and (1.2) we have

$$u^{(n+1)} = \mathscr{S}_\omega u^{(n)} + \omega(2 - \omega)(I - \omega U)^{-1}(I - \omega L)^{-1}c \tag{1.4}$$

where

$$\mathscr{S}_\omega = \mathscr{U}_\omega \mathscr{L}_\omega \tag{1.5}$$

Evidently

$$\mathscr{S}_\omega = I - \omega(2-\omega)(I-\omega U)^{-1}(I-\omega L)^{-1}D^{-1}A \tag{1.6}$$

and hence $I - \mathscr{S}_\omega$ is nonsingular if $0 < \omega < 2$ and if A is nonsingular. Moreover, since $\omega(2-\omega)(I-\omega U)^{-1}(I-\omega L)^{-1}c = (I - \mathscr{S}_\omega)A^{-1}b$ it follows by Theorems 3-2.2 and 3-2.4 that the SSOR method is completely consistent.

In Section 15.2 we shall study the convergence properties of the SSOR method under the assumption that A is real and symmetric with positive diagonal elements. It will be shown that convergence holds if and only if A is positive definite and $0 < \omega < 2$. Moreover, it will be shown that $\mathscr{S}_\omega$ has real and positive eigenvalues; hence, one can accelerate the convergence using semi-iterative methods. Thus, even though the SSOR method with the best value of ω is usually slower than the SOR method with the best ω, nevertheless the convergence of the SSOR method can often be accelerated by the use of semi-iterative methods so that it converges faster than the SOR method (without semi-iteration). Moreover, as we have seen, it does not appear to be possible to accelerate the convergence of the SOR method by semi-iteration.

In the special case where the matrix A has the form (7-1) there is a relation between the eigenvalues of $\mathscr{S}_\omega$ and those of B. However, in this case the optimum value of ω is unity and the optimum semi-iterative method based on $\mathscr{S}_1$ is no more effective than the optimum semi-iterative method based on $\mathscr{L}$.

In Section 15.3 we shall discuss the choice of ω when A is positive definite. It will be shown that a unique optimum value of ω exists, and that, in certain cases, a value of ω can be determined which, though not optimum, is sufficiently good so that when semi-iteration is used the resulting method is better than the SOR method by an order-of-magnitude. In Section 15.4 it will be shown that for the discrete Dirichlet problem one can choose an ω so that the SSOR method with semi-iteration requires on the order of $h^{-1/2}$ iterations as compared to h^{-1} for the SOR method.

In Section 15.5 we shall consider group SSOR methods. It will be shown that for the discrete Dirichlet problem if the line SSOR method is used, a significant improvement can be obtained over point SSOR.

In Section 15.6 and 15.7 we shall consider methods related to the SSOR method including the unsymmetric SOR method and, for the case where A has the form (7-1), the symmetric and unsymmetric MSOR methods.

For all of these methods we shall show that if A has the form (7-1), then the eigenvalues are the same as those of the MSOR method for suitable relaxation factors.

15.2. CONVERGENCE ANALYSIS

Theorem 2.1. Let A be a symmetric matrix with positive diagonal elements. For any real ω the eigenvalues of $\mathscr{S}_\omega$ are real and nonnegative. Moreover, if $0 < \omega < 2$ and if A is positive definite, then

$$\| \mathscr{S}_\omega \|_{A^{1/2}} = S(\mathscr{S}_\omega) = \| \mathscr{L}_\omega \|^2_{A^{1/2}} < 1 \tag{2.1}$$

Conversely, if $S(\mathscr{S}_\omega) < 1$, then $0 < \omega < 2$ and A is positive definite.

Proof. By (1.5) and (1.3) and by the symmetry of A, $\mathscr{S}_\omega$ is similar to the matrix $(D-\omega C_L)^{-1}(\omega C_L+(1-\omega)D)[(D-\omega C_L)^{-1}(\omega C_L+(1-\omega)D)]^T$ which, by Theorem 2-2.6, is nonnegative definite and hence has real and nonnegative eigenvalues. Moreover, since $0 < \omega < 2$ and since A is positive definite, we have $\| \mathscr{L}_\omega \|^2_{A^{1/2}} < 1$, by Theorem 4-3.1, and (2.1) follows from Theorem 7-5.1.

Let us assume now that $S(\mathscr{S}_\omega) < 1$. By (1.5) and (1.3) we have

$$\det \mathscr{S}_\omega = (1 - \omega)^{2N} \tag{2.2}$$

and hence the product of the eigenvalues of $\mathscr{S}_\omega$ is $(1 - \omega)^{2N}$. Therefore $S(\mathscr{S}_\omega) \geq | 1 - \omega |^2$ and $0 < \omega < 2$. From (1.6) it follows that $\mathscr{S}_\omega$ is similar to

$$\mathscr{S}_\omega^* = I - \omega(2 - \omega)D^{1/2}(D - \omega C_L)^{-1}A(D - \omega C_U)^{-1}D^{1/2} \tag{2.3}$$

If A is not positive definite, then there exists $v \neq 0$ and $\alpha \leq 0$ such that $Av = \alpha v$. If we let $w = D^{-1/2}(D - \omega C_U)v$, then we have

$$\begin{aligned}(w, \mathscr{S}_\omega^* w)/(w, w) &= [(w, w) - \omega(2 - \omega)(v, \alpha v)]/(w, w) \\ &= 1 - \omega(2 - \omega)\alpha(v, v)/(w, w) \geq 1 \end{aligned}\tag{2.4}$$

provided $0 < \omega < 2$. By Theorem 2-2.2 we have $S(\mathscr{S}_\omega^*) \geq 1$. This contradiction proves that A is positive definite and the proof of Theorem 2.1 is complete.

From Theorem 7-5.2 it follows that if A has the form (7-1), then $S(\mathscr{S}_\omega)$ is minimized when $\omega = 1$ and in this case

$$S(\mathscr{S}_1) = \bar{\mu}^2 = S(\mathscr{L}) \tag{2.5}$$

Thus we have

Theorem 2.2. If A is a positive definite matrix of the form (7-1), then

$$S(\mathscr{S}_\omega) = S(\mathscr{L}_{\omega(2-\omega)}) \leq 1 - \tfrac{1}{2}\omega^2(2-\omega)^2(1-\bar{\mu})^2 \tag{2.6}$$

and unless $\omega = 1$, we have

$$S(\mathscr{S}_\omega) > S(\mathscr{S}_1) = S(\mathscr{L}) = \bar{\mu}^2 \tag{2.7}$$

Thus in this case the SSOR method with the optimum ω is no better (based on the spectral radius) than the GS method even though more work is required per iteration. Clearly there is no advantage in using the SSOR method either with or without semi-iteration.

15.3. CHOICE OF RELAXATION FACTOR

We now study the problem of determining a relaxation factor for use with the SSOR method. We are primarily concerned with the case where A does *not* have the form (7-1).

Theorem 3.1. Let A be a positive definite matrix.

(a) There exists a unique value of ω, say ω_0, such that

$$0 < \omega_0 < 2 \tag{3.1}$$

and such that unless $\omega = \omega_0$ we have

$$S(\mathscr{S}_\omega) > S(\mathscr{S}_{\omega_0}) \tag{3.2}$$

(b) If

$$\omega_1 = \frac{2}{1 + (1 - 2\bar{\mu} + 4\gamma)^{1/2}} \tag{3.3}$$

where

$$\gamma = \max(S(LU), \tfrac{1}{4}) \tag{3.4}$$

then

$$S(\mathscr{S}_{\omega_1}) \leq \left[1 - \frac{1-\bar{\mu}}{(1 - 2\bar{\mu} + 4\gamma)^{1/2}}\right] \Big/ \left[1 + \frac{1-\bar{\mu}}{(1 - 2\bar{\mu} + 4\gamma)^{1/2}}\right] \tag{3.5}$$

(c) If $S(LU) \leq \frac{1}{4}$, then

$$\omega_1 = \frac{2}{1 + [2(1-\bar{\mu})]^{1/2}} \tag{3.6}$$

and

$$S(\mathscr{S}_{\omega_1}) \le \left[1 - \left(\frac{1-\bar{\mu}}{2}\right)^{1/2}\right] \Big/ \left[1 + \left(\frac{1-\bar{\mu}}{2}\right)^{1/2}\right] \tag{3.7}$$

Moreover, for any $\varepsilon > 0$ and for $1 - \bar{\mu}$ sufficiently small we have

$$R(\mathscr{S}_{\omega_1})/R(\mathscr{L}_{\omega_b}) \ge \tfrac{1}{2} - \varepsilon \tag{3.8}$$

where $R(\mathscr{S}_{\omega_1})$ and $R(\mathscr{L}_{\omega_b})$ are the rates of convergence of $\mathscr{S}_{\omega_1}$ and $\mathscr{L}_{\omega_b}$, respectively. Here $\bar{\mu} = S(B)$ and $\omega_b = 2(1 + [1 - \bar{\mu}^2]^{1/2})^{-1}$.

Proof. By (1.5) and Theorem 4-3.5 it is sufficient to consider the case where

$$A = I - L - U = I - B \tag{3.9}$$

where

$$L^{\mathrm{T}} = U \tag{3.10}$$

By Theorem 2.1 we consider only values of ω in the interval $0 < \omega < 2$. From (1.6) we have

$$\mathscr{S}_\omega = I - R(\omega)^{-1} \tag{3.11}$$

and

$$S(\mathscr{S}_\omega) = 1 - \frac{1}{S(R(\omega))} \tag{3.12}$$

where

$$R(\omega) = \frac{1}{\omega(2-\omega)} A^{-1}(I - \omega L)(I - \omega U) \tag{3.13}$$

Since $\mathscr{S}_\omega$ has real eigenvalues and $S(\mathscr{S}_\omega) < 1$ for $0 < \omega < 2$ it follows that $R(\omega)$ has real eigenvalues and that $1 \le S(R(\omega)) < \infty$ for $0 < \omega < 2$. Thus the problem of minimizing $S(\mathscr{S}_\omega)$ is equivalent to that of minimizing $S(R(\omega))$.

For each ω in the range $0 < \omega < 2$ let $v = v(\omega)$ be a real vector such that $(v, v) = 1$ and

$$R(\omega)v(\omega) = S(R(\omega))v(\omega) \tag{3.14}$$

By Theorem 2-1.10 the eigenvalues of $R(\omega)$, and hence $S(R(\omega))$, are continuous functions of ω. Moreover, we have

$$S(R(\omega)) = (v, R(\omega)v) \tag{3.15}$$

Let $\lambda = S(R(\omega))$, $v = v(\omega)$, $\lambda' = S(R(\omega'))$ and $v' = v(\omega')$. Evidently we have

$$Tv = \lambda Av, \qquad T'v' = \lambda' Av' \tag{3.16}$$

where

$$T = T(\omega) = AR(\omega), \qquad T' = T'(\omega') = AR(\omega') \tag{3.17}$$

Moreover,

$$(v', Tv) = \lambda(v', Av), \qquad (v, T'v') = \lambda'(v, Av') \tag{3.18}$$

Since T and T' are symmetric we have

$$(\lambda' - \lambda)(v', Av) = (v, T'v') - (v', Tv) \tag{3.19}$$

Letting $v' = v + \Delta v$ we have

$$\begin{aligned}(\lambda' - \lambda)(v, Av) &= -(\lambda' - \lambda)(\Delta v, Av) + (v, (T' - T)v) + (v, T'\,\Delta v) \\ &\quad -(\Delta v, Tv) \\ &= -\lambda'(\Delta v, Av) + (v, (T' - T)v) + (v, T'\,\Delta v) \\ &= -\lambda'(\Delta v, Av') + \lambda'(\Delta v, A\,\Delta v) + (v, (T' - T)v) \\ &\quad +(v', T'\,\Delta v) - (\Delta v, T'\,\Delta v) \\ &= -(\Delta v, T'v') + (\Delta v, (\lambda' A - T')\,\Delta v) \\ &\quad +(v, (T' - T)v) + (v', T'\,\Delta v)\end{aligned} \tag{3.20}$$

and hence

$$(\lambda' - \lambda)(v, Av) = (v, (T' - T)v) + (\Delta v, (\lambda' A - T')\,\Delta v) \tag{3.21}$$

Since λ' is the largest eigenvalue of $A^{-1}T'$, then λ' is the largest eigenvalue of $A^{-1/2}T'A^{-1/2}$. Hence by Theorem 2-2.2 we have

$$(z, A^{-1/2}T'A^{-1/2}z)/(z, z) \le \lambda' \tag{3.22}$$

for all $z \ne 0$, and hence

$$(y, T'y)/(A^{1/2}y, A^{1/2}y) \le \lambda' \tag{3.23}$$

and

$$(y, (\lambda' A - T')y) \ge 0 \tag{3.24}$$

for all $y \ne 0$. Thus the second term of (3.21) is nonnegative.

We now seek to represent $(v, (T' - T)v)$ in terms of $\Delta\omega$. Letting $\omega' = \omega + \Delta\omega$ we have

$$\begin{aligned}\frac{1}{\omega'(2-\omega')} &= \frac{1}{\omega(2-\omega)} + \frac{2(\omega-1)}{\omega^2(2-\omega)^2}\,\Delta\omega \\ &+ \frac{4-6\omega+3\omega^2}{\omega^3(2-\omega)^3}\,(\Delta\omega)^2 \\ &+ \left(-\frac{1}{\bar{\omega}_1{}^4} + \frac{1}{(2-\bar{\omega}_1)^4}\right)(\Delta\omega)^3 \end{aligned} \tag{3.25}$$

where $\bar{\omega}_1$ lies between ω and ω'. Similarly,

$$\begin{aligned}\frac{1}{2-\omega'} &= \frac{1}{2-\omega} + \frac{1}{(2-\omega)^2}\,\Delta\omega + \frac{1}{(2-\omega)^3}\,(\Delta\omega)^2 \\ &+ \frac{1}{(2-\bar{\omega}_2)^4}\,(\Delta\omega)^3 \end{aligned} \tag{3.26}$$

$$\begin{aligned}\frac{\omega'}{2-\omega'} &= \frac{\omega}{2-\omega} + \frac{2}{(2-\omega)^2}\,\Delta\omega + \frac{2}{(2-\omega)^3}\,(\Delta\omega)^3 \\ &+ \frac{2}{(2-\bar{\omega}_3)^4}\,(\Delta\omega)^3 \end{aligned} \tag{3.27}$$

Consequently, we have for some $\bar{a}_3$

$$(v, (T' - T)v) = a_1(\omega)\,\Delta\omega + a_2(\omega)(\Delta\omega)^2 + \bar{a}_3(\Delta\omega)^3 \tag{3.28}$$

where

$$a_1(\omega) = \frac{1}{\omega^2(2-\omega)^2}\,\{2(\omega-1) - \omega^2\alpha + 2\omega^2\beta\} \tag{3.29}$$

$$a_2(\omega) = \frac{1}{\omega^3(2-\omega)^3}\,\{4 - 6\omega + 3\omega^2 - \omega^3\alpha + 2\omega^3\beta\} \tag{3.30}$$

$$\alpha = (v, Bv), \qquad \beta = (v, LUv) \tag{3.31}$$

Since $I - 2L$ is nonsingular, it follows from Theorem 2-2.6 that $(I - 2L)(I - 2L)^{\mathrm{T}} = (I - 2L)(I - 2U) = I - 2B + 4LU$ is positive definite. Therefore

$$(v, (I - 2B + 4LU)v) = 1 - 2\alpha + 4\beta > 0$$

and

$$\begin{aligned}a_2(\omega) &= \omega^{-3}(2-\omega)^{-3}\{\tfrac{1}{2}\omega^3(1 - 2\alpha + 4\beta) + 4 - 6\omega + 3\omega^2 - \tfrac{1}{2}\omega^3\} \\ &= \tfrac{1}{2}\omega^{-3} + \tfrac{1}{2}(2-\omega)^{-3}(1 - 2\alpha + 4\beta) > 0\end{aligned}$$

If $a_1(\omega) = 0$, then for sufficiently small $\Delta\omega$ we have $\lambda' - \lambda \geq 0$ for all ω' such that $|\omega' - \omega| < \Delta\omega$. Thus if $a_1(\omega) = 0$, we have a relative minimum for $S(\mathscr{S}_\omega)$.

Since $S(R(0)) = S(R(2)) = \infty$ and since $0 < S(\mathscr{S}_\omega) < 1$ for $0 < \omega < 2$ it follows from the continuity of $S(R(\omega))$ that $S(R(\omega))$ assumes a minimum value, say ω_0, in the interval $0 < \omega < 2$. We show that the minimum is unique. If ω_0 and ω_1 were two relative minima, then in the open interval between ω_0 and ω_1 there would be a relative maximum. But this is impossible since the second term of the right member of (3.21) is nonnegative and since $a_2(\omega) \geq 0$. Therefore (a) follows.

By (3.13), (3.14), (3.31), and (3.9) we have

$$S(R(\omega)) = (1 - \omega\alpha + \omega^2\beta)/[\omega(2 - \omega)(1 - \alpha)] \tag{3.32}$$

Since

$$\beta = (v, LUv) \leq S(LU) \tag{3.33}$$

and since $1 - \omega\alpha + \omega^2\beta = ((I - \omega U)v, (I - \omega U)v) \geq 0$ we have, by (3.4),

$$S(R(\omega)) \leq (1 - \omega\alpha + \omega^2\gamma)/[\omega(2 - \omega)(1 - \alpha)] \tag{3.34}$$

where γ is given by (3.4). Since

$$\frac{d}{d\alpha}\left(\frac{1 - \omega\alpha + \omega^2\gamma}{1 - \alpha}\right) = \frac{\omega^2(\gamma - \frac{1}{4}) + \frac{1}{4}(\omega - 2)^2}{(1 - \alpha)^2} \tag{3.35}$$

it follows that the right member of (3.34) is an increasing function of α. Hence, since

$$\alpha = (v, Bv) \leq S(B) = \bar{\mu} \tag{3.36}$$

we have

$$S(R(\omega)) \leq (1 - \omega\bar{\mu} + \omega^2\gamma)/[\omega(2 - \omega)(1 - \bar{\mu})] \tag{3.37}$$

The derivative of the right member of (3.37) with respect to ω is

$$[\omega^2(2\gamma - \bar{\mu}) + 2(\omega - 1)]/[(1 - \bar{\mu})\omega^2(2 - \omega)^2] \tag{3.38}$$

which vanishes when

$$\omega^2(\bar{\mu} - 2\gamma) = 2(\omega - 1) \tag{3.39}$$

The root of (3.39) in the interval $0 < \omega < 2$ is ω_1, as given by (3.3).

Thus we have

$$\begin{aligned} S(R(\omega_1)) &\le [1 - \omega_1\bar{\mu} + \omega_1^2\gamma]/[\omega_1(2-\omega_1)(1-\bar{\mu})] \\ &= (1 - \tfrac{1}{2}\omega_1\bar{\mu})/[\omega_1(1-\bar{\mu})] \end{aligned} \tag{3.40}$$

The result (3.5) follows from (3.12) and (3.3).

If $S(LU) \le \frac{1}{4}$, then $\gamma = \frac{1}{4}$; hence (3.6) and (3.7) follow from (3.3) and (3.5), respectively. Moreover, by Theorem 4-1.2 we have $S(\mathscr{L}_{\omega_b}) \ge \omega_b - 1 = \bar{\mu}^2(1 + (1-\bar{\mu}^2)^{1/2})^{-2}$ and (3.8) follows by repeated application of L'Hôspital's rule. This completes the proof of Theorem 3.1.

Habetler and Wachspress [1961] proved (a) under the tacit assumption that there exists a vector $v = v(\omega)$ such that $\| v \| = 1$, $v(\omega)$ is continuous for $0 < \omega < 2$ and $\mathscr{S}_\omega v = S(\mathscr{S}_\omega)v$.† They show that ω_0 is uniquely determined by the condition

$$\omega_0 = \frac{2}{1 + [1 - 2(v, Bv) + 4(v, LUv)]^{1/2}} \tag{3.41}$$

Evans and Forrington [1963] developed an iterative scheme for determining the solution of (3.41). However, in the cases which they considered the condition $S(LU) \le \frac{1}{4}$ is satisfied, and a good estimate of ω_0 can be obtained by Theorem 3.1. Ehrlich [1963] proved (3.7) and (3.8) with ω_1 replaced by ω_0.

Theorem 3.2. (Ehrlich, 1963). If A is a positive definite matrix and if A_1 is a matrix obtained from A by deleting certain rows and the corresponding columns of A, then

$$S(\mathscr{S}_\omega[A_1]) \le S(\mathscr{S}_\omega[A]) \tag{3.42}$$

† That a continuous vector associated with the largest eigenvalue of a positive definite matrix need not exist can be seen from the example

$$A(\omega) = \begin{pmatrix} 2-\omega & 0 \\ 0 & \omega \end{pmatrix}$$

Here $v(\omega) = (1, 0)^{\mathrm{T}}$ for $0 < \omega < 1$ and $v(\omega) = (0, 1)^{\mathrm{T}}$ for $1 < \omega < 2$. Evidently $v(1)$ can be taken as $(a, b)^{\mathrm{T}}$ where $|a|^2 + |b|^2 = 1$. Thus $v(\omega)$ need not be continuous at $\omega = 1$. This by no means proves that $v(\omega)$ does not exist for $\mathscr{S}_\omega$, but it does indicate that the existence is not obvious.

Proof. From Theorem 4-3.5 it is sufficient to assume that diag $A = I$. Otherwise we can consider

$$\hat{A} = D^{-1/2}AD^{-1/2} \tag{3.43}$$

and

$$\begin{aligned}\mathscr{S}_\omega[\hat{A}] &= U_\omega[\hat{A}]\mathscr{L}_\omega[\hat{A}] = D^{1/2}U_\omega[A]\mathscr{L}_\omega[A]D^{-1/2}\\ &= D^{1/2}\mathscr{S}_\omega[A]D^{-1/2}\end{aligned} \tag{3.44}$$

where $D = \text{diag } A$. If $D_1 = \text{diag } A_1$ and $\hat{A}_1 = D_1^{-1/2}A_1D^{-1/2}$, then

$$\mathscr{S}_\omega[\hat{A}_1] = D_1^{1/2}\mathscr{S}_\omega[A_1]D_1^{-1/2} \tag{3.45}$$

By (1.6) we have

$$(I - \omega U[A_1])\mathscr{S}_\omega[A_1](I - \omega U[A_1])^{-1} = I - \omega(2 - \omega)Y[A_1] \tag{3.46}$$

where

$$Y[A_1] = (I - \omega L[A_1])^{-1}A_1(I - \omega U[A_1])^{-1} \tag{3.47}$$

By Theorem 2-2.2 we have

$$\begin{aligned}\underline{\lambda}(Y[A_1]) &= \min_{\substack{\hat{v}\in V_N' \\ \hat{v}\neq 0}} (\hat{v}, Y[A_1]\hat{v})/(\hat{v}, \hat{v})\\ &= \min_{\substack{\hat{v}\in V_N' \\ \hat{v}\neq 0}} ((I - \omega U[A_1])^{-1}\hat{v}, A_1(I - \omega U[A_1])^{-1}\hat{v})/(\hat{v}, \hat{v})\\ &= \min_{\substack{\hat{w}\in V_N' \\ \hat{v}\neq 0}} (\hat{w}, A_1\hat{w})/(\hat{v}, \hat{v})\end{aligned} \tag{3.48}$$

where $\hat{w} = (I - \omega U[A_1])^{-1}\hat{v}$. Here V_N' is the set of column vectors of order the same as the order of A_1. Let w be the column vector obtained from $\hat{w}$ by replacing the deleted elements by zeros. Evidently $(\hat{w}, A_1\hat{w}) = (w, Aw)$. Moreover, $(\hat{v}, \hat{v}) = ((I - \omega U[A_1])\hat{w}, (I - \omega U[A_1])\hat{w}) = ((I - \omega U[A])w, (I - \omega U[A])w)$ and hence

$$\begin{aligned}\underline{\lambda}(Y[A_1]) &= \min_{\substack{w\in V_N^* \\ w\neq 0}} (w, Aw)/((I - \omega U[A])w, (I - \omega U[A])w)\\ &\geq \min_{w\neq 0} (w, Aw)/((I - \omega U[A])w, (I - \omega U[A])w)\\ &= \underline{\lambda}(Y[A])\end{aligned} \tag{3.49}$$

where V_N^* is the set of column vectors of order N with the deleted elements equal to zero. The result (3.42) now follows.

15.4. SSOR SEMI-ITERATIVE METHODS: THE DISCRETE DIRICHLET PROBLEM

As in the case of the GS–SI method the eigenvalues of $\mathscr{S}_\omega$ are real and lie in the range

$$0 \leq \lambda \leq \bar{\lambda} = S(\mathscr{S}_\omega) \tag{4.1}$$

Consequently, the formulas for the SSOR–SI method are the same as those for the GS–SI method except that we replace $\bar{\mu}^2 = S(\mathscr{L})$ by $\bar{\lambda} = S(\mathscr{S}_\omega)$. Thus, (11-6.3) becomes

$$\begin{aligned} v^{(n+1)} &= \frac{\varrho_{n+1}}{1 - \frac{1}{2}\bar{\lambda}} \{(\mathscr{S}_\omega - \tfrac{1}{2}\bar{\lambda} I)v^{(n)} + k\} \\ &\quad + (1 - \varrho_{n+1})v^{(n-1)}, \qquad n = 0, 1, 2, \ldots \end{aligned} \tag{4.2}$$

where the ϱ_i are given by (11-2.29) with

$$z = (2/\bar{\lambda}) - 1 \tag{4.3}$$

Here

$$k = \omega(2 - \omega)(I - \omega U)^{-1}(I - \omega L)^{-1}c \tag{4.4}$$

Consider now the five-point discrete analog of the Dirichlet problem. If the natural ordering of the mesh points is used, then the matrices L and U correspond to the discrete operators

$$L[u](x, y) = \tfrac{1}{4}[u(x - h, y) + u(x, y - h)] \tag{4.5}$$

and

$$U[u](x, y) = \tfrac{1}{4}[u(x + h, y) + u(x, y + h)] \tag{4.6}$$

respectively. Therefore,

$$\| L \|_\infty \leq \tfrac{1}{2}, \qquad \| U \|_\infty \leq \tfrac{1}{2} \tag{4.7}$$

and, by Theorem 2-3.4,

$$S(LU) \leq \| LU \|_\infty \leq \| L \|_\infty \| U \|_\infty \leq \tfrac{1}{4} \tag{4.8}$$

If the region under consideration is a subset of the unit square, then

$$\bar{\mu} \leq \cos \pi h \tag{4.9}$$

and by Theorem 3.1 we have

$$S(\mathscr{S}_{\omega_1}) \leq \frac{1 - \sin(\pi h/2)}{1 + \sin(\pi h/2)} \sim 1 - \pi h \tag{4.10}$$

where

$$\omega_1 = \frac{2}{1 + 2\sin(\pi h/2)} \tag{4.11}$$

By (11-2.42) the virtual quasi-average rate of convergence for the SSOR–SI method is

$$(\overline{\overline{\mathscr{R}}}_n'(P_n(\mathscr{S}_{\omega_1})) \sim 2^{1/2}[\overline{\overline{\mathscr{R}}}_1'(P_1(\mathscr{S}_{\omega_1}))]^{1/2} \sim 2^{1/2}(-\log \sigma)^{1/2} \sim 2(\pi h)^{1/2} \tag{4.12}$$

since

$$\sigma = 1/z = S(\mathscr{S}_{\omega_1})/[2 - S(\mathscr{S}_{\omega_1})] \sim (1 - \pi h)/(1 + \pi h) \sim 1 - 2\pi h \tag{4.13}$$

Since $\overline{\overline{\mathscr{R}}}_n'(\mathscr{L}_{\omega_b}) \sim 2\pi h$ (see Section 6.3), we have a factor of improvement of $(\pi h)^{-1/2}$ over the SOR method. For example, if $h = \frac{1}{20}$, then the factor of improvement is about 2.5. This would probably not justify the additional work required since one must do at least twice as much work per iteration. Moreover there is extra storage and complexity involved in using the semi-iterative method. However, if $h = \frac{1}{80}$, then the factor of improvement is about 5, which is probably worthwhile.

The theoretical results are borne out by a number of numerical experiments carried out by Ehrlich [1963]. The results are given in the accompanying tabulation. For the SSOR method the value of ω was used which minimizes $S(\mathscr{S}_\omega)$ on the basis of numerical results. This value is close to ω_1. The observed factors of improvement are greater than the theoretical values. This is due to the fact that the SOR method converges more slowly than one would expect on the basis of the spectral radius (see Chapter 7). On the other hand, the theoretical numbers of

h^{-1}	SOR		SSOR–SI		Factor of improvement	
	ω	n	ω	n	Observed	Theoretical
20	1.75	58	1.76	17	3.4	2.5
40	1.86	117	1.87	24	4.9	3.5
80	1.93	236	1.94	34	6.9	5.0

iterations for the SSOR–SI method agree quite closely with the observed values. As a matter of fact, we note that for the SSOR method and any semi-iterative method based on the SSOR method, the $A^{1/2}$-norm is the same as the spectral radius.

We now show that for the difference equation Variant II corresponding to the self-adjoint equation

$$\frac{\partial}{\partial x}\left(A\frac{\partial u}{\partial x}\right)+\frac{\partial}{\partial y}\left(C\frac{\partial u}{\partial y}\right)=G \tag{4.14}$$

(see Section 2-2.8) we can obtain a rate of convergence of $O(h^{3/4})$ with the SSOR–SI method. Here we assume that in the domain of consideration $|A_x|$ and $|C_y|$ are bounded. By (2-8.11) the sum of the elements of the matrix U in the row corresponding to the point (x, y) is

$$\left(\frac{1}{A(x,y)+C(x,y)+O(h^2)}\right)\left[A\left(x+\frac{h}{2},y\right)+C\left(x,y+\frac{h}{2}\right)\right]$$
$$=\frac{1}{2}+\frac{h}{4}\left(\frac{A_x+C_y}{A+C}\right)+O(h^2) \tag{4.15}$$

Thus for h sufficiently small we have

$$\|U\|_\infty \leq \tfrac{1}{2}+Kh \tag{4.16}$$

and

$$S(LU)\leq \|LU\|_\infty \leq \|L\|_\infty\,\|U\|_\infty \leq \tfrac{1}{4}+K'h \tag{4.17}$$

for some constants K and K'.

By Theorem 3.1 if we let ω_1 be determined by (3.3) with $\gamma=\frac{1}{4}+K'h$, then since $\bar{\mu}=1-ch^2+O(h^4)$ we have for some constant K''

$$S(\mathscr{S}_{\omega_1})\sim\frac{1-K''h^{3/2}}{1+K''h^{3/2}}\sim 1-2K''h^{3/2} \tag{4.18}$$

Thus,

$$R(\mathscr{S}_{\omega_1})\sim 2K''h^{3/2} \tag{4.19}$$

and using semi-iteration we obtain

$$\overline{\mathscr{R}}_n'(P_n(\mathscr{S}_{\omega_1}))\sim 2(K'')^{1/2}h^{3/4} \tag{4.20}$$

so that we again have an order of magnitude improvement in the convergence rate.

We remark that we can use the results of Section 6.7 to obtain a bound on $\bar{\mu}$.

15.5. GROUP SSOR METHODS

As in Chapter 14 we can define, for any ordered grouping π, the group SSOR method by (1.4), replacing $\mathscr{S}_\omega$ by

$$\mathscr{S}_\omega^{(\pi)} = \mathscr{U}_\omega^{(\pi)} \mathscr{L}_\omega^{(\pi)} \tag{5.1}$$

where $\mathscr{U}_\omega^{(\pi)}$ and $\mathscr{L}_\omega^{(\pi)}$ are defined in Chapter 14. Most of the analysis in Sections 15.2–15.4 can be applied to group SSOR methods. Ehrlich [1963] considered the line SSOR method for the five-point discrete Dirichlet problem and was able to show that the convergence is faster than for point methods.

We now seek to show that if π corresponds to a partitioning by lines of mesh points (x, y) with y constant and if the ordering is with increasing y, then

$$\| U^{(\pi)} \|_\infty \leq \tfrac{1}{2} \tag{5.2}$$

where $U^{(\pi)} = (D^{(\pi)})^{-1} C_U^{(\pi)}$. Let us illustrate by a line with four points labeled 1, 2, 3, and 4 as indicated

5.	6.	7.	8.
1.	2.	3.	4.
9.	10.	11.	12.

If $v = U^{(\pi)} u$, then $D^{(\pi)} v = C_U^{(\pi)} u$ and we have

$$\begin{pmatrix} 4 & -1 & 0 & 0 \\ -1 & 4 & -1 & 0 \\ 0 & -1 & 4 & -1 \\ 0 & 0 & -1 & 4 \end{pmatrix} \begin{pmatrix} v_1 \\ v_2 \\ v_3 \\ v_4 \end{pmatrix} = \begin{pmatrix} u_5 \\ u_6 \\ u_7 \\ u_8 \end{pmatrix} \tag{5.3}$$

Clearly, it suffices to show that

$$\| (D^{(\pi)})^{-1} \|_\infty \leq \tfrac{1}{2} \tag{5.4}$$

This follows since $\| C_U^{(\pi)} \|_\infty \leq 1$ and hence $\| U^{(\pi)} \|_\infty \leq \| C_U^{(\pi)} \|_\infty \times \| (D^{(\pi)})^{-1} \|_\infty \leq \| (D^{(\pi)})^{-1} \|_\infty$. Let us consider the second column $(w_1, w_2, w_3, w_4)^T$ of $(D^{(\pi)})^{-1}$. Evidently

$$\begin{pmatrix} 4 & -1 & 0 & 0 \\ -1 & 4 & -1 & 0 \\ 0 & -1 & 4 & -1 \\ 0 & 0 & -1 & 4 \end{pmatrix} \begin{pmatrix} w_1 \\ w_2 \\ w_3 \\ w_4 \end{pmatrix} = \begin{pmatrix} 0 \\ 1 \\ 0 \\ 0 \end{pmatrix} \tag{5.5}$$

If we let α be the smallest root of

$$\alpha^2 - 4\alpha + 1 = 0 \tag{5.6}$$

then

$$12^{-1/2}\begin{pmatrix} 4 & -1 & 0 & 0 \\ -1 & 4 & -1 & 0 \\ 0 & -1 & 4 & -1 \\ 0 & 0 & -1 & 4 \end{pmatrix}\begin{pmatrix} \alpha \\ 1 \\ \alpha \\ \alpha^2 \end{pmatrix} = 12^{-1/2}\begin{pmatrix} \alpha^2 \\ 4-2\alpha \\ 0 \\ \alpha^3 \end{pmatrix} = \begin{pmatrix} 12^{-1/2}\alpha^2 \\ 1 \\ 0 \\ 12^{-1/2}\alpha^3 \end{pmatrix} \tag{5.7}$$

Since $D^{(\pi)}$ is an irreducible L-matrix with weak diagonal dominance, $(D^{(\pi)})^{-1} \geq 0$ by Theorems 2-7.1 and 2-7.3. Therefore

$$0 \leq \begin{pmatrix} w_1 \\ w_2 \\ w_3 \\ w_4 \end{pmatrix} \leq 12^{-1/2}\begin{pmatrix} \alpha \\ 1 \\ \alpha \\ \alpha^2 \end{pmatrix} \tag{5.8}$$

and

$$\sum_1^4 w_i = 12^{-1/2}(1 + 2\alpha + \alpha^2) = 6\alpha/12^{1/2} = (1 + 2/3^{1/2})^{-2} < \tfrac{1}{2} \tag{5.9}$$

This method can be applied to any column of $(D^{(\pi)})^{-1}$ and for any order of $D^{(\pi)}$. Thus in the general case $\sum w_i < 12^{-1/2}(1 + 2\alpha + 2\alpha^2 + \cdots)$ $\leq 12^{-1/2}(1 + 2\alpha(1-\alpha)^{-1}) = 12^{-1/2}(1+\alpha)(1-\alpha)^{-1} = \tfrac{1}{2}$. Since $D^{(\pi)}$ and $(D^{(\pi)})^{-1}$ are symmetric, we have (5.4), and as in the proof of (4.8) we have

$$S(L^{(\pi)}U^{(\pi)}) \leq \tfrac{1}{4} \tag{5.10}$$

We are now able to apply Theorem 3.1. By (3.6) and (3.7) if we use the relaxation factor

$$\omega_1^{(\pi)} = \frac{2}{1 + [2(1 - \bar{\mu}^{(\pi)})]^{1/2}} \tag{5.11}$$

then we have

$$S(\mathscr{S}_{\omega_1}^{(\pi)}) \leq \left\{1 - \left[\frac{1 - \bar{\mu}^{(\pi)}}{2}\right]^{1/2}\right\} \Big/ \left\{1 + \left[\frac{1 - \bar{\mu}^{(\pi)}}{2}\right]^{1/2}\right\} \tag{5.12}$$

By (14-4.6), in the case of the unit square we have

$$\bar{\mu}^{(\pi)} = S(B^{(\pi)}) = \cos \pi h/(2 - \cos \pi h) \sim 1 - \pi h^2 \tag{5.13}$$

and

$$S(\mathscr{S}_{\omega_1}^{(\pi)}) \sim 1 - 2^{1/2}\pi h \tag{5.14}$$

Thus for the unit square, or a subset thereof, we obtain an improvement in the rate of convergence of the line SSOR method over the point SSOR method. This improvement carries over to the corresponding semi-iterative methods. One can show that there is a gain of approximately a factor of $2^{1/4} \doteq 1.189$ in using line SSOR with semi-iteration as compared with point SSOR with semi-iteration. In order to achieve this relatively small improvement in terms of overall computational effort one should carry out the method using a normalized block iteration scheme such as that described at the end of Section 14.2. In certain cases treated by Ehrlich [1963, 1964] the factor of improvement was considerably greater than $2^{1/4}$. For the model problem the factor of improvement was closer to 2.

Habetler and Wachspress [1961] indicate that the SSOR theory is not applicable in certain cases involving variable coefficients. On the other hand, Ehrlich [1963] found the method to work quite well in a problem involving the differential equation

$$(e^{xy}u_x)_x + (e^{xy}u_y)_y - u = 0$$

The method was also found to work quite well for the equation

$$u_{xx} + u_{yy} - \frac{1}{y}u_y = 0$$

(see Ehrlich, 1964).

15.6. UNSYMMETRIC SOR METHOD

In this and the next section we consider variants of the SSOR method. First we consider the *unsymmetric SSOR method* (USSOR method) defined by

$$\begin{aligned} u^{(n+1/2)} &= \mathscr{L}_\omega u^{(n)} + (I - \omega L)^{-1}\omega c \\ u^{(n+1)} &= \mathscr{U}_{\tilde\omega} u^{(n+1/2)} + (I - \tilde\omega U)^{-1}\tilde\omega c \end{aligned} \tag{6.1}$$

or

$$u^{(n+1)} = \mathscr{C}_{\omega,\tilde\omega} u^{(n)} + (\omega + \tilde\omega - \omega\tilde\omega)(I - \tilde\omega U)^{-1}(I - \omega L)^{-1}\omega c \tag{6.2}$$

where

$$\mathscr{C}_{\omega,\tilde\omega} = \mathscr{U}_{\tilde\omega}\mathscr{L}_\omega \tag{6.3}$$

Since by (1.3)

$$\mathscr{E}_{\omega,\tilde{\omega}} = I - (\omega + \tilde{\omega} - \omega\tilde{\omega})(I - \tilde{\omega}U)^{-1}(I - \omega L)^{-1}D^{-1}A \qquad (6.4)$$

it follows that $I - \mathscr{E}_{\omega,\tilde{\omega}}$ is nonsingular provided

$$\omega + \tilde{\omega} - \omega\tilde{\omega} \neq 0 \qquad (6.5)$$

The USSOR method is thus seen to be completely consistent from (6.2) and (6.4).

We now prove the following convergence theorem:

Theorem 6.1. If A is a positive definite matrix and if $0 < \omega < 2$, $0 < \tilde{\omega} < 2$, then

$$S(\mathscr{E}_{\omega,\tilde{\omega}}) \leq \| \mathscr{E}_{\omega,\omega'} \|_{A^{1/2}} < 1 \qquad (6.6)$$

On the other hand if $S(\mathscr{E}_{\omega,\tilde{\omega}}) < 1$, then

$$0 < \omega + \tilde{\omega} - \omega\tilde{\omega} < 2 \qquad (6.7)$$

Proof. By (6.3) we have

$$\| \mathscr{E}_{\omega,\tilde{\omega}} \|_{A^{1/2}} \leq \| \mathscr{U}_{\tilde{\omega}} \|_{A^{1/2}} \| \mathscr{L}_{\omega} \|_{A^{1/2}} \qquad (6.8)$$

If A is positive definite and if $0 < \omega < 2$, $0 < \tilde{\omega} < 2$, then by Theorem 4-3.1 we have $\| \mathscr{L}_{\omega} \|_{A^{1/2}} < 1$ and $\| \mathscr{U}_{\tilde{\omega}} \|_{A^{1/2}} < 1$. Hence (6.6) follows.

Since the product of the eigenvalues of $\mathscr{E}_{\omega,\tilde{\omega}}$ is

$$\det \mathscr{E}_{\omega,\tilde{\omega}} = (1 - \omega)^N(1 - \tilde{\omega})^N \qquad (6.9)$$

it follows that

$$S(\mathscr{E}_{\omega,\tilde{\omega}}) \geq |(1 - \omega)(1 - \tilde{\omega})| \qquad (6.10)$$

Hence, if $S(\mathscr{E}_{\omega,\tilde{\omega}}) < 1$, we must have (6.7).

If A is a positive definite matrix of the form (7-1), then the eigenvalues of $\mathscr{E}_{\omega,\tilde{\omega}}$ are related to the eigenvalues of $\mathscr{L}_{\omega}$ for a suitable $\hat{\omega}$. Thus we have

Theorem 6.2. If A is a real symmetric matrix of the form (7-1) with positive diagonal elements, then the eigenvalues of $\mathscr{E}_{\omega,\tilde{\omega}}$ are the same as those of $\mathscr{L}_{\hat{\omega}}$ where

$$\hat{\omega} = \omega + \tilde{\omega} - \omega\tilde{\omega} \qquad (6.11)$$

If A is positive definite and if $0 < \hat{\omega} < 2$, then

$$S(\mathscr{C}_{\omega,\tilde{\omega}}) < 1 \qquad (6.12)$$

On the other hand, if (6.12) holds, then A is positive definite and

$$0 < \hat{\omega} < 2 \qquad (6.13)$$

Proof. By Lemma 8-5.2 the eigenvalues of $\mathscr{C}_{\omega,\tilde{\omega}} = \mathscr{U}_{\tilde{\omega},\tilde{\omega}}\mathscr{L}_{\omega,\omega}$ are the same as those of $\mathscr{L}_{\hat{\omega},\hat{\omega}} = \mathscr{L}_{\hat{\omega}}$. Thus, if A is positive definite and if $0 < \hat{\omega} < 2$, then (6.12) holds by Theorem 4-3.1. On the other hand, if (6.12) holds then, by Theorem 4-3.6, (6.13) holds and A is positive definite. The theorem follows.

From Theorem 6.2 and Theorem 6-2.3 it follows that the optimum values of ω and $\tilde{\omega}$ satisfy

$$\omega + \tilde{\omega} - \omega\tilde{\omega} = \omega_b \qquad (6.14)$$

and that in this case

$$S(\mathscr{C}_{\omega,\tilde{\omega}}) = \omega_b - 1 \qquad (6.15)$$

Thus, since twice as much work is required per iteration using the USSOR method as with the SOR method, and since the rate of convergence is no better, the USSOR method would appear to be mainly of academic interest, at least when A has the form (7-1).

15.7. SYMMETRIC AND UNSYMMETRIC MSOR METHODS

If A has the form (7-1), then analogous to the SSOR and USSOR methods we can consider the *symmetric MSOR method* (SMSOR method) and the *unsymmetric MSOR method* (USMSOR method). The iteration matrix for the USMSOR method is

$$\mathscr{W}_{\omega,\omega',\tilde{\omega},\tilde{\omega}'} = \mathscr{U}_{\tilde{\omega},\tilde{\omega}'}\mathscr{L}_{\omega,\omega'} \qquad (7.1)$$

where by (8-1.13) and (8-5.9) we have

$$\mathscr{L}_{\omega,\omega'} = \mathscr{L}_{0,\omega'}\mathscr{L}_{\omega,0}$$
$$\mathscr{U}_{\tilde{\omega},\tilde{\omega}'} = \mathscr{L}_{\tilde{\omega},0}\mathscr{L}_{0,\tilde{\omega}'} \qquad (7.2)$$

Thus one interation of the USMSOR method consists of a forward sweep and then a backward sweep. On the forward sweep we use the relaxation factor ω on the red equations and then ω' on the black equations. Then on the backward sweep we use $\tilde{\omega}'$ on the black equations and then $\tilde{\omega}$ on the red equations. Thus unprimed and primed relaxation factors are used for red and black equations, respectively.

In the case of the SMSOR method we have $\tilde{\omega} = \omega$, $\tilde{\omega}' = \omega'$ and the iteration matrix is

$$\mathscr{S}_{\omega,\omega'} = \mathscr{W}_{\omega,\omega',\omega,\omega'} = \mathscr{U}_{\omega,\omega'}\mathscr{L}_{\omega,\omega'} \tag{7.3}$$

We now prove

Theorem 7.1. If A is a symmetric matrix of the form (7.1) with positive diagonal elements, then the eigenvalues of $\mathscr{W}_{\omega,\omega',\tilde{\omega},\tilde{\omega}'}$ are the same as those of $\mathscr{L}_{\hat{\omega},\hat{\omega}'}$ where

$$\hat{\omega} = \omega + \tilde{\omega} - \omega\tilde{\omega}, \qquad \hat{\omega}' = \omega' + \tilde{\omega}' < \omega'\tilde{\omega}' \tag{7.4}$$

If A is positive definite and if $0 < \hat{\omega} < 2$ and $0 < \hat{\omega}' < 2$, then

$$\bar{S}(\mathscr{W}_{\omega,\omega'\tilde{\omega},\tilde{\omega}'}) \leq \|\overline{\mathscr{W}_{\omega,\omega',\tilde{\omega},\tilde{\omega}'}}\|_{A^{1/2}} - 1 \tag{7.5}$$

Here $\bar{S}(\mathscr{W})$ and $\|\overline{\mathscr{W}}\|_{A^{1/2}}$ denote the virtual spectral radius and virtual $A^{1/2}$ norms, respectively, of $\mathscr{W}$. On the other hand, if

$$\bar{S}(\mathscr{W}_{\omega,\omega',\tilde{\omega},\tilde{\omega}'}) < 1 \tag{7.6}$$

then A is positive definite and

$$0 < \hat{\omega} < 2, \qquad 0 < \hat{\omega}' < 2 \tag{7.7}$$

Proof. The eigenvalues relation holds by Lemma 8-5.2. If A is positive definite, then (7.5) follows by (7.1), (8-5.11), and Theorem 8-5.1. The rest of the proof follows from Theorem 8-3.2.

By Theorem 8-3.3 the optimum values of ω, ω', $\tilde{\omega}$, and $\tilde{\omega}'$ in the sense of minimizing $\bar{S}(\mathscr{W})$ satisfy

$$\omega + \tilde{\omega} - \omega\tilde{\omega} = \omega' + \tilde{\omega}' - \omega'\tilde{\omega}' = \omega_b \tag{7.8}$$

Moreover, the corresponding value of $\bar{S}(\mathscr{W})$ is $\omega_b - 1$. Thus we cannot obtain any better convergence rate based on the spectral radius than that of $\mathscr{L}_{\omega_b}$. Moreover, by Theorem 11-8.1 even if we use a semi-iterative method based on $\mathscr{W}$, if all of the eigenvalues of $\mathscr{W}$ are real, then we cannot obtain faster convergence. Thus, the USMSOR method would appear to be mainly of academic interest.

SUPPLEMENTARY DISCUSSION

The SSOR method was first considered by Sheldon [1955]. It is a generalization of a method proposed by Aitken [1950], which is equivalent to the SSOR method with $\omega = 1$. Sheldon also considered the acceleration of the SSOR method by use of semi-iteration. Ehrlich [1963, 1964] considered line SSOR and the use of semi-iterative methods.

Section 15.2. The proof of the sufficiency of the conditions of Theorem 2.1 is well known. The proof of the necessity and the relation of $S(\mathscr{S}_\omega)$ to $\| \mathscr{L}_\omega \|^2_{A^{1/2}}$ is given by Young [1969, 1970].

Section 15.3. The proof of (a), Theorem 3.1, is based on analysis of Habetler and Wachspress [1961] and Ehrlich [1963]. The results (b) and (c) of Theorem 3.1 were given by Young [1971a]. Similar results have been obtained by Habetler and Wachspress and by Ehrlich with ω_1 replaced by ω_0.

It can be shown (see Ehrlich, 1963; 1964) that one can always choose ω_1 so that the SSOR–SI method converges at least $3^{-1/2}$ times as fast as the SOR method.

Sections 15.4 and 15.5. The result (4.12) was given by Young [1971a]. Sharper inequalities than (4.8) and (5.10) can be obtained by more detailed analysis (see Ehrlich, 1963; 1964). The result that for line SSOR and for any region the rate of convergence is $O(h^{1/2})$ represents a generalization of the corresponding result of Ehrlich for the rectangle.

Sections 15.6 and 15.7. An analysis of the SSOR method for the case where A has the form (7.1) is given by D'Sylva and Miles [1963], who also considered the USSOR method. Lynn [1964] studied the relation between the SOR, SSOR, and USSOR methods. Much of the analysis given in Sections 15.6 and 15.7 can be found in Young [1969].

EXERCISES

Section 15.1

1. Verify (1.6) and show that (1.4) is completely consistent with $Au = b$ if $\omega \neq 0$ and $\omega \neq 2$ (even if A is singular).

2. Carry out two complete iterations of the SSOR method with $\omega = 1.5$ for the system

$$\begin{pmatrix} 4 & -1 & -1 & 0 \\ -1 & 4 & 0 & -1 \\ -1 & 0 & 4 & -1 \\ 0 & -1 & -1 & 4 \end{pmatrix} \begin{pmatrix} u_1 \\ u_2 \\ u_3 \\ u_4 \end{pmatrix} = \begin{pmatrix} 1000 \\ 1000 \\ 0 \\ 0 \end{pmatrix}$$

Section 15.2

1. Verify that

$$\mathscr{S}_\omega = (D - \omega C_U)^{-1}(\omega C_L + (1 - \omega)D)(D - \omega C_L)^{-1}(\omega C_U + (1 - \omega)D)$$

and that $\mathscr{S}_\omega$ is similar to SS^T where

$$S = (D - \omega C_L)^{-1}(\omega C_L + (1 - \omega)D)$$

2. Consider the model problem for Laplace's equation with $h = \frac{1}{3}$. Compute the eigenvalues of $\mathscr{S}_\omega$ both for the natural ordering and for the red–black ordering for the cases $\omega = 0.9$, 1.0, and 1.1. Also verify Theorem 2.2 for the case $\omega = 1.1$ using red–black ordering.

3. Show that if A is a positive definite matrix of the form (7-1) one can achieve the same rate of convergence for the SSOR method as with the SOR method provided one uses a complex value of ω for the SSOR method. Could one then effectively apply a semi-iterative procedure to the resulting method?

Section 15.3

1. Find $S(LU)$ for the matrices

$$A_1 = \begin{pmatrix} 2 & -1 & 0 \\ -1 & 2 & -1 \\ 0 & -1 & 2 \end{pmatrix}, \qquad A_2 = \begin{pmatrix} 2 & 0 & -1 \\ 0 & 2 & -1 \\ -1 & -1 & 2 \end{pmatrix}$$

In each case compute ω_1 by (3.3) and find a bound on $S(\mathscr{S}_{\omega_1})$ by (3.5) or (3.7). Also obtain the exact values of $S(\mathscr{S}_{\omega_1})$.

2. For the model problem with $h = \frac{1}{20}$ find ω_1 by (3.3) and a bound on $S(\mathscr{S}_{\omega_1})$. Assume that the natural ordering is used.

3. For the model problem for $h = \frac{1}{4}$ compute $S(\mathscr{S}_\omega)$ for $\omega = 0\ (0.1)\ 2.0$ for both the natural ordering and the red–black ordering. Also compute ω_1 and $S(\mathscr{S}_{\omega_1})$. Compare ω_1 with the observed optimum value for ω in each case.

4. Verify Theorem 3.2 for the case of the model problem with $h = \frac{1}{3}$ and the natural ordering and also for the case where the corner $\frac{2}{3} \le x \le 1$, $\frac{2}{3} \le y \le 1$ is removed. Consider the cases $\omega = 1$ and $\omega = 1.5$.

5. Let A be a positive definite matrix such that $S(LU) \le \frac{1}{4}$. Show that if $\bar{\mu}' \ge \bar{\mu} = S(B)$ and if

$$\omega_1' = \frac{2}{1 + [2(1 - \bar{\mu}')]^{1/2}}$$

then

$$S(\mathscr{S}_{\omega_1'}) \le \left[1 - \left(\frac{1 - \bar{\mu}'}{2}\right)^{1/2}\right] \Big/ \left[1 + \left(\frac{1 - \bar{\mu}'}{2}\right)^{1/2}\right]$$

6. Show that if A is positive definite and if γ is given by (3.4), then

$$S(\mathscr{S}_\omega) \le 1 - \omega(2 - \omega)(1 - \bar{\mu})/(1 - \omega\bar{\mu} + \omega^2\gamma)$$

7. For the matrix

$$A = \begin{pmatrix} 4 & -1 & -1 & 0 \\ -1 & 4 & 0 & -1 \\ -1 & 0 & 4 & -1 \\ 0 & -1 & -1 & 4 \end{pmatrix}$$

show that a continuous vector function $v(\omega)$ can be defined for $0 < \omega < 2$ such that $\| v(\omega) \| = 1$ and $\mathscr{S}_\omega v(\omega) = S(\mathscr{S}_\omega) v(\omega)$.

Section 15.4

1. Carry out two iterations of the SSOR–SI method for the system

$$\begin{pmatrix} 4 & -1 & -1 & 0 \\ -1 & 4 & 0 & -1 \\ -1 & 0 & 4 & -1 \\ 0 & -1 & -1 & 4 \end{pmatrix} \begin{pmatrix} u_1 \\ u_2 \\ u_3 \\ u_4 \end{pmatrix} = \begin{pmatrix} 1000 \\ 1000 \\ 0 \\ 0 \end{pmatrix}$$

using ω as determined by (3.3). Let $u^{(0)} = 0$.

2. For the model problem with $h = \frac{1}{20}$ estimate the number of iterations required to reduce the error by 10^{-6} of its original value using the SSOR–SI method.

3. Show that if A is the matrix corresponding to the difference equation (2-8.10) for solving (2-8.3), then

$$S(LU) \leq \tfrac{1}{4} + ch + O(h^2)$$

for some c. Show that one can choose ω_1 such that for small h we have

$$S(\mathscr{S}_{\omega_1}) \sim 1 - kh^{3/2}$$

for some constant k. Evaluate k for the equation $u_{xx} + u_{yy} - y^{-1}u_y = u$ in the region $0 < x < 1$, $1 < y < 2$.

4. Write a computer program to carry out the SSOR–SI method for the model problem. Allow h and ω to be input parameters. Let $u = 0$ on the boundary and let the starting values be unity. Apply the program to the cases $h^{-1} = 20, 40, 80$ where ω is determined by (3.3). Find the number of iterations required to reduce the maximum value of $|u_i|$ to 10^{-6}.

5. Show that for any semi-iterative method based on the SSOR method the $A^{1/2}$-norm is the same as the spectral radius.

Section 15.5

1. Carry out the argument used to prove (5.4) if the order of $D^{(\pi)}$ is 5 and one is seeking the second column of $(D^{(\pi)})^{-1}$. Also, give a proof in the case of any order.

2. Carry out two iterations of line SSOR for the system

$$\begin{pmatrix} 4 & -1 & -1 & 0 \\ -1 & 4 & 0 & -1 \\ -1 & 0 & 4 & -1 \\ 0 & -1 & -1 & 4 \end{pmatrix} \begin{pmatrix} u_1 \\ u_2 \\ u_3 \\ u_4 \end{pmatrix} = \begin{pmatrix} 1000 \\ 1000 \\ 0 \\ 0 \end{pmatrix}$$

with the ordered grouping $\{1, 2\}$, $\{3, 4\}$, and with $\omega = 1.5$.

3. Extend the computer program of Exercise 4, Section 5.4, to the case of line SSOR.

Section 15.6

1. Carry out two iterations of the USSOR method for the system

$$\begin{pmatrix} 4 & -1 & -1 & 0 \\ -1 & 4 & 0 & -1 \\ -1 & 0 & 4 & -1 \\ 0 & -1 & -1 & 4 \end{pmatrix} \begin{pmatrix} u_1 \\ u_2 \\ u_3 \\ u_4 \end{pmatrix} = \begin{pmatrix} 1000 \\ 1000 \\ 0 \\ 0 \end{pmatrix}$$

with $\omega = 1.4$, $\tilde{\omega} = 1.0$. Let $u^{(0)} = 0$. Also, carry out two iterations with $\mathscr{L}_{\hat{\omega}}$ where $\hat{\omega} = \omega + \tilde{\omega} - \omega\tilde{\omega}$. Find $\| \mathscr{E}_{\omega,\tilde{\omega}} \|_{A^{1/2}}$ and compare with $\| \mathscr{U}_{\tilde{\omega}} \|_{A^{1/2}} \| \mathscr{L}_{\omega} \|_{A^{1/2}}$.

2. If A is a positive definite matrix, if $\omega = 1.2$, and if the USSOR method converges, find bounds on $\hat{\omega}$.

3. Find the eigenvalues of $\mathscr{E}_{\omega,\tilde{\omega}}$ when $\omega = 1.5$, $\tilde{\omega} = 1.2$ and

$$A = \begin{pmatrix} 4 & 0 & -1 & -1 \\ 0 & 4 & -1 & -1 \\ -1 & -1 & 4 & 0 \\ -1 & -1 & 0 & 4 \end{pmatrix}$$

If $\omega = 1.2$, what value of $\tilde{\omega}$ will give the smallest value of $S(\mathscr{E}_{\omega,\tilde{\omega}})$?

4. Show that if A is positive definite and if $0 < \omega < 2$, then

$$\| \mathscr{U}_{\omega} \|_{A^{1/2}} < 1$$

5. Is it true that if A is a symmetric matrix with positive diagonal elements such that $S(\mathscr{E}_{\omega,\tilde{\omega}}) < 1$, then A is positive definite? Must ω and $\tilde{\omega}$ lie in the range $0 < \omega < 2$, $0 < \tilde{\omega} < 2$?

Section 15.7

1. Carry out two iterations with the USMSOR method for the system

$$\begin{pmatrix} 4 & 0 & -1 & -1 \\ 0 & 4 & -1 & -1 \\ -1 & -1 & 4 & 0 \\ -1 & -1 & 0 & 4 \end{pmatrix} \begin{pmatrix} u_1 \\ u_2 \\ u_3 \\ u_4 \end{pmatrix} = \begin{pmatrix} 1000 \\ 0 \\ 1000 \\ 0 \end{pmatrix}$$

with $\omega = 0.6$, $\omega' = 1.2$, $\tilde{\omega} = 0.8$, $\tilde{\omega}' = 1.6$.

2. Write the USMSOR method in the form

$$u^{(n+1)} = \mathscr{W}_{\omega, \omega', \tilde{\omega}, \tilde{\omega}'} u^{(n)} + k$$

and determine k.

3. Show that if A is a positive definite matrix of the form (7-1), then $\bar{S}(\mathscr{S}_{\omega, \omega'})$ is minimized if $\omega = \omega' = 1$.

4. Show that if A is a positive definite matrix and if $0 < \omega < 2$, $0 < \omega' < 2$, $0 < \tilde{\omega} < 2$, $0 < \tilde{\omega}' < 2$, then the USMSOR method is strongly convergent.

5. Show that if A is a positive definite matrix of the form (7-1), then the eigenvalues of $\mathscr{U}_{\omega' \omega} \mathscr{L}_{\omega, \omega'}$ are the same as those of $\mathscr{C}_{\omega, \omega'}$.

6. If A is a positive definite matrix of the form (7-1) and if $\bar{\mu} = 0.8$, find the value of $\tilde{\omega}'$ which minimizes $S(\mathscr{W}_{\omega, \omega', \tilde{\omega}, \tilde{\omega}'})$ if $\omega = 1.2$, $\omega' = 1.4$, $\tilde{\omega} = 1.6$.

7. Find the eigenvalues of $\mathscr{W}_{\omega, \omega', \tilde{\omega}, \tilde{\omega}'}$ if $\omega = 1.2$, $\omega' = 0.8$, $\tilde{\omega} = 0.6$, $\tilde{\omega}' = 1.5$ for the matrix

$$A = \begin{pmatrix} 2 & 0 & -1 \\ 0 & 2 & -1 \\ -1 & -1 & 2 \end{pmatrix}$$

Chapter 16 / SECOND-DEGREE METHODS

Let us consider the linear stationary iterative method of second degree defined by

$$u^{(n+1)} = Gu^{(n)} + Hu^{(n-1)} + k \tag{1}$$

where $u^{(0)}$ and $u^{(1)}$ are arbitrary. If the sequence defined converges to a limit $\hat{u}$, then $\hat{u}$ must satisfy the *related equation*

$$u = Gu + Hu + k \tag{2}$$

In order that whenever, for some n, $u^{(n-1)} = u^{(n)} = \bar{u}$, the exact solution of the system

$$Au = b \tag{3}$$

we have $u^{(n+1)} = u^{(n+2)} = \cdots = \bar{u}$, the *consistency condition*

$$(I - H - G)A^{-1}b = k \tag{4}$$

must be satisfied. Here and subsequently we assume that A is nonsingular. The method will be *completely consistent* if (4) holds and if $I - G - H$ is nonsingular.

Following an idea of Golub and Varga [1961] we observe that

$$\begin{pmatrix} u^{(n)} \\ u^{(n+1)} \end{pmatrix} = \begin{pmatrix} 0 & I \\ H & G \end{pmatrix} \begin{pmatrix} u^{(n-1)} \\ u^{(n)} \end{pmatrix} + \begin{pmatrix} 0 \\ k \end{pmatrix} \tag{5}$$

Hence, a necessary and sufficient condition that the method converge for all $u^{(0)}$ and $u^{(1)}$ is that

$$S(\hat{G}) < 1 \tag{6}$$

where†

$$\hat{G} = \begin{pmatrix} 0 & I \\ H & G \end{pmatrix} \tag{7}$$

Moreover, $S(\hat{G}) < 1$ if and only if all roots λ of

$$\det(\lambda^2 I - \lambda G - H) = 0 \tag{8}$$

are less than unity in modulus.

Given a linear stationary method of first degree

$$u^{(n+1)} = G_1 u^{(n)} + k_1 \tag{9}$$

the linear second-degree method

$$u^{(n+1)} = u^{(n)} + d(u^{(n)} - u^{(n-1)}) + e(G_1 u^{(n)} + k_1 - u^{(n)}) \tag{10}$$

is completely consistent for any constants d and e such that $e \neq 0$. We remark that (10) has a form similar to that of the semi-iterative method (11-2.28) which is given by

$$\begin{aligned} v^{(n+1)} &= v^{(n)} + (\varrho_{n+1} - 1)(v^{(n)} - v^{(n-1)}) \\ &\quad + \frac{2\varrho_{n+1}}{2 - (\alpha + \beta)} (G_1 v^{(n)} + k_1 - v^{(n)}) \end{aligned} \tag{11}$$

Thus if we replace $\varrho_{n+1} - 1$ by d and $2\varrho_{n+1}(2 - (\alpha + \beta))^{-1}$ by e we obtain (10).

Since

$$\begin{aligned} G &= (1 - e + d)I + eG_1 \\ H &= -dI \end{aligned} \tag{12}$$

we have from (8)

$$\det(\lambda^2 I - \lambda(eG_1 + (1 - e + d)I) + dI) = 0$$

† In this chapter we shall use the notation $\hat{G}$ as indicated. Normally we let $\hat{G} = D^{-1/2} G D^{-1/2}$ where D is a certain diagonal matrix. (See Chapter 4 and other previous chapters.)

and

$$\det\left(G_1 + \left(\frac{1-e+d}{e}\right)I - \left(\frac{\lambda^2+d}{e\lambda}\right)I\right) = 0 \tag{13}$$

Thus the eigenvalues λ of $\hat{G}$ are related to the eigenvalues μ of G_1 by

$$\mu + \frac{1-e+d}{e} = \frac{\lambda^2+d}{e\lambda} \tag{14}$$

The image of the circle $\lambda = \varrho e^{i\theta}$ is the ellipse

$$\frac{\left[\operatorname{Re}\mu + \left(\frac{1-e+d}{e}\right)\right]^2}{\left[\frac{1}{e}\left(\varrho + \frac{d}{\varrho}\right)\right]^2} + \frac{[\operatorname{Im}\mu]^2}{\left[\frac{1}{e}\left(\varrho - \frac{d}{\varrho}\right)\right]^2} = 1 \tag{15}$$

As in Section 6.4,† if the eigenvalues μ of G_1 are real and lie in the interval

$$\alpha \leq \mu \leq \beta < 1 \tag{16}$$

then the optimum choice of d and e satisfies the conditions

$$\begin{aligned} \varrho^2 &= d \\ (e-1-d)/e &= (\alpha+\beta)/2 \\ 2\varrho/e &= (\beta-\alpha)/2 \end{aligned} \tag{17}$$

Thus we have

$$\{(\beta-\alpha)/[2-(\beta+\alpha)]\}(1+\varrho^2) = 2\varrho \tag{18}$$

If we let

$$\bar{\sigma} = (\beta-\alpha)/[2-(\beta+\alpha)] \tag{19}$$

then we obtain

$$\bar{\sigma}(1+\varrho^2) = 2\varrho \tag{20}$$

and

$$1+\varrho^2 = \hat{\omega}_b = 2/[1+(1-\bar{\sigma}^2)^{1/2}] \tag{21}$$

Therefore,

$$d = \hat{\omega}_b - 1, \qquad e = 2\hat{\omega}_b\bar{\sigma}/(\beta-\alpha) = 2\hat{\omega}_b/[2-(\beta+\alpha)] \tag{22}$$

† See also Frankel [1950].

If we let

$$\hat{r} = \hat{\omega}_b - 1 \tag{23}$$

then we have

$$S(\hat{G}) = \varrho = (\hat{\omega}_b - 1)^{1/2} = \hat{r}^{1/2} \tag{24}$$

Hence it follows that by the use of a second-degree method we are able to accelerate the convergence of any iterative method which has real eigenvalues which are less than unity by an amount comparable to that attainable by using the optimum semi-iterative method.

As an example, if A is a positive definite matrix with Property A and if $G_1 = B$, then we have $\beta = -\alpha = \bar{\mu} = S(B)$, $\bar{\sigma} = \bar{\mu}$ and

$$\hat{\omega}_b = \omega_b = 2/[1 + (1 - \bar{\mu}^2)^{1/2}] \tag{25}$$

By (24) we have

$$S(\hat{G}) = (\omega_b - 1)^{1/2} \tag{26}$$

Hence, if A is consistently ordered the second-degree J method converges approximately half as fast as the SOR method.

If A is positive definite and consistently ordered and if $G_1 = \mathscr{L}$, corresponding to the GS method, then we have $\beta = \bar{\mu}^2$, $\alpha = 0$ and

$$\bar{\sigma} = \bar{\mu}^2/(2 - \bar{\mu}^2) \tag{27}$$

Therefore,

$$\hat{\omega}_b = \frac{2}{1 + (1 - \bar{\sigma}^2)^{1/2}} = 1 + \left(\frac{\bar{\mu}}{1 + (1 - \bar{\mu}^2)^{1/2}}\right)^4 = 1 + (\omega_b - 1)^2 \tag{28}$$

and

$$S(\hat{G}) = \omega_b - 1 \tag{29}$$

Thus the second-degree GS method converges approximately as fast as the SOR method.

If A is positive definite and if $G_1 = \mathscr{S}_\omega$, corresponding to the SSOR method, then $\beta = \bar{\lambda} = S(\mathscr{S}_\omega)$, $\alpha = 0$ and

$$\bar{\sigma} = \bar{\lambda}/(2 - \bar{\lambda}) \tag{30}$$

Therefore

$$\hat{\omega}_b = \frac{2}{1 + (1 - \bar{\sigma}^2)^{1/2}} = 1 + \left(\frac{\bar{\lambda}^{1/2}}{1 + (1 - \bar{\lambda})^{1/2}}\right)^4 \tag{31}$$

and

$$S(\hat{G}) = (\hat{\omega}_b - 1)^{1/2} = \left(\frac{\bar{\lambda}^{1/2}}{1 + (1 - \bar{\lambda})^{1/2}}\right)^2 \tag{32}$$

If, as is frequently the case, $\bar{\lambda}$ is substantially less than $\bar{\mu}^2$, then the second-degree SSOR method converges much faster than the SOR method.

If the choice of $u^{(1)}$ is specified, then we can give a more accurate assessment of the convergence properties of second-degree methods and a more direct comparison with the corresponding semi-iterative methods. It seems logical to let $u^{(1)}$ be determined as for the semi-iterative method. Thus, by (11-2.28) we let

$$u^{(1)} = \frac{1}{2 - (\beta + \alpha)} \{[2G_1 - (\beta + \alpha)I]u^{(0)} + 2k_1\} \tag{33}$$

If we let

$$G' = \frac{2}{2 - (\beta + \alpha)} G_1 - \frac{\beta + \alpha}{2 - (\beta + \alpha)} I \tag{34}$$

then by (10) and (22) we have, for $n = 1, 2, \ldots,$

$$u^{(n+1)} = \hat{\omega}_b G' u^{(n)} + (1 - \hat{\omega}_b) u^{(n-1)} + \frac{2\hat{\omega}_b}{2 - (\beta + \alpha)} k_1 \tag{35}$$

Thus we have

$$u^{(n+1)} - \bar{u} = Q_n(G')(u^{(0)} - \bar{u}) \tag{36}$$

where $\bar{u}$ is the exact solution of (3) and where the polynomials $Q_n(G')$ satisfy the recurrence relation

$$\begin{aligned} &Q_0(G') = I, \qquad Q_1(G') = G' \\ &Q_{n+1}(G') = \hat{\omega}_b G' Q_n(G') + (1 - \hat{\omega}_b) Q_{n-1}(G') \end{aligned} \tag{37}$$

These polynomials have been encountered before in Section 10.5 in connection with the modified Sheldon method. Since the eigenvalues γ of G' lie in the range

$$-\bar{\sigma} = -\frac{\beta - \alpha}{2 - (\beta + \alpha)} \le \gamma \le \frac{\beta - \alpha}{2 - (\beta + \alpha)} = \bar{\sigma} \tag{38}$$

it follows from (10-5.8) that

$$\max_{-\bar{\sigma}\le 0\le\bar{\sigma}} |Q_n(\gamma)| = Q_n(\bar{\sigma}) = \frac{2\hat{r}^{n/2}}{1+\hat{r}}\left[1+\left(\frac{n-1}{2}\right)(1-\hat{r})\right] \tag{39}$$

where $\hat{r}$ is given by (23). Therefore,

$$\bar{S}(Q_n(G')) = \frac{2\hat{r}^{n/2}}{1+\hat{r}}\left[1+\left(\frac{n-1}{2}\right)(1-\hat{r})\right] \tag{40}$$

as compared to

$$\bar{S}(P_n(G)) = \frac{2\hat{r}^{n/2}}{1+\hat{r}^n} \tag{41}$$

given by (11-2.37) for the corresponding semi-iterative method.

By the theory of Chebyshev polynomials we know that $\bar{S}(P_n(G)) \le \bar{S}(Q_n(G'))$. (A direct proof is given in Appendix C of Young and Kincaid, 1969.) Thus, the semi-iterative method converges faster than the second-degree method. This would be expected since in the second-degree method the three-term recurrence relation for $Q_n(G')$ has constant coefficients while variable coefficients are used for the semi-iterative method.

For the second-degree J method with A positive definite and having Property A we have by (33) and (35)

$$\begin{aligned} u^{(1)} &= Bu^{(0)} + c \\ u^{(n+1)} &= \omega_b(Bu^{(n)} + c) + (1-\omega_b)u^{(n-1)} \end{aligned} \tag{42}$$

For the second-degree GS method with A positive definite and consistently ordered we have

$$u^{(1)} = \frac{1}{2-\bar{\mu}^2}\{(2\mathscr{L} - \bar{\mu}^2 I)u^{(0)} + 2(I-L)^{-1}c\} \tag{43}$$

$$\begin{aligned} u^{(n+1)} = \hat{\omega}_b&\left\{\left(\frac{2}{2-\bar{\mu}^2}\mathscr{L} - \frac{\bar{\mu}^2}{2-\bar{\mu}^2}I\right)u^{(n)} + \frac{2}{2-\bar{\mu}^2}(I-L)^{-1}c\right\} \\ &+(1-\hat{\omega}_b)u^{(n-1)}, \qquad n = 1, 2, \ldots \end{aligned} \tag{44}$$

where $\hat{\omega}_b$ is given by (28). For the second degree SSOR method with A positive definite we have the same formulas as (43) with $\bar{\mu}^2$ replaced by $\bar{\lambda} = S(\mathscr{S}_\omega)$ and $\hat{\omega}_b$ given by (31).

As observed by Riley [1954], if A has the form (7-1), then one can use "compression" as was done in Section 11.5 in deriving the CCSI method from the J–SI method. Thus one needs compute only half the values of $u^{(n)}$ for each n. The method obtained in this way from the second-degree J method can be shown to be the same as the modified Sheldon method considered in Section 10.5.

If G_1, and hence G', are symmetric, then the $D^{1/2}$-norm and the $A^{1/2}$-norm of $Q_n(G')$ are the same as $\bar{S}(Q_n(G'))$. In particular, this is true for the case $G_1 = B$, where A is positive definite. In the case of the second-degree SSOR method with A positive definite, the $A^{1/2}$-norm is the same as the spectral radius.

Let us now assume that A is positive definite and has the form (7-1). We have already seen in Section 11.6 that the norms of the GS–SI method may be quite large. We were able to obtain smaller norms using the GS–SSI method. By analogy with the GS–SSI method we propose the following scheme defined by

$$u^{(1)} = \mathscr{L}u^{(0)} + (I - L)^{-1}c \tag{45}$$

$$u^{(2)} = \left(\frac{2}{2 - \bar{\mu}^2}\mathscr{L} - \frac{\bar{\mu}^2}{2 - \bar{\mu}^2}I\right)u^{(1)} + \frac{2}{2 - \bar{\mu}^2}(I - L)^{-1}c$$

and for $n \geq 2$, $u^{(n+1)}$ is determined by (44). Evidently for this method we have

$$u^{(n)} - \bar{u} = Q_{n-1}(\mathscr{L}')\mathscr{L}(u^{(n)} - \bar{u}) \tag{46}$$

for $n \geq 2$ where

$$\mathscr{L}' = \frac{2}{2 - \bar{\mu}^2}\mathscr{L} - \frac{\bar{\mu}^2}{2 - \bar{\mu}^2}I \tag{47}$$

and where the polynomials $Q_n(\mathscr{L}')$ are defined by (37). Thus, we have

$$\begin{aligned} S(Q_{n-1}(\mathscr{L}')\mathscr{L}) &= \frac{2\hat{r}^{(n-1)/2}}{1 + \hat{r}}\left[1 + \left(\frac{n-2}{2}\right)(1 - \hat{r})\right]\bar{\mu}^2 \\ &= \frac{8r^n}{(1+r^2)(1+r)^2}\left[1 + \left(\frac{n-2}{2}\right)(1 - r^2)\right] \end{aligned} \tag{48}$$

as compared with

$$\bar{S}(q_{n-1}(\mathscr{L})\mathscr{L}) = 8r^n/(1 + r^{2n-2})(1 + r)^2 \tag{49}$$

for the GS–SSI method [see (11-6.34)].

SUPPLEMENTARY DISCUSSION

Early work on the second-order Richardson method (the "second-degree Richardson method" in our terminology) was done by Frankel [1950] and Riley [1954]. See also Golub and Varga [1961]. The analysis given here appears in abbreviated form, in Young [1971a].

It does not seem to be generally recognized that in many cases, in particular for the SSOR–SI method, one can replace a semi-iterative method by a stationary second-degree method without a significant loss in convergence speed. It is possible that though this simplification is perhaps more apparent than real, nevertheless, it might lead to a more serious consideration of the SSOR method as a practical procedure in some cases.

EXERCISES

Chapter 16

1. Show that if $\hat{G}$ is given by (7), then $S(\hat{G}) < 1$ if and only if all roots λ of $\det(\lambda^2 I - \lambda G - H) = 0$ are less than unity in modulus.

2. Work out the derivation of (21) and (22) as follows: Show that if ϱ_1 and ϱ_2 are the largest roots of

$$-\frac{1-e+d}{e} - \frac{1}{e}\left(\varrho + \frac{d}{\varrho}\right) = \alpha$$

and

$$-\frac{1-e+d}{e} + \frac{1}{e}\left(\varrho + \frac{d}{\varrho}\right) = \beta$$

respectively, then, for fixed d, ϱ_1 is an increasing function of e and ϱ_2 is a decreasing function. Moreover, $\varrho^2 \geq |d|$ where $\varrho = \max(\varrho_1, \varrho_2)$. Then show that for given d we minimize ϱ by choosing

$$e = 2(1+d)/[2-(\beta+\alpha)]$$

Then show that for this value of e we minimize ϱ by letting $d = \hat{\omega}_b - 1$.

3. Find the optimum second-degree method based on the GS method for solving

$$\begin{pmatrix} 4 & -1 & -1 & 0 \\ -1 & 4 & 0 & -1 \\ -1 & 0 & 4 & -1 \\ 0 & -1 & -1 & 4 \end{pmatrix} \begin{pmatrix} u_1 \\ u_2 \\ u_3 \\ u_4 \end{pmatrix} = \begin{pmatrix} 1000 \\ 1000 \\ 0 \\ 0 \end{pmatrix}$$

Carry out two iterations with $u^{(0)} = 0$.

4. Let G be a matrix whose eigenvalues are real and lie in the interval $-\frac{1}{2} \leq \mu \leq \frac{3}{4}$. If $d = 0.5$, what value of e should be used to minimize $S(\hat{G})$. Find $S(\hat{G})$. Here

$$\hat{G} = \begin{pmatrix} 0 & I \\ -dI & (1 - e + d)I + eG \end{pmatrix}$$

5. Consider the application of the second degree SSOR method for the model problem with $h = \frac{1}{20}$. Estimate the number of iterations necessary to reduce the error to 10^{-6} of its original value.

6. Apply two iterations of the scheme which is analogous to the GS–SSI method to the problem of Exercise 3.

7. Show that if A has the form (7-1) and if one uses the second-degree J method with "compression" one obtains the modified Sheldon's method.

8. Show that if $0 \leq \hat{r} \leq 1$, then $\bar{S}(Q_n(G'))$ defined by (40) is not less than $\bar{S}(P_n(G))$ defined by (41).

9. Carry out an analysis of the second-degree RF method for the case where A is positive definite and the eigenvalues ν of A lie in the interval $0 < \bar{a} \leq \nu \leq \bar{b}$.

10. Consider the second-degree J method defined by

$$u^{(n+1)} = \omega_b(Bu^{(n)} + k) - (\omega_b - 1)u^{(n-1)}$$

Show that if A is a positive definite matrix such that diag $A = I$, then the eigenvalues λ of

$$\hat{G} = \begin{pmatrix} 0 & I \\ (1 - \omega_b)I & \omega_b B \end{pmatrix}$$

are related to the eigenvalues μ of B by

$$\lambda^2 + \omega_b - 1 = \omega_b \mu \lambda$$

Also, find an expression for $\| \hat{G} \|$ and $\| \hat{G}^m \|$.

11. Derive the conditions (21) and (22) as follows. Show that for fixed d the root radius of (14) considered as a quadratic in λ minimized when $e = 2(1 + d)[2 - (\alpha + \beta)]^{-1}$. Then find the best choice of d.

Chapter 17 / ALTERNATING DIRECTION IMPLICIT METHODS

17.1. INTRODUCTION: THE PEACEMAN–RACHFORD METHOD

In this chapter we consider a class of methods known as *alternating direction implicit* methods (ADI methods) for solving the system

$$Au = b \tag{1.1}$$

where A is nonsingular. We consider a splitting of A into the sum of three matrices

$$A = H_0 + V_0 + \Sigma \tag{1.2}$$

where Σ is a nonnegative diagonal matrix and where H_0, V_0, and Σ satisfy the following properties:

(a) $H_0 + \theta\Sigma + \varrho I$ and $V_0 + \theta\Sigma + \varrho I$ are nonsingular for all $\theta \geq 0$ and $\varrho > 0$.

(b) For any vectors s and t and for any $\theta \geq 0$, $\varrho > 0$ it is "convenient" to solve the systems

$$(H_0 + \theta\Sigma + \varrho I)x = s, \qquad (V_0 + \theta\Sigma + \varrho I)y = t \tag{1.3}$$

for x and y, respectively.

In a typical situation H and V would be tri-diagonal or could be made so by a permutation of the rows and corresponding columns. It would then be "convenient" to solve (1.3) in the sense that normally the work required would be small compared to that needed to solve (1.1).

By (1.2) we can write (1.1) in either of the forms

$$(H_0 + \theta\Sigma + \varrho I)u = b - (V_0 + (1 - \theta)\Sigma - \varrho I)u \tag{1.4}$$

$$(V_0 + \hat{\theta}\Sigma + \varrho' I)u = b - (H_0 + (1 - \hat{\theta})\Sigma - \varrho' I)u \tag{1.5}$$

In the *Peaceman–Rachford method* (PR method) the vector $u^{(n+1)}$ is determined from $u^{(n)}$ in two steps. Having chosen θ, $\hat{\theta}$, ϱ, and ϱ' one determines $u^{(n+1/2)}$ and $u^{(n+1)}$ successively by

$$\begin{aligned} (H_0 + \theta\Sigma + \varrho I)u^{(n+1/2)} &= b - (V_0 + (1 - \theta)\Sigma - \varrho I)u^{(n)} \\ (V_0 + \hat{\theta}\Sigma + \varrho' I)u^{(n+1)} &= b - (H_0 + (1 - \hat{\theta})\Sigma - \varrho' I)u^{(n+1/2)} \end{aligned} \tag{1.6}$$

By the assumption (b) above, it is convenient to determine $u^{(n+1/2)}$ and $u^{(n+1)}$. Frequently, the parameters ϱ and ϱ' are allowed to vary with n. In this case we have

$$\begin{aligned} (H_0 + \theta\Sigma + \varrho_{n+1} I)u^{(n+1/2)} &= b - (V_0 + (1 - \theta)\Sigma - \varrho_{n+1} I)u^{(n)} \\ (V_0 + \hat{\theta}\Sigma + \varrho'_{n+1} I)u^{(n+1)} &= b - (H_0 + (1 - \hat{\theta})\Sigma - \varrho'_{n+1} I)u^{(n+1/2)} \end{aligned} \tag{1.7}$$

A conventional choice of θ and $\hat{\theta}$ is given by $\theta = \hat{\theta} = \frac{1}{2}$. If we define

$$H = H_0 + \tfrac{1}{2}\Sigma, \qquad V = V_0 + \tfrac{1}{2}\Sigma \tag{1.8}$$

then (1.7) becomes

$$\begin{aligned} (H + \varrho_{n+1} I)u^{(n+1/2)} &= b - (V - \varrho_{n+1} I)u^{(n)} \\ (V + \varrho'_{n+1} I)u^{(n+1)} &= b - (H - \varrho'_{n+1} I)u^{(n+1/2)} \end{aligned} \tag{1.9}$$

Other choices have been used; for example, Wachspress [1966] used the choice $\theta = \hat{\theta} = 1$. (Actually, Wachspress did not assume that Σ is a diagonal matrix.)

If $\Sigma = \sigma I$, where σ is a constant, the choices of θ and $\hat{\theta}$ is immaterial in the following sense. By suitable choice of $\hat{\varrho}_{n+1}$ and $\hat{\varrho}'_{n+1}$ in (1.7), we can obtain (1.9). Thus, if $\hat{\varrho}_{n+1} = \varrho_{n+1} + (\theta - \frac{1}{2})\sigma$ and $\hat{\varrho}'_{n+1} = \varrho'_{n+1} + (\hat{\theta} - \frac{1}{2})\sigma$, then (1.7) becomes (1.9).

As an example, let us consider the problem of solving the five-point discrete analog of the generalized Dirichlet problem involving the unit square and the differential equation

$$u_{xx} + u_{yy} - u = 0 \tag{1.10}$$

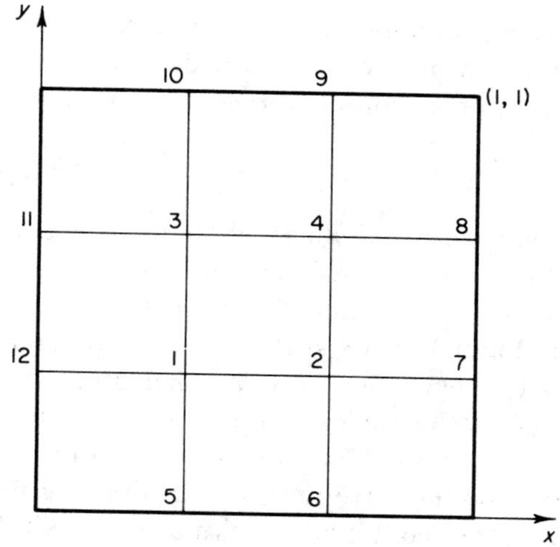

Figure 1.1.

With $h = \frac{1}{3}$ and the mesh points labeled as in Figure 1.1 we have, after replacing the derivatives by central difference quotients and multiplying by $-h^2$,

$$\begin{pmatrix} 4+h^2 & -1 & -1 & 0 \\ -1 & 4+h^2 & 0 & -1 \\ -1 & 0 & 4+h^2 & -1 \\ 0 & -1 & -1 & 4+h^2 \end{pmatrix} \begin{pmatrix} u_1 \\ u_2 \\ u_3 \\ u_4 \end{pmatrix} = \begin{pmatrix} u_5 + u_{12} \\ u_6 + u_7 \\ u_{10} + u_{11} \\ u_8 + u_9 \end{pmatrix} \tag{1.11}$$

This system of equations may be written in the form

$$Au = (H_0 + V_0 + \Sigma)u = b \tag{1.12}$$

where

$$H_0 = \begin{pmatrix} 2 & -1 & 0 & 0 \\ -1 & 2 & 0 & 0 \\ 0 & 0 & 2 & -1 \\ 0 & 0 & -1 & 2 \end{pmatrix}$$

$$V_0 = \begin{pmatrix} 2 & 0 & -1 & 0 \\ 0 & 2 & 0 & -1 \\ -1 & 0 & 2 & 0 \\ 0 & -1 & 0 & 2 \end{pmatrix} \tag{1.13}$$

$$\Sigma = h^2 I$$

Evidently H_0 and V_0 correspond to the discrete analogs of $-h^2 u_{xx}$ and $-h^2 u_{yy}$, respectively. Moreover, H_0 is tri-diagonal and V_0 can be made so by a permutation of its rows and columns. If we let

$$H = H_0 + \tfrac{1}{2}\Sigma, \qquad V = V_0 + \tfrac{1}{2}\Sigma \tag{1.14}$$

then H and V each have strong diagonal dominance and $H + \varrho I$, $V + \varrho I$ are nonsingular for any $\varrho > 0$ by Theorem 2-5.4. Hence, the PR method defined by (1.6) can be conveniently applied.

In Sections 17.2 and 17.3 we shall consider the stationary PR method where ϱ and ϱ' are independent of n. We shall study the convergence of the method as well as the choice of ϱ and ϱ' which maximizes the rate of convergence. In Sections 17.4–17.6 we shall study the nonstationary PR method in the commutative case, i.e., in the case where $HV = VH$ and where certain other conditions are satisfied. It will be shown that a very great improvement in the convergence rate can be attained by suitable choice of the iteration parameters. In Section 17.7 it will be shown that for the Helmholtz equation in the rectangle the commutative case holds and the optimum parameters can be explicitly determined. Next, it will be shown that for the stationary PR method for a system derived from a class of elliptic equations the rate of convergence is a monotone-decreasing function of the size of the region. (Unfortunately, this is not true in general for the nonstationary PR method.) In Section 17.9 we shall give necessary and sufficient conditions for the commutative case to hold. As will be indicated in Section 17.10, relatively little has been proved about the convergence properties of the nonstationary PR method in the noncommutative case even for many cases where the method works well in actual practice.

17.2. THE STATIONARY CASE: CONSISTENCY AND CONVERGENCE

Let us now consider the *stationary case* where

$$\varrho_n = \varrho, \qquad \varrho_n{}' = \varrho', \qquad n = 1, 2, \ldots \tag{2.1}$$

We show that if $H + \varrho I$ and $V + \varrho I$ are nonsingular for all $\varrho > 0$, then the PR method is completely consistent. Indeed, eliminating $u^{(n+1/2)}$ in (1.9) we obtain

$$u^{(n+1)} = T_{\varrho,\varrho'} u^{(n)} + k \tag{2.2}$$

where

$$T_{\varrho,\varrho'} = (V + \varrho'I)^{-1}(H - \varrho'I)(H + \varrho I)^{-1}(V - \varrho I) \tag{2.3}$$

and

$$k = (\varrho + \varrho')(V + \varrho'I)^{-1}(H + \varrho I)^{-1}b \tag{2.4}$$

But since

$$T_{\varrho,\varrho'} = I - (\varrho + \varrho')(V + \varrho'I)^{-1}(H + \varrho I)^{-1}A \tag{2.5}$$

it follows that

$$(I - T_{\varrho,\varrho'})A^{-1}b = k \tag{2.6}$$

Hence the method is consistent. Moreover, since $I - T_{\varrho,\varrho'}$ is nonsingular, the method is completely consistent.

Theorem 2.1. If $H + \frac{1}{2}(\varrho - \varrho')I$ and $V + \frac{1}{2}(\varrho' - \varrho)$ are positive definite, and if $\varrho > 0$, $\varrho' > 0$, then $S(T_{\varrho,\varrho'}) < 1$.

Proof. Let

$$\begin{aligned} \tilde{T}_{\varrho,\varrho'} &= (V + \varrho'I)T_{\varrho,\varrho'}(V + \varrho'I)^{-1} \\ &= (H - \varrho'I)(H + \varrho I)^{-1}(V - \varrho I)(V + \varrho'I)^{-1} \end{aligned} \tag{2.7}$$

Since $H + \frac{1}{2}(\varrho - \varrho')I$ is positive definite, it follows that $H + \frac{1}{2}(\varrho - \varrho')I$ and hence H are symmetric. If μ is any eigenvalue of H, then $(\mu - \varrho') \times (\mu+\varrho)^{-1}$ is an eigenvalue of $(H - \varrho'I)(H+\varrho I)^{-1}$. Since $H+\frac{1}{2}(\varrho - \varrho')I$ is positive definite, it follows that $\mu + \frac{1}{2}(\varrho - \varrho') > 0$; hence $\mu + \varrho > 0$ and

$$1 - \frac{\mu - \varrho'}{\mu + \varrho} = \frac{\varrho + \varrho'}{\mu + \varrho} > 0$$

$$\frac{\mu - \varrho'}{\mu + \varrho} - (-1) = \frac{2\mu + \varrho - \varrho'}{\mu + \varrho} > 0$$

and

$$\left| \frac{\mu - \varrho'}{\mu + \varrho} \right| < 1$$

Therefore,

$$\| (H - \varrho'I)(H + \varrho I)^{-1} \| = S((H - \varrho'I)(H + \varrho I)^{-1}) < 1 \tag{2.8}$$

Similarly,

$$\| (V - \varrho I)(V + \varrho'I)^{-1} \| < 1 \tag{2.9}$$

and

$$S(T_{\varrho,\varrho'}) = S(\tilde{T}_{\varrho,\varrho'}) \leq \| \tilde{T}_{\varrho,\varrho'} \| \leq \| (H - \varrho' I)(H + \varrho I)^{-1} \| \times \| (V - \varrho I)(V + \varrho' I)^{-1} \| < 1 \tag{2.10}$$

and the theorem follows.

For any ϱ, let us define

$$T_\varrho = T_{\varrho,\varrho} \tag{2.11}$$

Then we have

Corollary 2.2. If H and V are positive definite, then for any $\varrho > 0$ we have

$$S(T_\varrho) < 1 \tag{2.12}$$

As an extension of Theorem 2.1 we have

Theorem 2.3. If $\varrho > 0$ and $\varrho' > 0$ and if there exists a nonsingular matrix P such that $P^{-1}(H + \frac{1}{2}(\varrho - \varrho')I)P$ and $P^{-1}(V + \frac{1}{2}(\varrho' - \varrho)I)P$ are positive real or, equivalently, if there exists a positive definite matrix Q such that $Q(H + \frac{1}{2}(\varrho - \varrho')I)$ and $Q(V + \frac{1}{2}(\varrho' - \varrho)I)$ are positive real, then $S(T_{\varrho,\varrho'}) < 1$.

Proof. To show that the alternative hypotheses are equivalent we prove

Lemma 2.4. If W is a real matrix, then there exists a real nonsingular matrix P such that $P^{-1}WP$ is positive real if and only if there exists a real positive definite matrix Q such that QW is positive real.

Proof. If P exists, then $P^{-1}WP + (P^{-1}WP)^T$ is positive definite by Theorem 2-2.9. Hence by Corollary 2-2.8, $P(P^{-1}WP + (P^{-1}WP)^T)P^T = WR + RW^T$ is positive definite, where $R = PP^T$. Also, $R^{-1}(WR + RW^T) \times R^{-1} = R^{-1}W + W^TR^{-1} = (R^{-1}W) + (R^{-1}W)^T$ is positive definite and hence, by Theorem 2-2.9, QW is positive real where $Q = R^{-1}$.

On the other hand, if Q exists, then $QW + W^TQ$ is positive definite as is $Q^{-1/2}(QW + W^TQ)Q^{-1/2} = P^{-1}WP + (P^{-1}WP)^T$ where $P = Q^{-1/2}$. Hence $P^{-1}WP$ is positive real and the lemma follows.

Suppose now that P exists and let

$$\tilde{H} = P^{-1}HP, \qquad \tilde{V} = P^{-1}VP \tag{2.13}$$

Since

$$(\tilde{H} - \varrho' I)(\tilde{H} + \varrho I)^{-1} = I - (\varrho + \varrho')(\tilde{H} + \varrho I)^{-1} \tag{2.14}$$

we have

$$\begin{aligned}[(\tilde{H} - \varrho' I)(\tilde{H} + \varrho I)^{-1}][(\tilde{H} - \varrho' I)(\tilde{H} + \varrho I)^{-1}]^{\mathrm{T}} \\ = I - (\varrho + \varrho')(\tilde{H} + \varrho I)^{-1}(\tilde{H} + \tilde{H}^{\mathrm{T}} + (\varrho - \varrho')I)[(\tilde{H} + \varrho I)^{-1}]^{\mathrm{T}}\end{aligned} \tag{2.15}$$

Since $\tilde{H} + \frac{1}{2}(\varrho - \varrho')I$ is positive real, it follows by Theorem 2-2.9 that $\tilde{H} + \tilde{H}^{\mathrm{T}} + (\varrho - \varrho')I$ is positive definite, and

$$\| (\tilde{H} - \varrho' I)(\tilde{H} + \varrho I)^{-1} \| < 1 \tag{2.16}$$

Similarly,

$$\| (\tilde{V} - \varrho I)(\tilde{V} + \varrho' I)^{-1} \| < 1 \tag{2.17}$$

and hence

$$\begin{aligned}S(T_{\varrho,\varrho'}) &= S(P^{-1}(V + \varrho' I)T_{\varrho,\varrho'}(V + \varrho' I)^{-1}P) \\ &\leq \| P^{-1}(V + \varrho' I)T_{\varrho,\varrho'}(V + \varrho' I)^{-1}P \| \\ &= \| (\tilde{H} - \varrho' I)(\tilde{H} + \varrho I)^{-1}(\tilde{V} - \varrho I)(\tilde{V} + \varrho' I)^{-1} \| \\ &\leq \| (\tilde{H} - \varrho' I)(\tilde{H} + \varrho I)^{-1} \| \, \| (\tilde{V} - \varrho I)(\tilde{V} + \varrho' I)^{-1} \| < 1\end{aligned} \tag{2.18}$$

This completes the proof of Theorem 2.3.

Corollary 2.5. If there exists a nonsingular matrix P such that $P^{-1}HP$ and $P^{-1}VP$ are positive real, or, equivalently, if there exists a positive definite matrix Q such that QH and QV are positive real, then $S(T_\varrho) < 1$ for all $\varrho > 0$.

From (2.3) we have

$$I + T_{\varrho,\varrho'} = (V + \varrho' I)^{-1}(H + \varrho I)^{-1}(2HV + (\varrho - \varrho')(V - H) + 2\varrho\varrho' I) \tag{2.19}$$

Hence by (2.5) and Theorem 3-6.3, we have

Theorem 2.6. A necessary and sufficient condition that $S(T_{\varrho,\varrho'}) < 1$ is that

$$(H + V)^{-1}(HV + \tfrac{1}{2}(\varrho - \varrho')(V - H) + \varrho\varrho' I) \tag{2.20}$$

is N-stable.

Corollary 2.7. A necessary and sufficient condition that $S(T_\varrho) < 1$ is that

$$(H + V)^{-1}(HV + \varrho^2 I) \tag{2.21}$$

is N-stable.

Theorem 2.8. If H and V are nonsingular and if $S(T_\varrho) < 1$ for all $\varrho > 0$, then

$$H^{-1} + V^{-1} \tag{2.22}$$

is N-stable.

Proof. If $H^{-1} + V^{-1}$ is not N-stable, then $(H^{-1} + V^{-1})^{-1}$ is not N-stable. Moreover, $(H + V)^{-1}HV$, which is similar to $V(H + V)^{-1}H = (H^{-1} + V^{-1})^{-1}$, is also not N-stable. Hence by the continuity of the eigenvalues of $(H + V)^{-1}(HV + \varrho^2 I)$ as a function of ϱ, it follows that, for ϱ sufficiently small, $(H + V)^{-1}(HV + \varrho^2 I)$ is not N-stable. Therefore, by Corollary 2.7, it follows that $S(T_\varrho) \geq 1$.

Under certain circumstances, if A is positive definite, we can show that the $A^{1/2}$-norm of T_ϱ is less than one for all ϱ. As a matter of fact, by Theorem 3-5.3, we have

Theorem 2.9. If A is positive definite, then $\| T_\varrho \|_{A^{1/2}} < 1$ if and only if

$$HV + VH + 2\varrho^2 I \tag{2.23}$$

is positive definite.

Evidently the condition (2.23) is equivalent to requiring that $HV + \varrho^2 I$ should be positive real. Guilinger [1965] has proved that for the discrete analog of the Dirichlet problem in a convex region, HV and hence $HV + \varrho^2 I$ are positive real.

From the properties of the $A^{1/2}$-norm and from Theorem 2.9 we have, using Theorem 2-2.9,

Corollary 2.10. If A is positive definite and HV is positive real, then for any set of positive parameters, $\varrho_1, \varrho_2, \ldots, \varrho_m$ we have

$$S\left(\prod_{i=1}^{m} T_{\varrho_i}\right) < 1 \tag{2.24}$$

17.3. THE STATIONARY CASE: CHOICE OF PARAMETERS

In this section we shall assume that H and V are positive definite matrices and that the eigenvalues μ of H and ν of V lie in the ranges

$$0 < a \le \mu \le b, \qquad 0 < \alpha \le \nu \le \beta \tag{3.1}$$

Evidently, we have

$$\begin{aligned} S(T_{\varrho,\varrho'}) &= S[(V + \varrho' I)T_{\varrho,\varrho'}(V + \varrho' I)^{-1}] \\ &= S[(H - \varrho' I)(H + \varrho I)^{-1}(V - \varrho I)(V + \varrho' I)^{-1}] \\ &\le \| (H - \varrho' I)(H + \varrho I)^{-1} \| \, \| (V - \varrho I)(V + \varrho' I)^{-1} \| \\ &= \max_{a \le \mu \le b} \left| \frac{\mu - \varrho'}{\mu + \varrho} \right| \max_{\alpha \le \nu \le \beta} \left| \frac{\nu - \varrho}{\nu + \varrho'} \right| \\ &= \max_{\substack{a \le \mu \le b \\ \alpha \le \nu \le \beta}} \left| \left(\frac{\mu - \varrho'}{\mu + \varrho} \right) \left(\frac{\nu - \varrho}{\nu + \varrho'} \right) \right| \end{aligned} \tag{3.2}$$

We shall seek to choose ϱ and ϱ' to minimize the quantity

$$\psi(a, b; \alpha, \beta; \varrho, \varrho') = \max_{\substack{a \le \mu \le b \\ \alpha \le \nu \le \beta}} \left| \left(\frac{\mu - \varrho'}{\mu + \varrho} \right) \left(\frac{\nu - \varrho}{\nu + \varrho'} \right) \right| \tag{3.3}$$

We shall consider separately cases where $\varrho = \varrho'$ and/or where the ranges of μ and ν are the same.

We first show that we can assume that

$$\bar{a} \le \varrho \le \bar{b}, \qquad \bar{a} \le \varrho' \le \bar{b} \tag{3.4}$$

where

$$\bar{a} = \min(a, \alpha), \qquad \bar{b} = \max(b, \beta) \tag{3.5}$$

Evidently

$$\partial/\partial\varrho([\nu - \varrho]/[\mu + \varrho]) = -(\mu + \nu)/(\mu + \varrho)^2 < 0 \tag{3.6}$$

If $\varrho \ge \bar{b}$ then for $\bar{a} \le \nu \le \bar{b}$ we have

$$| (\nu - \varrho)/(\mu + \varrho) | = (\varrho - \nu)/(\mu + \varrho) \tag{3.7}$$

which is an increasing function of ϱ. Thus $| (\nu - \varrho)/(\mu + \varrho) |$ is reduced if ϱ is replaced by $\bar{b}$. Similarly, if $\varrho \le \bar{a}$, then $| (\nu - \varrho)/(\mu + \varrho) |$ is reduced if ϱ is replaced by $\bar{a}$.

Case Where $\varrho = \varrho'$: Same Range for μ and ν

Theorem 3.1. There exists a unique value of ϱ, namely,

$$\varrho^* = (ab)^{1/2} \tag{3.8}$$

such that

$$\psi(a, b; a, b; \varrho^*, \varrho^*) = \left(\frac{b^{1/2} - a^{1/2}}{b^{1/2} + a^{1/2}}\right)^2 \tag{3.9}$$

and unless $\varrho = \varrho^*$ we have

$$\psi(a, b; a, b; \varrho, \varrho) > \psi(a, b; a, b; \varrho^*, \varrho^*) \tag{3.10}$$

Proof. Evidently

$$\psi(a, b; a, b; \varrho, \varrho) = \phi(a, b; \varrho)^2 \tag{3.11}$$

where

$$\phi(a, b; \varrho) = \max_{a \le \mu \le b} |\,(\mu - \varrho)/(\mu + \varrho)\,| \tag{3.12}$$

Since

$$(d/d\mu)([\mu - \varrho]/[\mu + \varrho]) = 2\varrho/(\mu + \varrho)^2 \tag{3.13}$$

it follows that for given ϱ in the interval $a \le \varrho \le b$ the derivative is positive and the extreme values of $(\mu - \varrho)(\mu + \varrho)^{-1}$ occur at the end points of the interval $a \le \mu \le b$. Hence we have

$$\begin{aligned} \phi(a, b; \varrho) &= \max\{([\varrho - a]/[\varrho + a]), ([b - \varrho]/[b + \varrho])\} \\ &= \begin{cases} (b - \varrho)/(b + \varrho), & a \le \varrho \le (ab)^{1/2} \\ (\varrho - a)/(\varrho + a), & (ab)^{1/2} \le \varrho \le b \end{cases} \end{aligned} \tag{3.14}$$

Since

$$\begin{aligned} (d/d\varrho)([\varrho - a]/[\varrho + a]) &= 2a/(\varrho + a)^2 > 0 \\ (d/d\varrho)([b - \varrho]/[b + \varrho]) &= -2b/(b + \varrho)^2 < 0 \end{aligned} \tag{3.15}$$

it follows that $\phi(a, b; \varrho)$ is a decreasing function of ϱ for $a \le \varrho \le (ab)^{1/2}$ and an increasing function of ϱ for $(ab)^{1/2} \le \varrho \le b$. Evidently the minimum value is assumed when $\varrho = \varrho^*$ and in this case

$$\phi(a, b; \varrho^*) = \frac{b^{1/2} - a^{1/2}}{b^{1/2} + a^{1/2}} \tag{3.16}$$

The theorem now follows from (3.11).

Case Where the Ranges May Be Different; $\varrho = \varrho'$

Theorem 3.2. If

$$ab \leq \alpha\beta \tag{3.17}$$

then there exists a value of ϱ, namely

$$\varrho^* = \begin{cases} (ab)^{1/2} & \text{if } a \geq \alpha \text{ or } a \leq \alpha \text{ and } a\beta \geq b\alpha \\ (\alpha\beta)^{1/2} & \text{if } b \geq \beta \text{ or } b \leq \beta \text{ and } a\beta \leq b\alpha \end{cases} \tag{3.18}$$

such that

$$\begin{aligned} &\psi(a, b; \alpha, \beta; \varrho^*, \varrho^*) \\ &= \left(\frac{\varrho^* - a}{\varrho^* + a}\right)\left(\frac{\beta - \varrho^*}{\beta + \varrho^*}\right) \\ &= \begin{cases} \left(\dfrac{b^{1/2} - a^{1/2}}{b^{1/2} + a^{1/2}}\right)\left(\dfrac{\beta - (ab)^{1/2}}{\beta + (ab)^{1/2}}\right) & \text{if } \varrho^* = (ab)^{1/2} \\ \left(\dfrac{(\alpha\beta)^{1/2} - a}{(\alpha\beta)^{1/2} + a}\right)\left(\dfrac{\beta^{1/2} - \alpha^{1/2}}{\beta^{1/2} + \alpha^{1/2}}\right) & \text{if } \varrho^* = (\alpha\beta)^{1/2} \end{cases} \end{aligned} \tag{3.19}$$

Moreover, for all ϱ we have

$$\psi(a, b; \alpha, \beta; \varrho^*, \varrho^*) \leq \psi(a, b; \alpha, \beta; \varrho, \varrho) \tag{3.20}$$

and unless a, α, b, and β satisfy the conditions

$$a \leq \alpha, \qquad b \leq \beta, \qquad a\beta = b\alpha \tag{3.21}$$

then the strict inequality holds in (3.20) for $\varrho \neq \varrho^*$. If (3.21) holds, then we have the strict inequality in (3.20) unless $\varrho = (ab)^{1/2}$ or $\varrho = (\alpha\beta)^{1/2}$. Moreover, in this case

$$\psi(a, b; \alpha, \beta; (ab)^{1/2}, (ab)^{1/2}) = \psi(a, b; \alpha, \beta; (\alpha\beta)^{1/2}, (\alpha\beta)^{1/2}) \tag{3.22}$$

Proof. By (3.3) and (3.12) we have

$$\psi(a, b; a, b; \varrho, \varrho) = \phi(a, b; \varrho)\phi(\alpha, \beta; \varrho) \tag{3.23}$$

Evidently $\phi(a, b, \varrho)$ and $\phi(\alpha, \beta; \varrho)$ are both decreasing for $a \leq \varrho \leq (ab)^{1/2}$ and are both increasing for $(\alpha\beta)^{1/2} \leq \varrho \leq b$. Hence a minimum value of ψ, if it exists, must correspond to ϱ in the interval $(ab)^{1/2} \leq \varrho \leq (\alpha\beta)^{1/2}$.

But in that interval we have

$$\phi(a, b; \varrho) = (\varrho - a)/(\varrho + a), \qquad \phi(\alpha, \beta; \varrho) = (\beta - \varrho)/(\beta + \varrho) \tag{3.24}$$

and

$$\begin{cases} (d^2/d\varrho^2) \log \phi(a, b; \varrho) = -4a\varrho/(\varrho^2 - a^2)^2 \\ (d^2/d\varrho^2) \log \phi(\alpha, \beta; \varrho) = -4\beta\varrho/(\beta^2 - \varrho^2)^2 \end{cases} \tag{3.25}$$

so that

$$(d^2/d\varrho^2) \log \psi(a, b; \alpha, \beta; \varrho, \varrho) < 0 \tag{3.26}$$

On the other hand,

$$(d^2/d\varrho^2) \log \psi = [\psi\psi'' - (\psi')^2]/\psi^2 \tag{3.27}$$

Hence, if $\psi' = 0$, then $\psi'' < 0$ since $\psi > 0$ and $(\log \psi)'' < 0$. Thus there cannot be a local minimum for ψ corresponding to ϱ in the interior of the interval $(ab)^{1/2} \le \varrho \le (\alpha\beta)^{1/2}$. Therefore the extreme values must occur at the end points, i.e., for $\varrho = (ab)^{1/2}$ or $\varrho = (\alpha\beta)^{1/2}$ and we have

$$\min_{\varrho} \psi(a, b; \alpha, \beta; \varrho, \varrho)$$
$$= \min\left\{\left(\frac{b^{1/2} - a^{1/2}}{b^{1/2} + a^{1/2}}\right)\left(\frac{\beta - (ab)^{1/2}}{\beta + (ab)^{1/2}}\right), \left(\frac{(\alpha\beta)^{1/2} - a}{(\alpha\beta)^{1/2} + a}\right)\left(\frac{\beta^{1/2} - \alpha^{1/2}}{\beta^{1/2} + \alpha^{1/2}}\right)\right\} \tag{3.28}$$

Let x and y be defined by

$$\begin{aligned} S_1 &= \frac{1 - x}{1 + x} = \frac{b^{1/2} - a^{1/2}}{b^{1/2} + a^{1/2}} \frac{\beta - (ab)^{1/2}}{\beta + (ab)^{1/2}} \\ S_2 &= \frac{1 - y}{1 + y} = \left(\frac{(\alpha\beta)^{1/2} - a}{(\alpha\beta)^{1/2} + a}\right)\left(\frac{\beta^{1/2} - \alpha^{1/2}}{\beta^{1/2} + \alpha^{1/2}}\right) \end{aligned} \tag{3.29}$$

Thus

$$x = \frac{(a/b)^{1/2} + [(ab)^{1/2}/\beta]}{1 + (a/\beta)}, \qquad y = \frac{[a/(\alpha\beta)^{1/2}] + (\alpha/\beta)^{1/2}}{1 + (a/\beta)} \tag{3.30}$$

Evidently $S_1 \le S_2$ if and only if $x \ge y$, i.e., if and only if

$$(\beta a^{1/2} + ba^{1/2})/b^{1/2} \ge (a\beta^{1/2} + \alpha\beta^{1/2})/\alpha^{1/2} \tag{3.31}$$

i.e., if and only if

$$T_1 = (\beta/b)^{1/2} + (b/\beta)^{1/2} \ge (a/\alpha)^{1/2} + (\alpha/a)^{1/2} = T_2 \tag{3.32}$$

Suppose now that $a \geq \alpha$. Then $b \leq \beta$; otherwise, $ab \geq \alpha\beta$. Moreover, $a/\alpha \leq \beta/b$ since $ab \leq \alpha\beta$; hence† $T_1 \geq T_2$, $S_1 \geq S_2$, and $\varrho^* = (ab)^{1/2}$. Similarly, if $b \geq \beta$, then $a \leq \alpha$ and $\alpha/a \geq b/\beta$. Therefore, $T_2 \geq T_1$, $S_2 \geq S_1$, and $\varrho^* = (\alpha\beta)^{1/2}$.

Now suppose $a \leq \alpha$ and $b \leq \beta$. If $a\beta \geq b\alpha$, then $\beta/b \geq \alpha/a$ and $T_1 \geq T_2$, $S_1 \geq S_2$, $\varrho^* = (ab)^{1/2}$. On the other hand, if $a\beta \leq b\alpha$, then $\beta/b \leq \alpha/a$, and $T_1 \leq T_2$, $S_1 \leq S_2$, $\varrho^* = (\alpha\beta)^{1/2}$. This completes the proof of Theorem 3.2.

Corollary 3.3. If $a + b = \alpha + \beta$, and $ab \leq \alpha\beta$, then $\varrho^* = (\alpha\beta)^{1/2}$.

Proof. Let $s = a + b = \alpha + \beta$. Then $a(s - a) \leq \alpha(s - \alpha)$. Hence $0 \leq \alpha(s - \alpha) - a(s - a) = (\alpha - a)(s - (a + \alpha))$. But since $a \leq s/2$, $\alpha \leq s/2$, it follows that $\alpha \geq a$. Moreover, $b \geq \beta$. Hence $\alpha/a \geq 1 \geq \beta/b$, and $\alpha b \geq a\beta$. Thus by Theorem 3.2 it follows that $\varrho^* = (\alpha\beta)^{1/2}$.

Case Where the Ranges Are the Same: ϱ and ϱ' May Be Different

Theorem 3.4. If

$$\varrho^* = (ab)^{1/2} \tag{3.33}$$

then

$$\psi(a, b; a, b; \varrho^*, \varrho^*) = \left(\frac{b^{1/2} - a^{1/2}}{b^{1/2} + a^{1/2}}\right)^2 \tag{3.34}$$

and unless $\varrho = \varrho' = \varrho^*$, then

$$\psi(a, b; a, b; \varrho, \varrho') > \psi(a, b; a, b; \varrho^*, \varrho^*) \tag{3.35}$$

Proof. Evidently

$$\psi(a, b; a, b; \varrho, \varrho') = \Gamma(a, b; \varrho, \varrho')\Gamma(a, b; \varrho', \varrho) \tag{3.36}$$

where

$$\Gamma(a, b; \varrho, \varrho') = \max_{a \leq \mu \leq b} \left| \frac{\mu - \varrho'}{\mu + \varrho} \right| \tag{3.37}$$

Since $a \leq \varrho \leq b$, $a \leq \varrho' \leq b$ we have

$$\frac{d}{d\mu}\left(\frac{\mu - \varrho'}{\mu + \varrho}\right) = \frac{\varrho + \varrho'}{(\mu + \varrho)^2} > 0 \tag{3.38}$$

† Here we use the fact that if $w \geq 1$, $z \geq 1$, then $w + w^{-1} \leq z + z^{-1}$ if and only if $w \leq z$.

and it follows that for fixed $\varrho \geq 0$ we have

$$\Gamma(a, b; \varrho, \varrho') = \max\left[\left(\frac{\varrho' - a}{\varrho + a}\right), \left(\frac{b - \varrho'}{\varrho + b}\right)\right]$$

$$= \begin{cases} \dfrac{b - \varrho'}{b + \varrho} & \text{if} \quad a \leq \varrho' \leq g(\varrho) \\ \dfrac{\varrho' - a}{\varrho + a} & \text{if} \quad g(\varrho) \leq \varrho' \leq b \end{cases} \tag{3.39}$$

where

$$g(\varrho) = \frac{2ab + (b + a)\varrho}{2\varrho + b + a} \tag{3.40}$$

Moreover, we have

$$\psi(a, b; a, b; \varrho, \varrho') = \begin{cases} \left(\dfrac{b - \varrho'}{b + \varrho'}\right)\left(\dfrac{b - \varrho}{b + \varrho}\right) & \text{if} \quad (\varrho, \varrho') \in R_1 \\ \left(\dfrac{\varrho' - a}{\varrho' + a}\right)\left(\dfrac{\varrho - a}{\varrho + a}\right) & \text{if} \quad (\varrho, \varrho') \in R_2 \\ \left(\dfrac{b - \varrho'}{b + \varrho}\right)\left(\dfrac{\varrho - a}{\varrho' + a}\right) & \text{if} \quad (\varrho, \varrho') \in R_3 \\ \left(\dfrac{\varrho' - a}{\varrho + a}\right)\left(\dfrac{b - \varrho}{b + \varrho'}\right) & \text{if} \quad (\varrho, \varrho') \in R_4 \end{cases} \tag{3.41}$$

where we define the subsets R_1, R_2, R_3, and R_4 of the set $a \leq \varrho \leq b$, $a \leq \varrho' \leq b$ by

$$\begin{aligned} &(\varrho, \varrho') \in R_1 && \text{if} \quad \varrho \leq g(\varrho'), \quad \varrho' \leq g(\varrho) \\ &(\varrho, \varrho') \in R_2 && \text{if} \quad \varrho \geq g(\varrho'), \quad \varrho' \geq g(\varrho) \\ &(\varrho, \varrho') \in R_3 && \text{if} \quad \varrho \geq g(\varrho'), \quad \varrho' \leq g(\varrho) \\ &(\varrho, \varrho') \in R_4 && \text{if} \quad \varrho \leq g(\varrho'), \quad \varrho' \geq g(\varrho) \end{aligned}$$

(see Figure 3.1).

By (3.40) we have

$$g'(\varrho) = (b - a)^2/(2\varrho + b + a)^2 \geq 0 \tag{3.42}$$

$$g(\varrho) - \varrho = 2(ab - \varrho^2)/(2\varrho + b + a) \tag{3.43}$$

$$g(\varrho) - (ab)^{1/2} = (a+b - 2[ab]^{1/2})(\varrho - [ab]^{1/2})/(2\varrho+b+a) \tag{3.44}$$

If $(\varrho, \varrho') \in R_1$ and if $\varrho' \leq \varrho$, then since $\varrho \leq g(\varrho')$ we have by (3.42)

$$\varrho \leq g(\varrho') \leq g(\varrho) \tag{3.45}$$

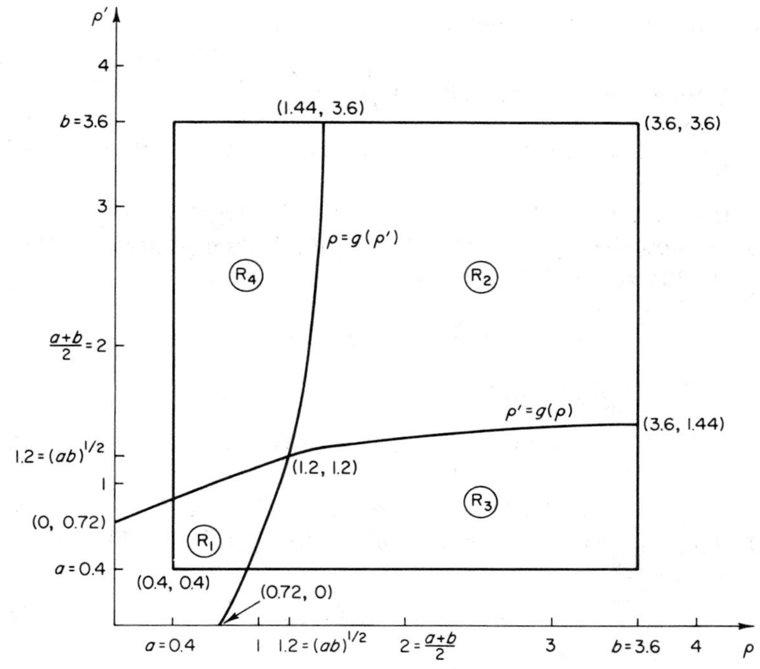

Figure 3.1. Case a $= \alpha = 0.4$. Case b $= \beta = 3.6$.

Hence $(\varrho, \varrho) \in R_1$. Moreover,

$$d/d\varrho'([b - \varrho']/[b + \varrho']) = -2b/(b + \varrho')^2 \leq 0 \qquad (3.46)$$

and, by (3.41),

$$\psi(a, b; a, b; \varrho, \varrho') \geq ([b - \varrho]/[b + \varrho])^2 = \psi(a, b; a, b; \varrho, \varrho) \quad (3.47)$$

Similarly, if $\varrho' \geq \varrho$, then we have

$$\psi(a, b; a, b; \varrho, \varrho') \geq \psi(a, b; a, b; \varrho', \varrho') = ([b - \varrho']/[b+\varrho'])^2 \quad (3.48)$$

Let us now determine the minimum value of $\psi(a, b; a, b; \varrho, \varrho)$ when $\varrho \leq g(\varrho)$. Since

$$\psi(a, b; a, b; \varrho, \varrho) = \psi(a, b; \varrho, \varrho)^2 = ([b - \varrho]/[b + \varrho])^2 \quad (3.49)$$

is a decreasing function of ϱ for $\varrho \leq g(\varrho)$ and since the largest value of ϱ for which $\varrho \leq g(\varrho)$ is $\varrho = (ab)^{1/2}$, by (3.43), it follows that the optimum pair (ϱ, ϱ') in R_1 is

$$\varrho = \varrho' = (ab)^{1/2} \qquad (3.50)$$

and the corresponding value of ψ is

$$(\psi(a, b; (ab)^{1/2}, (ab)^{1/2}))^2 = \left(\frac{b^{1/2} - a^{1/2}}{b^{1/2} + a^{1/2}}\right)^2$$
$$= \psi(a, b; a, b; (ab)^{1/2}, (ab)^{1/2}) \qquad (3.51)$$

A similar argument can be used to show that this result is also true for R_2.

Suppose now that $(\varrho, \varrho') \in R_3$. If $\varrho \geq (ab)^{1/2}$, then $(\varrho, g(\varrho)) \in R_3 \cap R_2$. For by (3.43) and (3.44) we have

$$(ab)^{1/2} \leq g(\varrho) \leq \varrho \qquad (3.52)$$

since $a + b \geq 2(ab)^{1/2}$, and

$$(ab)^{1/2} \leq g(g(\varrho)) \leq g(\varrho) \leq \varrho \qquad (3.53)$$

On the other hand, if $\varrho \leq (ab)^{1/2}$, then $(g(\varrho'), \varrho') \in R_3 \cap R_1$. For by (3.44), $g(\varrho) \leq (ab)^{1/2}$ and hence $\varrho' \leq g(\varrho) \leq (ab)^{1/2}$. Therefore, by (3.43) and (3.44)

$$\varrho' \leq g(\varrho') \leq (ab)^{1/2} \qquad (3.54)$$

and

$$\varrho' \leq g(\varrho') \leq g(g(\varrho')) \leq (ab)^{1/2} \qquad (3.55)$$

By (3.40) we have, since $\varrho \geq g(\varrho')$ and $\varrho' \leq g(\varrho)$,

$$(b + a)(\varrho - \varrho') \geq 2 \,|\, ab - \varrho\varrho' \,|\, \geq 0 \qquad (3.56)$$

and hence $\varrho \geq \varrho'$.

Since

$$\begin{aligned} \frac{\partial}{\partial \varrho}\left[\left(\frac{b - \varrho'}{b + \varrho}\right)\left(\frac{\varrho - a}{\varrho' + a}\right)\right] &= \left(\frac{b - \varrho'}{\varrho' + a}\right)\frac{(b + a)}{(b + \varrho)^2} \geq 0 \\ \frac{\partial}{\partial \varrho'}\left[\left(\frac{b - \varrho'}{b + \varrho}\right)\left(\frac{\varrho - a}{\varrho' + a}\right)\right] &= -\left(\frac{\varrho - a}{b + \varrho}\right)\frac{(a + b)}{(\varrho' + a)^2} \leq 0 \end{aligned} \qquad (3.57)$$

we can reduce $\psi(a, b; a, b; \varrho, \varrho')$ by either decreasing ϱ or increasing ϱ'. Thus if $\varrho \geq (ab)^{1/2}$, we decrease ψ by replacing ϱ' by $g(\varrho)$, while if $\varrho \leq (ab)^{1/2}$, we decrease ψ by replacing ϱ by $g(\varrho')$. Since in the first case $(\varrho, g(\varrho)) \in R_2$ and in the second case $(g(\varrho'), \varrho') \in R_1$ it follows that we can reduce ψ still further by letting $\varrho = \varrho' = (ab)^{1/2}$. The theorem now follows.

Case Where the Ranges May Be Different: ϱ and ϱ' May Be Different

We now assume that μ and ν vary over different ranges $a \leq \mu \leq b$, $\alpha \leq \nu \leq \beta$, and that ϱ and ϱ' may be different. We first describe how by a change of variables proposed by Wachspress and Jordan (see Wachspress, 1966) one can reduce to the case of a single range.

We seek to introduce new variables $\hat{\mu}$ and $\hat{\nu}$ such that

$$\mu = \frac{p + q\hat{\mu}}{1 + s\hat{\mu}}, \qquad \nu = \frac{p' + q'\hat{\nu}}{1 + s'\hat{\nu}} \tag{3.58}$$

so that for some $\hat{\varrho}$ and $\hat{\varrho}'$ we have

$$\left(\frac{\mu - \varrho'}{\mu + \varrho}\right)\left(\frac{\nu - \varrho}{\nu + \varrho'}\right) = \left(\frac{\hat{\mu} - \hat{\varrho}'}{\hat{\mu} + \hat{\varrho}}\right)\left(\frac{\hat{\nu} - \hat{\varrho}}{\hat{\nu} + \hat{\varrho}'}\right) \tag{3.59}$$

and where $\hat{\mu}$ and $\hat{\nu}$ both vary over the ranges

$$r \leq \hat{\mu} \leq R, \qquad r \leq \hat{\nu} \leq R \tag{3.60}$$

From (3.58) we have

$$\frac{\mu - \varrho'}{\mu + \varrho} = \left(\frac{q - s\varrho'}{q + s\varrho}\right)\frac{\hat{\mu} - \left(\dfrac{\varrho' - p}{q - \varrho' s}\right)}{\hat{\mu} + \left(\dfrac{\varrho + p}{q + \varrho s}\right)} \tag{3.61}$$

$$\frac{\nu - \varrho}{\nu + \varrho'} = \left(\frac{q' - s'\varrho}{q' + s'\varrho'}\right)\left(\frac{\hat{\nu} - \left(\dfrac{\varrho - p'}{q' - \varrho s'}\right)}{\hat{\nu} + \left(\dfrac{\varrho' + p'}{q' + s'\varrho'}\right)}\right) \tag{3.62}$$

Evidently (3.59) will hold for all ϱ and ϱ' provided the following conditions are satisfied

$$\left(\frac{q - s\varrho'}{q + s\varrho}\right)\left(\frac{q' - s'\varrho}{q' + s'\varrho'}\right) = 1 \tag{3.63}$$

$$\frac{\varrho' - p}{q - \varrho' s} = \frac{\varrho' + p'}{q' + s'\varrho'} \tag{3.64}$$

$$\frac{\varrho + p}{q + s\varrho} = \frac{\varrho - p'}{q' - \varrho s'} \tag{3.65}$$

The first condition (3.63) implies

$$(\varrho + \varrho')(qs' + q's) = 0 \tag{3.66}$$

which holds for all ϱ and ϱ' provided

$$qs' + q's = 0 \tag{3.67}$$

The second condition (3.64) implies

$$(\varrho')^2(s' + s) + \varrho'(q' - ps' + p's - q) = qp' + pq' \tag{3.68}$$

This will be satisfied for all ϱ' provided

$$\begin{aligned} s' &= -s \\ q' - ps' &= q - p's \\ qp' + pq' &= 0 \end{aligned} \tag{3.69}$$

But since $qs' + q's = 0$ we have $q' = q$, and hence $p' = -p$. Summarizing, we have

$$p' = -p, \qquad q' = q, \qquad s' = -s \tag{3.70}$$

It is easy to verify that if the conditions (3.70) hold, then (3.65) is satisfied. Thus (3.58) becomes

$$\mu = \frac{p + q\hat{\mu}}{1 + s\hat{\mu}}, \qquad \nu = \frac{-p + q\hat{\nu}}{1 - s\hat{\nu}} \tag{3.71}$$

To determine p, q, s, r, and R we use (3.60). We require that $\mu = a$ corresponds to $\hat{\mu} = r$, that $\mu = b$ corresponds to $\hat{\mu} = R$, etc. Thus we have

$$\begin{aligned} a &= \frac{p + qr}{1 + sr}, & b &= \frac{p + qR}{1 + sR} \\ \alpha &= \frac{-p + qr}{1 - sr}, & \beta &= \frac{-p + qR}{1 - sR} \end{aligned} \tag{3.72}$$

Evidently we have

$$\begin{aligned} p + qr &= a + asr, & p + qR &= b + bsR \\ -p + qr &= \alpha - \alpha sr, & -p + qR &= \beta - \beta sR \end{aligned} \tag{3.73}$$

hence

$$\begin{aligned} 2qr &= (a + \alpha) + (a - \alpha)rs, & 2qR &= (b + \beta) + (b - \beta)Rs \\ 2p &= (a - \alpha) + (a + \alpha)rs, & 2p &= (b - \beta) + (b + \beta)Rs \end{aligned} \tag{3.74}$$

Thus, eliminating q and p, we have

$$\frac{a+\alpha}{rs} + a - \alpha = \frac{b+\beta}{Rs} + b - \beta$$
$$(a+\alpha)rs + a - \alpha = (b+\beta)Rs + b - \beta \tag{3.75}$$

or

$$\frac{a+\alpha}{rs} + \beta - \alpha = \frac{b+\beta}{Rs} + b - a$$
$$(a+\alpha)rs + \beta - \alpha = (b+\beta)Rs + b - a \tag{3.76}$$

Solving each equation for s we have

$$s = \frac{[(b+\beta)/R] - [(a+\alpha)/r]}{(\beta-\alpha)-(b-a)} = \frac{(\beta-\alpha)-(b-a)}{(b+\beta)R-(a+\alpha)r} \tag{3.77}$$

Equating the two values of s we obtain

$$c + (1/c) = 2(1+\theta) \tag{3.78}$$

where

$$c = r/R \tag{3.79}$$

$$\theta = 2(\beta-\alpha)(b-a)/[(a+\alpha)(b+\beta)] \tag{3.80}$$

Hence

$$c = \frac{1}{1+\theta+[\theta(2+\theta)]^{1/2}} \tag{3.81}$$

and

$$Rs = \frac{(\beta-\alpha)-(b-a)}{(b+\beta)-(a+\alpha)c} \tag{3.82}$$

$$Rq = \frac{(b+\beta)+(b-\beta)Rs}{2} \tag{3.83}$$

$$p = \frac{(b-\beta)+(b+\beta)Rs}{2} \tag{3.84}$$

Therefore,

$$\mu = \frac{p+(Rq)(\hat{\mu}/R)}{1+(Rs)(\hat{\mu}/R)}, \qquad \nu = \frac{-p+(Rq)(\hat{\nu}/R)}{1-(Rs)(\hat{\nu}/R)} \tag{3.85}$$

$$\varrho = \frac{-p+(Rq)(\hat{\varrho}/R)}{1-(Rs)(\hat{\varrho}/R)}, \qquad \varrho' = \frac{p+(Rq)(\hat{\varrho}'/R)}{1+(Rs)(\hat{\varrho}'/R)} \tag{3.86}$$

Evidently we can choose any positive value for R, for example, $R = 1$.

The optimum parameters $\hat{\varrho}_b$ and $\hat{\varrho}_b'$ for the transformed problems are $\hat{\varrho}_b = \hat{\varrho}_b' = (rR)^{1/2}$. Thus by (3.86) the actual parameters to be used are

$$\varrho_b = \frac{-p + Rqc^{1/2}}{1 - Rsc^{1/2}}, \qquad \varrho_b' = \frac{p + Rqc^{1/2}}{1 + Rsc^{1/2}} \tag{3.87}$$

One determines θ by (3.80), c by (3.81), Rs by (3.82), Rq by (3.83), and p by (3.84). The corresponding value of ψ is

$$\psi(a, b; \alpha, \beta; \varrho_b, \varrho_b') = \left(\frac{1 - c^{1/2}}{1 + c^{1/2}}\right)^2 \tag{3.88}$$

As an example, let us consider the case $a = 1$, $b = 3$, $\alpha = 2$, $\beta = 5$. For simplicity we let $R = 1$. By (3.80) we have $\theta = \frac{1}{2}$ and, from (3.81),

$$c = 2/(3 + 5^{1/2}) \doteq 0.382$$

From (3.82), (3.83), and (3.84) we obtain

$$\begin{aligned} s &= 1/(8 - 3c) \doteq 0.146 \\ q &= 4 - s \doteq 3.854 \\ p &= -1 + 4s \doteq -0.416 \end{aligned}$$

The optimum parameters $\hat{\varrho}_b$ and $\hat{\varrho}_b'$ for the common range $0.382 \le \hat{\mu} \le 1$, $0.382 \le \hat{\nu} \le 1$ are

$$\hat{\varrho}_b = \hat{\varrho}_b' = c^{1/2} \doteq 0.618$$

Consequently, the actual parameters to be used are, by (3.87),

$$\varrho_b \doteq 3.076, \qquad \varrho_b' \doteq 1.804$$

Evidently, by (3.85)

$$\psi(a, b; \alpha, \beta; \varrho_b, \varrho_b') = \left(\frac{1 - c^{1/2}}{1 + c^{1/2}}\right)^2 = 0.0557$$

17.4. THE COMMUTATIVE CASE

We shall now consider the "commutative case", as defined by Birkhoff *et al.* [1962], where the matrices H_0, V_0, and Σ of (1.2) satisfy the conditions

$$\begin{aligned} &H_0 V_0 = V_0 H_0 \\ &\Sigma = \sigma I, \quad \text{where } \sigma \text{ is a nonnegative constant} \\ &H_0 \text{ and } V_0 \text{ are similar to nonnegative diagonal matrices} \end{aligned} \tag{4.1}$$

Evidently, if these conditions are satisfied and if $H = H_0 + \frac{1}{2}\Sigma$, $V = V_0 + \frac{1}{2}\Sigma$, then

$$HV = VH \tag{4.2}$$

and

$$H \text{ and } V \text{ are similar to nonnegative diagonal matrices} \tag{4.3}$$

Moreover $H + \varrho I$ and $V + \varrho I$ are nonsingular for any $\varrho > 0$.

The following theorem relates conditions (4.1) to another set of conditions which are sometimes used:

Theorem 4.1. If H_0 and V_0 are similar to nonnegative diagonal matrices and if $H_0 + V_0$ is irreducible, then the conditions (4.1) are equivalent to the conditions

$$H_0V_0 = V_0H_0, \qquad H_0\Sigma = \Sigma H_0, \qquad V_0\Sigma = \Sigma V_0 \tag{4.4}$$

Proof. If conditions (4.1) hold, then conditions (4.4) are trivially satisfied. If conditions (4.4) hold, then $(H_0 + V_0)\Sigma = \Sigma(H_0 + V_0)$. Thus for each i, j we have

$$a_{i,j}^{(0)}\sigma_j = \sigma_i a_{i,j}^{(0)}$$

where $H_0 + V_0 = A^{(0)} = (a_{i,j}^{(0)})$ and the diagonal elements of Σ are $\sigma_1, \sigma_2, \ldots, \sigma_N$. For any pair i, j such that $a_{i,j}^{(0)} \neq 0$ and $i \neq j$, it follows that $\sigma_i = \sigma_j$. By the irreducibility of $A^{(0)}$ it follows that all σ_i are equal. Hence $\Sigma = \sigma I$ and the theorem is proved.

The analysis of the PR method in the commutative case depends on the following theorem of Frobenius. (See, for instance, Thrall and Tornheim, 1957, page 190.)

Theorem 4.2. There exists a nonsingular matrix W such that $W^{-1}HW$ and $W^{-1}VW$ are diagonal matrices if and only if $HV = VH$ and H and V are similar to diagonal matrices.

Proof. If such a matrix exists, then

$$W^{-1}HW = \Lambda_H, \qquad W^{-1}VW = \Lambda_V$$

where Λ_H and Λ_V are diagonal matrices. Therefore

$$\begin{aligned} HV &= (W\Lambda_H W^{-1})(W\Lambda_V W^{-1}) = W\Lambda_H\Lambda_V W^{-1} \\ &= W\Lambda_V\Lambda_H W^{-1} = (W\Lambda_V W^{-1})(W\Lambda_H W^{-1}) = VH \end{aligned}$$

Suppose now that $HV = VH$ and that H and V are similar to diagonal matrices. Then there exists a nonsingular matrix P such that $P^{-1}VP$ is a diagonal matrix. We can assume without loss of generality that $P^{-1}VP$ has the form

$$P^{-1}VP = \begin{pmatrix} \lambda_1 I_1 & 0 & 0 & \cdots & 0 \\ 0 & \lambda_2 I_2 & 0 & \cdots & 0 \\ \vdots & \vdots & \vdots & \cdots & 0 \\ 0 & 0 & 0 & \cdots & \lambda_p I_p \end{pmatrix}$$

where $I_1, I_2, \ldots, I_p$ are square identity matrices and where $\lambda_1, \lambda_2, \ldots, \lambda_p$ are distinct eigenvalues of V. For simplicity, let us assume $p = 4$. Then†

$$\hat{H} = P^{-1}HP = \begin{pmatrix} \hat{H}_{1,1} & \hat{H}_{1,2} & \hat{H}_{1,3} & \hat{H}_{1,4} \\ \hat{H}_{2,1} & \hat{H}_{2,2} & \hat{H}_{2,3} & \hat{H}_{2,4} \\ \hat{H}_{3,1} & \hat{H}_{3,2} & \hat{H}_{3,3} & \hat{H}_{3,4} \\ \hat{H}_{4,1} & \hat{H}_{4,2} & \hat{H}_{4,3} & \hat{H}_{4,4} \end{pmatrix}$$

and

$$P^{-1}(HV - VH)P$$
$$= \begin{pmatrix} 0 & \hat{H}_{1,2}(\lambda_2 - \lambda_1) & \hat{H}_{1,3}(\lambda_3 - \lambda_1) & \hat{H}_{1,4}(\lambda_4 - \lambda_1) \\ \hat{H}_{2,1}(\lambda_1 - \lambda_2) & 0 & \hat{H}_{2,3}(\lambda_3 - \lambda_2) & \hat{H}_{2,4}(\lambda_4 - \lambda_2) \\ \hat{H}_{3,1}(\lambda_1 - \lambda_3) & \hat{H}_{3,2}(\lambda_2 - \lambda_3) & 0 & \hat{H}_{3,4}(\lambda_4 - \lambda_3) \\ \hat{H}_{4,1}(\lambda_1 - \lambda_4) & \hat{H}_{4,2}(\lambda_2 - \lambda_4) & \hat{H}_{4,3}(\lambda_3 - \lambda_4) & 0 \end{pmatrix}$$

Hence, since the above matrix vanishes we have $\hat{H}_{1,2} = \hat{H}_{1,3} = \cdots = 0$ and

$$\hat{H} = \begin{pmatrix} \hat{H}_{1,1} & 0 & 0 & 0 \\ 0 & \hat{H}_{2,2} & 0 & 0 \\ 0 & 0 & \hat{H}_{3,3} & 0 \\ 0 & 0 & 0 & \hat{H}_{4,4} \end{pmatrix}$$

Let Q_i be the nonsingular matrix which reduces $\hat{H}_{ii}$ to Jordan canonical form J_i. Then if

$$Q = \begin{pmatrix} Q_1 & 0 & 0 & 0 \\ 0 & Q_2 & 0 & 0 \\ 0 & 0 & Q_3 & 0 \\ 0 & 0 & 0 & Q_4 \end{pmatrix}$$

† Here $\hat{H}$ and $\hat{H}_{i,j}$ are defined as indicated rather than as in Chapter 4.

we have

$$Q^{-1}\hat{H}Q = \begin{pmatrix} J_1 & 0 & 0 & 0 \\ 0 & J_2 & 0 & 0 \\ 0 & 0 & J_3 & 0 \\ 0 & 0 & 0 & J_4 \end{pmatrix}$$

Evidently, this is a Jordan canonical form for H or can be made so by permuting the rows and columns. Hence each J_i must be diagonal as otherwise the Jordan canonical form of H would not be diagonal.

Since

$$Q^{-1}(P^{-1}VP)Q = \begin{pmatrix} \lambda_1 I_1 & 0 & 0 & 0 \\ 0 & \lambda_2 I_2 & 0 & 0 \\ 0 & 0 & \lambda_3 I_3 & 0 \\ 0 & 0 & 0 & \lambda_4 I_4 \end{pmatrix}$$

it follows that $W = PQ$ simultaneously reduces H and V to diagonal form. This completes the proof of Theorem 4.2.

Suppose now that H and V satisfy conditions (4.2) and (4.3). Then by Theorem 4.2 there exist a set of N linearly independent vectors $v_1, v_2, \ldots, v_N$ which are eigenvectors both of H and V. Let v be any such vector and let

$$Hv = \mu v, \qquad Vv = \nu v \tag{4.5}$$

Then for any positive ϱ and ϱ' we have

$$T_{\varrho,\varrho'}v = (V + \varrho' I)^{-1}(H - \varrho' I)(H + \varrho I)^{-1}(V - \varrho I)v = \left(\frac{(\mu - \varrho')(\nu - \varrho)}{(\mu + \varrho)(\nu + \varrho')}\right)v \tag{4.6}$$

This follows since, for instance, $(H+\varrho I)v = (\mu+\varrho)v$; hence $(H+\varrho I)^{-1}v = (\mu + \varrho)^{-1}v$, etc. Therefore, all eigenvalues of $\prod_{i=1}^{m} T_{\varrho_i,\varrho_i'}$ are given by

$$\prod_{i=1}^{m} \frac{(\mu - \varrho_i')(\nu - \varrho_i)}{(\mu + \varrho_i)(\nu + \varrho_i')}$$

for some eigenvalue μ of H and some eigenvalue ν of V. Therefore, if the eigenvalues of H and V lie in the ranges $a \le \mu \le b$, $\alpha \le \nu \le \beta$, respectively, then

$$S\left(\prod_{i=1}^{m} T_{\varrho_i,\varrho_i'}\right) \le \psi_m(a, b; \alpha, \beta; \boldsymbol{\rho}, \boldsymbol{\rho}') \tag{4.7}$$

where

$$\boldsymbol{\rho} = (\varrho_1, \varrho_2, \ldots, \varrho_m), \qquad \boldsymbol{\rho}' = (\varrho_1', \varrho_2', \ldots, \varrho_m') \tag{4.8}$$

and

$$\psi_m(a, b; \alpha, \beta; \boldsymbol{\rho}, \boldsymbol{\rho}') = \max_{\substack{a \le \mu \le b \\ \alpha \le \nu \le \beta}} \prod_{i=1}^{m} \left| \frac{(\mu - \varrho_i')(\nu - \varrho_i)}{(\mu + \varrho_i)(\nu + \varrho_i')} \right| \tag{4.9}$$

By the transformation (3.85) and (3.86) we have

$$\psi_m(a, b; \alpha, \beta; \boldsymbol{\rho}, \boldsymbol{\rho}') = \psi_m(r, R; r, R; \hat{\boldsymbol{\rho}}, \hat{\boldsymbol{\rho}}') \tag{4.10}$$

where

$$\psi_m(r, R; r, R; \hat{\boldsymbol{\rho}}, \hat{\boldsymbol{\rho}}') = \max_{\substack{r \le \hat{\mu} \le R \\ r \le \hat{\nu} \le R}} \prod_{i=1}^{m} \left| \frac{(\hat{\mu} - \hat{\varrho}_i')(\hat{\nu} - \hat{\varrho}_i)}{(\hat{\mu} + \hat{\varrho}_i)(\hat{\nu} + \hat{\varrho}_i')} \right| \tag{4.11}$$

Here R is arbitrary and r is determined by (3.79), (3.80), and (3.81).

We have already shown in Section 16.3 that for the case $m = 1$, there is a unique pair of values ϱ and ϱ' which minimize $\Gamma_1(r, R; \varrho, \varrho')$. Wachspress [1968] has shown that this is true for any m, and, moreover, that $\boldsymbol{\rho} = \boldsymbol{\rho}'$, i.e.,

$$\varrho_i = \varrho_i', \qquad i = 1, 2, \ldots, m \tag{4.12}$$

We are thus led to consider the "m-parameter problem" of minimizing

$$\phi_m(a, b; \boldsymbol{\rho}) = \max_{a \le \gamma \le b} \prod_{i=1}^{m} \left| \frac{\gamma - \varrho_i}{\gamma + \varrho_i} \right| \tag{4.13}$$

In Section 17.5 we shall describe how the optimum parameters can be obtained. In Section 17.6 we shall describe how one can obtain "good parameters" which, though not optimum, nevertheless are easy to work with and frequently are close enough to optimal for practical purposes.

17.5. OPTIMUM PARAMETERS

Wachspress and Habetler [1960] showed that if m is even, one can solve the m-parameter problem by solving the $m/2$-parameter problem. (See also Gastinel, 1962.) They also gave an algorithm for solving the m-parameter problem if $m = 2^p$ for some nonnegative integer p.

Suppose m is even. As stated in Section 17.4, $\phi_m(a, b; \boldsymbol{\rho})$ is minimized for a vector $\boldsymbol{\rho}^* = (\varrho_1^*, \varrho_2^*, \ldots, \varrho_m^*)$ which is unique except for the order of the components. Since

$$\frac{ab/\gamma - \varrho}{ab/\gamma + \varrho} = \frac{ab/\varrho - \gamma}{ab/\varrho + \gamma} \tag{5.1}$$

and since γ varies over the same range as $\tilde{\gamma}$, where

$$\tilde{\gamma} = ab/\gamma \tag{5.2}$$

we have

$$\begin{aligned} \max_{a \le \gamma \le b} \prod_{i=1}^{m} \left| \frac{\gamma - \varrho_i}{\gamma + \varrho_i} \right| &= \max_{a \le \tilde{\gamma} \le b} \prod_{i=1}^{m} \left| \frac{\tilde{\gamma} - (ab/\varrho_i)}{\tilde{\gamma} + (ab/\varrho_i)} \right| \\ &= \max_{a \le \gamma \le b} \prod_{i=1}^{m} \left| \frac{\gamma - (ab/\varrho_i)}{\gamma + (ab/\varrho_i)} \right| \end{aligned} \tag{5.3}$$

Thus if $\varrho_1^*, \varrho_2^*, \ldots, \varrho_m^*$ are optimum parameters, then

$$ab/\varrho_1^*, ab/\varrho_2^*, \ldots, ab/\varrho_m^*$$

are also optimum parameters though perhaps in a different order. Let us assume that the ϱ_i^* have been labeled so that

$$\varrho_1^* \le \varrho_2^* \le \cdots \le \varrho_m^* \tag{5.4}$$

Then we have

$$ab/\varrho_m^* \le ab/\varrho_{m-1}^* \le \cdots \le ab/\varrho_1^* \tag{5.5}$$

Hence

$$\varrho_i^* \varrho_{m+1-i}^* = ab, \qquad i = 1, 2, \ldots, m/2 \tag{5.6}$$

Moreover,

$$\begin{aligned} \prod_{i=1}^{m} \left(\frac{\gamma - \varrho_i^*}{\gamma + \varrho_i^*} \right) &= \prod_{i=1}^{m/2} \left(\frac{\gamma - \varrho_i^*}{\gamma + \varrho_i^*} \right) \left(\frac{\gamma - \varrho_{m+1-i}^*}{\gamma + \varrho_{m+1-i}^*} \right) \\ &= \prod_{i=1}^{m/2} \left(\frac{\hat{\gamma} - \hat{\varrho}_i^*}{\hat{\gamma} + \hat{\varrho}_i^*} \right) \end{aligned} \tag{5.7}$$

where

$$\hat{\gamma} = \frac{1}{2}\left(\gamma + \frac{ab}{\gamma}\right), \quad \hat{\varrho}_i^* = \frac{1}{2}\left(\varrho_i^* + \frac{ab}{\varrho_i^*}\right), \quad i = 1, 2, \ldots, \frac{m}{2} \tag{5.8}$$

Since

$$\frac{d}{d\gamma}\left(\gamma + \frac{ab}{\gamma}\right) = 1 - \frac{ab}{\gamma^2}$$

it follows that as γ varies from a to b, $\hat{\gamma}$ decreases from $(a + b)/2$ at $\gamma = a$ to $(ab)^{1/2}$ at $\gamma = (ab)^{1/2}$ and back up to $(a + b)/2$ at $\gamma = b$. Therefore, we have

$$\max_{a \le \gamma \le b} \prod_{i=1}^{m} \left| \frac{\gamma - \varrho_i^*}{\gamma + \varrho_i^*} \right| = \max_{(ab)^{1/2} \le \hat{\gamma} \le (a+b)/2} \prod_{i=1}^{m/2} \left| \frac{\hat{\gamma} - \hat{\varrho}_i^*}{\hat{\gamma} + \hat{\varrho}_i^*} \right| \tag{5.9}$$

Thus the $\hat{\varrho}_i^*$ are the solution of the $m/2$-parameter problem with the range $(ab)^{1/2} \le \hat{\gamma} \le (a + b)/2$. Having determined the $\hat{\varrho}_i^*$ we can obtain the ϱ_i^* from

$$\begin{aligned} \varrho_{m+1-i} &= \hat{\varrho}_{m/2+1-i} + (\hat{\varrho}_{m/2+1-i} - ab)^{1/2}, \qquad & i &= 1, 2, \ldots, m/2 \\ \varrho_i &= ab/\varrho_{m+1-i}, & i &= 1, 2, \ldots, m/2 \end{aligned} \tag{5.10}$$

If $m = 2^p$, we can repeat this process, eventually obtaining a one-parameter problem which can be solved directly. Let

$$F_m(a, b) = \min_{\boldsymbol{\rho}} \phi_m(a, b; \boldsymbol{\rho}) \tag{5.11}$$

and let

$$\begin{aligned} a_0 &= a, \qquad & b_0 &= b \\ a_{k+1} &= (a_k b_k)^{1/2}, & b_{k+1} &= (a_k + b_k)/2, \qquad k = 1, 2, \ldots, p - 1 \end{aligned} \tag{5.12}$$

Then we have

$$F_m(a, b) = F_{m/2}([ab]^{1/2}, (a + b)/2) \tag{5.13}$$

and

$$\begin{aligned} F_{2^p}(a, b) &= F_{2^{p-1}}(a_1, b_1) = \cdots = F_1(a_p, b_p) \\ &= (b_p^{1/2} - a_p^{1/2})/(b_p^{1/2} + a_p^{1/2}) \end{aligned} \tag{5.14}$$

The iteration parameters are determined as follows. Let

$$\varrho_1^{(p)} = (a_p b_p)^{1/2} \tag{5.15}$$

Then determine $\varrho_1^{(p-1)}$, $\varrho_2^{(p-1)}$ from (5.10) (replacing ϱ_i by $\varrho_i^{(p-1)}$, and $\hat{\varrho}_i$ by $\varrho_i^{(p)}$). Then determine $\boldsymbol{\rho}^{(p-2)}, \boldsymbol{\rho}^{(p-3)}, \ldots, \boldsymbol{\rho}^{(0)} = \boldsymbol{\rho}$.

As an example, suppose $m = 4$, $a = 0.5$, $b = 3.5$. Then $p = 2$ and we have

$$\begin{aligned} a_0 &= 0.5, \qquad & b_0 &= 3.5 \\ a_1 &= 1.3229, & b_1 &= 2.0 \\ a_2 &= 1.6266, & b_2 &= 1.6614 \end{aligned}$$

Hence,

$$\varrho_1^{(2)} = 1.6439, \qquad \varrho_2^{(1)} = 1.8820, \qquad \varrho_4^{(0)} = 3.2207$$
$$\varrho_1^{(1)} = 1.4058, \qquad \varrho_3^{(0)} = 0.5433$$
$$\varrho_2^{(0)} = 1.8814$$
$$\varrho_1^{(0)} = 0.9301$$

and

$$F_4(0.5, 3.5) = (b_2^{1/2} - a_2^{1/2})/(b_2^{1/2} + a_2^{1/2}) \doteq 0.005301$$
$$S\left(\prod_{i=1}^{4} T_{\varrho_i^*}\right) \doteq 0.0000281$$

It is easy to show that as p increases, then

$$[F_{2^p}(a, b)]^{1/2^p}$$

decreases (see Gaier and Todd, 1967). Hence, in some sense the average rate of convergence is increased if p is increased. To show this, it is clearly sufficient to show that

$$F_{2^{p+1}}(a, b) \leq [F_{2^p}(a, b)]^2 \tag{5.16}$$

But

$$F_{2^p}(a, b) = F_1(a_p, b_p) = (b_p^{1/2} - a_p^{1/2})/(b_p^{1/2} + a_p^{1/2}) = z \tag{5.17}$$

$$\begin{aligned} F_{2^{p+1}}(a, b) &= F_1(a_{p+1}, b_{p+1}) \\ &= (b_{p+1}^{1/2} - a_{p+1}^{1/2})/(b_{p+1}^{1/2} + a_{p+1}^{1/2}) = w \end{aligned} \tag{5.18}$$

where $a_{p+1} = (a_p b_p)^{1/2}$, $b_{p+1} = (a_p + b_p)/2$. Let $\theta = a_p/b_p$. Then

$$z = (1 - \theta^{1/2})/(1 + \theta^{1/2}), \qquad w = (1 - x^{1/2})/(1 + x^{1/2}) \tag{5.19}$$

where

$$x = 2\theta^{1/2}/(1 + \theta) \tag{5.20}$$

But

$$z^2 = \frac{1 - [2\theta^{1/2}/(1 + \theta)]}{1 + [2\theta^{1/2}/(1 + \theta)]} = \frac{1 - x}{1 + x} \tag{5.21}$$

Since $x < 1$ it follows that $x \leq x^{1/2}$ and hence $z^2 \geq w$.

As $p \to \infty$, the values of a_p and b_p converge to a common limit, say ζ, which is sometimes known as the *arithmetic–geometric mean* of a and b. The sequence $a_1, a_2, \ldots$ increases to ζ and the sequence $b_1, b_2, \ldots$ decreases to ζ. There appears to be little point in choosing p larger than is necessary so that a_p and b_p are reasonably close to ζ. Certainly, one should not choose p so large that 2^p will exceed the number of iterations required to solve the problem.

Solution of the m-Parameter Problem by Elliptic Functions

We now summarize, without proofs, the solution to the m-parameter problem in terms of elliptic functions. This solution is due to W. B. Jordan and is described by Wachspress [1966]. Todd [1967, 1967a] and Gaier and Todd [1967] relate the solution to work of Zolotareff [1877] and Cauer [1958]. We describe the solution and also approximations proposed by Wachspress and by Todd.

The optimum parameters are given by

$$\varrho_i^* = b\, dn\left(\frac{2(m-i)+1}{2m} K, k\right), \qquad i = 1, 2, \ldots, m \tag{5.22}$$

where the elliptic function $dn(u, k)$ is defined by

$$dn(u, k) = (1 - k^2x^2)^{1/2} \tag{5.23}$$

Here,

$$x = \sin \phi \tag{5.24}$$

and ϕ is defined in terms of u by

$$\int_0^\phi (1 - k^2 \sin^2 \theta)^{-1/2}\, d\theta = u \tag{5.25}$$

The modulus k is given by

$$k = (1 - c^2)^{1/2} \tag{5.26}$$

where

$$c = a/b \tag{5.27}$$

Here K is the complete elliptic integral

$$K = \int_0^{\pi/2} (1 - k^2 \sin^2 \theta)^{-1/2}\, d\theta \tag{5.28}$$

As an example, let $a = 0.5$, $b = 3.5$, $m = 4$. Evidently, $c = 1/7 \doteq 0.14286$.

$$k = [1 - (1/7)^2]^{1/2} \doteq 0.98974$$

Using a table of complete elliptic integrals we obtain, using linear interpolation, $K = 3.3460$. Also using (double) inverse interpolation in a table of incomplete elliptic integrals we have the values shown in the accompanying tabulation. Thus we obtain results which are close to those obtained with the previous method. The differences are due to the inaccuracy of the interpolation in the tables.

i	$\frac{2(m-i)+1}{2m}$	u	ϕ	$x = \sin\phi$	$(1 - k^2x^2)^{1/2}$	$\varrho_i^* = b(1-k^2x^2)^{1/2}$
1	$\frac{7}{8}$	2.92779	86.39°	0.99801	0.15585	0.5455
2	$\frac{5}{8}$	2.09128	76.77°	0.97345	0.26780	0.9373
3	$\frac{3}{8}$	1.25477	58.37°	0.85145	0.53835	1.8842
4	$\frac{1}{8}$	0.41826	23.29°	0.39538	0.92024	3.2208

Wachspress [1966] gave the following approximate formula for ϱ_i

$$\varrho_i \doteq b\,\frac{2(c/4)^{t_i}(1 + (c/4)^{2(1-t_i)})}{1 + (c/4)^{2t_i}}, \qquad i = 1, 2, \ldots, m \tag{5.29}$$

where

$$t_i = [2(m - i) + 1]/(2m) \tag{5.30}$$

Thus in the above example we have, since $c = \frac{1}{7}$,

$$\varrho_1 = 0.5424, \qquad \varrho_2 = 0.9294, \qquad \varrho_3 = 1.8828, \qquad \varrho_4 = 3.2263$$

Wachspress showed that with the optimum parameters the value of $\phi_m(a, b; \boldsymbol{\rho})$ is given by

$$\phi_m(a, b; \boldsymbol{\rho}) = \frac{1 - (k'm)^{1/2}}{1 + (k'm)^{1/2}} \tag{5.31}$$

Here $k_m' = (1 - k_m^2)^{1/2}$ and k_m is determined so that

$$K(k_m')/K(k_m) = mK(k')/K(k) \tag{5.32}$$

Here $K(k)$ is defined by (5.28), k is given by (5.26) and $k' = c$.

Thus in our example we have

$$K(k) \doteq 3.359, \qquad K(k') \doteq 1.5787$$

hence

$$K(k')/K(k) = 1.5787/3.359 \doteq 0.470$$

Therefore,

$$mK(k')/K(k) = 4(0.470) = 1.880$$

We thus seek k_4 and $k_4' = (1 - k_4^2)^{1/2}$ so that

$$K(k_4')/K(k_4) = 1.880$$

From a table of elliptic integrals we obtain

$$k_4 \doteq 0.208, \qquad k_4' \doteq 0.978$$

Hence,

$$\frac{1 - (k_4')^{1/2}}{1 + (k_4')^{1/2}} \doteq \frac{1 - 0.989}{1 + 0.989} = \frac{0.011}{1.989} = 0.00552$$

which agrees reasonably well with the value 0.005301 obtained previously.

Wachspress also gave the following approximate formula for $\phi_m(a, b; \rho^*)$,

$$\phi_m(a, b; \rho^*) \doteq 2q_1/(1 + 2q_1^4) \tag{5.33}$$

where

$$q_1 = q^m = (e^{-\pi K'/K})^m \tag{5.34}$$

where $K' = K(k')$, $K = K(k)$. This formula is valid for small k' and for small ϕ_m.

Thus in our example we have $K'/K = 0.470$ and

$$q = e^{-\pi(0.470)} \doteq e^{-1.475} = 0.229$$

$$q_1 = q^4 = 0.00275$$

Hence,

$$\phi_4(a, b; \rho^*) \doteq 2(0.00275) = 0.00550$$

which agrees closely with the previous result.

Todd [1967] has shown that

$$\eta_m = [\phi_m(a, b; \boldsymbol{\rho}^*)]^{1/m} \tag{5.35}$$

is a decreasing function of m and, as $m \to \infty$, and

$$\log \eta_m = \log q + (\log 2)/m + O(m^{-2}) \tag{5.36}$$

where, as before,

$$q = e^{-\pi K'/K} \tag{5.37}$$

Thus,

$$\eta_m = -\pi(K'/K) + (\log 2)/m + O(m^{-2}) \tag{5.38}$$

and in our example above

$$\eta_4 \doteq -\pi(0.470) + (0.693/4) + O(m^{-2}) \doteq -1.302 + O(m^{-2})$$

The actual value of η_4 is

$$\tfrac{1}{4}\log(0.00552) \doteq \tfrac{1}{4}(-5.21) = -1.3025$$

Thus the approximate formula (5.31) gives very good results.

17.6. GOOD PARAMETERS

In this section we describe the use of "good" parameters, which though not optimal, nevertheless are nearly as good as the optimal parameters in many cases, and are much easier to determine. We shall consider the Peaceman–Rachford parameters (Peaceman and Rachford, 1955)

$$\varrho_i^{(P)} = b(a/b)^{(2i-1)/(2m)}, \qquad i = 1, 2, \ldots, m \tag{6.1}$$

and the Wachspress parameters (Wachspress, 1957),

$$\varrho_i^{(W)} = b(a/b)^{(i-1)/(m-1)}, \qquad m \geq 2, \quad i = 1, 2, \ldots, m \tag{6.2}$$

Theorem 6.1. For the Peaceman–Rachford parameters we have

$$\phi_m \leq (1 - z)/(1 + z) \tag{6.3}$$

where

$$z = c^{1/(2m)}, \qquad c = a/b \tag{6.4}$$

and

$$\frac{4}{m} z \leq -\frac{2}{m} \log \phi_m$$
$$\leq -\frac{2}{m} \log\left(\frac{1-z}{1+z}\right) + \frac{4}{m} \frac{z^3}{(1-z)^2} = \frac{4}{m} z + O(z^2) \tag{6.5}$$

as $z \to 0$. Here $\phi_m = \phi_m(a, b; \boldsymbol{\rho})$ is defined by (4.13) with $\boldsymbol{\rho} = \boldsymbol{\rho}^{(P)}$.

Proof. Each factor of (4.13) is less than one. Moreover, $(\gamma - \varrho)(\gamma+\varrho)^{-1}$ is an increasing function of γ. Clearly, we have $\varrho_i = bz^{2i-1}$ and

$$b > \varrho_1 > \varrho_2 > \cdots > \varrho_m > a$$

In the interval $a \leq \gamma \leq \varrho_m$, we have

$$\left|\frac{\gamma - \varrho_m}{\gamma + \varrho_m}\right| \leq \frac{\varrho_m - a}{\varrho_m + a} = \frac{(a/z) - a}{(a/z) + a} = \frac{1-z}{1+z}$$

The same inequality holds in the interval $\varrho_1 \leq \gamma \leq b$. In the interval $\varrho_{i+1} \leq \gamma \leq \varrho_i$ the function

$$\left(\frac{\gamma - \varrho_{i+1}}{\gamma + \varrho_{i+1}}\right)\left(\frac{\gamma - \varrho_i}{\gamma + \varrho_i}\right)$$

has an extreme value at $\gamma = (\varrho_i \varrho_{i+1})^{1/2}$ since

$$\frac{d}{d\gamma}\left(\frac{\gamma - \varrho_{i+1}}{\gamma + \varrho_{i+1}}\right)\left(\frac{\gamma - \varrho_i}{\gamma + \varrho_i}\right) = \frac{2(\varrho_i + \varrho_{i+1})(\gamma^2 - \varrho_i \varrho_{i+1})}{(\gamma + \varrho_{i+1})^2(\gamma + \varrho_i)^2} \tag{6.6}$$

This extreme value has modulus

$$\left(\frac{(\varrho_i)^{1/2} - (\varrho_{i+1})^{1/2}}{(\varrho_i)^{1/2} + (\varrho_{i+1})^{1/2}}\right)^2 = \left(\frac{1-z}{1+z}\right)^2$$

and in the interval

$$\left|\left(\frac{\gamma - \varrho_{i+1}}{\gamma + \varrho_{i+1}}\right)\left(\frac{\gamma - \varrho_i}{\gamma + \varrho_i}\right)\right| \leq \left(\frac{1-z}{1+z}\right)^2$$

Hence (6.3) follows.

Since

$$-\log\frac{1-z}{1+z} = -\log(1-z) + \log(1+z)$$
$$= \left(z + \frac{z^2}{2} + \frac{z^3}{3} + \cdots\right) + \left(z - \frac{z^2}{2} + \frac{z^3}{3} + \cdots\right)$$
$$= 2\left(z + \frac{z^3}{3} + \cdots\right) \geq 2z$$

the first inequality of (6.5) holds. To prove the second inequality we have

$$\phi_m \geq \prod_{i=1}^{m} \frac{b - \varrho_i}{b + \varrho_i} = \prod_{i=1}^{m} \frac{1 - z^{2i-1}}{1 + z^{2i-1}} = \frac{1-z}{1+z} \prod_{i=2}^{m} \frac{1 - z^{2i-1}}{1 + z^{2i+1}} \tag{6.7}$$

Hence†

$$-\log \phi_m \leq -\log\frac{1-z}{1+z} + \frac{2z^3}{(1-z)^2} \tag{6.8}$$

and (6.5) follows. This completes the proof of Theorem 6.1.

In contrast to the situation with the optimum parameters, the function $(\phi_m)^{1/m}$ decreases with m up to a certain point and then increases. This critical value of m can be estimated as follows. We seek to minimize

$$\eta_m = \left(\frac{1-z}{1+z}\right)^{2/m} \tag{6.9}$$

Evidently,

$$-\log \eta_m = -\frac{2}{m}\log\frac{1-z}{1+z} \tag{6.10}$$

If we let

$$\delta = \frac{1-z}{1+z} \tag{6.11}$$

then

$$z = \frac{1-\delta}{1+\delta}, \qquad m = \frac{1}{2}\,\frac{\log c}{\log((1-\delta)/(1+\delta))} \tag{6.12}$$

so that

$$-\log \eta_m = -\frac{4 \log \delta \log((1-\delta)/(1+\delta))}{\log c} \tag{6.13}$$

† See Birkhoff *et al.* [1962], p. 214.

Equating to zero the first derivative with respect to δ we obtain

$$\frac{1-\delta^2}{2}\log\frac{1-\delta}{1+\delta} = \delta\log\delta \tag{6.14}$$

which has the solution

$$\bar{\delta} = 2^{1/2} - 1 \doteq 0.414 \tag{6.15}$$

Thus, the estimated value of the optimum m is determined by finding the smallest integer such that

$$(\bar{\delta})^{2m} \leq c \tag{6.16}$$

The corresponding value of $-\log \eta_m$ is

$$-\log \eta_m \doteq \frac{4(\log \bar{\delta})^2}{-\log c} \doteq \frac{3.11}{-\log c} \tag{6.17}$$

As an example, let us assume $a = 0.1$, $b = 3.9$. Then $c = 0.0257$. To determine m we solve (6.16) obtaining

$$(0.414)^{2m} \leq 0.0257$$

and $m = 2.08$. If we accept $m = 3$, we then compute the parameters ϱ_i by (6.1), obtaining

$$\varrho_i = (3.9)(0.0257)^{(2i-1)/10}, \qquad i = 1, 2, 3$$

The Wachspress parameters are usually, but not always, better than the Peaceman–Rachford parameters. We have

Theorem 6.2. For the Wachspress parameters

$$\phi_m \leq \left(\frac{1-y}{1+y}\right)^2 \tag{6.18}$$

where

$$y = c^{1/[2(m-1)]} \tag{6.19}$$

Moreover,

$$\begin{aligned} \frac{8}{m}y &\leq -\frac{2}{m}\log\phi_m \\ &\leq -\frac{4}{m}\log\left(\frac{1-y}{1+y}\right) + \frac{4}{m}\frac{y^3}{(1-y)^2} = \frac{8}{m}y + O(y^2) \end{aligned} \tag{6.20}$$

as $y \to 0$.

Proof. As in the proof of Theorem 6.1 the maximum value of $|(\gamma - \varrho_{i+1})(\gamma - \varrho_i)(\gamma + \varrho_{i+1})^{-1}(\gamma + \varrho_i)^{-1}|$ in the interval $\varrho_{i+1} \leq \gamma \leq \varrho_i$ is

$$\left(\frac{\varrho_i^{1/2} - \varrho_{i+1}^{1/2}}{\varrho_i^{1/2} + \varrho_{i+1}^{1/2}}\right)^2 = \left(\frac{1 - y^2}{1 + y}\right)^2$$

Hence (6.18) holds. Moreover, $-\log((1 - y)(1 + y)^{-1}) \geq 2y$; hence the first inequality of (6.20) holds.

To prove the second inequality we have

$$\begin{aligned}\phi_m(a, b; \boldsymbol{\rho}) &\geq \prod_{i=1}^{m} \left| \frac{(\varrho_1\varrho_2)^{1/2} - \varrho_i}{(\varrho_1\varrho_2)^{1/2} + \varrho_i} \right| \\ &= \left(\frac{1 - y}{1 + y}\right)^2 \prod_{i=2}^{m} \left| \frac{by - \varrho_i}{by + \varrho_i} \right| \\ &= \left(\frac{1 - y}{1 + y}\right)^2 \prod_{i=2}^{m} \frac{1 - y^{2i-1}}{1 + y^{2i-1}} \end{aligned} \tag{6.21}$$

Hence[†]

$$-\frac{2}{m} \log \phi_m \leq -\frac{4}{m} \log\left(\frac{1 - y}{1 + y}\right) + \frac{4}{m} \frac{y^3}{(1 - y)^2} \tag{6.22}$$

and (6.20) follows.

As in the case of the Peaceman–Rachford parameters, there is an optimum number of Wachspress parameters to use. We estimate this number as follows. We seek to maximize

$$\zeta_m = -\frac{4}{m} \log\left(\frac{1 - y}{1 + y}\right) = -\frac{4}{m} \log \frac{1 - c^{1/[2(m-1)]}}{1 + c^{1/[2(m-1)]}} \tag{6.23}$$

As an approximation we replace $m - 1$ by m in the above formula and have

$$\zeta_m{}' = -\frac{4}{m} \log \frac{1 - c^{1/2m}}{1 + c^{1/2m}} = -\frac{4}{m} \log \frac{1 - z}{1 + z} \tag{6.24}$$

As in the case of the Peaceman–Rachford parameters, the optimum value of m is, approximately, the smallest integer such that (6.16) is satisfied. The corresponding value of

$$-\frac{4}{m} \log\left(\frac{1 - y}{1 + y}\right)$$

[†] See Birkhoff *et al.* [1962], p. 217.

is, approximately,

$$\frac{8(\log \bar{\delta})^2}{-\log c} \doteq \frac{6.22}{-\log c}$$

which appears to be twice as good as the result (6.17) for the Peaceman–Rachford parameters. Of course, it must be noted that an approximate method was used to determine m.

Table 6.1 summarizes some numerical experiments reported by Birkhoff *et al.* [1962] for the model problem with $h = \frac{1}{40}$. It can be seen that on the basis of the theoretical numbers of iterations and on the

TABLE 6.1.

m	N_t^{P} [a]	N_t^{W} [a]	N_t^{B} [a]	N_c^{P} [b]	N_c^{W} [b]	N_O^{P} [c]	N_O^{W} [c]	N_O^{B} [c]
1	82		86	82		91		91
2	33	82	24	34	82	36	73	26
3	26	25	17	29	26	27	22	18
4	23	18	15	29	20	27	15	18
5	22	16		30	18	25	14	
6	21	15		31	18	24	14	
7	21	15		33	18	26	11	
8	20	15		34	18	24	13	
9	20	15		36	17	26	14	
10	20	15		38	20	27	11	

[a] N_t^{P}, N_t^{W}, N_t^{B} are theoretical numbers of iterations for the Peaceman–Rachford, the Wachspress parameters and the optimum parameters, respectively. In each case

$$N_t = -\log 10^{-6}/(-2 \log \phi_m).$$

[b] N_c^{P}, N_c^{W} are estimated theoretical numbers of iterations based on the use of (6.3) and (6.18), respectively.

[c] N_O^{P}, N_O^{W}, N_O^{B} are the observed numbers of iterations.

observed numbers, the Wachspress parameters are substantially better than the Peaceman–Rachford parameters. Moreover, slightly fewer iterations were required in the given case with the Wachspress parameters than with the optimum parameters. This was apparently because of the choice of starting values. However, it appears that the Wachspress parameters are nearly as good as the optimum parameters except for small m.

17.7. THE HELMHOLTZ EQUATION IN A RECTANGLE

Let us consider the solution of the Helmholtz equation

$$\frac{\partial^2 u}{\partial x^2} + \frac{\partial^2 u}{\partial y^2} - cu = G(x, y) \tag{7.1}$$

in the rectangle $0 \leq x \leq L_x$, $0 \leq y \leq L_y$, where c is a nonnegative constant. The five-point difference equation analog of (7.1) is given by

$$\mathscr{H}_0[u](x, y) + \mathscr{V}_0[u](x, y) + h^2 cu(x, y) = -h^2 G \tag{7.2}$$

where

$$\begin{aligned} \mathscr{H}_0[u](x, y) &= 2u(x, y) - u(x + h, y) - u(x - h, y) \\ \mathscr{V}_0[u](x, y) &= 2u(x, y) - u(x, y + h) - u(x, y - h) \end{aligned} \tag{7.3}$$

Evidently (7.3) is equivalent to the system

$$(H_0 + V_0 + \Sigma)u = b \tag{7.4}$$

where

$$\Sigma = \sigma I = ch^2 I \tag{7.5}$$

and where H_0 and V_0 correspond to $\mathscr{H}_0[u]$ and $\mathscr{V}_0[u]$. By (1.8), $H = H_0 + \frac{1}{2}\Sigma$ and $V = V_0 + \frac{1}{2}\Sigma$ correspond to

$$\begin{aligned} \mathscr{H}[u](x, y) &= \mathscr{H}_0[u](x, y) + \tfrac{1}{2}\sigma u(x, y) \\ \mathscr{V}[u](x, y) &= \mathscr{V}_0[u](x, y) + \tfrac{1}{2}\sigma u(x, y) \end{aligned} \tag{7.6}$$

To determine the eigenvalues and eigenvectors of H we seek a function $v(x, y)$, vanishing on S_h such that

$$\mathscr{H}[v](x, y) = \mu v(x, y) \tag{7.7}$$

Such a function is

$$v_{p,q}(x, y) = \sin(p\pi x/L_x) \sin(q\pi y/L_y) \tag{7.8}$$

where p and q are integers such that $1 \leq p \leq I - 1$, $1 \leq q \leq J - 1$ where $Ih = L_x$, $Jh = L_y$. Evidently we have

$$\begin{aligned} \mathscr{H}[v_{p,q}](x, y) &= \mu_p v_{p,q}(x, y) \\ \mathscr{V}[v_{p,q}](x, y) &= \nu_q v_{p,q}(x, y) \end{aligned} \tag{7.9}$$

where

$$\mu_p = 4\sin^2\frac{\pi p}{2I} + \frac{1}{2}\sigma, \qquad \nu_q = 4\sin^2\frac{\pi q}{2J} + \frac{1}{2}\sigma$$

Therefore, the eigenvalue bounds a, b, α, and β are given by

$$\begin{aligned} a = 4\sin^2\frac{\pi}{2I} + \frac{1}{2}\sigma \le \mu \le 4\cos^2\frac{\pi}{2I} + \frac{1}{2}\sigma = b \\ \alpha = 4\sin^2\frac{\pi}{2J} + \frac{1}{2}\sigma \le \nu \le 4\cos^2\frac{\pi}{2J} + \frac{1}{2}\sigma = \beta \end{aligned} \tag{7.10}$$

It can be shown (see Section 4.6) that the $v_{p,q}(x, y)$ are orthogonal over R_h; hence they are linearly independent and form a basis for the $(I-1)\times(J-1)$-dimensional space of functions defined on R_h and vanishing on S_h. Thus, we have accounted for all eigenvectors of H and of V. Consequently since H and V have a common basis of eigenvectors, then by Theorem 4.2 they commute. In Section 17.9 we shall give a direct proof of the commutativity.

If $L_x \ge L_y$, then $I \ge J$ and $a \le \alpha$, $b \ge \beta$. By Corollary 3.3 it follows that the value of ϱ which minimizes $\psi(a, b; \alpha, \beta; \varrho, \varrho)$, as defined by (3.3), is

$$\varrho^* = (\alpha\beta)^{1/2} = \left[\left(4\sin^2\frac{\pi}{2J} + \frac{1}{2}\sigma\right)\left(4\cos^2\frac{\pi}{2J} + \frac{1}{2}\sigma\right)\right]^{1/2} \tag{7.11}$$

This follows since $a + b = \alpha + \beta = 4 + \sigma$.

For the Dirichlet problem over the unit square we have, by (7.10),

$$a = \alpha = 4\sin^2(\pi h/2), \qquad b = \beta = 4\cos^2(\pi h/2) \tag{7.12}$$

Therefore, if we use the Peaceman–Rachford parameters, we have, by Theorem 6.1,

$$S\left(\prod_{i=1}^{m} T_{\varrho_i}\right) \le \left(\frac{1-(\tan(\pi h/2))^{1/m}}{1+(\tan(\pi h/2))^{1/m}}\right)^2 \sim 1 - 4\left(\frac{\pi h}{2}\right)^{1/m} \tag{7.13}$$

for small h and hence the average rate of convergence is given by

$$R_m\left(\prod_{i=1}^{m} T_{\varrho_i}\right) = -\frac{1}{m}\log S\left(\prod_{i=1}^{m} T_{\varrho_i}\right) \sim \frac{4}{m}\left(\frac{\pi h}{2}\right)^{1/m} \tag{7.14}$$

Thus, the average rate of convergence is asymptotically proportional to $h^{1/m}$ for small h. A similar result holds for the Wachspress parameters.

One can also show, using (7.12), that if one uses the Peaceman–Rachford parameters and chooses m according to (6.16), where $c = \tan^2(\pi h/2)$, then

$$R_m\left(\prod_{i=1}^{m} T_{\varrho_i}\right) \sim \frac{3.11}{-\log \tan^2(\pi h/2)} \sim \frac{1.55}{-\log(\pi h/2)} \tag{7.15}$$

Thus the average rate of convergence is proportional to $\log(\pi h/2)$, a quantity which decreases very slowly with h. Thus, the number of iterations required to achieve a given degree of convergence increases very slowly indeed.

For the Wachspress parameters, we have a similar result. Indeed, by Theorem 6.2 we have

$$R_m\left(\prod_{i=1}^{m} T_{\varrho_i}\right) \sim \frac{6.22}{-\log \tan^2(\pi h/2)} \sim \frac{3.11}{-\log(\pi h/2)}$$

Table 7.1 gives the results obtained using the PR method for the model problem for various values of h (see Birkhoff *et al.*, 1962). The numbers of iterations are given using the Peaceman–Rachford parameters and the Wachspress parameters for several values of m. For comparison some results for the SOR method are also given. It can be seen that the improvement in using the PR method is substantial and that, as h decreases, the number of iterations increases only very slightly with the Wachspress parameters.

TABLE 7.1. Numerical Results with the PR and SOR Methods for the Model Problem

h^{-1}	SOR		Peaceman–Rachford parameters				Wachspress parameters				
	ω	n	$m=1$	2	3	4	$m=1$	2	3	4	5
5	1.27	12	12	10	9	8	—	5	5	5	5
10	1.54	28	23	16	15	15	—	18	8	9	7
20	1.74	53	46	24	21	20	—	37	14	11	11
40	1.86	117	91	36	27	27	—	73	22	15	14
80	1.93	236	183	49	37	31	—	146	32	21	18
120	—	351 (est.)	274	61	44	38	—	—	41	23	19
160	—	472 (est.)	—	71	47	39	—	—	47	27	22

17.8. MONOTONICITY

We now seek upper and lower bounds for the eigenvalues of H and V corresponding to the difference equation Variant II (see Section 2.8) for the differential equation

$$\frac{\partial}{\partial x}\left(A\frac{\partial u}{\partial x}\right)+\frac{\partial}{\partial y}\left(C\frac{\partial u}{\partial y}\right)+Fu=G \tag{8.1}$$

where $A>0$, $C>0$, $F\le 0$ in the region under consideration. Let us consider two mesh regions R_h and R_h' such that $R_h\subseteq R_h'$ and let the matrices be H, V, and H', V', respectively. Let the eigenvalues of H and V lie in the ranges

$$a\le\mu\le b,\qquad \alpha\le\nu\le\beta \tag{8.2}$$

for R_h and

$$a'\le\mu\le b',\qquad \alpha'\le\nu\le\beta' \tag{8.3}$$

for R_h'. Since H, H', V, V' are positive definite, and since H can be obtained from H' by deleting certain rows and the corresponding columns it follows from Theorem 2-2.2 that

$$a'\le a,\qquad \alpha'\le\alpha,\qquad b'\ge b,\qquad \beta'\ge b \tag{8.4}$$

Next, following Birkhoff *et al.* [1962] we let

$$\begin{aligned}\bar{A}&=\max A(x,y), & \underline{A}&=\min A(x,y)\\ \bar{C}&=\max C(x,y), & \underline{C}&=\min C(x,y)\\ \bar{\sigma}&=\max(-h^2F), & \underline{\sigma}&=\min(-h^2F)\end{aligned} \tag{8.5}$$

Let $\bar{H}'$ and $\underline{H}'$ be matrices obtained from H' by replacing $A(x,y)$ by $\bar{A}$ and $\underline{A}$, respectively, and by replacing $-h^2F$ by $\bar{\sigma}$ and $\underline{\sigma}$, respectively. Evidently $\bar{H}'-H'$ is a nonnegative definite matrix, say, D. Hence the largest eigenvalue of $\bar{H}'$ is given by

$$\begin{aligned}\bar{b}'&=\max_{v\ne 0}\frac{(v,\bar{H}'v)}{(v,v)}=\max_{v\ne 0}\left\{\frac{(v,H'v)}{(v,v)}+\frac{(v,Dv)}{(v,v)}\right\}\\ &\ge\max_{v\ne 0}\frac{(v,H'v)}{(v,v)}=b'\end{aligned} \tag{8.6}$$

since $(v, Dv) \geq 0$. Similarly, we have $H' - \underline{H}'$ is a positive definite matrix, say, E, and

$$a' = \min_{v \neq 0} \frac{((\underline{H}' + E)v, v)}{(v, v)} = \min_{v \neq 0}\left[\frac{(\underline{H}'v, v)}{(v, v)} + \frac{(Ev, v)}{(v, v)}\right]$$
$$\geq \min_{v \neq 0} \frac{(\underline{H}'v, v)}{(v, v)} = \underline{a}' \tag{8.7}$$

Here $\underline{a}'$ is the smallest eigenvalue of $\underline{H}'$. Thus we have

$$\begin{aligned} \underline{a}' \leq a' \leq a, \qquad \bar{b}' \geq b' \geq b \\ \underline{\alpha}' \leq \alpha' \leq \alpha, \qquad \bar{\beta}' \geq \beta' \geq \beta \end{aligned} \tag{8.8}$$

But we can show, as in Section 17.7, that

$$\begin{aligned} \underline{a}' = 4\underline{A} \sin^2 \frac{\pi}{2I} + \frac{\sigma}{2}, \qquad \bar{b}' = 4\bar{A} \cos^2 \frac{\pi}{2I} + \frac{\bar{\sigma}}{2} \\ \underline{\alpha}' = 4\underline{C} \sin^2 \frac{\pi}{2J} + \frac{\sigma}{2}, \qquad \bar{\beta}' = 4\bar{C} \cos^2 \frac{\pi}{2J} + \frac{\bar{\sigma}}{2} \end{aligned} \tag{8.9}$$

We can then apply the results of Section 17.3 to determine values of ϱ and ϱ' which, though not necessarily optimum, yield spectral radii for the stationary case which are no larger than the values indicated.

It must be noted, of course, that while we can obtain bounds on the eigenvalues of H and V, which are useful in the stationary case, we cannot handle the nonstationary case, rigorously, unless $HV = VH$.

17.9. NECESSARY AND SUFFICIENT CONDITIONS FOR THE COMMUTATIVE CASE

Let us consider the discrete analogue of the generalized Dirichlet problem for the self-adjoint differential equation

$$\frac{\partial}{\partial x}\left(A \frac{\partial u}{\partial x}\right) + \frac{\partial}{\partial y}\left(C \frac{\partial u}{\partial y}\right) + Fu = G \tag{9.1}$$

We consider the symmetric difference equation (Variant II—see Section 2.8)

$$\mathscr{H}_0[u](x, y) + \mathscr{V}_0[u](x, y) + \mathscr{E}[u](x, y) = -h^2G \tag{9.2}$$

where

$$\mathscr{H}_0[u](x, y) = [A(x + \tfrac{1}{2}h, y) + A(x - \tfrac{1}{2}h, y)]u(x, y) \\ -A(x+\tfrac{1}{2}h, y)u(x+h, y) - A(x - \tfrac{1}{2}h, y)u(x-h, y) \quad (9.3)$$

$$\mathscr{V}_0[u](x, y) = [C(x, y + \tfrac{1}{2}h) + C(x, y - \tfrac{1}{2}h)]u(x, y) \\ -C(x, y+\tfrac{1}{2}h)u(x, y+h) - C(x, y - \tfrac{1}{2}h)u(x, y-h) \quad (9.4)$$

$$\mathscr{E}[u](x, y) = -h^2F(x, y)u(x, y) \quad (9.5)$$

Evidently (9.2) corresponds to the matrix equation

$$(H_0 + V_0 + \Sigma)u = b \quad (9.6)$$

where H_0, V_0, and Σ correspond to the operators $\mathscr{H}_0[u]$, $\mathscr{V}_0[u]$, and $\mathscr{E}[u]$, respectively. We seek to determine necessary and sufficient conditions that there exists a diagonal matrix P with positive diagonal elements such that the commutative case theory can be applied to the system

$$(PH_0 + PV_0 + P\Sigma)u = Pb \quad (9.7)$$

Since H_0 and V_0 are positive definite, it follows that H_0 and V_0 are similar to nonnegative diagonal matrices. Moreover, the same is true of PH_0 and PV_0 since they are similar to the positive definite matrices $P^{-1/2}(PH_0)P^{1/2} = P^{1/2}H_0P^{1/2}$ and $P^{-1/2}(PV_0)P^{1/2} = P^{1/2}V_0P^{1/2}$, respectively. Thus, in order that (4.1) be satisfied, we must have

$$(PH_0)(PV_0) = (PV_0)(PH_0) \quad (9.8)$$

and

$$P\Sigma = \sigma I \quad (9.9)$$

where σ is a nonnegative constant.

We remark that if a matrix P with the desired properties exists, then we can let $v = P^{-1/2}u$ and obtain from (9.6)

$$(H_0' + V_0' + \Sigma')v = P^{1/2}b \quad (9.10)$$

where

$$H_0' = P^{1/2}H_0P^{1/2}, \qquad V_0' = P^{1/2}V_0P^{1/2}, \qquad \Sigma' = P\Sigma \quad (9.11)$$

Evidently, if $H_0'' = PH_0$, $V_0'' = PV_0$ and Σ' satisfy (4.1), then so do H_0', V_0', and Σ'. Moreover, H_0' and V_0' are symmetric and, in fact, positive definite.

Evidently, the operators $\mathscr{H}_0'[u](x, y)$ and $\mathscr{V}_0'[u](x, y)$ corresponding to H_0' and V_0', respectively, are given by

$$\begin{aligned}\mathscr{H}_0'[u](x, y) &= A_0(x, y)u(x, y) - A_1(x, y)u(x + h, y)\\ &\quad -A_3(x, y)u(x - h, y)\\ \mathscr{V}_0'[u](x, y) &= C_0(x, y)u(x, y) - C_2(x, y)u(x, y + h)\\ &\quad -C_4(x, y)u(x, y - h)\end{aligned} \tag{9.12}$$

$$\begin{aligned}A_0(x, y) &= P(x, y)[A(x + \tfrac{1}{2}h, y) + A(x - \tfrac{1}{2}h, y)]\\ A_1(x, y) &= P(x, y)^{1/2}P(x + h, y)^{1/2}A(x + \tfrac{1}{2}h, y)\\ A_3(x, y) &= P(x, y)^{1/2}P(x - h, y)^{1/2}A(x - \tfrac{1}{2}h, y)\\ C_0(x, y) &= P(x, y)[C(x, y + \tfrac{1}{2}h) + C(x, y - \tfrac{1}{2}h)]\\ C_2(x, y) &= P(x, y)^{1/2}P(x, y + h)^{1/2}C(x, y + \tfrac{1}{2}h)\\ C_4(x, y) &= P(x, y)^{1/2}P(x, y - h)^{1/2}C(x, y - \tfrac{1}{2}h)\end{aligned} \tag{9.13}$$

Actually, the matrices H_0' and V_0' correspond more directly to the operators $\tilde{\mathscr{H}}_0'[u]$ and $\tilde{\mathscr{V}}_0'[u]$, respectively, which are defined for functions $u(x, y)$ vanishing on S_h and are such that $\tilde{\mathscr{H}}_0'[u](x, y)$ and $\tilde{\mathscr{V}}_0'[u](x, y)$ vanish on S_h. Thus we have, for any $u(x, y)$ vanishing on S_h

$$\begin{aligned}\tilde{\mathscr{H}}_0'[u](x, y) &= \Gamma(x, y)\mathscr{H}_0'[u](x, y)\\ \tilde{\mathscr{V}}_0'[u](x, y) &= \Gamma(x, y)\mathscr{V}_0'[u](x, y)\end{aligned} \tag{9.14}$$

TABLE 9.1 Coefficients of Product Operators

	Coefficient in $\mathscr{V}_0'\mathscr{H}_0'[u](x, y)$	Coefficient in $\mathscr{H}_0'\mathscr{V}_0'[u](x, y)$
$u(x, y)$	$A_0(x, y)C_0(x, y)$	$A_0(x, y)C_0(x, y)$
$u(x+h, y)$	$-C_0(x, y)A_1(x, y)$	$-\Gamma(x+h, y)A_1(x, y)C_0(x+h, y)$
$u(x, y+h)$	$-\Gamma(x, y+h)C_2(x, y)A_0(x, y+h)$	$-A_0(x, y)C_2(x, y)$
$u(x - h, y)$	$-C_0(x, y)A_3(x, y)$	$-\Gamma(x - h, y)A_3(x, y)C_0(x - h, y)$
$u(x, y - h)$	$-\Gamma(x, y - h)C_4(x, y)A_0(x, y - h)$	$-A_0(x, y)C_4(x, y)$
$u(x+h, y+h)$	$\Gamma(x, y+h)C_2(x, y)A_1(x, y+h)$	$\Gamma(x+h, y)A_1(x, y)C_2(x+h, y)$
$u(x - h, y+h)$	$\Gamma(x, y+h)C_2(x, y)A_3(x, y+h)$	$\Gamma(x - h, y)A_3(x, y)C_2(x - h, y)$
$u(x+h, y - h)$	$\Gamma(x, y - h)C_4(x, y)A_1(x, y - h)$	$\Gamma(x+h, y)A_1(x, y)C_4(x+h, y)$
$u(x - h, y - h)$	$\Gamma(x, y - h)C_4(x, y)A_3(x, y - h)$	$\Gamma(x - h, y)A_3(x, y)C_4(x - h, y)$

where

$$\Gamma(x, y) = \begin{cases} 1 & \text{if} \quad (x, y) \in R_h \\ 0 & \text{if} \quad (x, y) \notin R_h \end{cases} \tag{9.15}$$

The product matrix $V_0'H_0'$ corresponds to the operator $\mathscr{V}_0'\mathscr{H}_0'[u](x, y)$ which, for $(x, y) \in R_h$, can be determined by

$$\begin{aligned}
\mathscr{V}_0'\mathscr{H}_0'[u](x, y) &= C_0(x, y)\mathscr{H}_0'[u](x, y) - C_2(x, y)\mathscr{H}_0'[u](x, y + h) \\
&\quad -C_4(x, y)\mathscr{H}_0'[u](x, y - h) \\
&= C_0(x, y)\mathscr{H}_0'[u](x, y) \\
&\quad -C_2(x, y)\Gamma(x, y + h)\mathscr{H}_0'[u](x, y + h) \\
&\quad -C_4(x, y)\Gamma(x, y - h)\mathscr{H}_0'[u](x, y - h) \\
&= A_0(x, y)C_0(x, y)u(x, y) - C_0(x, y)A_1(x, y)u(x + h, y) \\
&\quad -\Gamma(x, y + h)C_2(x, y)A_0(x, y + h) \\
&\quad -C_0(x, y)A_3(x, y)u(x - h, y) \\
&\quad -\Gamma(x, y - h)C_4(x, y)A_0(x, y - h)u(x, y - h) \\
&\quad +\Gamma(x, y + h)C_2(x, y)A_1(x, y + h)u(x + h, y + h) \\
&\quad +\Gamma(x, y + h)C_2(x, y)A_3(x, y + h)u(x - h, y + h) \\
&\quad +\Gamma(x, y - h)C_4(x, y)A_1(x, y - h)u(x + h, y - h) \\
&\quad +\Gamma(x, y - h)C_4(x, y)A_3(x, y - h)u(x - h, y - h)
\end{aligned} \tag{9.16}$$

A similar expression can be obtained for $\mathscr{H}_0'\mathscr{V}_0'[u](x, y)$. Indeed, we can summarize the results in Table 9.1.

We first show that for commutativity we must have a rectangular region, assuming that R_h is connected and that $A > 0$, $C > 0$, $F \leq 0$. Suppose $(x + h, y + h)$ as well as (x, y) is in R_h. Then by equating the coefficients of $u(x + h, y + h)$ we see that $\Gamma(x, y + h) = \Gamma(x + h, y)$. Thus, $(x + h, y)$ belongs to R_h if and only if $(x, y + h)$ does. In this way one can show that, if *any* three of the four points (x, y), $(x + h, y)$, $(x, y + h)$, $(x + h, y + h)$ lie in R_h the fourth does also. From this and the fact that R_h is connected, one can show that R_h is a rectangular set of points or else R_h is a line of points.

Equating coefficients of $u(x \pm h, y)$, $u(x, y \pm h)$ we have

$$\begin{aligned}
C_0(x - h, y) &= C_0(x, y) = C_0(x + h, y) \\
A_0(x, y + h) &= A_0(x, y) = A_0(x, y - h)
\end{aligned} \tag{9.17}$$

Thus for all points of R_h it follows that $A_0(x, y)$ is independent of y and $C_0(x, y)$ is independent of x.

From (9.13) it follows that the operators $\mathscr{H}_0'[u]$ and $\mathscr{V}_0'[u]$ are *symmetric* since

$$\begin{aligned} A_1(x, y) &= A_3(x + h, y) \\ C_2(x, y) &= C_4(x, y + h) \end{aligned} \tag{9.18}$$

Equating coefficients of $u(x + h, y + h)$ we have

$$C_2(x, y)A_1(x, y + h) = A_1(x, y)C_2(x + h, y) \tag{9.19}$$

Equating coefficients of $u(x^* - h, y + h)$ in the equation for (x^*, y), where $x^* = x + h$ we have

$$C_2(x + h, y)A_3(x + h, y + h) = A_3(x + h, y)C_2(x, y) \tag{9.20}$$

or by (9.18)

$$C_2(x + h, y)A_1(x, y + h) = A_1(x, y)C_2(x, y) \tag{9.21}$$

But this implies that

$$A_1(x, y + h) = A_1(x, y) \tag{9.22}$$

We can therefore show that $A_1(x, y)$ and $A_3(x, y)$ are independent of y for all $(x, y) \in R_h$ and, similarly, $C_2(x, y)$ and $C_4(x, y)$ are independent of x for all $(x, y) \in R_h$. Thus, in summary, we see that the operators $\mathscr{H}_0'[u]$ and $\mathscr{V}_0'[u]$ are symmetric and commute if and only if the $A_i(x, y)$ are independent of y and the $C_i(x, y)$ are independent of x for all $(x, y) \in R_h$.

Suppose now that there exist positive functions $E_1(x)$, $E_2(x)$, $F_1(y)$, $F_2(y)$ and a constant $c \geq 0$ such that

$$\begin{aligned} A(x, y) &= E_1(x)F_1(y) \\ C(x, y) &= E_2(x)F_2(y) \\ F(x, y) &= -cE_2(x)F_1(y) \end{aligned} \tag{9.23}$$

If we let

$$P(x, y) = 1/(E_2(x)F_1(y)) \tag{9.24}$$

we obtain, by (9.13)

$$\begin{aligned}
A_0(x, y) &= \frac{E_1(x + \frac{1}{2}h) + E_1(x - \frac{1}{2}h)}{E_2(x)} \\
A_1(x, y) &= \frac{E_1(x + \frac{1}{2}h)}{[E_2(x)]^{1/2}[E_2(x + h)]^{1/2}} \\
A_3(x, y) &= \frac{E_1(x - \frac{1}{2}h)}{[E_2(x)]^{1/2}[E_2(x - h)]^{1/2}} \\
C_0(x, y) &= \frac{F_2(y + \frac{1}{2}h) + F_2(y - \frac{1}{2}h)}{F_1(y)} \\
C_2(x, y) &= \frac{F_2(y + \frac{1}{2}h)}{[F_2(y)]^{1/2}[F_2(y + h)]^{1/2}} \\
C_4(x, y) &= \frac{F_2(y - \frac{1}{2}h)}{[F_2(y)]^{1/2}[F_2(y - h)]^{1/2}}
\end{aligned} \tag{9.25}$$

Moreover,

$$\mathscr{E}[u](x, y) = ch^2u(x, y) \tag{9.26}$$

Hence the conditions required for the commutative case hold.

We remark that in actual practice one can simply multiply (9.1) by $P(x, y)$ obtaining

$$\frac{1}{E_2(x)} \frac{\partial}{\partial x}\left(E_1(x) \frac{\partial u}{\partial x}\right) + \frac{1}{F_1(y)} \frac{\partial}{\partial y}\left(F_2(y) \frac{\partial u}{\partial y}\right) - cu = \frac{G}{E_2(x)F_1(y)} \tag{9.27}$$

We can use the difference equation

$$\mathscr{H}_0''[u](x, y) + \mathscr{V}_0''[u](x, y) + \mathscr{E}[u](x, y) = -h^2G/(E_2(x)F_1(y)) \tag{9.28}$$

where

$$\begin{aligned}
\mathscr{H}_0''[u](x, y) = &\left(\frac{E_1(x + \frac{1}{2}h) + E_1(x - \frac{1}{2}h)}{E_2(x)}\right)u(x, y) \\
&- \frac{E_1(x + \frac{1}{2}h)}{E_2(x)} u(x + h, y) - \frac{E_1(x - \frac{1}{2}h)}{E_2(x)} u(x - h, y)
\end{aligned} \tag{9.29}$$

$$\begin{aligned}
\mathscr{V}_0''[u](x, y) = &\left(\frac{F_2(y + \frac{1}{2}h) + F_2(y - \frac{1}{2}h)}{F_1(y)}\right)u(x, y) \\
&- \frac{F_2(y + \frac{1}{2}h)}{F_1(y)} u(x, y + h) - \frac{F_2(y - \frac{1}{2}h)}{F_1(y)} u(x, y - h)
\end{aligned} \tag{9.30}$$

and

$$\mathscr{E}[u](x, y) = h^2cu(x, y) \tag{9.31}$$

It is readily seen that $\mathscr{H}_0''[u]$ and $\mathscr{V}_0''[u]$ commute, but that they are not symmetric. However, if H_0'' and V_0'' are the corresponding matrices and if P is a diagonal matrix obtained from the function $P(x, y)$, we know that $H_0'' = PH_0$, $V_0'' = PV_0$ where H_0 and V_0 are positive definite. Thus $P^{-1/2}H_0''P^{1/2} = P^{1/2}H_0P^{1/2}$ and $P^{-1/2}V_0''P^{1/2} = P^{1/2}V_0P^{1/2}$ are positive definite. Hence H_0'' and V_0'' are both similar to positive definite matrices and hence are similar to nonnegative diagonal matrices.

As an example, let us consider the equation

$$\frac{\partial}{\partial x}\left(\frac{1}{y}\frac{\partial u}{\partial x}\right) + \frac{\partial}{\partial y}\left(\frac{1}{y}\frac{\partial u}{\partial y}\right) - \frac{2}{y}u = e^x \tag{9.32}$$

Evidently the conditions (9.23) are satisfied with

$$E_1(x) = 1, \quad F_1(y) = 1/y, \quad E_2(x) = 1, \quad F_2(y) = 1/y, \quad c = 2 \tag{9.33}$$

Letting

$$P(x, y) = y \tag{9.34}$$

and multiplying by y we have

$$\frac{\partial^2 u}{\partial x^2} + y\frac{\partial}{\partial y}\left(\frac{1}{y}\frac{\partial u}{\partial y}\right) - 2u = ye^x \tag{9.35}$$

The difference equation (9.28) becomes

$$\begin{aligned}\{2u(x, y) - u(x+h, y) - u(x-h, y)\} &+ \left\{\left[\frac{y}{y+\frac{1}{2}h} + \frac{y}{y-\frac{1}{2}h}\right]u(x, y)\right. \\ &\left. - \frac{y}{y+\frac{1}{2}h}u(x, y+h) - \frac{y}{y-\frac{1}{2}h}u(x, y-h)\right\} \\ &+ 2h^2u(x, y) = -h^2ye^x\end{aligned} \tag{9.36}$$

This leads to the linear system

$$(H_0'' + V_0'' + \Sigma'')u = b'' \tag{9.37}$$

where $H_0''V_0'' = V_0''H_0''$ and Σ'' is a constant times the identity matrix. Moreover, V_0'' is not symmetric though it is similar to a positive definite matrix, namely,

$$P^{-1/2}V_0''P^{1/2} = P^{1/2}V_0P^{1/2} \tag{9.38}$$

A difference equation which leads to a system

$$(H_0' + V_0' + \Sigma'')u = b' \tag{9.39}$$

where $H_0'V_0' = V_0'H_0'$ and where H_0' and V_0' are positive definite is given by

$$\begin{aligned}&\{2u(x, y) - u(x + h, y) - u(x - h, y)\} + \Big\{y\Big[\frac{1}{y+\frac{1}{2}h} + \frac{1}{y-\frac{1}{2}h}\Big]u(x, y)\\ &\quad -[y(y+h)]^{1/2}\Big(\frac{1}{y+\frac{1}{2}h}\Big)u(x, y+h) - [y(y-h)]^{1/2}\Big(\frac{1}{y-\frac{1}{2}h}\Big)u(x, y-h)\Big\}\\ &\quad +2h^2u(x, y) = -h^2y^{1/2}e^x\end{aligned} \tag{9.40}$$

Let us now seek *necessary* conditions that a continuous function $P(x, y)$ exist which is positive for all (x, y) and such that the operators $\mathscr{H}_0'[u]$ and $\mathscr{V}_0'[u]$ given by (9.14), (9.12), and (9.13) commute, and such that $P(x, y)F(x, y)$ is a constant. We have already seen that for commutativity R_h must be a rectangular grid. Hence, let us assume that we are dealing with a rectangle $R + S$ and that for a sequence of mesh sizes, $h_1, h_2, \ldots,$ tending to zero the sides of the rectangle are multiples of each h_i. Moreover, let us assume that h_1 is an integral multiple of h_2, that h_2 is an integral multiple of h_3, etc. Thus, if the point (x, y) is a mesh point for some h_i it is also a mesh point for $h_{i+1}, h_{i+2}, \ldots.$

If we require that $A_1(x, y)$ be independent of y for all $x \in R_h$, and that this be true for all h, we must have, by (9.13) and the continuity of P and A,

$$P(x, y)A(x, y) = X(x) \tag{9.41}$$

for some continuous function $X(x)$. Similarly, we must have

$$P(x, y)C(x, y) = Y(y) \tag{9.42}$$

Actually, (9.41) and (9.42) have been shown to be true only for points which are mesh points for some h_i, but since the set of such points is dense in $R + S$, it follows by the continuity of A, C, and P that (9.41) and (9.42) hold for all points of $R + S$.

Since $A_1(x, y)$ is a function of x we have by (9.41) and (9.13)

$$A(x + \tfrac{1}{2}h, y)/\{[A(x, y)]^{1/2}[A(x + h, y)]^{1/2}\} = \phi(x) \tag{9.43}$$

for some function $\phi(x)$ of x. If we let

$$\zeta(x, y) = \log A(x, y) \tag{9.44}$$

we have

$$\zeta(x + \tfrac{1}{2}h, y) - \tfrac{1}{2}\zeta(x, y) - \tfrac{1}{2}\zeta(x + h, y) = \log \phi(x) = \psi(x) \quad (9.45)$$

The general solution of this difference equation is[†]

$$\zeta(x, y) = \sigma(y) + xY_1(y) + \omega(x) \quad (9.46)$$

for some functions $\sigma(y)$, $\omega(x)$, and $Y_1(y)$, and hence

$$A(x, y) = E_1(x)F_1(y)e^{xY_1(y)} \quad (9.47)$$

Similarly, we have

$$C(x, y) = E_2(x)F_2(y)e^{yX_1(x)} \quad (9.48)$$

for suitable $E_2(x)$, $F_2(y)$, and $X_1(x)$.

But since $A(x, y)/C(x, y) = X(x)/Y(y)$ we have

$$\frac{\partial^2}{\partial x\, \partial y}\left(\log \frac{A}{C}\right) = 0 \quad (9.49)$$

Hence for some constant a we have

$$X_1'(x) = Y_1'(y) = a \quad (9.50)$$

and therefore

$$A(x, y) = E_1(x)F_1(y)e^{axy}, \qquad C(x, y) = E_2(x)F_2(y)e^{axy} \quad (9.51)$$

By (9.41) and (9.42) we have

$$P(x, y)E_2(x)F_1(y)e^{axy} = X(x)E_2(x)/E_1(x) = Y(y)F_1(y)/F_2(y) = q \quad (9.52)$$

for some constant q and hence

$$P(x, y) = q/[E_2(x)F_1(y)e^{axy}] \quad (9.53)$$

Let us now impose the requirement that $A_0(x, y)$ be independent of y. By (9.12), (9.51), and (9.53), this implies that

$$\frac{q}{E_2(x)F_1(y)e^{axy}} \times \{E_1(x+\tfrac{1}{2}h)F_1(y)\exp[a(x+\tfrac{1}{2}h)y]+E_1(x-\tfrac{1}{2}h)F_1(y)\exp[a(x-\tfrac{1}{2}h)y]\}$$

[†] Consider the difference equation $w_{n+1} - 2w_n + w_{n-1} = f(x)$. Evidently if $z_{n+1} = w_{n+1} - w_n$, then $z_n = z_0 + \sum_{i=0}^{n-1} f(x_i)$ and $w_n = w_0 + nz_0 + \sum_{j=0}^{n} \sum_{i=0}^{j-1} f(x_i)$.

is independent of y, i.e.,

$$[e^{ahy/2}E_1(x+\tfrac{1}{2}h)+e^{-ahy/2}E_1(x-\tfrac{1}{2}h)]/E_2(x)$$

is independent of y. For this to be true for all y and all h we must have $a=0$, since $E_1(x)$ is a positive function. Therefore, we have

$$\begin{aligned} A(x,y)&=E_1(x)F_1(y),\\ C(x,y)&=E_2(x)F_2(y)\\ P(x,y)&=q/[E_2(x)F_1(y)],\\ F(x,y)&=-cE_2(x)F_1(y) \end{aligned} \tag{9.54}$$

for some constant $c>0$. But these conditions are identical to the conditions (9.23).

We remark that if we do not require that the difference equation correspond to a symmetric matrix we can obtain commutativity even when the above conditions do not hold. Thus, for example, consider the equation

$$\frac{\partial^2 u}{\partial x^2}+\frac{\partial^2 u}{\partial y^2}+\frac{2}{x+y}\frac{\partial u}{\partial x}+\frac{2}{x+y}\frac{\partial u}{\partial y}=0 \tag{9.55}$$

in the unit square. If we consider the difference equation

$$\mathscr{H}[u](x,y)+\mathscr{V}[u](x,y)=0 \tag{9.56}$$

where

$$\begin{aligned} \mathscr{H}[u](x,y)&=2u(x,y)-\left(1+\frac{h}{x+y}\right)u(x+h,y)\\ &\quad-\left(1-\frac{h}{x+y}\right)u(x-h,y)\\ \mathscr{V}[u](x,y)&=2u(x,y)-\left(1+\frac{h}{x+y}\right)u(x,y+h)\\ &\quad-\left(1-\frac{h}{x+y}\right)u(x,y-h) \end{aligned} \tag{9.57}$$

one can easily show that $\mathscr{H}\mathscr{V}[u]=\mathscr{V}\mathscr{H}[u]$. Hence the corresponding matrices H and V commute. Moreover, it can be shown that H and V are similar to positive definite matrices (see Birkhoff *et al.*, 1962).

17.10. THE NONCOMMUTATIVE CASE

We have seen that in the commutative case, the PR method, with an appropriate choice of parameters, converges extremely rapidly in comparison with other methods such as the SOR method. For the Dirichlet problem the number of iterations increases as $|\log h|$. As a matter of fact, if one chooses the initial vector $u^{(0)}$ on the basis of a sufficiently smooth function satisfying the boundary conditions, then the number of iterations required for convergence does not increase with h (see Guilinger, 1965, and Lynch and Rice, 1968).

In the stationary case, the convergence is known and, in fact, there is a monotonicity theorem which enables one to get a bound on $S(T_\varrho)$ in terms of a corresponding bound for a circumscribing rectangle. As a matter of fact, as shown by Guilinger [1965], if the initial guess is sufficiently smooth and if the region is convex, then the number of iterations is independent of the mesh size.

The theory for the nonstationary PR method in the noncommutative case is in an unsatisfactory state. Even the convergence of the method for an arbitrary choice of positive parameters is not guaranteed as shown by the following example due to Price and Varga [1962]. It is easy to verify that the matrices H and V given below are nonnegative definite and that $A = H + V$ is positive definite. Nevertheless we have

$$S(T_{1.0}T_{3.0}) \doteq 1.369$$

The matrices were derived from the differential equation (9.1) with mixed boundary conditions where $A(x, y)$ and $C(x, y)$ are discontinuous. The question as to whether this phenomenon of divergence can occur if $A(x, y)$ and $C(x, y)$ are continuous remains open.

$$H = \frac{1}{2}\begin{pmatrix} 101 & -1 & 0 & 0 & 0 & 0 & 0 & 0 & 0 \\ -1 & 2 & -1 & 0 & 0 & 0 & 0 & 0 & 0 \\ 0 & -1 & 101 & 0 & 0 & 0 & 0 & 0 & 0 \\ 0 & 0 & 0 & 202 & -101 & 0 & 0 & 0 & 0 \\ 0 & 0 & 0 & -101 & 202 & -101 & 0 & 0 & 0 \\ 0 & 0 & 0 & 0 & -101 & 202 & 0 & 0 & 0 \\ 0 & 0 & 0 & 0 & 0 & 0 & 101 & -100 & 0 \\ 0 & 0 & 0 & 0 & 0 & 0 & -100 & 200 & -100 \\ 0 & 0 & 0 & 0 & 0 & 0 & 0 & -100 & 101 \end{pmatrix}$$

$$V = \frac{1}{2}\begin{pmatrix} 101 & 0 & 0 & -101 & 0 & 0 & 0 & 0 & 0 \\ 0 & 2 & 0 & 0 & -2 & 0 & 0 & 0 & 0 \\ 0 & 0 & 101 & 0 & 0 & -101 & 0 & 0 & 0 \\ -101 & 0 & 0 & 202 & 0 & 0 & -101 & 0 & 0 \\ 0 & -2 & 0 & 0 & 202 & 0 & 0 & -200 & 0 \\ 0 & 0 & -101 & 0 & 0 & 202 & 0 & 0 & -101 \\ 0 & 0 & 0 & -101 & 0 & 0 & 101 & 0 & 0 \\ 0 & 0 & 0 & 0 & -200 & 0 & 0 & 200 & 0 \\ 0 & 0 & 0 & 0 & 0 & -101 & 0 & 0 & 101 \end{pmatrix}$$

Guilinger [1965] proved that for the Dirichlet problem for any convex region

$$(Hv, Vv) \geq 0 \tag{10.1}$$

for all real v. From this it follows that HV is positive real since for any real v we have

$$(v, (HV + VH)v) = 2(Hv, Vv) \geq 0 \tag{10.2}$$

Thus by Theorem 2.9 and Corollary 2.10 the nonstationary PR method is convergent for any set of positive parameters.

On the other hand, HV need not be positive real for a nonconvex region. For example, for the 13-point cross

$$x = 0, \qquad y = 0, \pm h, \pm 2h, \pm 3h$$
$$y = 0, \qquad x = 0, \pm h, \pm 2h, \pm 3h$$

the matrix $HV + VH$ corresponding to the Dirichlet problem is not positive definite. As a matter of fact, $HV + VH$ has an eigenvalue of -0.267. Thus even though HV, which is similar to the positive definite matrix $H^{1/2}VH^{1/2}$ and VH have positive eigenvalues, $HV + VH$ need not have positive eigenvalues. Nevertheless, on the basis of numerical evidence it seems clear in this case that for any $\varrho_1 > 0$, $\varrho_2 > 0$

$$S(T_{\varrho_1} T_{\varrho_2}) < 1$$

Numerical experiments described by Young and Ehrlich [1960] and by Birkhoff *et al.* [1962] showed that for the Dirichlet problem and for a variety of regions the commutative case theory holds reasonably well. In most cases considered, the number of iterations was a monotonic function of the size of the region, though there were some exceptions.

Even in this relatively restricted class of problems there does not exist a proof of the observed behavior of the method.

Pearcy [1962] showed that for any given problem one can be assured of convergence in the nonstationary case by using m parameters $\bar{b} \geq \varrho_1, \geq \varrho_2 \geq \cdots \geq \varrho_m \geq \bar{a}$ where all eigenvalues of H and V lie in the interval $\bar{a} \leq \gamma \leq \bar{b}$, and where m is sufficiently large. Habetler [1959] and Birkhoff and Varga [1959] proved similar results. None of the techniques, however, appears to be useful in explaining the very rapid convergence of the PR method which has been observed in many situations involving the noncommutative case.

Widlund [1966] showed that for the generalized Dirichlet problem for the rectangle, if the coefficients are sufficiently smooth, then iteration parameters can be chosen so that the convergence is as large as for the comparable commutative cases. The iteration parameters are required to vary with the independent variables. (Thus in (1.6) we replace ϱI by ϱE where E is a diagonal matrix with positive diagonal elements.) As shown by Widlund [1969] such a choice of iteration parameters is equivalent to using the ordinary PR method on "scaled equations." Thus, for example, in Section 17.9 we "scaled" the difference equation (9.6), corresponding to (9.1) with $A = E_1(x)F_1(y)$, $C = E_2(x)F_2(y)$, $F = kE_2(x)F_1(y)$ by a matrix P corresponding to $F_1(y)^{-1}E_2(x)^{-1}$ and thereby obtained (9.7) for which the commutative case theory holds. Of course, a different scaling must be used in the more general case considered by Widlund.

SUPPLEMENTARY DISCUSSION

Section 17.1. The PR method was introduced by Peaceman and Rachford [1955] and is closely related to a method of Douglas [1955] which is applicable to initial value problems. Another ADI method is the method of Douglas and Rachford [1956]. One variant of the Douglas–Rachford method is defined by

$$(H_0 + \Sigma + \varrho_{n+1}I)u^{(n+1/2)} = k - (V_0 - \varrho_{n+1}I)u^{(n)}$$
$$(V_0 + \Sigma + \varrho_{n+1}I)u^{(n+1)} = (V_0 + \tfrac{1}{2}\Sigma)u^{(n)} + (\tfrac{1}{2}\Sigma + \varrho_{n+1}I)u^{(n+1/2)}$$

The Douglas–Rachford method is somewhat slower than the PR method in many cases (see the discussion of Birkhoff *et al.*, 1962). However, it can be generalized to the case where A is split into the sum of several matrices, e.g., $A = H_1 + H_2 + \cdots + H_p + \Sigma$. (Such splittings naturally occur in higher dimensional problems.) Other ADI methods are considered by Guittet [1967] and by D'Jakonov [1969].

Section 17.2. The first general convergence proof for the stationary PR method (Corollary 2.2) is due to Sheldon and Wachspress (see Wachspress, 1957). Theorem 2.1 was proved by Birkhoff *et al.* [1962]. Theorem 2.3 is a slight extension of a result in that paper. Corollaries 2.5 and 2.7 were proved by Wachspress and Habetler [1960]. Theorem 2.9 and Corollary 2.10 follow from results of Wachspress [1966].

Section 17.3. Proofs of Theorems 3.1 and 3.2 can be found in Birkhoff *et al.* [1962]. Theorem 3.4 is a special case proved by elementary methods of a result of Wachspress [1968].

Section 17.5. A still further improvement in the convergence rate can be obtained by using semi-iteration based on the PR method with optimum parameters (see Gourlay, 1968; 1970).

Section 17.8. The results in this section are given by Birkhoff *et al.* [1962].

Section 17.9. The necessary and sufficient conditions for the commutative case were given by Birkhoff *et al.* [1962]. Earlier, Birkhoff and Varga [1958] had shown that for the case $P = I$ and $A \equiv C$ the region must be a rectangle and A and C must be constant. The commutative case holds when the differential equation is, in a sense, separable and the region is rectangular. Methods based on the separation of variables can then be used to solve either the differential equation or the difference equation analytically. As pointed out by Widlund [1969], very efficient noniterative methods exist for such cases; see, for instance, Hockney [1965], and Buzbee *et al.* [1969].

The results of numerical studies on the convergence of the Peaceman–Rachford method are given in a report by Mouradoglou [1967].

EXERCISES

Section 17.1

1. Carry out two iterations of the stationary PR method for the five-point discrete analog of the Dirichlet problem in the unit square with $h = \frac{1}{3}$ and $u = 1$ for $y = 0$ and $u = 0$ elsewhere on the boundary. Take zero starting values and let $\varrho = \varrho' = 1$.

2. Consider the problem of solving the linear system $Au = b$ where A is a nonsingular matrix. Suppose that $A = H + V$ where for any $\varrho > 0$ the systems $(H + \varrho I)w = s$ and $(V + \varrho I)z = t$ can be conveniently and uniquely solved for w and z, respectively, for any s and t.

(a) Determine k_1 and k_2 so that each of the two methods

$$u^{(n+1)} = G_1 u^{(n)} + k_1$$
$$u^{(n+1)} = G_2 u^{(n)} + k_2$$

will be completely consistent where

$$G_1 = I - (H + \varrho I)^{-1} A$$
$$G_2 = I - (V + \varrho I)^{-1} A$$

(b) Suppose $u^{(n+1)} = G_1 u^{(n)} + k_1$ if n is even and $u^{(n+1)} = G_2 u^{(n)} + k_2$ if n is odd. Let $v^{(n)} = u^{(2n)}$, $n = 0, 1, 2, \ldots$. Find G and k so that $v^{(n+1)} = Gv^{(n)} + k$ and show that the method is completely consistent. Also, show that

$$G = (V + \varrho I)^{-1}(H - \varrho I)(H + \varrho I)^{-1}(V - \varrho I)$$
$$k = 2\varrho(V + \varrho I)^{-1}(H + \varrho I)^{-1} b$$

Section 17.2

1. For the model problem show that

$$\min_{\varrho > 0} S(T_\varrho) = \frac{1 - \sin \pi h}{1 + \sin \pi h}$$

2. Show that the stationary PR method with optimum ϱ and the point SOR method with ω_b have the same asymptotic rates of convergence for the model problem. (Use the result of Exercise 1.)

3. For the model problem with $h = \frac{1}{4}$ for what values of ϱ and ϱ' is $S(T_{\varrho,\varrho'}) < 1$?

Section 17.3

1. Find ϱ and ϱ' to minimize $\psi(a, b; \alpha, \beta; \varrho, \varrho')$ for the following cases:

a	b	α	β
0.1	1	0.5	4
0.1	3.9	0.2	3.8
0.1	1	2	4
0.5	2.0	0.4	3.0

Also find the optimum ϱ which minimizes $\psi(a, b; \alpha, \beta; \varrho, \varrho)$.

2. Find the optimum values of ϱ and ϱ' to be used in solving the discrete Dirichlet problem for the 2×1 rectangle with $h = \frac{1}{20}$ by the stationary Peaceman–Rachford method.

Section 17.4

1. Verify that for the five-point discrete analog of the Dirichlet problem for the unit square with $h = \frac{1}{3}$ that H and V commute. Construct a set of eigenvectors common to both H and V.

2. Verify that for the five-point discrete analog of the Dirichlet problem for the L-shaped region consisting of the unit square with the subregion $\frac{2}{3} \leq x \leq 1$, $\frac{2}{3} \leq y \leq 1$ removed and $h = \frac{1}{3}$ that H and V do not commute. Show that $S(\prod_{i=1}^{m} T_{\varrho_i}) < 1$ for any set of positive parameters.

Section 17.5

1. If the eigenvalues μ of H and ν of V lie in the ranges $0.5 \leq \mu \leq 3.5$, $0.1 \leq \nu \leq 5.9$, and $m = 4$, compute the optimum parameters $\varrho_i{}^*$ and $\varrho_i'{}^*$, $i = 1, 2, 3, 4$. Also, compute

$$\psi_4(a, b; \alpha, \beta; \boldsymbol{\rho}^*, (\boldsymbol{\rho}')^*)$$

and

$$\lambda_4^{1/4} \quad \text{and} \quad -\tfrac{1}{4} \log \psi_4$$

2. Determine the optimum parameters for the cases $m = 3$ and $m = 4$ for the case of the model problem for $h = \frac{1}{20}$. Use (5.29).

Section 17.6

1. Consider the use of the Peaceman–Rachford method for solving the model problem with $h = \frac{1}{60}$. What is the optimum number of Peaceman–Rachford parameters? Give the values of the parameters to three decimals. How many double sweeps would be required to reduce an initial error by a factor of 10^{-5}?

2. For the model problem with $h = \frac{1}{80}$, determine the optimum number of Peaceman–Rachford parameters. Estimate the number of double sweeps required using these parameters to reduce an initial error by a factor of 10^{-4}. Determine the optimum relaxation factor to be used with the SOR method and estimate the number of iterations required to reduce the initial error by a factor of 10^{-4}. Which method would be faster on a computer?

Section 17.7

1. For the five-point discrete analog of the generalized Dirichlet problem with the differential equation $u_{xx} + u_{yy} - u = 0$ in the rectangle $0 \leq x \leq 2$, $0 \leq y \leq 1$ compute the optimum parameters for the PR method if $h = \frac{1}{20}$, $m = 4$. Also compute the Peaceman–Rachford and the Wachspress parameters. In each case compute a bound for

$$S\left(\prod_{i=1}^{4} T_{\varrho_i, \varrho_i'}\right)$$

Section 17.8

1. For the five-point discrete analog of the Dirichlet problem, verify that the eigenvalues of H and V for any subregion of the unit square with $h = \frac{1}{3}$ are bounded by the eigenvalues of H and V for the square. Compute a bound for $\min_{\varrho>0} S(T_\varrho)$ for all these regions. (Use the result of Problem 1, Section 17.2.)

2. For the five-point discrete analog of the generalized Dirichlet problem in the rectangle $1 \leq x \leq 2$, $1 \leq y \leq \frac{5}{3}$ with $h = \frac{1}{3}$ and with the differential equation

$$\frac{\partial}{\partial x}\left(x \frac{\partial u}{\partial x}\right) + \frac{\partial}{\partial y}\left(y \frac{\partial u}{\partial y}\right) = 0$$

verify (8.8).

Section 17.9

1. Consider the five-point discrete analog of the generalized Dirichlet problem in a rectangle involving the differential equation

$$u_{xx} + u_{yy} + \frac{1}{x} u_x + \frac{1}{y} u_y = 0$$

Describe how you would set up a difference equation leading to a linear system of the form $(H + V)u = k$ where $HV = VH$. Assume that the rectangle is contained in the region: $x \geq 1$, $y \geq 1$.

2. Show that (9.55) is essentially self-adjoint but when one uses Variant II (see Section 2.8) with the self-adjoint form one does not obtain the commutative case. (Consider the case of the unit square with $h = \frac{1}{3}$.)

Section 17.10

1. (Pearcy) Show that if H and V are positive definite and if their eigenvalues lie in the interval $a \leq \gamma \leq b$, then there exists N such that if $m > N$ and $b \geq \varrho_1 \geq \varrho_2 \geq \cdots \geq \varrho_m \geq a$, then $S(\prod_{i=1}^{m} T_{\varrho_i}) < 1$.

2. Show that $H + V$ in the example of Price and Varga is positive definite and that H and V are nonnegative definite.

3. Show that if H and V are positive definite, then $S(\prod_{i=1}^{m} T_{\varrho_i}) < 1$ if $\varrho_1 \geq \varrho_2 \geq \cdots \geq \varrho_m > 0$ and $\varrho_i - \varrho_{i+1} < 2\gamma_{\min}$ for $i = 1, 2, \ldots, m-1$ where $\gamma_{\min}$ is the smallest eigenvalue of V.

Chapter 18 / SELECTION OF ITERATIVE METHOD

We now give a brief discussion of considerations in the selection of an iterative method to be used in solving the large linear system

$$Au = b \tag{1}$$

when A is a large sparse positive definite matrix. We shall give special attention to the case where the system (1) corresponds to the finite difference solution of a self-adjoint elliptic partial differential equation of the form (2-8.3).

We assume that $\bar{a}$ and $\bar{b}$, the lower and upper bounds, respectively, of the eigenvalues ν of A are known or can be estimated to a reasonable accuracy. We also assume that similar bounds $\hat{a}$ and $\hat{b}$, respectively, are available for the eigenvalues of the scaled matrix $\hat{A} = D^{-1/2}AD^{-1/2}$ where $D = \operatorname{diag} A$. Frequently, reasonable bounds can be found by the methods of Chapter 6 (Sections 6.5–6.9). If nothing else is known, one can use the RF method with $p = -2/(\bar{a} + \bar{b})$ obtaining

$$u^{(n+1)} = \left(I - \frac{2}{\bar{a} + \bar{b}}\right)u^{(n)} - \left(\frac{2}{\bar{a} + \bar{b}}\right)b \tag{2}$$

The spectral radius of this method is

$$S\left(I - \frac{2}{\bar{a} + \bar{b}} A\right) = \frac{\bar{b} - \bar{a}}{\bar{b} + \bar{a}} < 1 \tag{3}$$

We can then improve the convergence by an order-of-magnitude using the RF–SI method, as described in Chapter 11. Also, even though the J method need not converge, nevertheless, the eigenvalues of B are real and less than unity, and hence semi-iteration can be applied. In general, we do not know whether this procedure, the J–SI method, is better than the RF–SI method. However, we do know that if A has Property A, then the J–SI method is superior (see Section 11.3).

It is also known that the GS method converges and the SOR method converges for $0 < \omega < 2$, but as shown in the example in Section 12.1, the SOR may not be effective no matter what choice of ω is used. Furthermore, since the eigenvalues of $\mathscr{L}$ and $\mathscr{L}_\omega$ may be complex, the use of semi-iterative methods does not appear promising.

If A is an L-matrix, then it is known (see Section 4.5) that the GS method converges at least as fast as the J method. However, the eigenvalues of $\mathscr{L}$ may still be complex—hence one would not expect a significant improvement using semi-iteration. Nevertheless, in this case $\bar{\mu} = S(B) < 1$ and, if we can estimate $\bar{\mu}$, then we can obtain a substantial improvement over the GS method by using $\mathscr{L}_{\omega_b}$ where $\omega_b = 2(1 + [1 - \bar{\mu}^2]^{1/2})^{-1}$ (see Section 12.2). Indeed, the spectral radius of $\mathscr{L}_{\omega_b}$ does not exceed that of the J–SI method.

If A has Property A, then by a suitable rearrangement of the rows and corresponding columns of A we can obtain a consistently ordered matrix for which the SOR theory is rigorously applicable. In some cases an ordered grouping π exists such that A has Property $A^{(\pi)}$ and the theory can also be applied (see Chapter 14). One can also permute the rows and columns of A to obtain the form (8-1.2) and then apply any variant of the MSOR method with variable relaxation factors including the CCSI method (see Chapter 10). Depending on whether one is more concerned about the $D^{1/2}$-norm or the $A^{1/2}$-norm one would choose either the CCSI method or the GS–SSI method. From the standpoint of simplicity, however, one might be inclined to use Sheldon's method.

As a matter of fact, it is recommended that before carrying out any iterative method one first perform one iteration using the GS method. We have seen in Chapter 10 that if A has the form

$$A = \begin{pmatrix} D_1 & H \\ K & D_2 \end{pmatrix} \tag{4}$$

where D_1 and D_2 are square diagonal matrices, and if one performs this initial iteration followed by the SOR method with $\omega = \omega_b$, the resulting method, Sheldon's method, has a smaller $D^{1/2}$-norm than does the regular SOR method. Since the two methods have nearly the same $A^{1/2}$-norm, it seems that there can be no harm, and there may be some gain, in this initial iteration. We have also seen in Section 11.6 that using this initial iteration followed by the GS–SI method results in a method, the GS–SSI method, which has considerably better norm properties than the GS–SI method.

It seems worthwhile to investigate the possibility of using the SSOR method combined with semi-iteration. If the system is derived from an elliptic equation with slowly varying coefficients, the chances of the method working effectively are good. However, rather than use the ordering of the mesh points corresponding to (4) one should instead use the "natural" ordering (see Section 5.6). One can compute ω_1 on the basis of $\gamma = \frac{1}{4}$ in (15-3.3). For this value of ω, and possibly for a few slightly larger values, one can iterate with $\mathscr{S}_\omega$ for a problem with zero boundary values and with constant starting values to estimate $S(\mathscr{S}_\omega)$ (see Section 6.6). If an ω can be found so that $S(\mathscr{S}_\omega)$ is small enough to afford a gain by the use of semi-iteration, then the SSOR–SI method can be applied. It would seem reasonable to perform this investigation for line SSOR as well as point SSOR in a given case (see Chapter 15).

Under ideal conditions, the PR method gives much faster convergence than any of the other methods considered. Unfortunately, except in special cases there does not exist any systematic procedure to tell in advance whether or not the method will be effective. In the special case considered in Section 17.9, a scaling of the equations was suggested. Widlund [1966, 1969] has suggested a procedure for scaling the equations and has proved that for a rectangular region the PR method will be effective provided the coefficients of the differential equation are sufficiently regular. On the other hand, there exist instances such as the example of Price and Varga [1962] (see Section 17.10) where the PR method does not converge for some choice of parameters. Clearly, a considerable amount of research is needed to provide a solid theoretical foundation for this important method.

BIBLIOGRAPHY

Pages on which a reference is cited are listed in italic numerals at the end of each reference.

Aitken, A. C. [1950], Studies in practical mathematics V. On the iterative solution of a system of linear equations, *Proc. Roy. Soc. Edinburgh Sec. A*63, 52–60; *480*.

Albrecht, J. [1966], Zur verallgemeinerten Iteration in Einzel schritten bein Mehrstellen verfahren, *Z. Angew. Math. Mech.* **46**, 322–324; *228*.

Angel, Edward [1970], Inverse boundary-value problems: elliptic equations, *J. Math. Anal. Appl.* **30**, 86–98; *6*.

Arms, R. J., and Gates, L. D. [1956], Iterative methods of solving linear systems, comparisons between point and block iteration, unpublished memorandum from the U. S. Naval Proving Grounds, Dahlgren, Virginia; *454*.

Arms, R. J., Gates, L. D., and Zondek, B. [1956], A method of block iteration, *J. Soc. Indust. Appl. Math.* **4**, 220–229; *451, 456*.

Barnard, S., and Child, J. M. [1952], "Higher Algebra." MacMillan, New York; *54*.

Bilodeau, G. G., Cadwell, W. R., Dorsey, J. P., Fairey, J. M., and Varga, R. S. [1957], PDQ—An ABM-704 code to solve the two-dimensional few-group neutron-diffusion equations, Report WAPD-TM-70, Bettis Atomic Power Laboratory, Westinghouse Electric Corp., Pittsburgh, Pennsylvania; *211*.

Birkhoff, G., and MacLane, S. [1953]. "A Survey of Modern Algebra." MacMillan, New York; *7, 53*.

Birkhoff, G., and Varga, R. S. [1959], Implicit alternating direction methods, *Trans. Amer. Math. Soc.* **92**, 13–24; *547*.

Birkhoff, G., Varga, R. S., and Young, D. [1962], Alternating direction implicit methods, *Advances in Computers*, **3**, 189–273; *514, 527, 529, 530, 533, 534, 544, 546, 547, 548*.

Blair, A., Metropolis, N., von Neumann, J., Taub, A. H., and Tsingori, M. [1959], A study of a numerical solution of a two-dimensional hydrodynamical problem, *Math. Tables Aids Comput.* **13**, 145–184; *386*.

Bramble, J. H., and Hubbard, B. E. [1964], On a finite difference analogue of an elliptic boundary problem which is neither diagonally dominant nor of non-negative type, *J. Math. Phys.* **43**, 117–132; *54*.

Broyden, C. G. [1964], Some generalisations of the theory of successive overrelaxation, *Numer. Math.* **6**, 269–284; *419, 429.*

Broyden, C. G. [1968], Some aspects of consistent ordering, *Numer. Math.* **12**, 47–56; *163, 406, 429.*

Buzbee, B. L., Golub, G. H., and Nielson, C. W. [1969], The method of odd/even reduction and factorization with application to Poisson's equation, Computer Sciences Department Tech. Report No. 128, Stanford University, Stanford, California; *548.*

Carré, B. A. [1961], The determination of the optimum accelerating factor for successive over-relaxation, *The Computer Journal* **4**, 73–78; *210.*

Cauer, W. [1958], "Synthesis of Linear Communications Networks." McGraw-Hill, New York; *522.*

Collatz, L. [1952], Aufgaben monotoner Art, *Arch. Math.* **3**, 366–376; *54.*

Cuthill, E. H., and Varga, R. S. [1959], A method of normalized block iteration, *J. Assoc. Comput. Mach.* **6**, 236–244; *442, 456.*

De Vogelaere, R. [1958], Over-relaxations, Abstract No. 539-53, *Amer. Math. Soc. Notices* **5**, 147; *273.*

D'Jakonov, E. G. [1969], On certain iterative methods for solving nonlinear difference equations, *Proc. Conf. Numer. Sol. Diff. Eqs., Dundee, Scotland, June 23–27, 1969*, Springer-Verlag, Heidelberg, Germany; *548.*

Douglas, J., Jr. [1955], On the numerical integration of $\partial^2 u/\partial x^2 + \partial^2 u/\partial y^2 = \partial u/\partial t$ by implicit methods, *J. Soc. Ind. Appl. Math.* **3**, 42–65; *547.*

Douglas, J., Jr., and Rachford, H. [1956], On the numerical solution of heat conduction problems in two and three space variables, *Trans. Amer. Math. Soc.* **82**, 421–439; *547.*

D'Sylva, E., and Miles, G. A. [1963], The SSOR iteration scheme for equations with σ_1 ordering, *Comput. J.* **6**, 271–273; *480.*

Ehrlich, L. W. [1963], The block symmetric successive overrelaxation method, doctoral thesis, Univ. Texas, Austin, Texas; *469, 472, 474, 476, 480.*

Ehrlich, L. W. [1964], The block symmetric successive overrelaxation method, *J. Soc. Indust. Appl. Math.* **12**, 807–826; *476, 480.*

Evans, D. J., and Forrington, C. V. D. [1963], An iterative process for optimizing symmetric overrelaxation, *Comput. J.* **6**, 271–273; *469.*

Faddeev, D. K., and Faddeeva, V. N. [1963], "Computational Methods of Linear Algebra." Freeman, San Francisco, California; *7, 53, 60, 132, 206.*

Flanders, D., and Shortley, G. [1950], Numerical determination of fundamental modes, *J. Appl. Phys.* **21**, 1326–1332; *304, 385.*

Forsythe, G. E. [1953], Solving linear algebraic equations can be interesting, *Bull. Amer. Math. Soc.* **59**, 299–329; *94.*

Forsythe, G. E., and Moler C. B. [1967], "Computer Solution of Linear Algebraic Systems." Prentice-Hall, Englewood Cliffs, New Jersey; *38, 54.*

Forsythe, G. E., and Straus, E. G. [1955], On best conditioned matrices, *Proc. Amer. Math. Soc.* **6**, 340–345; *214.*

Forsythe, G. E., and Wasow, W. R. [1960], "Finite Difference Methods for Partial Differential Equations." Wiley, New York; *94.*

Frank, W. [1960], Solution of linear systems by Richardson's method, *J. Assoc. Comput. Mach.* **7**, 274–286; *386.*

Frankel, S. P. [1950], Convergence rates of iterative treatments of partial differential equations, *Math. Tables Aids Comput.* **4**, 65–75; *94, 488, 493.*

Friedman, B. [1957], The iterative solution of elliptic difference equations, A. E. C. Research and Development Report NYO-7698, Institute of Mathematical Sciences, New York University, New York; *163, 456.*

Frobenius, G. [1908], Über Matrizen aus positiven Elementen, *S.-B. Preuss Akad. Wiss. Berlin,* 471–476; *18.*

Frobenius, G. [1912], Über Matrizen aus nicht negativen Elementen, *S.-B. Preuss. Akad. Wiss. Berlin,* 456–477; *54.*

Gaier, D., and Todd, J. [1967], On the rate of convergence of optimal ADI processes, *Numer. Math.* **9**, 452–459; *521, 522.*

Garabedian, P. R. [1956], Estimation of the relaxation factor for small mesh size, *Math. Tables Aids Comput.* **10**, 183–185; *228.*

Gastinel, N. [1962], Sur le meilleur choix des paramètres de surrelaxation, *Chiffres* **5**, 109–126; *518.*

Geiringer, H. [1949], On the solution of systems of linear equations by certain iterative methods, *Reissner Anniversary Volume,* Contributions to Applied Mechanics, 365–393. Edwards, Ann Arbor, Michigan; *54, 94, 456.*

Golub, G. H. [1959], The use of Chebyshev matrix polynomials in the iterative solution of linear systems compared with the method of successive overrelaxation, doctoral thesis, University of Illinois, Urbana, Illinois; *265.*

Golub, G. H., and Varga, R. S. [1961], Chebyshev semi-iterative methods, successive over-relaxation iterative methods, and second-order Richardson iterative methods, *Numer. Math.*, Parts I and II **3**, 147–168; *265, 319, 330, 340, 365, 486, 493.*

Gourlay, A. R. [1968], The acceleration of the Peaceman–Rachford method by Chebyshev polynomials, *Comput. J.* **10**, 378–382; *548.*

Gourlay, A. R. [1970], On Chebychev acceleration procedures for alternating direction iterative methods, *J. Inst. Math. Applic.* **6**, 1–11; *548.*

Guilinger, W. H., Jr. [1965], The Peaceman–Rachford method for small mesh increments, *J. Math. Anal. Appl.* **11**, 261–277; *502, 545, 546.*

Guittet, J. [1967], Une nouvelle méthode de directions alternées à q variables, *J. Math. Anal. Appl.* **17**, 199–213; *548.*

Habetler, G. J. [1959], Concerning the implicit alternating-direction method, Report KAPL-2040, Knolls Atomic Power Laboratory, Schenectady, New York; *547.*

Habetler, G. J., and Wachspress, E. L. [1961], Symmetric successive overrelaxation in solving diffusion difference equations, *Math. Comp.* **15**, 356–362; *469, 476, 480.*

Habetler, G. J., Wachspress, E. L., and Baldwin, B. D. [1959], Symmetric successive overrelaxation in solving diffusion difference equations, Knolls Atomic Power Laboratory Report 2038, Schenectady, New York.

Hageman, L. A. [1967], The Chebyshev polynomial method of iteration, WAPD-TM-537 (available from the Clearinghouse for Federal Scientific and Technical Information, National Bureau of Standards, U.S. Department of Commerce, Springfield, Virginia).

Hageman, L. A., and Kellogg, R. B. [1966], Estimating optimum acceleration parameters for use in the successive overrelaxation and Chebyshev polynomial methods of iteration, WAPD-TM-592 (available from the Clearinghouse for Federal Scientific and Technical Information, National Bureau of Standards, U.S. Department of Commerce, Springfield, Virginia); *211.*

Hageman, L. A., and Kellogg, R. B. [1968], Estimating optimum overrelaxation parameters, *Math. Comp.* **22**, 60–68; *211.*

Heller, J. [1960], Simultaneous, successive and alternating direction iteration schemes, *J. Soc. Indust. Appl. Math.* **8**, 150–173; *454.*

Henrici, P. [1960], Estimating the best over-relaxation factor, Report NN-144, Applied Math. Dept., Space Technology Laboratories, Los Angeles, California; *95.*

Henrici, P. [1962], Bounds for iterates, inverses, spectral variation, and fields of values of non-normal matrices, *Numer. Math.* **4**, 24–40; *222.*

Hockney, R. W. [1965], A fast solution of Poisson's equation using Fourier analysis, *J. Assoc. Comput. Mach.* **12**, 95–113; *548.*

Householder, A. S. [1958], Unitary triangularization of a nonsymmetric matrix, *J. Assoc. Comput. Mach.* **5**, 339–342; *17.*

Householder, A. S. [1964], "The Theory of Matrices in Numerical Analysis." Blaisdell, New York; *33, 54, 95.*

Jahnke, E., and Emde, F. [1945], "Tables of Functions." Dover, New York; *224.*

Kahan, W. [1958], Gauss–Seidel methods of solving large systems of linear equations, doctoral thesis, University of Toronto, Toronto, Canada; *107, 126, 133, 393, 401.*

Kalähne, A. [1907], Über die Wurzeln einiger Zylinderfunktionen und gewisser aus ihen gebildeter Gleichungen, *Z. Math. Phys.* **54**, 55–86; *224.*

Keller, H. B. [1960], Special block iterations with applications to Laplace and biharmonic difference equations, *SIAM Rev.* **2**, 277–287; *456.*

Keller, H. B. [1965], On the solution of singular and semidefinite linear systems by iteration, *SIAM J. Numer. Anal.* **2**, 281–290; *94.*

Kincaid, David R. [1971], An analysis of a class of norms of iterative methods for systems of linear equations, doctoral thesis, Univ. Texas, Austin, Texas; *291.*

Kjellberg, G. [1958], On the convergence of successive over-relaxation applied to a class of linear systems of equations with complex eigenvalues, *Ericsson Technics Stockholm* **2**, 245–258.

Kjellberg, G. [1961], On the successive over-relaxation method for cyclic operators, *Numer. Math.* **3**, 87–91; *227.*

Kredell, B. [1962], On complex successive overrelaxation, *BIT* **2**, 143–152; *227.*

Kulsrud, H. E. [1961], A practical technique for the determination of the optimum relaxation factor of the successive over-relaxation method, *Comm. Assoc. Comput. Mach.* **4**, 184–187; *210.*

Lanczos, C. [1952], Solution of systems of linear equations by minimized iterations, *J. Res. Nat. Bur. of Standards* **49**, 33–53; *386.*

Loizou, G. [1969], Nonnormality and Jordan condition numbers of matrices, *J. Assoc. Comput. Mach.* **16**, 580–584; *95.*

Lynch, R. E., and Rice, J. R. [1968], Convergence rates of ADI methods with smooth initial error, *Math. Comp.* **22**, 311–335; *545.*

Lynn, M. S. [1964], On the equivalence of SOR, SSOR, and USSOR as applied to σ_1-ordered systems of linear equations, *Comput. J.* **7**, 72–75; *480.*

Markoff, W. [1892], Über Polynome, die in einem gegeben Intervalle möglichst wenig von Null abweichen, *Math. Ann.* **77** (1916): 213–258 (translation and condensation by J. Grossman of Russian article published in 1892); *304.*

Maybee, J. S. [1965], Some structural theorems for partial difference operators, *Numer. Math.* **7**, 66–72.

McDowell, L. K. [1967], Variable successive overrelaxation, Report No. 244, Dept. Computer Sciences, University of Illinois, Urbana, Illinois; *273, 291.*

Miles, G. A., Stewart, K. L., and Tee, G. J. [1964], Elementary divisors of the Liebmann process, *Comput. J.* **6**, 352–355; *265, 374.*

Mouradoglou, A. J. [1967], Numerical studies on the convergence of the Peaceman–Rachford alternating direction implicit method, M. A. thesis, Univ. Texas, Austin, Texas; also Report TNN-67, Computation Center, Univ. Texas, Austin, Texas; *548.*

Nichols, N. K., and Fox, L. [1969], Generalized consistent ordering and optimum successive over-relaxation factor, *Numer. Math.* **13**, 425–433; *413, 414.*

Oldenburger, R. [1940], Infinite powers of matrices and characteristic roots, *Duke Math. J.* **6**, 357–361; *18.*

Ortega, J., and Rheinboldt, W. [1970], "The Numerical Solution of Nonlinear Systems of Equations." Academic Press, New York; *33.*

Ostrowski, A. M. [1954], On the linear iteration procedures for symmetric matrices, *Rend. Mat. e Appl.* **14**, 140–163; *118, 132.*

Ostrowski, A. M. [1955], Über Normen von Matrizen, *Math. Z.* **63**, 2–18; *54.*

Ostrowski, A. M. [1960], "Solution of Equations and Systems of Equations." Academic Press, New York; *15.*

Ostrowski, A., and Schneider, H. [1962], Some theorems on the inertia of general matrices, *J. Math. Anal. Appl.* **4**, 72–84; *95.*

Parter, S. V. [1959], On "two-line" iterative methods for the Laplace and biharmonic difference equations, *Numer. Math.* **1**, 240–252; *456.*

Parter, S. V. [1961], "Multi-line" iterative methods for elliptic difference equations and fundamental frequencies, *Numer. Math.* **3**, 305–319; *453, 456.*

Parter, S. V. [1965], On estimating the "rates of convergence" of iterative methods for elliptic difference equations, *Trans. Amer. Math. Soc.* **114**, 320–354; *228, 456.*

Peaceman, D. W., and H. H., Rachford, Jr. [1955], The numerical solution of parabolic and elliptic differential equations, *J. Soc. Indus. Appl. Math.* **3**, 28–41; *525, 547.*

Pearcy, C. [1962], On convergence of alternating direction procedures, *Numer. Math.* **4**, 172–176; *547.*

Price, H., and Varga, R. S. [1962], Recent numerical experiments comparing successive overrelaxation iterative methods with implicit alternating direction methods, Report No. 91, Gulf Research and Development Co., Pittsburgh, Pennsylvania; *545, 555.*

Randall, T. J. [1968], A note on the estimation of the optimum successive overrelaxation parameter for Laplace's equation, *Comput. J.* **10**, 400–401.

Reich, E. [1949], On the convergence of the classical iterative method of solving linear simultaneous equations, *Ann. Math. Statist.* **20**, 448–451; *132.*

Reid, J. K. [1966], A method for finding the optimum successive over-relaxation parameter, *The Computer Journal* **9**, 200–204; *210.*

Rheinboldt, W. C. [1969], An elementary proof of the basic convergence theorem for powers of a matrix, unpublished manuscript; *17, 34.*

Richardson, L. F. [1910], The approximate arithmetical solution by finite differences of physical problems involving differential equations with an application to the stresses in a masonry dam, *Philos. Trans. Roy. Soc. London Ser. A* **210**, 307–357; *74, 94.*

Rigler, A. K. [1965], Estimation of the successive over-relaxation factor, *Math. of Comp.* **19**, 302–307; *210.*

Riley, J. D. [1954], Iteration procedures for the Dirichlet difference problem, *Math. Tables Aids Comput.* **8**, 125–131; *491, 493.*

Russell, D. B. [1963], On obtaining solutions to the Navier-Stokes equations with automatic digital computers, Aeronautical Research Council Report R & M 3331, Engineering Laboratory, Oxford; *227, 291.*

Schur, I. [1909], Über die charakteristischen Wurzeln einer linearen Substitution mit einer Anwendung auf die Theorie der Integralgleichungen, *Math. Ann.* **66**, 488–510; *17.*

Sheldon, J. [1955], On the numerical solution of elliptic difference equations, *Math. Tables Aids Comput.* **9**, 101–112; *386, 480.*

Sheldon, J. W. [1959], On the spectral norms of several iterative processes, *J. Assoc. Comput. Mach.* **6**, 494–505; *265, 315, 340, 367, 372, 386.*

Shortley, G. [1953], Use of Tschebyscheff polynomial operators in the numerical solution of boundary value problems, *J. Appl. Phys.* **24**, 392–396; *386.*

Smith, R. A. [1967], The condition numbers of the matrix eigenvalue problem, *Numer. Math.* **10**, 232–240; *95.*

Southwell, R. V. [1946], "Relaxation Methods in Theoretical Physics." Oxford Univ. Press, New York; *456.*

Stein, P. [1952], Some general theorems on iterants, *J. Res. Nat. Bur. Standards* **48**, 82–83; *80.*

Stein, P., and Rosenberg, R. [1948], On the solution of linear simultaneous equations by iteration, *J. London Math. Soc.* **23**, 111–118; *54, 120.*

Stiefel, E. [1956], On solving Fredholm integral equations, *J. Soc. Indus. Appl. Math.* **4**, 63–88; *386.*

Taussky, O. [1949], A recurring theorem on determinants, *Amer. Math. Monthly* **56**, 672–676; *54, 60.*

Taussky, O. [1961], A remark on a theorem of Lyapunov, *J. Math. Anal. Appl.* **2**, 105–107; *95.*

Taylor, A. E. [1963], "Introduction to Functional Analysis." Wiley, New York; *54.*

Taylor, P. J. [1969], A generalisation of systematic relaxation methods for consistently ordered matrices, *Numer. Math.* **13**, 377–395; *273, 291.*

Tee, G. J. [1963], Eigenvectors of the successive over-relaxation process, and its combination with Chebyshev semi-iteration, *Comput. J.* **6**, 250–263; *374, 386.*

Tee, G. J. [1964], An application of p-cyclic matrices, for solving periodic parabolic problems, *Numer. Math.* **6**, 142–159; *428.*

Thomas, L. H. [1949], Elliptic problems in linear difference equations over a network, Watson Scientific Computing Lab., Columbia Univ., New York; *441.*

Thrall, R. M., and Tornheim, L. [1957], "Vectors, Spaces and Matrices." Wiley, New York; *515.*

Todd, J. [1967], Inequalities of Chebyshev, Zolotareff, Cauer, and W. B. Jordan, Proc. of Conf. at Dayton, Ohio, on Inequalities, pp. 321–328. Academic Press, New York; *522.*

Todd, J. [1967a], Optimal ADI-parameters, *Funktional. Anal. Approx. Theor. Numer. Math.* **7**, 58–70; *522, 525.*

Varga, R. S. [1957], A comparison of the successive overrelaxation method and semi-iterative methods using Chebyshev polynomials, *J. Soc. Indus. Appl. Math.* **5**, 39–46; *375, 386.*

Varga, Richard S. [1957a], On estimating rates of convergence in multigroup diffusion problems, AEC Research and Development Report WAPD-TM-41, Office of Technical Services, U.S. Dept. of Commerce, Washington 25, D.C.; *211.*

Varga, R. S. [1959], "p-cyclic matrices: a generalization of the Young–Frankel successive overrelaxation scheme, *Pacific J. Math.* **9**, 617–628; *413, 414, 429.*

Varga, R. S. [1959a], Orderings of the successive overrelaxation scheme, *Pacific J. Math.* **9**, 925–939; *401.*

Varga, R. S. [1960], Factorization and normalized iterative methods, "Boundary Problems in Differential Equations" (R. E. Langer, ed.), pp. 121–142. Univ. Wisconsin Press, Madison, Wisconsin; *77, 453, 454, 456.*

Varga, Richard S. [1961], Numerical methods for solving multi-dimensional multigroup diffusion equations, "*Proc. Symp. Appl. Math., Amer. Math. Soc.* (Providence, Rhode Island) **11**, 164–189.

Varga, R. S. [1962], "Matrix Iterative Analysis." Prentice-Hall, Englewood Cliffs, New Jersey; *6, 18, 54, 95, 113, 162, 393, 408, 414, 418, 429, 449, 450.*

Verner, J. H., and Bernal, M. J. M. [1968], On generalizations of the theory of consistent orderings for successive over-relaxation methods, *Numer. Math.* **12**, 215–222; *429, 456.*

Wachspress, E. L. [1957], CURE: A generalized two-space-dimension multigroup coding for the IBM-704, Report KAPL-1724, Knolls Atomic Power Lab., Schenectady, New York; *548.*

Wachspress, E. L. [1966], "Iterative Solution of Elliptic Systems and Applications to the Neutron Diffusion Equations of Reactor Physics." Prentice-Hall, Englewood Cliffs, New Jersey; *24, 54, 94, 95, 118, 265, 401, 496, 511, 522, 523, 546.*

Wachspress, E. L. [1968], Solution of the ADI minimax problems, doctoral thesis, Rensselaer Polytechnic Inst., Troy, New York; *548.*

Wachspress, E. L., and Habetler, G. J. [1960], An alternating-direction-implicit iteration technique, *J. Soc. Indust. Appl. Math.* **8**, 403–424; *518, 548.*

Warlick, C. H. [1955], Convergence rates of numerical methods for solving $\partial^2 u/\partial x^2 + k/\varrho\ \partial u/\partial \varrho + \partial^2 u/\partial \varrho^2 = 0$, M. A. thesis, Univ. Maryland, College Park, Maryland; *222, 224, 228.*

Warlick, C. H. [1955a], Convergence rates of numerical methods for solving $\partial^2 f/\partial x^2 = \partial^2 f/\partial y^2 + k/y\ \partial f/\partial y = 0$ (further results), Mathematical Applications Development Memorandum No. 19, DM 60-191, Evandale Computations, Flight Propulsion Division, General Electric Co., Cincinnati, Ohio; *222, 224, 228.*

Warlick, C. H., and Young, D. M. [1970], *A priori* methods for the determination of the optimum relaxation factor for the successive overrelaxation method, TNN-105, Computation Center, Univ. Texas, Austin, Texas; *225.*

Widder, D. V. [1947], "Advanced Calculus." Prentice-Hall, Englewood Cliffs, New Jersey; *345.*

Widlund, O. B. [1966], On the rate of convergence of an alternating direction implicit method in a noncommutative case, *Math. Comp.* **20**, 500–515; *547, 555.*

Widlund, O. B. [1969], On the effects of scaling of the Peaceman–Rachford method, *Proc. Conf. Numer. Sol. Diff. Eqs. Dundee, Scotland, June 23–27, 1969*, Springer-Verlag, Heidelberg, Germany; *547, 548, 555.*

Wrigley, E. E. [1962], On accelerating the Jacobi method for solving simultaneous equations by Chebyshev extrapolation when the eigenvalues of the iteration matrix are complex, AEEW-R224, Atomic Energy Establishment, Winfrith, Dorchester, Dorset, England (also, *The Computer Journal* **6** (1963), 169–176); *227, 386.*

Young, D. M. [1950], Iterative methods for solving partial difference equations of elliptic type, doctoral thesis, Harvard Univ., Cambridge, Massachusetts; *54, 94, 95, 133, 163, 227, 238, 265, 401.*

Young, D. M. [1954], Iterative methods for solving partial difference equations of elliptic type, *Trans. Amer. Math. Soc.* **76**, 92–111; *54, 94, 95, 163, 227, 234, 265, 365.*

Young, D. M. [1954a], On Richardson's method for solving linear systems with positive definite matrices, *J. Math. Phys.* **XXXII**, 243–255; *386.*

Young, D. M. [1955], Ordvac solutions of the Dirichlet problem, *J. Assoc. Comput. Mach.* **2**, 137–161; *210, 227.*

Young, D. M. [1956], On the solution of linear systems by iteration, *Proc. Sixth Symp. in Appl. Math. Amer. Math. Soc.*, Vol. VI, pp. 283–298. McGraw-Hill, New York; *365, 387.*

Young, D. M. [1962], The numerical solution of elliptic and parabolic partial differential equations, "Survey of Numerical Analysis" (J. Todd, ed.), pp. 380–438. McGraw-Hill, New York; *228.*

Young, D. M. [1969], Convergence properties of the symmetric and unsymmetric successive overrelaxation methods and related methods, Report TNN-96, Computation Center, Univ. Texas, Austin, Texas; also to appear in *Math. Comp.*; *265, 291, 340, 376, 480.*

Young, D. M. [1970], Generalizations of Property A and consistent orderings, Report CNA-6, Center for Numerical Analyzis, Univ. Texas, Austin, Texas; *265, 429, 480.*

Young, D. M. [1971], A bound on the optimum relaxation factor for the successive overrelaxation method, to appear in *Numer. Math*; *228.*

Young, D. M. [1971a], Second-degree iterative methods for the solution of large linear systems, to appear in *J. of Approximation Theory*; *480, 493.*

Young, D. M. [1971b], On the consistency of linear stationary iterative methods, to appear in *SIAM J. Numer. Anal*; *94.*

Young, D. M., and Ehrlich, L. [1960], Some numerical studies of iterative methods for solving elliptic difference equations, "Boundary Problems in Differential Equations" (R. E. Langer, ed.), pp. 143–162. Univ. Wisconsin Press, Madison, Wisconsin; *546.*

Young, D. M., and Eidson, H. E. [1970], On the determination of the optimum relaxation factor for the SOR method when the eigenvalues of the Jacobi method are complex Report CNA-1, Center for Numerical Analysis Univ. Texas, Austin, Texas; *200.*

Young, D. M., and Kincaid, D. R. [1969], Norms of the successive overrelaxation method and related methods, Report TNN-94, Computation Center, Univ. Texas, Austin, Texas; *265, 329, 335, 340, 372, 386, 491.*

Young, D. M., and Shaw, H. [1955], Ordvac solutions of $\partial^2 u/\partial x^2 + \partial^2 u/\partial y^2 + k/y \times \partial u/\partial y = 0$ for boundary value problems and problems of mixed type, Interim Tech. Report No. 14, Office of Ordnance Research Contract DA-36-034-ORD-1486, Univ. Maryland, College Park, Maryland; *210, 227.*

Young, D. M., and Warlick, C. H. [1953], On the use of Richardson's method for the numerical solution of Laplace's equation on the ORDVAC, Ballistic Research Labs. Memorandum Report No. 707, Aberdeen Proving Ground, Maryland; *365.*

Young, D. M., and Wheeler, M. F. [1964], Alternating direction methods for solving partial difference equations, "Nonlinear Problems in Engineering" (W. F. Ames, ed.), pp. 220–246. Academic Press, New York.

Young, D. M., Wheeler, M. F., and Downing, J. A. [1965], On the use of the modified successive overrelaxation method with several relaxation factors, *Proc. of IFIP* **65**, 177–182; *265, 273, 291, 309, 340.*

Zolotareff, E. [1877], Anwendung der elliptischen Funktionen auf Probleme über Funktionen, die von Null am weiningsten oder am meisten abweichen. *Abh. St. Petersburg* **30**; *522.*

INDEX

A

$A^{1/2}$-norm, 109, 118–119
  of MSOR method, fixed parameters, 288–291
  of SOR method, 255–265
a priori methods for choosing ω_b, 216–224
ADI methods, 494–552
Alternating direction implicit methods, *see* ADI methods
Arithmetic-geometric mean, 522
Associated integers, 144
Augmented matrix, 14

B

Basic eigenvalue relation, 239–245
Basic iterative methods, 65–74
  convergence of, 77–84, 106–139
  generation of, 75–77
Basis, 11
Bessel's functions, 224
Biharmonic equation, 454
Black equations, 271
Block diagonal matrix, 429
Block iterative methods, *see* Group iterative methods
Block Jacobi method, 429
Block tridiagonal matrix, 141–144

C

Canonical forms
  for CO(q, r)-matrices, 422–428
  for matrices with Property $A_{q,r}$, 422–428
CCSI method, 321–326, 365–367, 492, 554
Central difference quotient, 3
Cesàro summable sequence, 345
Chain, 419–420
  closed, 419
  increment of, 420
  length of, 419
  simple, 419
Characteristic equation, 14
Chebyshev polynomials, 211, 301–303, 491
  three-term recurrence relation for, 349
Closed set, 27
Closure of set, 27
CO*(q, r)-matrix, 424
Column iteration, 440
Column matrix, 8
Column vector, 8
Commensurable numbers, 127
Commutative case
  conditions for, 535–544
  optimum parameters, 518–525
  for PR method, 514–518
Completely consistent iterative method, 64–70

Compatible norms, 31
Composite method, 98, 300
Compression, 491
Condition numbers of matrix, 85, 213
Conjugate of matrix, 9
Conjugate transpose of matrix, 9
Consistent iterative method, 64–70
  generation of, 75
Consistent norms, 31–32
Consistently ordered matrix, 106, 140–168
  alternative definition of, 162
Convergence
  of basic iterative methods, 106–139
  of sequence of matrices, 34
  of sequence of vectors, 34
Cramer's rule, 13
Cuthill–Varga procedure, 442, 458, 476
Cycle
  even, 12
  odd, 12
Cyclic Chebyshev semi-iterative method, *see* CCSI method
Cyclic matrix of index p, 162

D

$D^{1/2}$-norm
  of MSOR method, fixed parameters, 283–288
  of SOR method, 245–255, 264–265
Determinant of matrix, 12
Diagonal matrix, 8
Difference equation
  five-point, 2
  nine-point, 454
  nonsymmetric, 224
  symmetric, 224
  thirteen-point, 454
  Variant I, 51
  Variant II, 51
Dimension, 11
Direct methods, 6
Directed graph, 38
  connected, 38
Dirichlet problem, 2, 49
  discrete generalized, 49
  generalized, 49
Douglas–Rachford method, 547

E

Eigenvalue of matrix, 14
Eigenvalue spectrum of matrix, 15
Eigenvector of matrix, 14
Elliptic functions, 522
Elliptic integral, 522
Elliptic partial differential equation, 48, 553
Euclidean length, 11

F

Factorization techniques, 453
Five-diagonal system, 454
Forsythe and Straus theorem, 214
Frobenius, theorem of, 515

G

Gauss elimination method, 2, 435
Gauss–Seidel method, 5, 13, *see also* GS method
GCO-matrix, 404–433
GCO(q, r)-matrix
  alternative definition of, 429
  relation to CO(q, r)-matrix, 419–422
Generalized consistently ordered matrix, *see* GCO-matrix
Generalized Dirichlet problem, discrete analog, 496–498
Generalized π-consistently-ordered matrix, *see* π-GCO-matrix
Generalized (q, r) consistently ordered matrix, *see* GCO(q, r)-matrix
Good parameters for PR method, 525–530
GRF method, 74, 109, 117
Group, 436
Group iterative methods, 434–460
  construction of, 435–441
  convergence, 445–452
  GS method, 454
  J method, 455
  JOR method, 455
  SOR method, 438
  SSOR method, 474–476
GS method, 72, 109, 554
  rate of convergence for model problem, 127–132

GS–SI method, 367–372, 554
GS–SSI method, 372–374, 554

H

Helmholtz equation, 498, 531–533
Hermitian matrix, 9, 18
Homogeneous system, 40

I

Identity matrix, 8
Implicit function theorem, 263
Induced matrix norm, 30
Inner product, 11
Integrating factor, 49
Inverse matrix, 10
Inversions, 12
Irreducible matrix, 36–41, 107–108
Iterative method, 2, 63
  choosing ω_b for, 209–211
  convergence, 77
  degree of, 64
  linear stationary, 63–105
  nonlinear, 64
  nonstationary, 295–305
  point, 454
  selection of, 553–555
  strong convergence, 171
  weak convergence, 77

J

J method, 71, 140–168, 554
  bounds for spectral radius, 211–216
  convergence, 107–111
  eigenvectors for, 234
  rate of convergence for model problem, 127–132
Jacobi method, 5, *see also* J method
JOR method, 72
  convergence, 107–111
JOR–SI method, 357–358
Jordan canonical form, 17
Jordan condition number of matrix, 85, 99
  for 2 × 2 matrix, 89–94
J–SI method, 355–357, 554

K

k-line SOR method, 453

L

L-matrix, 42–48, 120–127, 554
Line GS method, 453
Line iteration, 452
Line J method, 452
Line SOR method, 453, 555
Linear systems, 1
Lower triangular matrix, 8
  strictly lower-triangular matrix, 8
Lyapunov's theorem, 82

M

M-matrix, 42–48, 120–127
M-parameter problem, 528
Matrix norm, 25
  β-norm, 25
  β,L-norm, 25
  L-norm, 32
Matrix product, 8
Matrix sum, 8
Mesh points
  adjacent, 49
  connected subset of, 52
  properly adjacent, 49
  regular, 49, 55
Metric space, 27
  distance function for, 27
Modified Sheldon method, 319–321, 490
Modified SOR method, *see* MSOR method
Monotone matrix, 42–48, 120–127
Monotonicity theorems, 206
  for PR method, 534–535
Model problem, 2
  discrete analog, 3
  numerical experiments for, 132
MSOR method
  backward, 289
  fixed parameters, 271–294
    $A^{1/2}$-norm, 288–291
    convergence, 277–283, 307
    $D^{1/2}$-norm, 283–288

semi-iterative methods based on, 376–382
variable parameters, 306–343, 554
alternative optimum parameter sets, 311–314
convergence, 307
norms, 315–340
optimum choice of relaxation factors, 307–310

N

N-stable matrix, 82, 502
Natural ordering, 130, 159, 555
Navier–Stokes equation, 291
Negative definite matrix, 21
Negative stable matrix, *see* *N*-stable matrix
Neumann problem, 94
Noncommutative case
for convex regions, 545
PR method for, 545–547
Nonmigratory permutation, 153
Nonnegative definite matrix, 21
Nonnegative matrix, 18
Nonnegative real matrix, 24
Nonpositive definite matrix, 21
Nonsingular matrix, 10
Nonstationary iterative methods
complete consistency, 296
consistency, 296
convergence, 297
reciprocal consistency, 296
weak convergence, 297
Normal matrix, 9
Normalized block iteration, *see* Cuthill–Varga procedure
Null space, 13
Nullity, 14

O

Ordered grouping, 435, 554
Ordering by diagonals, 159
Ordering vector, 144–148
compatible, 145
compatible (q, r)-ordering vector, 406
complete (q, r)-ordering vector, 424
(q, r)-ordering vector, 406
Orthogonal matrix, 9
Orthogonal vectors, 11
pairwise, 11
Orthonormal vectors, 11

P

P-cyclic matrix, 162, 428–429
Partitioning of matrix, 434
Peaceman–Rachford method, *see* PR method
Peaceman–Rachford parameters, 525
Periodic method, 300–301
Permutation
even, 12
function, 10
migratory, 422–426
nonmigratory, 422–426
odd, 12
Permutation matrix, 10
Perron–Frobenius theory, 18
π-consistently ordered matrix, *see* π-CO-matrix
π-CO-matrix, 445
π-GCO-matrix, 449
Poisson's equation, 2
Positive definite matrix, 18–25, 108–118
Positive matrix, 18
Positive real matrix, 24, 500
Positive stable matrix, 95
Power method, 206
PR method, 496–552, 555
$A^{1/2}$-norm, 502
good parameters, 525–530
optimum parameters, 518–525
semi-iteration, 548
stationary case, 498–514
Prescaling of matrix, 74
Price–Varga example, 545–546, 552, 555
Primitive matrix, 162
Principal vector, 17
grade of, 17
Property A, 41, 140–168, 554
Property A_p, 406
Property A_p^*, 425
Property $A^{(\pi)}$, 445, 554
Property $A_{q,r}$, 405
Property $A_{q,r}^*$, 424

Q

(q, r)-consistently-ordered matrix, *see* CO (q, r)-matrix
Ouadratic form, 114
Quasi $T(q, r)$-matrix, 423

R

Range, 13
Rank, 14
  column, 14
  row, 14
Rate of convergence, 188–190
  asymptotic, 88
  asymptotic average, 88, 299
  asymptotic quasi-average, 299
  average, 87
  for group SSOR method, 476
  for linear stationary iterative methods, 84
  quasi-average, 299
  virtual asymptotic average, 300
  virtual asymptotic quasi-average, 300, 352
  virtual quasi-average, 300, 352
Reciprocally consistent iterative method, 64
Rectangular matrix, 7
Red–black ordering, 158
Red equation, 271
Related linear system, 64
Relaxation factor
  choice if B has complex eigenvalues, 192–200
    B has real eigenvalues, 171–175
  optimum, 169–232
RF method, 74, 109, 117, 553
RF–SI method, 358–361, 554
Richardson's method, 94, 361–365
Root radius, 176
Row iteration, 438
Row matrix, 8
Row vector, 8

S

Scaled equations, 547
Scaling, 555
Schur's theorem, 17, 34
Schwarz inequality, 11, 27
Second-degree methods
  complete consistency, 486
  consistency, 486
  convergence, 487
  GS, 489
  J, 489
  RF, 493
  related equation, 486
  relation to semi-iterative methods, 487
  Richardson, 393
  SSOR, 489
Second-order Richardson method, 493
Self-adjoint differential equation, 49
  essentially self-adjoint, 49
Semi-iterative methods, 344–390
  MSOR, 376–382
  SOR, 374–376
Separation of variables, method of, 131
Sheldon's method, 315–319, 554
Similar matrices, 15
Simultaneous overrelaxation method, *see* JOR method
Singular linear system, 94
Singular matrix, 13
SMSOR method, 478–480
Solvable linear system, 68
SOR method, 73, 109, 113, 140–270, 554
  $A^{1/2}$-norm, 255–265
  backward, 265
  CO(q, r)-matrices for, 411–418
  convergence, 106–127
  $D^{1/2}$-norm, 245–255, 264–265
  eigenvalues, 140–168, 203–206
  extensions of, 391–403
  forward, 461
  Jordan canonical form, 234–239
  norms, 233–270
  principal vectors, 237
  semi-iterative methods based on, 374–376
  varying relaxation factors, 118–120
SOR theory, limitations of, 391–395
Sparse matrix, 1, 553
Spectral condition number of matrix, 85
Spectral norm, 31
Spectral radius, 15
  asymptotic average, 299

average, 297
virtual asymptotic average, 300
virtual average, 300
Spectrum of matrix, *see* Eigenvalue spectrum of matrix
Splitting of matrix, 75, 495
regular, 77, 123
square matrix, 8
Square root of matrix, 22
SSOR method, 100, 136, 273, 555
choice of relaxation factors, 464–470
convergence, 463
SSOR–SI method, 471–473, 555
Stable matrix, 82, 95
Stationary generalized Richardson method, *see* GRF method
Stationary Richardson method, *see* RF method
Stein–Rosenberg theory, 120–127
Stieltjes matrix, 42–48, 120–123, 395–401
Strong diagonal dominance, 41
Strong property A_p, *see* Property A_p^*
Strong property $A_{q,r}$, *see* Property $A_{q,r}^*$
Strongly (q, r)-consistently-ordered matrix, *see* CO*(q, r)-matrix
Subordinate matrix norm, 32
Successive overrelaxation method, *see* SOR method
Summability method, 344–346
regular, 345
Symmetric matrix, 9
Symmetric MSOR method, *see* SMSOR method
Symmetric SOR method, *see* SSOR method

T

Theoretical number of iterations, 327
Thirteen-point cross, 546
Three-eigenvalue problem, 197
T-matrix, 141
Trace of matrix, 8
Transpose of matrix, 9
Transpositions, 12
Tridiagonal matrix, 495–498
solution of linear system with, 441–444
Trivial solution, 13
Two-cyclic matrix, 54
Two-eigenvalue problem, 196
Two-line iterative methods, 453
two-line SOR method, 453

U

Unitary matrix, 9
Unsymmetric MSOR method, *see* USMSOR method
Unsymmetric SOR method, *see* USSOR method
Upper-triangular matrix, 8
strictly upper-triangular matrix, 8
USMSOR method, 478–480
USSOR method, 476–478

V

Vector, 10
Vector norm, 25
α-norm, 25
α,L-norm, 32
Virtual E-norm, 233
Virtual norm, 233
Virtual spectral radius, 170

W

Wachspress parameters, 525
Weak diagonal dominance, 36, 107
Weakly cyclic matrix of index p, 162

Computer Science and Applied Mathematics

A SERIES OF MONOGRAPHS AND TEXTBOOKS

Editor

Werner Rheinboldt

University of Pittsburgh

HANS P. KÜNZI, H. G. TZSCHACH, and C. A. ZEHNDER. Numerical Methods of Mathematical Optimization: With ALGOL and FORTRAN Programs, Corrected and Augmented Edition

AZRIEL ROSENFELD. Picture Processing by Computer

JAMES ORTEGA AND WERNER RHEINBOLDT. Iterative Solution of Nonlinear Equations in Several Variables

AZARIA PAZ. Introduction to Probabilistic Automata

DAVID YOUNG. Iterative Solution of Large Linear Systems

ANN YASUHARA. Recursive Function Theory and Logic

JAMES M. ORTEGA. Numerical Analysis: A Second Course

G. W. STEWART. Introduction to Matrix Computations

CHIN-LIANG CHANG AND RICHARD CHAR-TUNG LEE. Symbolic Logic and Mechanical Theorem Proving

C. C. GOTLIEB AND A. BORODIN. Social Issues in Computing

ERWIN ENGELER. Introduction to the Theory of Computation

F. W. J. OLVER. Asymptotics and Special Functions

DIONYSIOS C. TSICHRITZIS AND PHILIP A. BERNSTEIN. Operating Systems

ROBERT R. KORFHAGE. Discrete Computational Structures

PHILIP J. DAVIS AND PHILIP RABINOWITZ. Methods of Numerical Integration

A. T. BERZTISS. Data Structures: Theory and Practice, Second Edition

N. CHRISTOPHIDES. Graph Theory: An Algorithmic Approach

ALBERT NIJENHUIS AND HERBERT S. WILF. Combinatorial Algorithms

AZRIEL ROSENFELD AND AVINASH C. KAK. Digital Picture Processing

SAKTI P. GHOSH. Data Base Organization for Data Management

DIONYSIOS C. TSICHRITZIS AND FREDERICK H. LOCHOVSKY. Data Base Management Systems

JAMES L. PETERSON. Computer Organization and Assembly Language Programming

WILLIAM F. AMES. Numerical Methods for Partial Differential Equations, Second Edition

ARNOLD O. ALLEN. Probability, Statistics, and Queueing Theory: With Computer Science Applications

ELLIOTT I. ORGANICK, ALEXANDRA I. FORSYTHE, AND ROBERT P. PLUMMER. Programming Language Structures

ALBERT NIJENHUIS AND HERBERT S. WILF. Combinatorial Algorithms. Second edition.

JAMES S. VANDERGRAFT. Introduction to Numerical Computations

AZRIEL ROSENFELD. Picture Languages, Formal Models for Picture Recognition

ISAAC FRIED. Numerical Solution of Differential Equations

ABRAHAM BERMAN AND ROBERT J. PLEMMONS. Nonnegative Matrices in the Mathematical Sciences

BERNARD KOLMAN AND ROBERT E. BECK. Elementary Linear Programming with Applications

CLIVE L. DYM AND ELIZABETH S. IVEY. Principles of Mathematical Modeling

ERNEST L. HALL. Computer Image Processing and Recognition

ALLEN B. TUCKER, JR., Text Processing: Algorithms, Languages, and Applications

MARTIN CHARLES GOLUMBIC. Algorithmic Graph Theory and Perfect Graphs

In preparation

GABOR T. HERMAN. Image Reconstruction from Projections: The Fundamentals of Computerized Tomography

WEBB MILLER AND CELIA WRATHALL. Software from Roundoff Analysis of Matrix Algorithms